Modern Physics: Concepts and Applications

Modern Physics: Concepts and Applications

Editor: Aydan Bailey

NY RESEARCH PRESS

New York

Published by NY Research Press
118-35 Queens Blvd., Suite 400,
Forest Hills, NY 11375, USA
www.nyresearchpress.com

Modern Physics: Concepts and Applications
Edited by Aydan Bailey

© 2017 NY Research Press

International Standard Book Number: 978-1-63238-543-7 (Hardback)

Cataloging-in-Publication Data

Modern physics : concepts and applications / edited by Aydan Bailey.
    p. cm.
Includes bibliographical references and index.
ISBN 978-1-63238-543-7
1. Physics. I. Bailey, Aydan.
QC21.3 .M63 2017
530--dc23

Printed in the United States of America.

Contents

Preface

Physics studies in detail the varied aspects of matter with respect to force, space, energy and time. Particle physics, nuclear physics, bio physics, optical physics etc. are some of the significant branches of physics which have applications across multiple industries. This book on modern physics deals with the breakthroughs that have occurred in this field and the various modern applications that derive their propositions on the basis of these theories. The book includes advanced topics that are relevant for those seeking insights into this field. It elucidates the concepts and innovative models around prospective developments with respect to modern physics. The book will be of great help for students and researchers in the field of quantum mechanics, quantum theory, particle physics, and other related areas of study.

This book unites the global concepts and researches in an organized manner for a comprehensive understanding of the subject. It is a ripe text for all researchers, students, scientists or anyone else who is interested in acquiring a better knowledge of this dynamic field.

I extend my sincere thanks to the contributors for such eloquent research chapters. Finally, I thank my family for being a source of support and help.

Editor

A polymer scaffold for self-healing perovskite solar cells

Yicheng Zhao[1,2,*], Jing Wei[1,2,*], Heng Li[1,2,*], Yin Yan[1,2], Wenke Zhou[1,2], Dapeng Yu[1,2] & Qing Zhao[1,2]

Advancing of the lead halide perovskite solar cells towards photovoltaic market demands large-scale devices of high-power conversion efficiency, high reproducibility and stability via low-cost fabrication technology, and in particular resistance to humid environment for long-time operation. Here we achieve uniform perovskite film based on a novel polymer-scaffold architecture via a mild-temperature process. These solar cells exhibit efficiency of up to ~16% with small variation. The unencapsulated devices retain high output for up to 300 h in highly humid environment (70% relative humidity). Moreover, they show strong humidity resistant and self-healing behaviour, recovering rapidly after removing from water vapour. Not only the film can self-heal in this case, but the corresponding devices can present power conversion efficiency recovery after the water vapour is removed. Our work demonstrates the value of cheap, long chain and hygroscopic polymer scaffold in perovskite solar cells towards commercialization.

[1] State Key Laboratory for Mesoscopic Physics, School of Physics, Peking University, Beijing, 100871, China. [2] Collaborative Innovation Center of Quantum Matter, Beijing, 100084, China. * These authors contribute equally to this work. Correspondence and requests for materials should be addressed to Q.Z. (email: zhaoqing@pku.edu.cn).

The rapid rising of the lead halide perovskite solar cells (PSCs), typically based on $CH_3NH_3PbI_3$ ($MAPbI_3$) or its analogues, demonstrate great potential for market applications[1-13]. The worldwide power conversion efficiency (PCE) race for the perovskite solar cells has stimulated the highest record of $>20\%$ update via solvent engineering[8], interface engineering[5] and composition engineering[10], which approaches the best of polycrystalline silicon modules. The excellent performance of PSCs is attributed to the high absorption coefficient, weak exciton binding energy, and long diffusion length of the perovskite phase[12,14-17]. Two dominant device architectures, namely mesoporous scaffold[1-3] and planar heterojunction structures[4,5] have been developed. The PCE in mesoporous scaffold architecture depends on the pore size, porosity and morphology of the metal oxide nanoparticles, which dominantly determine the coverage, morphology variation of the perovskite layer and the carrier lifetime. Moreover, the fabrication of inorganic metal oxide mesoporous scaffold is complicated and needs sintering at high temperature of $>450\,^\circ C$ (refs 2,3), which makes the fabrication more expensive. The planar heterojunction PSCs fabricated layer by layer is either fabricated via high vacuum deposition[4] or solution-based method[5,18]. However, the low-cost solution process is rather challenging to form homogeneous film due to the dewetting process and sensitivity to the atmosphere[18-20]. Recently, progress has been reported to control the nucleation rate by changing dissolvent or temperature of the substrate[21,22]. Besides high efficiency and high reproducibility, long-term stability is also a crucial barrier to commercialize the PSCs, because the lead halide perovskite materials are very sensitive to ambient humidity and easy to dissolve and degrade in humid environment[5,23,24]. Therefore, it is very challenging and pressing to develop ultra stable and high PCE devices resistant to hostile operation environment.

Here we report a novel perovskite solar cell architecture based on an insulating polymer scaffold structure. The polymer-scaffold perovskite layer is simply fabricated via a single one-step process under mild temperature of $105\,^\circ C$. The as-made polymer-scaffold perovskite solar cell (PPSC) devices demonstrate high PCE up to $\sim 16\%$ under standard AM1.5 illumination and very small efficiency variation compared with that without the insulating polymer. Remarkably, unsealed PPSCs exhibit excellent performance in stability test in highly humid environment (70% relative humidity), which can endure the moisture for over 300 h. Most strikingly, the PPSCs show strong self-healing or water resistant effect under vapour spray. The perovskite film, as well as the corresponding devices can recover rapidly to its original phase and performance after being vapour sprayed and dried in ambient air, respectively.

Results

Design, fabrication and characterization of PPSCs. The PPSC architecture inherits the basic structure of PSCs with inorganic mesoporous scaffold[1-3], instead, the inorganic nanocrystal scaffolds are replaced with long- and flexible-polymer chain networks, which can be processed under mild temperature (Fig. 1a–c). The fabrication of PPSC is similar to that of planar heterojunction PSCs, except for the precursor solution preparation (Fig. 1d). Long-chain insulating polymers, dissolved in precursor solution, assemble the polymer scaffold simultaneously as the precursor solution is spin coated on the substrate. The selected polymer should be chemically inert to ingredients for perovskite synthesis; it should also act as an electrical insulator to guarantee the effective transport of photo-generated carriers. Here polyethylene glycol (PEG), which

meets the requirements, is chosen to be added in the precursor solution of $PbCl_2$ and CH_3NH_3I (MAI; 1:3 molar ratio) in dimethylformamide. The molar ratio of PEG monomers (C_2H_4O) to ultimate product $MAPbI_3$ is close to 1:1 (PEG in 40 mg ml^{-1}). The PEG scaffold perovskite absorber layer was prepared on the TiO_2 (~ 40 nm thickness)-coated fluorine doped tin oxide (FTO) glass by spin coating followed by annealing at $105\,^\circ C$ for 70 min. 2,29,7,79-tetrakis-(N,N-di-p-methoxyphenylamine) 9,99-spirobifluorene (Spiro-OMeTAD) and gold were used as the hole transport layer and electrode, respectively.

Figure 1e,f present top-view scanning electron microscope (SEM) images of the perovskite films without and with polymer scaffold. Those without scaffold exhibited large pinholes and bare conducting substrates, increasing the risk of short circuits. Those with scaffold demonstrated continuous and complete coverage. The cross-section SEM image showed the perovskite layer without polymer to have considerable variation in thickness and many voids (Fig. 1g). The film with polymer was uniform in thickness over large area (Fig. 1h). The PEG scaffold can also work in another precursor solution PbI_2:MAI (1:1) (Supplementary Fig. 1). PEG plays an important role in improvement of the film quality for the following reasons. First, the simultaneously formed PEG scaffold, which is formed by entangled long-chain molecules, acts as a three-dimensional skeleton to support perovskite crystals, causing it to cover the substrate more uniformly (Fig. 1c), demonstrating an advantage over planar heterojunction PSCs. Uniformly distributed PEG scaffolds function as Al_2O_3 scaffolds[3], and they can be produced in a much simpler and less costly manner. Second, the wetting properties of the precursor solution have been improved after introducing PEG (Supplementary Fig. 2), which can improve film morphology[19]. Third, PEG molecules in precursor solution can slow down the crystallization of perovskite from the X-ray diffraction characterization by tracking the intermediate phase in the crystallization process, which is shown below.

To confirm that PEG acts as scaffold in the architecture, detailed energy dispersive spectroscopy (EDS) analysis in a Cs-corrected scanning transmission electron microscope (STEM) was conducted for element mapping investigation (Fig. 2a–c). The high-angle annular dark field (HAADF) STEM image (Fig. 2a) reveals that the perovskite crystalline grains are surrounded by polymer material. The profiles of perovskite grains are highlighted with dashed lines according to the EDS mapping of Pb component (Fig. 2b). The amorphous material surrounding the perovskite grains is PEG in accordance with the EDS mapping of O element because only PEG contains O. The O mapping demonstrates that PEG not only covers the perovskite grains, but also cross-links the neighbouring perovskite grains via self-assembly forming a polymer-scaffold network to support the perovskite grains (Fig. 2c). Above conclusion is also evidenced by the fact that O mapping covers larger area than Pb and extends from crystal boundaries, indicating that the perovskite grains are imbedded in the PEG networks, corresponding well to the HAADF–STEM image. In addition, the O element mapping on the cross-sectional area of the PPSC film confirms a homogeneous distribution of PEG molecules in the perovskite film (Supplementary Fig. 3). High-resolution transmission electron microscope image of perovskite grain demonstrates good crystal lattice (Supplementary Fig. 4), suggesting that PEG molecules do not damage the perovskite crystal structure.

The phase evolution of PEG scaffold perovskite film during annealing was characterized by X-ray diffraction analysis (Fig. 2d). Over the course of 30–50 min, the film turned from orange to yellow, the (110) and (220) peaks of perovskite phase[25,26] became more pronounced and some new peaks appeared between 10 and 20°. The intermediate phase

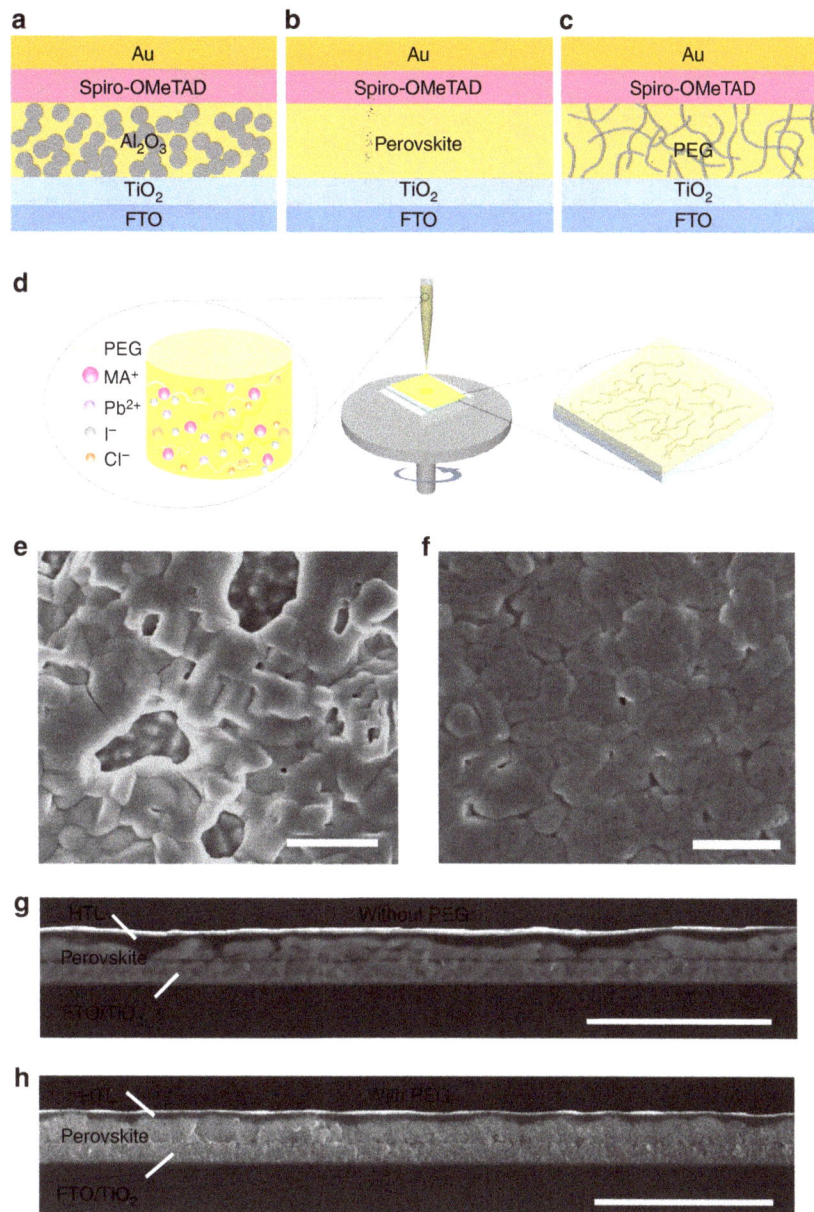

Figure 1 | Scheme and morphology of polymer-scaffold perovskite solar cells. Schematic diagram of mesoporous-scaffold-structured PSC (**a**), planar-heterojunction-structured PSC (**b**) and polymer-scaffold structured PSC (**c**). (**d**) Schematic diagram showing the fabrication process of perovskite film with polymer scaffold using one-step spin coating method. (**e,f**) Top-view SEM images of the perovskite films (**e**) without PEG and (**f**) with PEG. Scale bar, 1 μm. (**g,h**) Cross-sectional SEM images of the perovskite solar cells (**g**) without PEG and (**h**) with PEG scaffold. Scale bar, 3 μm.

disappeared after annealing for 60 min, and the sample became deep black at 105 °C after 70 min of annealing, with stronger (110) and (220) peaks. X-ray diffraction patterns of perovskite film without PEG and with PEG under the same 100 °C sintering temperature are also shown in Supplementary Fig. 5. Note that intermediate phase disappears after 70 min in film without PEG. While in film with PEG, intermediate phase disappears after 90 min (Supplementary Note 1), which strongly suggests a slower crystallization process in PEG scaffold perovskite film.

Photoluminescence (PL) spectra were measured to verify whether excess recombination sites and defect states are induced in the PEG scaffold perovskite film[27,28]. Compared with the PL spectrum of pristine perovskite film, the intensity is not influenced after being coated with PEG (Fig. 3b), indicating the insulating nature of PEG, as shown in Fig. 3a. Figure 3c shows the time-resolved PL spectra with and without PEG polymer scaffold,

with the similar carrier lifetime by using single exponential fitting (110.8 ns with PEG, 120.1 ns without PEG). This result indicates ∼1 μm diffusion length if diffusion coefficient is supposed to be 0.1 cm^2 s^{-1}, which is sufficient to guarantee effective collection efficiency. On the other hand, no visible shift (peak position of 760 nm) and broadening (half width of 40 nm) of peaks associated with perovskite layers with and without PEG scaffold is observed, indicating that PEG does not induce any excessive recombination (Supplementary Fig. 6).

Photovoltaic performance of PPSCs. The photovoltaic properties of PPSCs are characterized systematically. The PEG molecular weight and its concentration in precursor solution were found to play an important role in the performance of PPSCs. Figure 4a shows the current density–voltage (J–V) curves of samples with PEG in different molecule weights (12,000/20,000/

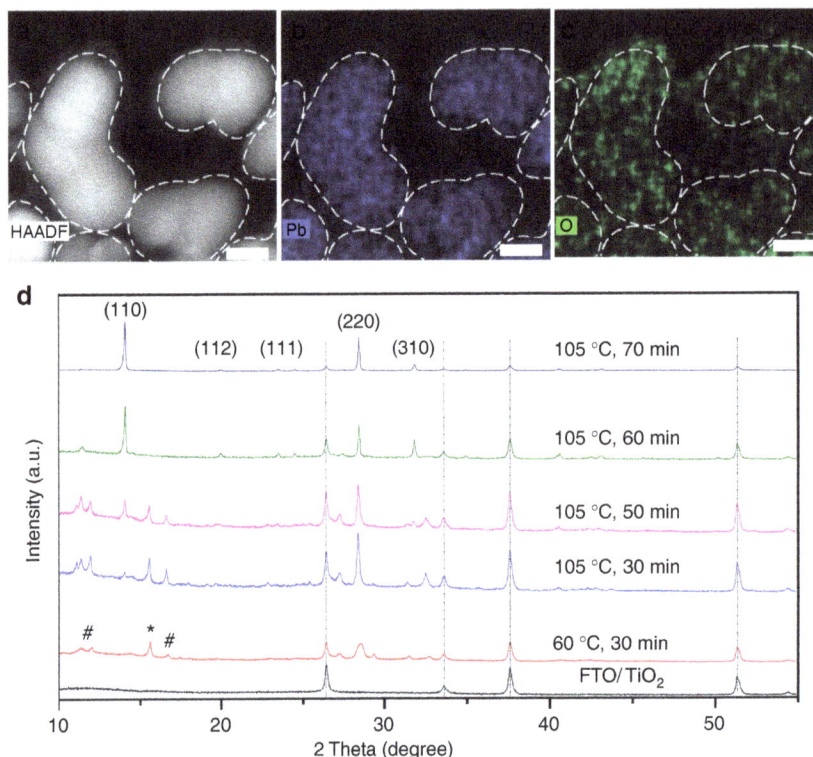

Figure 2 | Characterization and structure evolution of polymer-scaffold solar cells. (**a**) HAADF image of the perovskite film with PEG scaffold in STEM mode. Scale bar, 10 nm. (**b**) EDS mapping of Pb element of the peroskite film. Scale bar, 10 nm. (**c**) EDS mapping of O element of the peroskite film. O signal corresponds to PEG distribution because only the PEG contains O element among the materials used in perovskite thin film synthesis. Scale bar, 10 nm. (**d**) X-ray diffraction patterns of the perovskite film (PEG molecule weight: 20,000 Da; PEG concentration: 40 mg ml^{-1}) evolution with annealing time. '#' symbols and '*' symbols denote the intermediate phase and MAPbCl$_3$ phase, respectively.

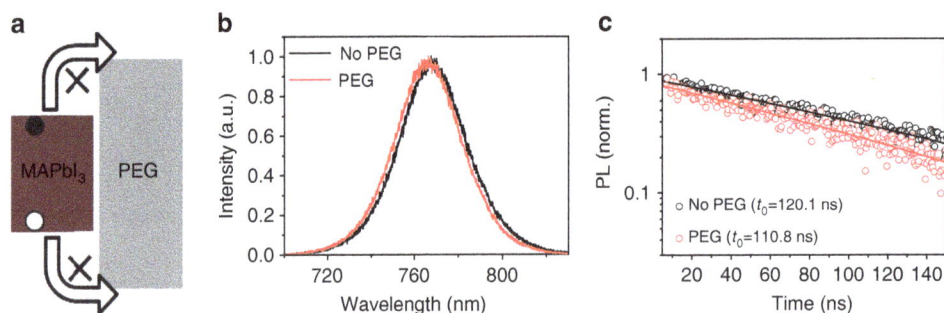

Figure 3 | Photoluminescence and energy-level alignment of polymer-scaffold perovskite films. (**a**) Schematic diagram of the energy offsets between CH$_3$NH$_3$PbI$_3$ and PEG. (**b**) PL spectra of pristine perovskite film (control) and the corresponding perovskite film after being coated with PEG. (**c**) Characteristics of PL transient spectra of PEG scaffold perovskite films and perovskite film without PEG, indicating the similar lifetime of carriers in perovskite with and without PEG scaffold.

100,000 Da) measured under simulated AM1.5 illumination of 100 mW cm^{-2}. The molecular weight of PEG, proportional to the length of polymer, has a significant effect on the performance of PPSCs, and the optimal molecular weight was found to be > 20,000 Da because longer PEG molecules facilitate the formation of three-dimensional molecular network in the perovskite film, and guarantee uniform morphology. PEG concentration is also critical to devices because it is related to the series resistance. As demonstrated in Fig. 4b, the device shows much better performance with 20 and 40 mg ml^{-1} PEG concentration, and its PCE decreases at higher concentrations (80 mg ml^{-1}) because the excess PEG in the film increases the series resistance and lowers the efficiency. Note that viscosity of the solution can be

strongly influenced by PEG concentration and molecule weight, so the r.p.m. value of spin coating is optimized at different PEG concentration and molecule weight (see Methods for details) to guarantee ~ 400 nm film thickness. The improved efficiency was also consistent with the increased recombination time constant[28] in PPSCs, compared with that of control devices, as is shown in Fig. 4c. Owing to the improved coverage, the recombination time is almost one order of magnitude larger than that without polymer scaffold at zero bias. The highest PCE of 16% was obtained for PPSCs with optimal PEG molecule weight of 20,000 Da and concentration of 20 mg ml^{-1}, and showed an open voltage (V_{oc}) of 0.98 V, short-circuit current (J_{sc}) of 22.5 mA cm^{-2}, and fill factor of 0.72 (Supplementary Fig. 7a).

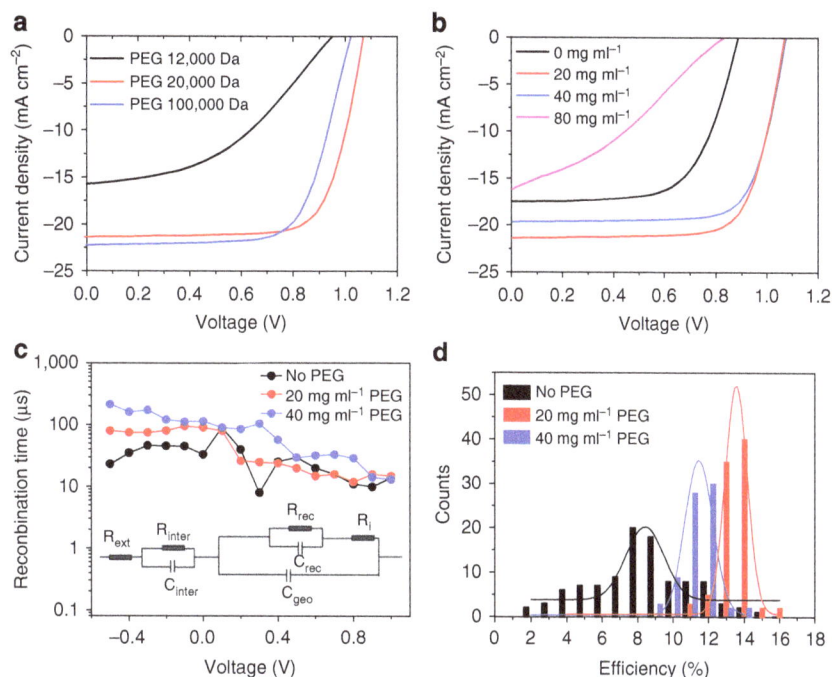

Figure 4 | Photovoltaic characterization of polymer-scaffold perovskite solar cells. (a) $J-V$ curves of the reverse scan based on precursor solution with different PEG molecular weight (concentration: 20 mg ml^{-1}) measured from 1.5 to -0.2 V under 500 mV s^{-1}. **(b)** $J-V$ curves of the reverse scan based on precursor solution with different PEG concentration (molecular weight: 20,000 Da) measured from 1.5 to -0.2 V under 500 mV s^{-1}. **(c)** Recombination time constant $R_{rec} \cdot C_{rec}$ (R_{rec}, recombination resistance, C_{rec}, capacitance) under different applied voltage can be extracted by fitting the Nyquist plot using the inset circuit. **(d)** PCE distribution of perovskite solar cells with (20 or 40 mg ml^{-1}) and without PEG scaffold. PCE values are extracted from reverse scan curve measured from 1.5 to -0.2 V under 500 mV s^{-1}. The active area for testing is 9 mm^2.

Another representative device shows a steady-state PCE of 15.4% at first 10 s, and the PCE gradually drops down and eventually stabilized at 13.5% in 1,000 s (0.7 V bias, Supplementary Fig. 7b). Parallel devices present H value from 10 to 30% under 500 mV s^{-1} if hysteresis index is defined $H = (PCE_{reverse} - PCE_{forward})/PCE_{reverse} \times 100\%$ to characterize the hysteresis effect. $PCE_{reverse}$ and $PCE_{forward}$ represent PCE obtained from J to V measurement in reverse scan and forward scan, respectively. The statistical PCE distributions of >100 samples without and with (for two concentrations) polymer scaffolds are shown in Fig. 4d. The PPSCs exhibited a narrow PCE distribution for 20 and 40 mg ml^{-1}, with an average efficiency of 14 and 12.5%, respectively, but the PCE distribution of the PSCs without polymer scaffold was much broader, varying from 2 to 14%, averaging at $\sim8\%$. The significant increase in PCE and high reproducibility of the PPSC devices are direct reflections of the uniform thickness and homogeneous morphology of the polymer-scaffold perovskite layer.

Improved stability in highly humid environment. More importantly, the insulating PEG scaffold structure can stabilize the PSCs in humid environment due to the strong hygroscopicity property of PEG. Here three groups of unsealed devices were prepared without and with PEG (two concentrations), respectively. Figure 5a shows the PCE evolution as a function of time of the three groups of devices in the dark under highly humid environment (relative humidity 70%) without any sealing. In contrast to the fast PCE decay from 12 to 0.5% within 50 h of the pristine PSCs, the PPSCs retained 65% of its PCE after 300 h aging. Similarly, the J_{sc}, V_{oc}, and fill factor of the pristine sample decreased quickly during the first 50 h, while those of the PPSCs retained relatively high values for up to 300 h (Supplementary Fig. 8a,b). Note that the device with PEG concentration of

40 mg ml^{-1} showed better performance than the one with 20 mg ml^{-1} because more PEG can form a more compact layer, enhancing the moisture resistance of devices. We also studied the stability under continuous light illumination in 70% humidity for unsealed devices (Supplementary Fig. 8c,d). Photovoltaic performance degraded quickly within 10 h in the device without PEG scaffold, while PPSC retained 83% of its PCE during the same aging period. Similarly, V_{oc}, J_{sc}, and fill factor suffered substantial decrease in the first 10 h for pristine sample while they remained relatively high values in the presence of PEG. X-ray diffraction patterns of the perovskite film with PEG scaffold showed no structure destruction after 72 h continuous light irradiation, while the samples without PEG degraded into PbI$_2$ quickly within 10 h (Supplementary Fig. 9). Evidence above strongly suggests that PEG can greatly protect perovskite film from decomposition under moisture attack and significantly improve the stability of PPSCs at very humid environment.

Self-healing property of PPSCs after vapour spray. In addition to its improved stability, the PPSC devices can demonstrate a self-healing behaviour and humidity resistance effect on exposure to water vapour. Figure 5b shows a comparison of the changes in colour of the perovskite layer with PEG (top) and the perovskite layer without PEG (bottom) after both were sprayed with water vapour for 60 s. The perovskite film without PEG decomposed into PbI$_2$ and turned yellow irreversibly. In contrast, the PEG scaffold perovskite film showed yellow in colour at first and recovered to black in ~45 s after removing from the spray (Fig. 5b). This amazing self-healing behaviour is further vividly demonstrated in 120 s videos (Supplementary Movies 1 and 2). The self-healing capability of the PPSC device was also manifested in the $J-V$ characterization (Fig. 5c). The PCE degrades with smaller V_{oc} and J_{sc} after exposure to water vapour

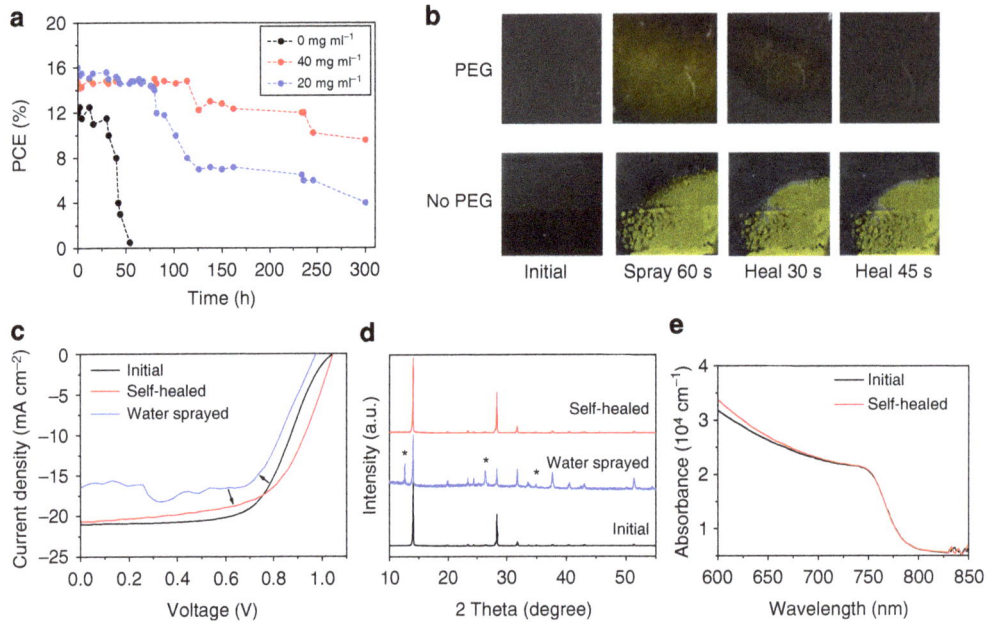

Figure 5 | Device stability and self-healing demonstration. (**a**) PCE evolution as a function of time of perovskite solar cells with 20 or 40 mg ml^{-1} PEG and without PEG scaffold exposed in high humid (70% relative humidity) dark environment without any sealing. (**b**) Photographs of perovskite films with and without PEG showing colour change evolution after water-spraying for 60 s and kept in ambient air in 45 s. (**c**) J-V curves of PPSCs before and after water spray, revealing a complete recover of the cells in one minute when it puts back to ambient air. (**d**) X-ray diffraction evolution revealing the self-healing process: X-ray diffraction pattern of initial perovskite film with PEG, after vapour sprayed and after 10 min from removing from water vapour, respectively. The symbol '*' represents for the peaks of PbI$_2$. (**e**) Absorption coefficient as function of wavelength for perovskite film before and after vapour spray and self-healing.

for 60 s. Surprisingly, results showed that the J–V curve can almost regain its original value in 45 s, indicating that the PPSCs can heal itself when it is returned to ambient air after being sprayed with water. This unique humidity-resist characteristic is very suitable for practical applications because once the devices are exposed to very humid environment; they can self-heal to high PCE again in a short time when they return to sunlight again. Moreover, to further prove the self-healing property, perovskite film with PEG was sprayed with water vapour for 60 s, then after self-healing, it was coated by Spiro-MeOTAD and Au electrode to fabricate into corresponding device, which showed photovoltaic performance comparable to its original PPSC devices without water spray (Supplementary Fig. 10).

The self-healing process was recorded by X-ray diffraction analysis (Fig. 5d). The peak at 12.7° reveals that PbI$_2$ phase formed when water vapour was sprayed on the film. After self-healing, the PbI$_2$ peak disappeared and the perovskite film fully recovered to its original crystal phase. The self-healing can also be demonstrated by the absorption spectra (Fig. 5e), which presented the same absorbance and similar Urbach energy ($\sim$40 meV; ref. 29). The decomposition of perovskite phase is caused by the chemical reaction with water molecules[30–32], which is manifested by the X-ray diffraction analysis of the perovskite film without PEG after water spray (Supplementary Fig. 11).

According to the observations in Fig. 2a–c, we conjecture that these entangled PEG molecules are anchored on the surface of perovskite grains by hydrogen bonding, and the formation is illustrated in Fig. 6a. The –OH I–interaction had been evidenced in the work by Li et al[32]. We use liquid state ^{1}H NMR measurements to prove such interactions. The proton NMR spectra of three samples were compared (Fig. 6b): deuterated DMSO solution with MAPbI$_3$ (sample 1), with MAPbI$_3$ + PEG (sample 2) and with PEG (sample 3). In the signals of sample 3, double methylene group linked to oxygen on both ends ($-[O-CH_2-CH_2]_n-$) is characterized by the peak at

$\delta = 3.48$ p.p.m. (Fig. 6b). An upfield chemical shift of $\Delta\delta \sim -0.18$ p.p.m. with several splitting peaks is observed in sample 2. Such chemical shift can be explained by the hydrogen bond between MA$^+$ and O in PEG chain, which weakens the influence of O on protons of methylene in $-[O-CH_2-CH_2]_n-$. Furthermore, the proton resonance signals of $-NH_3^+$ in sample 1 (peak at $\delta = 7.33$ p.p.m.) shift towards upfield with $\Delta\delta \sim -0.13$ p.p.m. in sample 2 (Fig. 6b), which can also be attributed to the hydrogen bonds discussed above. Such interaction is also evidenced by a dissolution experiment devised by us (Supplementary Fig. 12).

As presented in Fig. 6c, the improved stability and self-healing effect of the PPSC devices can be ascribed to the excellent hygroscopicity of the PEG molecules and their strong interaction with the perovskite. On one hand, the omnipresent PEG molecules can absorb water efficiently to form a compact moisture barrier around perovskite crystal grains with little water penetrating into the film. On the other hand, on water spray, the black perovskite film with PEG turned light yellow (PbI$_2$ forms) at first. However, due to the strong interaction between MAI and PEG, (NMR measurements and a dissolution experiment in Supplementary Fig. 12), the MAI molecules were anchored by the nearby PEG molecules rather than escape away (process 1, 2 in Fig. 6c). After being kept away from water vapour, PbI$_2$ in the film reacted with nearby MAI to regenerate the perovskite MAPbI$_3$ phase, very similar to the two-step synthesis[2] (process 3 in Fig. 6c). The instant decomposition-regeneration mechanism explains the fast self-healing process in the PEG scaffold perovskite film.

Discussion

The hygroscopic PEG scaffold can stabilize the perovskite film, rendering the devices resistant to moisture with strong self-healing property. Hygroscopic polymer-scaffold architecture paved a new effective way in perovskite solar cell, to solve the

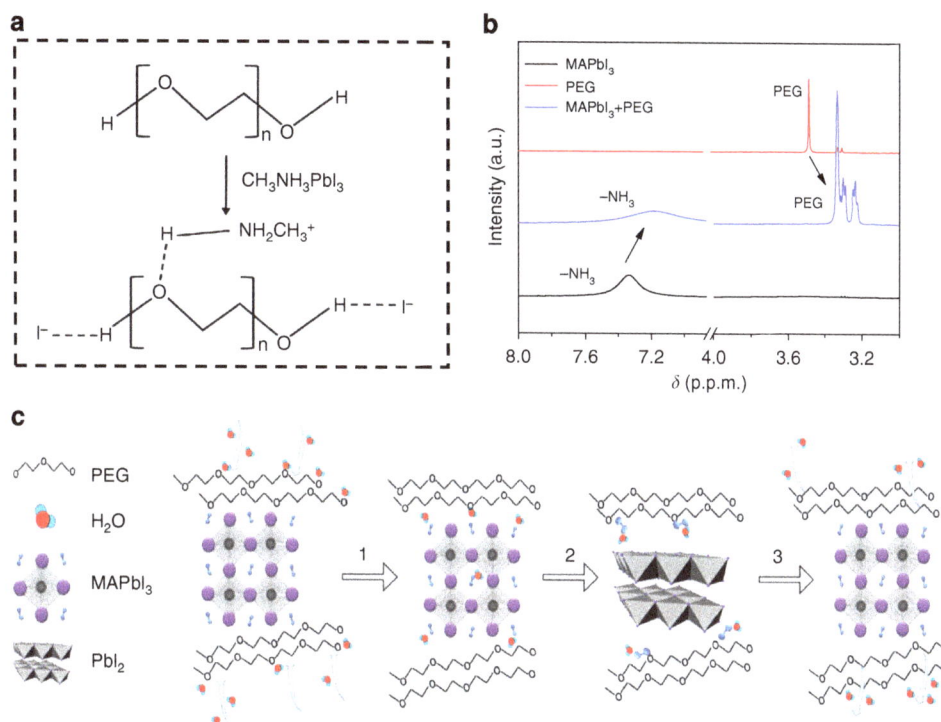

Figure 6 | Mechanism demonstration of self-healing. (**a**) Schematic diagram of the hydrogen bonding formation between PEG molecules and MAPbI$_3$. (**b**) A comparison of NMR spectra from 3.0 to 8.0 p.p.m. among three samples: deuterated DMSO solutions with MAPbI$_3$, mixture of MAPbI$_3$ + PEG and PEG, respectively. (**c**) Schematic diagram to show mechanisms for the self-healing properties in PPSCs: (1) Water absorb on perovskite; (2) Perovskite hydrolysis into PbI$_2$ and MAI•H$_2$O by water; (3). Restrained MAI by PEG react with nearby PbI$_2$ to form perovskite again after water evaporates. PEG has a strong interaction with MAI, preventing it from evaporating, subsequently MAI and PbI$_2$ react *in situ* to form perovskite after the film was removed away from the vapour source.

hydrolysis problem in ambient air with greatly decreased package cost. The polymer-scaffold perovskite layer reinforces the architecture and improves strain tolerance. It may also facilitate production of flexible, wearable devices in the future. Future work will be focused on looking for cheap long-chain polymers which have stronger hygroscopicity and binding effect with MAI to protect the perovskite from decomposition in ambient environment. Moreover, by comparing the stability test result under light illumination (Supplementary Fig. 8) and in dark (Fig. 5a), we found that light has a strong destructive influence on stability performance. One would expect that this influence comes from decomposition of perovskite film accelerated by light with the aid of hydrolysis by water molecules under illumination (reactive equation in Supplementary Fig. 11). However, after 72 h light soaking, the perovskite film with PEG was not destructed compared with that without PEG from X-ray diffraction spectra (Supplementary Fig. 9). This indicated another important factor influencing the efficiency in addition to material decomposition. Hence, we speculate that I$^-$ ion migration[33–36] may be induced or enhanced by light. Ion migration may answer for the poor stability under illumination, no matter in the traditional architecture without PEG scaffold or in our PPSCs, which needs further investigation. Further improvement needs be focused on revealing the mechanism of ion migration in the long-term stability issue, as well as how to inhibit it in PPSCs, which would lead to much better performance in stability test under illumination with high humidity.

Methods

Device fabrication and device characterization. All chemicals were purchased from Sigma-Aldrich or J&K Scientific Ltd. unless otherwise stated. The photovoltaic devices were fabricated on FTO-coated glass (Pilkington, Nippon Sheet Glass). First, laser-patterned, FTO-coated glass substrates were cleaned by ultrasonic cleaned in

deionized water, acetone and ethanol, followed by an ultraviolet treatment for 5 min. Compact layers were deposited on the substrates by spin coating titanium diiso-propoxide bis(acetylacetonate) solution (75% in 2-propanol) diluted in ethanol (1:20, volume ratio) for two times and annealed at 450 °C for 30 min. After cooling to room temperature, the substrates were transferred to a hot plate at 90 °C before spin coating. CH$_3$NH$_3$I was synthesized according to the reported procedure[1]. In a typical synthesis, 33.77 ml methylamine (33% in methanol) and 30 ml of hydroiodic acid (57% in water) were reacting in a 250 ml flask at 0 °C for 2 h under stirring. The precipitate was formed by evaporation at 50 °C for 1 h. The product, methylammonium iodide CH$_3$NH$_3$I, was washed with diethyl ether by stirring the solution for 30 min, repeated three times, and then finally dried at 60 °C in vacuum oven for 24 h. The prepared CH$_3$NH$_3$I, PbCl$_2$ (Aldrich) (3:1) for 430 and 250 mg ml^{-1} solution was mixed with PEG (Sigma-Aldrich) and stirred in dimethylformamide at 60 °C for 12 h. The molar ratio of PEG monomers (C$_2$H$_4$O) to ultimate product MAPbI$_3$ is ~1:1 (PEG in 40 mg ml^{-1}). The resulting solution was then coated onto the TiO$_2$/FTO substrate by spin coating at 500 r.p.m. for 10 s, and then 3,500/4,500/5,500 r.p.m. for 30 s, for the solution with 20/40/80 mg ml^{-1} PEG, respectively. For different molecular weight 12,000/20,000/100,000 Da. with 20 mg ml^{-1}, we use the optimized 3.500/3.500/4.500 r.p.m. to prepare the film. Then the substrate was dried on a hot plate at 60 °C for 45 min, then sintered at 105 °C for 70 min. For the preparation of peroskite film without PEG, the optimized sintering temperature is 100 °C. After cooling to room temperature, the hole transport material was spin coated onto the perovskite film at 3,000 r.p.m. for 40 s. The spin coating formulation was prepared by dissolving 72.3 mg 2,2',7,7'-Tetrakis(N,N-p-dimethoxy-phenylamino)-9,9'-spirobifluorene(spiro-MeOTAD), purchased from Yingkou OPV Tech New Energy Co. Ltd., 30 μl 4-tert-butylpyri-dine and 20 μl of a stock solution of 520 mg ml^{-1} lithium bis (trifluoromethylsulphonyl) imide in acetonitrile in 1 ml chlorobenzene. Finally, 90-nm-thick gold electrodes were deposited on top of the devices by evaporation at ~10^{-4} mbar. The active area of the electrode was fixed at 9 mm^2.

The surface morphology and EDS mapping of the perovskite thin film was characterized by SEM (Nano430, FEI). The instrument uses an electron beam accelerated at 15 kV, enabling operation at a variety of currents. Considering that only PEG polymer contains oxygen element among those materials used in perovskite thin film synthesis, here oxygen stands for PEG polymer, lead for perovskite. HAADF–STEM is examined by Cs-corrected FEI (Titan G2 80–200) transmission electron microscope operating at an accelerating voltage of 300 kV, equipped with ChemiSTEM EDS detector, which enables very quick high-resolution element mapping. The sample is prepared from perovskite precursor solution spin coated on TEM grid (3,000 r.p.m.), then annealed at 105 °C for 70 min.

For NMR measurement, it is examined by Bruker 600 UltraShield using pulse signal. We use deuterated DMSO as solvent. The $CH_3NH_3PbI_3$ solution is prepared from CH_3NH_3I, PbI_2 (Aldrich; 1:1) for 53 and 153 mg ml^{-1} solution and stirred in DMSO at 60 °C for 12 h. $CH_3NH_3PbI_3$ solution mixed with PEG (Sigma-Aldrich) is prepared by adding 40 mg PEG (molecule weight: 20,000 Da) to 1 ml $CH_3NH_3PbI_3$ solution.

For X-ray diffraction measurement, flat PbI_2 and TiO_2/PbI_2 nanocomposites were deposited on glass slides using the above-mentioned procedures. X-ray powder diagrams were recorded on an X'PertMPD PRO from PANalytical equipped with a ceramic tube (Cu anode, $\lambda = 1.5406$ Å), a secondary graphite (002) monochromator and a RTMS X'Celerator detector, and operated in BRAGG-BRENTANO geometry. The samples were mounted without further modification, and the automatic divergence slit and beam mask were adjusted to the dimensions of the thin films. A step size of 0.008 deg was chosen and an acquisition time of up to 7.5 min per deg. A baseline correction was applied to all X-ray powder diagrams to remove the broad diffraction peak arising from the amorphous glass slide.

Steady-state PL spectra were measured using a He–Cd laser (325 nm in wavelength) and green laser (514 nm in wavelength) guided by a micro-zone confocal Raman spectroscope (Renishaw inVia microRaman system) as the laser beam with a spot size diameter of 2 μm. The collected duration is ~5 ms. The time-resolved fluorescence spectra were recorded with a high-resolution streak camera system (Hamamatsu C10910). We used an amplified mode-lock Ti: Sapphire femtosecond laser system (Legend, Coherent) and a two-stage optical parametric amplifier (OperA Solo, Coherent) to generate the pump beam with a repetition rate of 1 KHz. All the samples were excited by 517 nm at room temperature with 110 nJ cm^{-2}.

The electrochemical impedance spectrum and cyclic voltammograms were measured using a potentiostat/galvanostat (SP-150, Bio-Logic, France). The frequency can be tuned from 0.1 Hz to 1 MHz. All samples were measured under 10 mW cm^{-2}.

The $J-V$ characteristics were obtained using an Agilent B2900 Series precision source/measure unit, and the cell was illuminated by a solar simulator (Solar IV-150A, Zolix) under AM1.5 irradiation (100 mW cm^{-2}). Light intensity was calibrated with a Newport calibrated KG5-filtered Si reference cell. We use black mask to define the cells' area, and the masking effect is confirmed by testing the J_{sc} with and without it, which has a 5% J_{sc} difference. The $J-V$ curves are tested from 1.5 to -0.2 V with a scan velocity 500 mV s^{-1}. The masked active area is 9 mm^2.

Stability test. Unsealed PPSCs and PSCs were put near a humidifier to control its relative humidity ~70%. For the stability test under dark, the cells were stored in a black box with 70% relative humidity, and were tested every 8 h. For the stability tests under continuous light illumination, light source ranges from 300 to 800 nm wavelength with 70 mW cm^{-2} was used. $J-V$ curves were recorded under AM1.5 light irradiation. The temperature is controlled ~30 °C. Three devices were measured in this way and demonstrate excellent repeatability. For the self-healing behaviour, we sprayed the water vapour for 60 s onto the resulting devices by humidifier. $J-V$ curves were collected before and right after water vapour exposure under standard illumination AM1.5.

References

1. Kim, H. S. et al. Lead iodide perovskite sensitized all-solid-state submicron thin film mesoscopic solar cell with efficiency exceeding 9%. Sci. Rep. **2**, 591 (2012).
2. Burschka, J. et al. Sequential deposition as a route to high-performance perovskite-sensitized solar cells. Nature **499**, 316–319 (2013).
3. Lee, M. M. et al. Efficient hybrid solar cells based on meso-superstructured organometal halide perovskites. Science **338**, 643–647 (2012).
4. Liu, M., Johnston, M. B. & Snaith, H. J. Efficient planar heterojunction perovskite solar cells by vapour deposition. Nature **501**, 395–398 (2013).
5. Zhou, H. et al. Interface engineering of highly efficient perovskite solar cells. Science **345**, 542–546 (2014).
6. Mei, A. et al. A hole-conductor–free, fully printable mesoscopic perovskite solar cell with high stability. Science **345**, 295–298 (2014).
7. Liu, D. & Kelly, T. L. Perovskite solar cells with a planar heterojunction structure prepared using room-temperature solution processing techniques. Nat. Photon. **8**, 133–138 (2014).
8. Jeon, N. J. et al. Solvent engineering for high-performance inorganic-organic hybrid perovskite solar cells. Nat. Mater. **13**, 897–903 (2014).
9. Chiang, C.H, Tseng, Z. L. & Wu, C. G. Planar heterojunction perovskite/PC$_{71}$BM solar cells with enhanced open-circuit voltage via a (2/1)-step spin-coating process. J. Mater. Chem. A **2**, 15897–15903 (2014).
10. Jeon, N. J. et al. Compositional engineering of perovskite materials for high-performance solar cells. Nature **517**, 476–480 (2015).
11. Wei, J. et al. Hysteresis analysis based on the ferroelectric effect in hybrid perovskite solar cells. J. Phys. Chem. Lett. **5**, 3937–3945 (2014).
12. Kojima, A., Teshima, K., Shirai, Y. & Miyasaka, T. Organometal halide perovskites as visible-light sensitizers for photovoltaic cells. J. Am. Chem. Soc. **131**, 6050–6051 (2009).
13. Xiao, Z. et al. Giant switchable photovoltaic effect in organometal trihalide perovskite devices. Nat. Mater. **14**, 193–198 (2015).
14. Stranks, S. D. et al. Electronhole diusion lengths exceeding 1 micrometer in an organometal trihalide perovskite absorber. Science **342**, 341–344 (2013).
15. Lin, Q. et al. Electro-optics of perovskite solar cells. Nat. Photon.s **9**, 106–112 (2015).
16. Nie, W. Y. et al. High-efficiency solution-processed perovskite solar cells with millimeter-scale grains. Science **347**, 522–525 (2015).
17. Dong, Q. F. et al. Electron-hole diffusion lengths > 175 μm in solution-grown $CH_3NH_3PbI_3$ single crystals. Science **347**, 967–970 (2015).
18. Eperon, G. E. et al. Morphological control for high performance, solution-processed planar heterojunction perovskite solar cells. Adv. Funct. Mater. **24**, 151–157 (2014).
19. Thompson, C. V. Solid-state dewetting of thin films. Annu. Rev. Mater. Res. **42**, 399–434 (2012).
20. Giles, E. Eperon et al. The importance of moisture in hybrid lead halide perovskite thin film fabrication. ACS Nano **10**, 1021 (2015).
21. Jeon, Y. J. et al. Planar heterojunction perovskite solar cells with superior reproducibility. Sci. Rep. **4**, 6953 (2014).
22. Xie, F. et al. Vacuum-assisted thermal annealing of $CH_3NH_3PbI_3$ for highly stable and efficient perovskite solar cells. ACS Nano **9**, 639–646 (2015).
23. Ahn, N. et al. Highly reproducible perovskite solar cells with average efficiency of 18.3% and best efficiency of 19.7% fabricated via lewis base adduct of lead(II) iodide. J. Am. Chem. Soc. **137**, 8696–8699 (2015).
24. Kaltenbrunner, M. et al. Flexible high power-per-weight perovskite solar cells with chromium oxide-metal contacts for improved stability in air. Nat. Mater. **14**, 1032–1039 (2015).
25. Song, T. B. et al. Unraveling film transformations and device performance of planar perovskite solar cells. Nano Energy **12**, 494–500 (2015).
26. Yu, H. et al. The role of chlorine in the formation process of "$CH_3NH_3PbI_{3-x}Cl_{(x)}$" perovskite. Adv. Funct. Mater. **24**, 7102–7108 (2014).
27. Quilettes, D. W. et al. Impact of microstructure on local carrier lifetime in perovskite solar cells. Science **348**, 683–686 (2015).
28. Shao, Y. C. et al. Origin and elimination of photocurrent hysteresis by fullerene passivation in $CH_3NH_3PbI_3$ planar heterojunction solar cells. Nat. Commun. **5**, 5784 (2014).
29. Snaith, H. J. et al. Metal-halide perovskites for photovoltaic and light-emitting devices. Nat. Nanotechnol. **10**, 391–402 (2015).
30. Baikie, T. et al. Synthesis and crystal chemistry of the hybrid perovskite (CH_3NH_3) PbI_3 for solid-state sensitised solar cell applications. J. Mater. Chem. A **1**, 5628–5641 (2013).
31. Dualeh, A. et al. Thermal behavior of methylammonium lead-trihalide perovskite photovoltaic light harvesters. Chem. Mater. **26**, 6160–6164 (2014).
32. Li, X. et al. Improved performance and stability of perovskite solar cells by crystal crosslinking with ω-ammonium-alkylphosphonic acid chlorides. Nat. Chem. **7**, 703–711 (2015).
33. Zhang, Y. et al. Charge selective contacts, mobile ions and anomalous hysteresis in organic–inorganic perovskite solar cells. Mater. Horiz **2**, 315 (2015).
34. Eames, C. et al. Ionic transport in hybrid lead iodide perovskite solar cells. Nat. Commun. **6**, 7497 (2015).
35. Yang, T.-Y et al. The Significance of ion conduction in a hybrid organic-inorganic lead-iodide-based perovskite photosensitizer. Angew. Chem. Int. Ed. **54**, 7905–7910 (2015).
36. Azpiroz, J. M. et al. Defect migration in methylammonium lead iodide and its role in perovskite solar cell operation. Energy Environ. Sci. **8**, 2118 (2015).

Acknowledgements

We thank Jinlong Pan, Professor Dongsheng Xu, Dr Zhenxuan Zhao, Professor Jianjun Tian, Professor Guozhong Cao and Professor Han Zhang for experimental help. We thank Yu Li and Professor Shufeng Wang for time-resolved PL measurements. We thank Professors Jixue Li and Ze Zhang for STEM observations. We also thank Professors Jun Xu, Liping You and Li Chen for electron microscope help. This work was supported by National 973 projects (2013CB932602, 2011CB707601, MOST) from Ministry of Science and Technology, China, National Natural Science Foundation of China (NSFC51272007, 61571015, 11234001, 91433102, 11327902). Q.Z. acknowledges Beijing Nova Program (XX2013003) and the Program for New Century Excellent Talents in University of China.

Author contributions

Y.Z., Q.Z. and D.Y. conceived the idea and designed the experiment. Y.Z., Y.Y., J.W. and W.Z. performed the experiment and analysed the data. Y.Z., Y.Y. and J.W. optimized the PCE to 15%. H.L., Y.Z. and Q.Z. contributed to the experiment discussion, schematic diagram and image processing. D.Y. and Q.Z. supervised the project during the whole process. D.Y., Q.Z., H.L. and Y.Z. co-wrote the manuscript. All authors read and comment on the manuscript.

Additional information

2

High-performance thermoelectric nanocomposites from nanocrystal building blocks

Maria Ibáñez[1,2,3], Zhishan Luo[3], Aziz Genç[4], Laura Piveteau[1,2], Silvia Ortega[3], Doris Cadavid[3], Oleksandr Dobrozhan[3], Yu Liu[3], Maarten Nachtegaal[5], Mona Zebarjadi[6], Jordi Arbiol[4,7], Maksym V. Kovalenko[1,2] & Andreu Cabot[3,7]

The efficient conversion between thermal and electrical energy by means of durable, silent and scalable solid-state thermoelectric devices has been a long standing goal. While nanocrystalline materials have already led to substantially higher thermoelectric efficiencies, further improvements are expected to arise from precise chemical engineering of nanoscale building blocks and interfaces. Here we present a simple and versatile bottom–up strategy based on the assembly of colloidal nanocrystals to produce consolidated yet nanostructured thermoelectric materials. In the case study on the PbS–Ag system, Ag nanodomains not only contribute to block phonon propagation, but also provide electrons to the PbS host semiconductor and reduce the PbS intergrain energy barriers for charge transport. Thus, PbS–Ag nanocomposites exhibit reduced thermal conductivities and higher charge carrier concentrations and mobilities than PbS nanomaterial. Such improvements of the material transport properties provide thermoelectric figures of merit up to 1.7 at 850 K.

[1] Department of Chemistry and Applied Biosciences, Institute of Inorganic Chemistry, ETH Zürich, Vladimir Prelog Weg 1, CH-8093 Zurich, Switzerland. [2] Laboratory for Thin Films and Photovoltaics, Empa-Swiss Federal Laboratories for Materials Science and Technology, Dübendorf, Überlandstrasse 129, CH-8600 Dübendorf, Switzerland. [3] Advanced Materials Department, Catalonia Energy Research Institute - IREC, Sant Adria de Besos, Jardins de les Dones de Negre n.1, Pl. 2, 08930 Barcelona, Spain. [4] Department of Advanced Electron Nanoscopy, Catalan Institute of Nanoscience and Nanotechnology (ICN2), CSIC and The Barcelona Institute of Science and Technology, Campus UAB, Bellaterra, 08193 Barcelona, Spain. [5] Paul Scherrer Institute, 5232 Villigen PSI, Switzerland. [6] Department of Mechanical and Aerospace Engineering, Rutgers University, 98 Brett Rd, Piscataway, New Jersey 08854-8058, USA. [7] Institució Catalana de Recerca i Estudis Avançats, ICREA, Passeig de Lluís Companys, 23 08010 Barcelona, Spain. Correspondence and requests for materials should be addressed to A.C. (email: acabot@irec.cat).

Thermoelectric devices allow direct conversion of heat into electricity and *vice versa*, holding great potential for heat management, precise temperature control and energy harvesting from ubiquitous temperature gradients. The efficiency of thermoelectric devices is primarily governed by three interrelated material parameters: the electrical conductivity, σ, the Seebeck coefficient or thermopower, S, and the thermal conductivity, κ. These parameters are grouped into a dimensionless figure of merit, ZT, defined as $ZT = \sigma S^2 T \kappa^{-1}$ where T is the absolute temperature. While there is no known limitation to the maximum thermoelectric energy conversion efficiency other than the Carnot limit, current thermoelectric materials struggle to simultaneously display high σ and S, and low κ, which prevents their widespread implementation[1].

Control of the chemical composition and crystallinity of thermoelectric materials at the nanoscale via engineering of multicomponent nanomaterials (nanocomposites) has proven to be effective for the reduction of thermal conductivity by promoting phonon scattering at grain boundaries[2–7]. The remaining major challenge facing the next generation of high-efficiency thermoelectric materials is the enhancement of the thermoelectric power factor ($PF = S^2\sigma$) while keeping a low thermal conductivity. Strategies to accomplish this goal have focused on increasing the average energy per carrier through energy filtering[8], carrier localization in narrow bands in quantum confined structures[9], or the introduction of resonant levels[10]. At the same time, the electrical conductivity must be optimized by properly adjusting the concentration of charge carriers and maximizing their mobility.

These improvements of thermoelectric properties have been mainly demonstrated at the thin-film level, oftentimes using precise but expensive vacuum-based materials growth techniques[11]. However, practical applications of thermoelectric materials demand inexpensive and, for withstanding relatively high temperature gradients, macroscopic devices. In this regard, current approaches to produce bulk nanocomposites, such as ball-milling or the precipitation of secondary phases from metastable solid solutions, lack precision control over the distribution of phases and/or are limited in compositional versatility. Thus, novel cost-effective and general strategies to produce bulk nanocomposites with high accuracy and versatility need to be developed.

Here we demonstrate that a simple route to engineering nanocomposites by the assembly of precisely designed nanocrystal building blocks is able to reach high thermoelectric efficiencies. We propose to blend semiconductor nanocrystals with metallic nanocrystals forming Ohmic contact with the host semiconductor. In such nanocomposites, metallic nanocrystals control charge carrier concentration through charge spill over to the host semiconductor. The goal of this configuration is to reach large charge carrier concentrations without deteriorating the mobility of charge carriers[12–16]. In this three-dimensional (3D) modulation doping strategy[17–20], composition, size and distribution of semiconductor and metal nanodomains control the nanocomposite transport properties. Consider a slab of a semiconductor sandwiched between two metallic plates. Transfer of charges at the interfaces will cause band bending, extending over charge-screening length. If the width of a semiconductor slab is on the order of this screening length, there will be an overlap of the bended bands and therefore carriers will not be confined to the interfaces, but they will be able to travel through the bulk of a semiconducting region. Similarly, in the 3D case, the size of the semiconducting nanocrystals should be on the order of the screening length, enabling charge transport with minimum scattering. The position of the quasi Fermi level relative to the conduction band of the semiconductor at the

metal–semiconductor interface should be adjusted to align the bands and minimize scattering rates. This can be done by selecting metallic nanocrystals of the appropriate material and size and adjusting their volume fraction. The bottom-up approach presented in this work (Fig. 1) offers sufficient materials versatility to harness the key benefits of such 3D modulation doping by selecting materials with appropriate Fermi levels and allowing a facile control of nanocrystals size and volume fraction. For this study, PbS was selected as inexpensive host semiconductor that comprises Earth-abundant elements (that is, contrary to tellurides), and holds great potential for reaching high thermoelectric efficiencies; with reported ZT values of up to 1.3 at 923 K in PbS–CdS[21]. Silver is chosen as a nanoscopic metallic dopant owing to its low work function (4.26–4.9 eV) (refs 22–24) needed for efficient injection of electrons into the PbS conduction band. The PbS–Ag nanocomposites derived from colloidal PbS and Ag nanocrystals exhibit high electrical conductivities due to (i) injection of electrons from Ag nanoinclusions to the host PbS and (ii) improved charge carrier mobility. The simultaneous combination of high electrical conductivity, relatively large Seebeck coefficients, and reduced thermal conductivities provides thermoelectric figures of merit up to 1.7 at 850 K.

Results

PbS–Ag nanocomposites. PbS–Ag nanocomposites were produced by combining cubic PbS nanocrystals (*ca.* 11 nm, Fig. 2a) and spherical Ag nanocrystals (*ca.* 3 nm Fig. 2b), followed by the evaporation of a solvent. Thereby obtained powdered nanocrystal blend was annealed to remove residual organic compounds and then hot-pressed into pellets. This simple procedure yielded nanocomposites with a highly homogeneous distribution of metallic Ag nanodomains at the interfaces of PbS grains, as evidenced by high-resolution transmission electron microscopy (HRTEM), high-angle annular dark field scanning transmission electron microscopy and energy-dispersive X-ray spectroscopy (Fig. 2c,d). Further atomistic insights into bonding and chemical identities of constituents were obtained by X-ray absorption

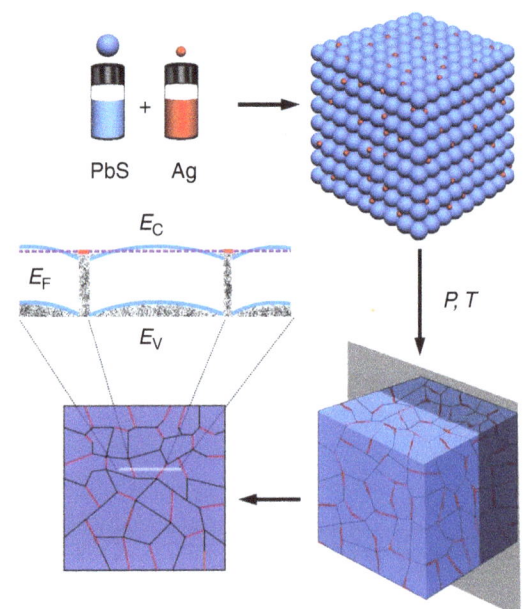

Figure 1 | Bottom-up Design. Bottom-up assembly process to produce PbS–Ag TE nanocomposites from the assembly of PbS (blue) and Ag (red) NCs, and the corresponding band alignment of the resulting nanocomposite.

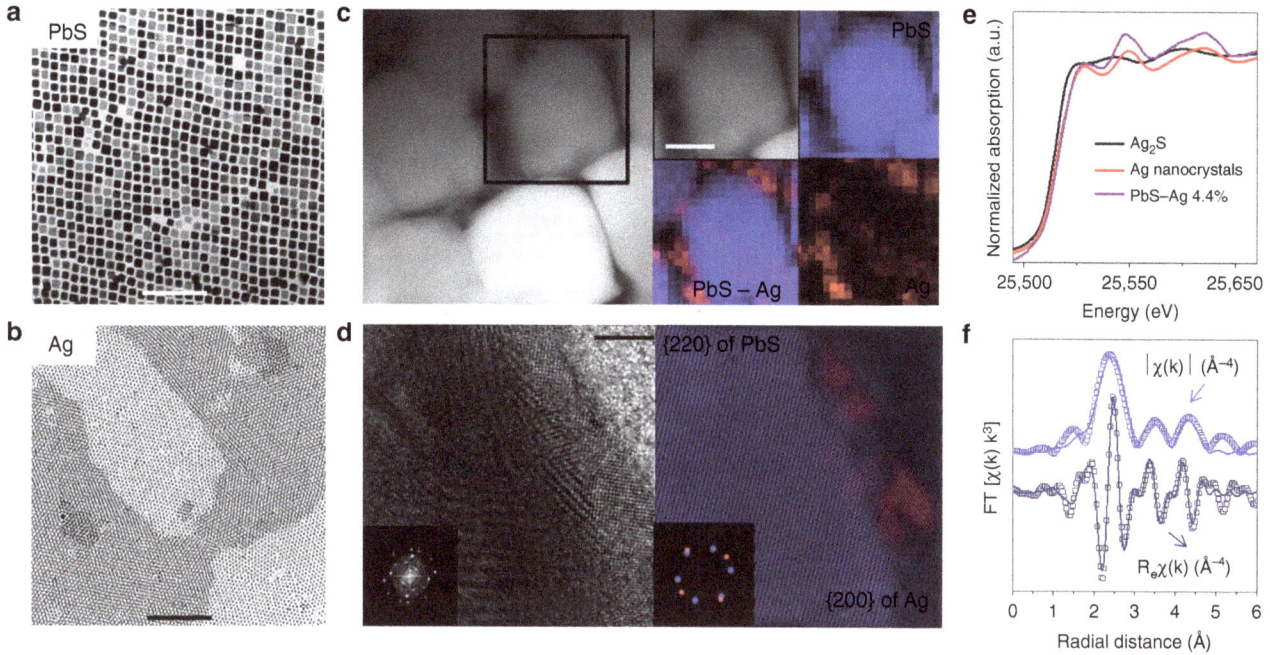

Figure 2 | Structural and compositional characterization of initial NCs and resulting PbS–Ag 4.4 mol% nanocomposite. TEM micrographs of (**a**) PbS and (**b**) Ag NCs; (**c**) HAADF-STEM micrograph and elemental EDX maps; and (**d**) HRTEM micrograph of the PbS–Ag interface with the corresponding power spectrum (inset) and filtered colourful composite image for the {220} family of planes of PbS (blue) and {200} family of planes of Ag (red). In the filtered images, PbS and Ag are visualized along their [111] and [001] zone axes, respectively (Supplementary Note 1); (**e**) Ag K-edge XANES spectra of Ag_2S reference (black), Ag NCs (red) and PbS–Ag 4.4 mol% nanocomposite (purple); (**f**) Fourier transform magnitude, $|\chi_{(k)}|$, and real part, $_{Re}\chi_{(k)}$, of the Pb L_3-edge EXAFS spectrum, the experimental data are given by the dotted line, the best fit by the solid line. Scale bars, 100 nm (**a**), 100 nm (**b**), 50 nm (**c**), 5 nm (**d**).

spectroscopy (XAS) at the Ag K edge (25515 eV) and Pb L_3 edge (13035 eV), respectively. Linear combination fitting of the X-ray absorption near edge structure (XANES) around the Ag absorption K edge, using references of Ag_2S and Ag metal, confirmed that at least 97% of Ag retained its metallic state (Fig. 2e, Supplementary Fig. 1 and Table 1). Only a small percentage of Ag may had diffused as Ag^+ within the PbS matrix. In addition, fitting of the Pb L_3 edge extended X-ray absorption fine structure spectra indicated interatomic distances and coordination numbers characteristics of PbS[25] (Fig. 2f). Metallic lead, lead oxide or lead sulfate species did not noticeably contribute to the spectra, suggesting that the presence of these phases can be disregarded (Supplementary Table 2).

Thermoelectric properties. To determine the effect of the Ag content on thermoelectric performance, a series of nanocomposites with Ag concentrations up to 5 mol% were prepared and analysed. Ag-free PbS nanomaterials exhibited relatively low electrical conductivities, which greatly increased from 0.07 S cm^{-1} at room temperature (RT) up to 46 S cm^{-1} at 850 K due to band to band charge carrier thermal excitation (Fig. 3a). This increase was accompanied by a sign inversion of the Seebeck coefficient from positive to negative at around 470 K (Fig. 3b). On the contrary, PbS–Ag nanocomposites possess significantly higher, Ag concentration-dependent, electrical conductivities. At RT, electrical conductivities of up to 660 S cm^{-1} were measured for Ag concentrations above 4 mol% (Fig. 3a). Furthermore, over the whole studied temperature range of 300–850 K, PbS–Ag nanocomposites exhibit negative Seebeck coefficients. Unlike to electrical conductivity, Seebeck coefficient decreases with increasing Ag content. Hence an optimal concentration of *ca.* 4.4–4.6 mol% Ag nanocrystals was established for maximizing the PF (Fig. 3c). Overall, around 20 PbS–Ag pellets were produced, all showing PFs above 1 mW m^{-1} K^{-2} at 850 K, with a champion

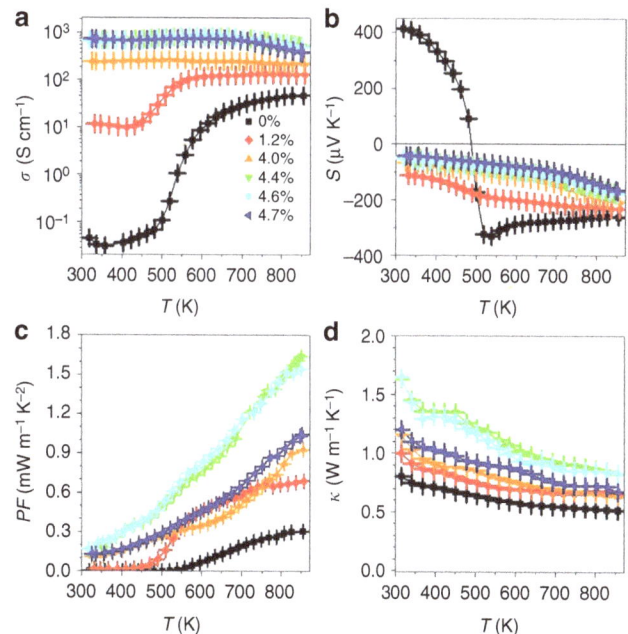

Figure 3 | Thermoelectric characterization of PbS–Ag nanocomposites. Temperature dependence of the (**a**) electrical conductivity, σ; (**b**) Seebeck coefficient, S; (**c**) thermal conductivity, κ; and (**d**) power factor, PF. Error bars were estimated from the repeatability of the experimental result; 3–5 measurements were carried out for each material.

value of 1.68 mW m^{-1} K^{-2} at 850 K for a PbS–Ag 4.4 mol%. This is a sixfold increase over identically prepared Ag-free samples, and a significant improvement by 23–47% over previously reported PbS-based nanocomposites[21,26].

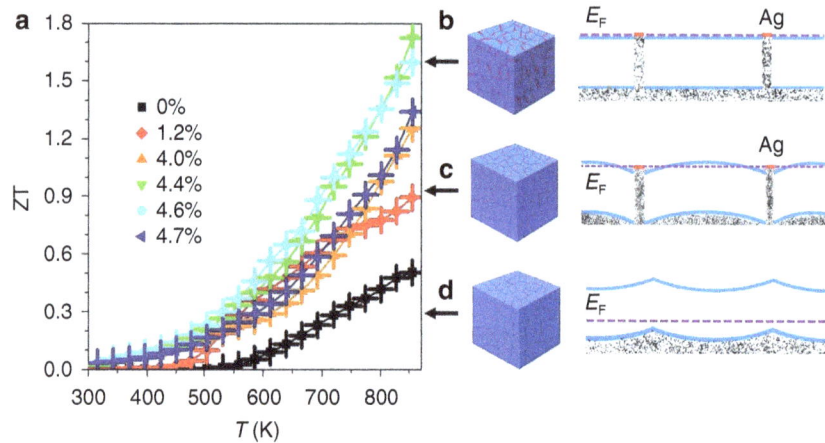

Figure 4 | Figure of merit and schematic representation of the electron energy band alignment. (**a**) Figure of merit, ZT, of PbS and PbS–Ag pellets; (**b**) band alignment in PbS–Ag nanocomposite with a larger amount of Ag and flat bands across the whole PbS domains; (**c**) band aligment in PbS–Ag nanocomposite with a low Ag volume fraction, showing electron energy wells in between PbS domains; (**d**) band alignment in bare PbS with an upward band-bending at the PbS intergrains. Error bars were estimated from the repeatability of the experimental result; 3-5 measurements were carried out for each material.

The thermal conductivities of the PbS–Ag nanocomposites monotonically increased with the Ag content (Fig. 3d) due to the increase of the electronic contribution (Supplementary Fig. 2 and Methods), yet not exceeding in the high-temperature range the values reported for PbS-based nanocomposites produced by co-precipitation of secondary phases[21,26,27]. Thus, beyond injecting charge carriers and facilitating charge transport between PbS nanocrystals, Ag nanodomains also assisted in blocking phonon propagation. The relatively low thermal conductivities may also in part result from phonon scattering at nanodomains within the large PbS grains as observed by extensive HRTEM analysis (Supplementary Figs 3 and 4 and Note 2).

Overall, the outstanding electrical properties along with low thermal conductivities of PbS–Ag samples resulted in thermoelectric figures of merit of up to ZT = 1.7 at 850 K (Fig. 4a). This value corresponds to a 30% increase over the highest figure of merit obtained for PbS to date (ZT = 1.3, $Pb_{0.975}Na_{0.025}S + 3\%$ CdS)[21] and is comparable to the highest thermoelectric figures of merit reported for other lead chalcogenide materials (Supplementary Table 3)[3,21,26-30].

Discussion
The high PFs displayed by PbS–Ag nanocomposites are at the origin of their outstanding thermoelectric figures of merit. Such high PFs could not be explained by a simple weighed sum of the properties of two randomly distributed compounds (Supplementary Table 4 and Supplementary Discussion). Neither can we explain the high electrical conductivity by percolation transport through Ag domains, as much lower Seebeck coefficients would be expected for a metallic conductor. It must be also pointed out that doping of PbS with Ag^+ ions cannot explain these transport properties either, since previous studies have demonstrated Ag^+ to be a p-type dopant for PbS[31-34]. Simultaneous combination of high electrical conductivity and relatively large Seebeck coefficients can be explained by an efficient injection of electrons from the metal to the conduction band of the semiconductor (Fig. 4b–d). In this regard, RT Hall charge carrier concentration measurements evidenced an increase in concentration of majority charge carriers from $p = 1 \times 10^{16}$ in the bare PbS nanomaterial to $n = 3 \times 10^{19}$ in PbS–Ag 4.4 mol% samples (Fig. 5 and Supplementary Table 5). This is consistent with the initial Ag Fermi level above that of the intrinsic PbS.

What is surprising is that the obtained charge carrier mobilities also increased with the Ag introduction, from $20\,cm^2\,V^{-1}\,s^{-1}$ for bare PbS to $90\,cm^2\,V^{-1}\,s^{-1}$ for PbS–Ag 4.4 mol% (Supplementary Table 5). In a simple modulation doping scenario, one would expect the injection of charge carriers from Ag nanodomains into PbS to have little negative impact on the charge carrier mobility. However, in PbS–Ag nanocomposites, the actual effect of Ag was found to facilitate charge transport through the material. We attribute this to a reduction of the energy barriers between PbS crystal domains (Fig. 4b–d). To determine the band alignment, we used the Anderson model to align the vacuum levels of Ag and PbS and then solve the Poisson equation self consistently assuming a parabolic two-band model for PbS (Fig. 5a). We used the Ag work function, which depends on its crystallographic surface and domain size, as a fitting parameter to fit the experimentally measured Hall data (Fig. 5b). The obtained fitted parameter was 4.4 eV, which is in the correct range (4.26–4.9 eV) (refs 22,23).

At a bulk metal–semiconductor junction, the Fermi level is pinned by the metallic layer due to the large carrier density in the metallic layer (Fig. 5a). However, for small nanocrystals, the number of electrons is limited and at low metal concentrations it may be not enough to completely pin the Fermi level. To simulate charge transfer from Ag nanocrystals to PbS, we assumed Ag nanocrystals were spheres of radius 1.5 nm (to replicate the 3-nm size of Ag nanodomains observed) embedded within another sphere made out of PbS. The radius of PbS sphere was determined by the volume fraction of each sample (Fig. 5b, inset). Figure 5c shows the Fermi level with respect to the bottom of the conduction band. This is an indicator for the effective well depth for the electrons as marked in Fig. 5a. As can be seen in Fig. 5c, when the Ag fraction increases, the Fermi level also increases, lowering the effective well depth in between PbS grains, which reaches zero at around a 0.5 % Ag volume fraction, that is, 1.5 mol%. At this point and beyond, there is no effective well in the path of the electrons (Fig. 4b). Therefore, as the Ag fraction increases, the electron-nanocrystal scattering decreases, which is consistent with the observed enhancement in electron mobility.

Note that the high PFs found in PbS–Ag nanocomposites cannot be obtained by a conventional doping strategy, where the introduction of ionic impurities would lead to increased scattering of charge carriers. This is shown in Fig. 6a,b, where the thermoelectric properties of PbS nanomaterials with different

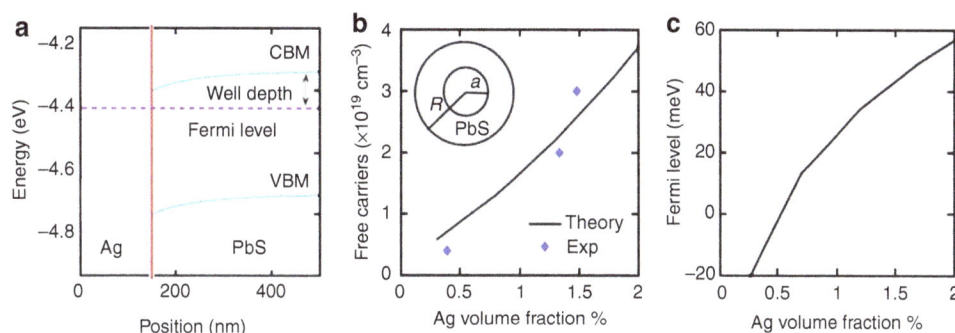

Figure 5 | Theoretical calculations. (**a**) Band alignment at the Ag-PbS interface considering bulk Ag and bulk PbS at 300 K. The Fermi level is pinned at -4.4 eV (silver work function). Electrons in PbS experience a well, as marked, when going to the silver side. (**b**) Carrier concentration versus silver volume fraction. Experimentally measured Hall data are shown by blue dots. Free carrier concentration, calculated using Anderson model and Poisson solver for the geometry shown in the inset of the figure, is shown by solid line. The inner sphere represents a silver NC ($a = 1.5$ nm) and the outer sphere represent PbS host matrix with radius R (silver volume fraction $= (a/R)^3$. (**c**) Fermi level plotted with respect to the conduction band minimum of PbS (far away from the interface). When negative, this value corresponds to the well depth experienced by electrons when moving between PbS grains.

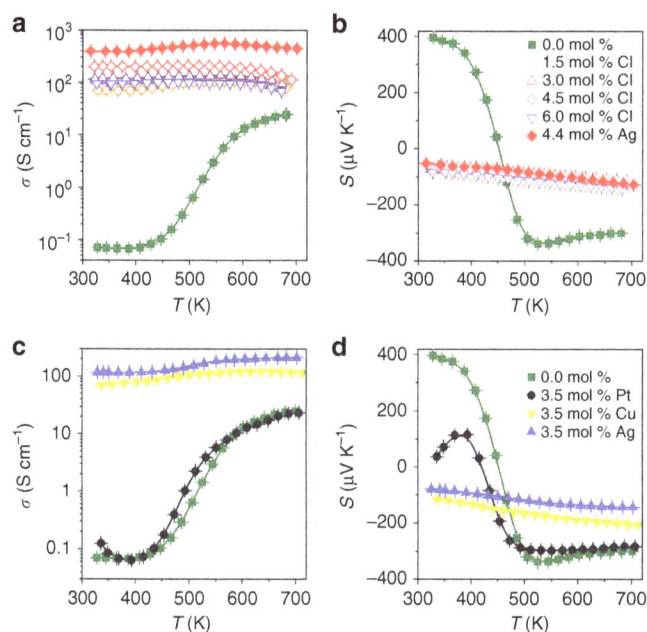

Figure 6 | Electrical properties of PbS nanomaterials with different doping strategies. (**a**) σ and (**b**) S of x mol% PbS ($x = 1.5, 3.0, 4.5$ and 6.0) doped with Cl (open symbols) compared with pure PbS (0.0 mol %) and PbS–Ag (4.4 mol%). (**c**) σ and (**d**) S of PbS (0.0 mol%) and PbS-X (3.5 mol%) with X = Pt, Cu and Ag.

chlorine concentrations are presented. By introducing Cl⁻ ions, a common dopant used in n-type PbS, the electron density can be increased, which translates into an enhancement of the electrical conductivity[35]. However, the highest electrical conductivities reached by halide doping remained below the values obtained for PbS–Ag nanocomposites. Thus, despite that PbS:Cl (4.5 mol%) nanomaterials had charge carrier concentrations in the same order of magnitude and similar Seebeck coefficients as the PbS–Ag (4.4 mol%) nanocomposites, the later showed significantly higher electrical conductivities and, therefore, higher PFs.

Besides PbS–Ag nanocomposites, a variety of other compositions were accomplished from colloidal nanocrystal building blocks, often showing similar synergistic effects on charge transport. For instance, we have carefully examined composites of PbS with nanoscopic Cu (Supplementary Fig. 5) and Pt (Supplementary Fig. 6). Like Ag, Cu has a low work function and hence is able to inject charges into the conduction band of PbS.

On the contrary, Pt has a much higher work function and does not exhibit efficient charge transfer (Fig. 6c,d).

Colloidal nanocrystals can be prepared with unmatched control over size, composition, shape, crystal phase and surface chemistry and benefit from facile handling and mixing in stable dispersions[36]. The current availability of a rich palette of such building blocks lends the opportunity to create a plethora of different nanocomposites by simply blending nanocrystals of various materials in the appropriate proportions and, afterward, consolidating them into arbitrarily shaped composites. Thus the facile bottom–up approach used here allows engineering a nearly endless variety of nanocomposites, which will allow a high-throughput screening of materials in the effort to maximize the thermoelectric energy conversion efficiency of durable, silent and scalable thermoelectric devices.

Methods

Nanocomposite preparation. *Chemicals.* Lead(II) oxide (PbO, 99.9%), copper(I) acetate (CuOAc, 97%), silver nitrate (AgNO₃, ≥99.8%), iron(III) nitrate non-ahydrate (Fe(NO₃)₃·9H₂O, 99.99%), platinum acetylacetonate (Pt(acac)₂, 97%), manganese(0) carbonyl (Mn₂(CO)₁₀, 98%), elemental sulfur (99.998%), oleic acid (OA, tech. 90%), 1-octadecene (ODE, 90%), oleylamine (OLA, tech. 70%) and benzyl ether (BE, 98%) were purchased from Aldrich. Tri-n-octylamine (TOA, 97%) was purchased from Across. Tetradecylphosphonic acid (TDPA, 97%) was purchased from PlasmaChem. Hexane, toluene, ethanol, anhydrous chloroform and anhydrous methanol were obtained from various sources. All chemicals, except OLA, were used as received without further purification. OLA was distilled to remove impurities.

Synthesis of nanocrystals. All syntheses were carried out using standard air-free techniques; a vacuum/dry argon Schlenk line was used for synthesis and an argon-filled glove box for storing and handling air and moisture-sensitive chemicals.

PbS nanocrystals with a mean edge size of 11 nm were prepared similarly to previously reported procedures[35]. In a typical synthesis, PbO (4.46 g, 20 mmol) and OA (50 ml, 0.158 mol) were mixed with 100 ml of ODE. This mixture was degassed at RT and 100 °C for 0.5 h each to form the lead oleate complex. Then the solution was flushed with Ar, and the temperature was raised to 210 °C. At this temperature, a sulfur precursor, prepared by dissolving elemental sulfur (0.64 g, 20 mmol) in distilled OLA (20 ml, 0.061 mol), was rapidly injected. The reaction mixture was maintained between 195 °C and 210 °C for 5 min and then quickly cooled down to RT using a water bath. Ag nanocrystals with an average diameter of 2–3 nm were produced using a modified approach of that reported by Wang *et al.*[37] In a typical reaction, AgNO₃ (0.17 g, 1 mmol), Fe(NO₃)₃·9H₂O (0.04 g, 0.01 mmol), OA (10 ml, 0.0316 mol) and OLA (10 ml, 0.0305 mol) were mixed and placed under Ar at RT for 30 min. Afterwards the reaction mixture was heated to 120 °C at the rate of 5 °C min⁻¹ and kept at this temperature for an additional 60 min. Cu nanocrystals with an average diameter of 5–6 nm were prepared following the approach developed by Yang *et al.*[38] In a typical synthesis, TOA (50 ml, 0.114 mol) was heated in a 100 ml three-neck flask to 130 °C for 30 min under Ar atmosphere. After cooling to RT, CuOAc (613 mg, 5 mmol) and TDPA (696 mg, 2.5 mmol) were added to the flask. The mixture was heated to 180 °C and maintained at this temperature for 30 min. Then, the reaction temperature was further increased to 270 °C and held for another 30 min. Cu nanocrystals are highly air sensitive and easily oxidized. To avoid any possible oxidation, the nanocrystals were purified in an Ar filled glove box. Pt nanocrystals with an average diameter of 6 nm were prepared using the method developed by Murray et al.[39] In a typical synthesis, $Pt(acac)_2$ (80 mg, 0.20 mmol) was dissolved in BE (10 ml, 52.6 mmol), OLA (7.36 ml, 22.37 mmol) and OA (1.25 ml, 3.94 mmol) under Ar atmosphere for 30 min at 60 °C. The precursor mixture was heated to 160 °C and a solution of 80 mg $Mn_2(CO)_{10}$ in 1 ml of chloroform was rapidly injected. Afterwards, the temperature was heated to 200 °C and held for an additional 30 min at this reaction temperature. Finally, the crude solution was cooled to RT.

Blending of nanocrystals. In this work we prepared PbS-metal semiconductor nanocomposites with different metal concentrations. The blending of nanocrystals was performed by wetting 1 g of dried PbS nanocrystals (a powder) with different amounts of a 0.093 M solution of metallic nanocrystals in anhydrous chloroform. Subsequently, the solvent was allowed to evaporate under Ar atmosphere. The concentration of metallic nanocrystals in chloroform was initially estimated by mass (considering a 30% of organic ligand) and later verified by inductively coupled plasma (ICP) spectroscopy. The values were found to differ slightly between the estimation from the weight and from ICP. The final quantities reported in this work correspond to the values obtained from ICP measurements.

Pellet fabrication. Before powder consolidation in a hot press, the nanocrystal powders were treated thermally to decompose the remaining organic ligands present at the nanocrystal surface. All nanocrystal powders were heated to 450 °C at 10 °C min $^{-1}$ and held at this temperature for 1 h under an Ar flow. After cooling to RT, the nanocrystal powders were pressed using a custom-made hot press. In this system, heat is provided by an induction coil operated in the RF range and applied directly to a graphite die acting as a susceptor. Before hot pressing, coarse powders were ground into fine powder using a mortar inside the glove box and then loaded into a 10-mm diameter graphite die lined with 0.13-mm thick graphite paper. The filled die was placed in the hot press system. The densification profile applied an axial pressure of 45 MPa before heating the die to between 420 and 440 °C. The temperature was held between 420 and 440 °C for 4 min. The pressure was then removed and the die cooled to RT. The resulting pellets were >85% dense compared with theoretical maximum into air-stable monoliths measuring ~1-mm thick by 10 mm in diameter. The density of the pressed pellets was measured by the Archimedes method.

Structural characterization. The size and shape of the initial nanocrystals were examined by TEM using a ZEISS LIBRA 120 instrument, operating at 120 kV. Structural and compositional characterizations of the nanocomposites were examined after thermoelectrical characterization. TEM Samples were mechanically thinned to 20–30 μm and further thinned to electron transparency by Ar^+ polishing using a Gatan Precision Ion Polishing System. HRTEM and scanning transmission electron microscopy studies were conducted by using a FEI Tecnai F20 field emission gun microscope operated at 200 kV, which is equipped high-angle annular dark field and energy-dispersive X-ray spectroscopy detectors. ICP atomic emission spectrometry was used for elemental analysis of the nanocomposites, especially to determine the ratio between Pb and Ag, Pt or Cu. ICP atomic emission spectrometry measurements were carried out using Perkin–Elmer Optima instrument, model 3200RL, under standard operating conditions. Samples were prepared by microwave-assisted digestion of the dried materials in a mixture of HNO_3 and H_2O_2 in a closed container. X-ray powder diffraction analyses were collected directly on the as-synthesized nanocrystals and final pellets using a Bruker AXS D8 Advance X-ray diffractometer with Ni-filtered (2 μm thickness) Cu K_α radiation ($\lambda = 1.5406$ Å) operating at 40 kV and 40 mA (Supplementary Fig. 7). A LynxEye linear position-sensitive detector was used in reflection geometry. XAS measurements were carried out at the X10DA (SuperXAS) beamline at the Swiss Light Source, Villigen, Switzerland, which operated with a ring current of ~400 mA in top-up mode. The polychromatic radiation from the superbend magnet, with a magnetic field of 2.9 T and critical energy of 11.9 keV, was monochromatized using a channel cut Si(311) crystal monochromator. Spectra were collected on pressed pellets optimized to 1 absorption length at the Ag K edge (25515 eV) and Pb L_3 edge (13035 eV) in transmission mode. XAS data were treated with the Demeter software suite[40]. For all samples, three spectra were acquired and merged. These averaged XAS data were background-subtracted and normalized. Linear combination fitting of the Ag K edge X-ray absorption near edge structure spectra was performed over the energy range from 25,498 to 25,598 eV. The goal was to determine composition of Ag-species present in the PbS–Ag sample. For this, reference spectra of metallic Ag NPs and Ag_2S were combined linearly. Fourier transformation of the Pb L_3 edge extended X-ray absorption fine structure spectra was performed over the k range of $3 - 9$ Å $^{-1}$ yielding a pseudo radial structure function. The R-range from 1 to 4.8 Å was fitted using theoretical single scattering paths of bulk PbS (S6)[25] based on crystallographic data.

Thermoelectric characterization. *Electric properties.* The pressed samples were polished, maintaining the disk-shape morphology. Final pellets had a 10-mm diameter and were ~1 mm thick. The Seebeck coefficient was measured using a static DC method. Electrical resistivity data were obtained by a standard four-probe method. Both the Seebeck coefficient and the electrical resistivity were simultaneously measured with accuracies better than 1% in a LSR-3 LINSEIS system from RT to 850 K, under helium atmosphere. Samples were held between two alumel

electrodes and two probe thermocouples with spring-loaded pressure contacts. A resistive heater on the lower electrode created temperature differentials in the sample to determine the Seebeck coefficient. Samples measured up to 850 K were spray coated with boron nitride to minimize out-degassing except where needed for electrical contact with the thermocouples, heater and voltage probes. In addition, before high-temperature measurements, samples were heated within the LINSEIS system in a He atmosphere up to 850 K at 3 K min $^{-1}$ and hold at this temperature for 10 min with the boron nitride coating. Such preliminary treatment warrants sample stability for all the cycles tested (Supplementary Figs 8 and 9 and Discussion). Carrier concentration and mobility were estimated from Hall Effect measurements, which were performed at RT using an Ecopia HMS-3000 set-up with golden spring-loaded contacts positioned at the edges of plates in the Van der Pauw configuration.

Thermal properties. An XFA 600 Xenon Flash Apparatus was used to determine the thermal diffusivities of all samples with an accuracy of ca. 6%. Total thermal conductivity (κ) was calculated using the relation $\kappa = DC_p\rho$, where D is the thermal diffusivity, C_p is the heat capacity and ρ is the mass density of the pellet. The ρ values were calculated using the Archimedes method. The specific heat (C_p) of the samples was measured using a Differential Scanning Calorimeter DSC 204 F1 Phoenix from NETZSCH (Supplementary Fig. 10). The electronic contribution of the thermal conductivity was calculated using the Wiedemann–Franz law (Supplementary Fig. 11 and Methods).

References

1. Bell, L. E. Cooling, heating, generating power, and recovering waste heat with thermoelectric systems. *Science* **321**, 1457–1461 (2008).
2. Vineis, C. J., Shakouri, A., Majumdar, A. & Kanatzidis, M. G. Nanostructured thermoelectrics: big efficiency gains from small features. *Adv. Mater.* **22**, 3970–3980 (2010).
3. Biswas, K. et al. High-performance bulk thermoelectrics with all-scale hierarchical architectures. *Nature* **489**, 414–418 (2012).
4. Ibáñez, M. et al. Core-shell nanoparticles as building blocks for the bottom-up production of functional nanocomposites: PbTe-PbS thermoelectric properties. *ACS Nano* **7**, 2573–2586 (2013).
5. Hsu, K. F. et al. Cubic $AgPb_mSbTe_{2+m}$: bulk thermoelectric materials with high figure of merit. *Science* **303**, 818–821 (2004).
6. Kim, S. I. et al. Dense dislocation arrays embedded in grain boundaries for high-performance bulk thermoelectrics. *Science* **348**, 109–114 (2015).
7. Ibáñez, M. et al. Crystallographic control at the nanoscale to enhance functionality: polytypic Cu_2GeSe_3 nanoparticles as thermoelectric materials. *Chem. Mater.* **24**, 4615–4622 (2012).
8. Bahk, J.-H., Bian, Z. & Shakouri, A. Electron energy filtering by a nonplanar potential to enhance the thermoelectric power factor in bulk materials. *Phys. Rev. B* **87**, 075204 (2013).
9. Ohta, H. et al. Giant thermoelectric Seebeck coefficient of a two-dimensional electron gas in SrTiO3. *Nat. Mater.* **6**, 129–134 (2007).
10. Heremans, J. P. et al. Enhancement of thermoelectric efficiency in PbTe by distortion of the electronic density of states. *Science* **321**, 554–557 (2008).
11. Harman, T. C., Taylor, P. J., Walsh, M. P. & LaForge, B. E. Quantum dot superlattice thermoelectric materials and devices. *Science* **297**, 2229–2232 (2002).
12. García de Arquer, F. P., Lasanta, T., Bernechea, M. & Konstantatos, G. Tailoring the electronic properties of colloidal quantum dots in metal–semiconductor nanocomposites for high performance photodetectors. *Small* **11**, 2636–2641 (2015).
13. Zebarjadi, M., Esfarjani, K., Dresselhaus, M. S., Ren, Z. F. & Chen, G. Perspectives on thermoelectrics: from fundamentals to device applications. *Energ. Environ. Sci.* **5**, 5147–5162 (2012).
14. Dingle, R., Störmer, H. L., Gossard, A. C. & Wiegmann, W. Electron mobilities in modulation-doped semiconductor heterojunction superlattices. *Appl. Phys. Lett.* **33**, 665–667 (1978).
15. Hicks, L. D. & Dresselhaus, M. S. Thermoelectric figure of merit of a one-dimensional conductor. *Phys. Rev. B* **47**, 16631–16634 (1993).
16. Moon, J., Kim, J.-H., Chen, Z. C. Y., Xiang, J. & Chen, R. Gate-modulated thermoelectric power factor of hole gas in Ge–Si core–shell nanowires. *Nano Lett.* **13**, 1196–1202 (2013).
17. Zebarjadi, M. et al. Power factor enhancement by modulation doping in bulk nanocomposites. *Nano Lett.* **11**, 2225–2230 (2011).
18. Yu, B. et al. Enhancement of thermoelectric properties by modulation-doping in silicon germanium alloy nanocomposites. *Nano Lett.* **12**, 2077–2082 (2012).
19. Pei, Y.-L., Wu, H., Wu, D., Zheng, F. & He, J. High thermoelectric performance realized in a BiCuSeO system by improving carrier mobility through 3D modulation doping. *J. Am. Chem. Soc.* **136**, 13902–13908 (2014).
20. Wu, D. et al. Significantly enhanced thermoelectric performance in n-type heterogeneous BiAgSeS composites. *Adv. Funct. Mater.* **24**, 7763–7771 (2014).
21. Zhao, L.-D. et al. Raising the thermoelectric performance of p-type PbS with endotaxial nanostructuring and valence-band offset engineering using CdS and ZnS. *J. Am. Chem. Soc.* **134**, 16327–16336 (2012).
22. Chelvayohan, M. & Mee, C. H. B. Work function measurements on (110), (100) and (111) surfaces of silver. *J. Phys. C* **15**, 2305 (1982).

23. Dweydari, A. W. & Mee, C. H. B. Work function measurements on (100) and (110) surfaces of silver. *Phys. Status Solidi A* **27**, 223–230 (1975).
24. Knapp, R. A. Photoelectric Properties of Lead Sulfide in the Near and Vacuum Ultraviolet. *Phys. Rev.* **132**, 1891–1897 (1963).
25. Noda, Y., Ohba, S., Sato, S. & Saito, Y. Charge distribution and atomic thermal vibration in lead chalcogenide crystals. *Acta Crystallogr. Sect. B* **39**, 312–317 (1983).
26. Zhao, L.-D. *et al.* High performance thermoelectrics from earth-abundant materials: enhanced figure of merit in PbS by second phase nanostructures. *J. Am. Chem. Soc.* **133**, 20476–20487 (2011).
27. Zhao, L.-D. *et al.* Thermoelectrics with earth abundant elements: high performance p-type PbS nanostructured with SrS and CaS. *J. Am. Chem. Soc.* **134**, 7902–7912 (2012).
28. Wu, H. J. *et al.* Broad temperature plateau for thermoelectric figure of merit ZT > 2 in phase-separated $PbTe_{0.7}S_{0.3}$. *Nat. Commun.* **5**, 4515 (2014).
29. Korkosz, R. J. *et al.* High ZT in p-Type (PbTe)1−2x(PbSe)x(PbS)x thermoelectric materials. *J. Am. Chem. Soc.* **136**, 3225–3237 (2014).
30. Girard, S. N. *et al.* High performance Na-doped PbTe–PbS thermoelectric materials: electronic density of states modification and shape-controlled nanostructures. *J. Am. Chem. Soc.* **133**, 16588–16597 (2011).
31. Yun, Z. *et al.* Thermoelectric transport properties of p-type silver-doped PbS with in situ Ag 2 S nanoprecipitates. *J. Phys. D Appl. Phys* **47**, 115303 (2014).
32. Shanyu, W. *et al.* Exploring the doping effects of Ag in p-type PbSe compounds with enhanced thermoelectric performance. *J. of Phys. D Appl. Phys* **44**, 475304 (2011).
33. Dow, H. S. *et al.* Effect of Ag or Sb addition on the thermoelectric properties of PbTe. *J. Appl. Phys.* **108**, 113709 (2010).
34. Ryu, B. *et al.* Defects responsible for abnormal n-type conductivity in Ag-excess doped PbTe thermoelectrics. *J. Appl. Phys.* **118**, 015705 (2015).
35. Ibáñez, M. *et al.* Electron doping in bottom-up engineered thermoelectric nanomaterials through HCl-mediated ligand displacement. *J. Am. Chem. Soc.* **137**, 4046–4049 (2015).
36. Kovalenko, M. V. *et al.* Prospects of nanoscience with nanocrystals. *ACS Nano* **9**, 1012–1057 (2015).
37. Li, L., Hu, F., Xu, D., Shen, S. & Wang, Q. Metal ion redox potential plays an important role in high-yield synthesis of monodisperse silver nanoparticles. *Chem. Commun.* **48**, 4728–4730 (2012).
38. Hung, L.-I., Tsung, C.-K., Huang, W. & Yang, P. Room-temperature formation of hollow Cu2O nanoparticles. *Adv. Mater.* **22**, 1910–1914 (2010).
39. Kang, Y. *et al.* Shape-controlled synthesis of Pt nanocrystals: the role of metal carbonyls. *ACS Nano* **7**, 645–653 (2012).
40. Ravel, B. & Newville, M. ATHENA, ARTEMIS, HEPHAESTUS: data analysis for X-ray absorption spectroscopy using IFEFFIT. *J. Synchrotron Radiat.* **12**, 537–541 (2005).

Acknowledgements

At IREC, work was supported by European Regional Development Funds and the Framework 7 program under project UNION (FP7-NMP 310250). M.I. and S.O. thank AGAUR for their Beatriu i Pinós post-doctoral grant (2013 BP-A00344) and the PhD grant, respectively. A.G. thanks to the Turkish Ministry of National Education for the PhD grant. A.G. and J.A. acknowledge the Spanish MINECO MAT2014-51480-ERC (e-ATOM) and the ICN2 Severo Ochoa Excellence Program. Z.L. and Y.L. thanks the China Scholarship Council for their PhD grant. IREC and ICN2 groups acknowledge the funding from Generalitat de Catalunya 2014SGR1638. The work performed at Rutgers was supported by NSF grant number 1400246. M.V.K. acknowledges partial financial support by the European Union (EU) via FP7 ERC Starting Grant 2012 (Project NANOSOLID, GA No. 306733). We thank Prof. Yaroslav Romanyuk for the use of his Ecopia HMS-3000 set-up for Hall Effect measurements and Dr Nicholas Stadie for reading the manuscript. The Swiss Light Source is thanked for the provision of beamtime at the SuperXAS beamline.

Author contributions

The manuscript was prepared through the contribution of all authors. M.I., Z.L., S.O., O.D. and D.C. produced the nanocomposite and performed the thermoelectric characterization. A.G., J.A., L.P. and M.N. performed structural nanocomposite characterization. M.Z. performed band alignment calculations. M.I., M.K. and A.C. planned and supervised the work and had major input in the writing of the manuscript.

Additional information

3

Towards do-it-yourself planar optical components using plasmon-assisted etching

Hao Chen[1], Abdul M. Bhuiya[2], Qing Ding[2], Harley T. Johnson[1] & Kimani C. Toussaint Jr[1]

In recent years, the push to foster increased technological innovation and basic scientific and engineering interest from the broadest sectors of society has helped to accelerate the development of do-it-yourself (DIY) components, particularly those related to low-cost microcontroller boards. The attraction with DIY kits is the simplification of the intervening steps going from basic design to fabrication, albeit typically at the expense of quality. We present herein plasmon-assisted etching as an approach to extend the DIY theme to optics, specifically the table-top fabrication of planar optical components. By operating in the design space between metasurfaces and traditional flat optical components, we employ arrays of Au pillar-supported bowtie nanoantennas as a template structure. To demonstrate, we fabricate a Fresnel zone plate, diffraction grating and holographic mode converter—all using the same template. Applications to nanotweezers and fabricating heterogeneous nanoantennas are also shown.

[1] Department of Mechanical Science and Engineering, University of Illinois Urbana-Champaign, Urbana, Illinois 61801, USA. [2] Department of Electrical and Computer Engineering, University of Illinois Urbana-Champaign, Urbana, Illinois 61801, USA. Correspondence and requests for materials should be addressed to K.C.T. (email: ktoussai@illinois.edu).

The maker movement has gained momentum in recent years thanks in large part to the reduction in cost of three-dimensional printers and the concomitant rise of inexpensive, do-it-yourself (DIY) microcontroller boards such as those made by Arduino and Raspberry Pi[1,2]. A strong theme with this movement is that reducing the number of steps in the manufacturing process, for example, from original equipment manufacturing to actual end product, could help spur learning and innovation, as well as potentially transform existing industries or usher in new ones, thereby leading to economic growth. This concept has even been extended to the production of low-cost DIY atomic force microscopes[3]. Notwithstanding the impressive range of activities and projects that have resulted from DIY kits, this trend has yet to lead to the realization of basic DIY optical components such as lenses or diffraction gratings. A major reason is because inexpensive additive manufacturing approaches result in effectively non-functional optical elements due to the inherent surface roughness between the various added layers. Interestingly, a hint towards realizable DIY optical components possessing basic functionality is to use planar structures based on either diffractive optical elements (DOEs) or metasurfaces[4–8]. DOEs are the state-of-the-art and utilize surface features on the order of the wavelength of light, in typically millimetre-thick plastic or quartz, to impart the desired phase onto an optical field. Moreover, advances in fabrication have made the design iteration step increasingly practical. DOEs are primarily limited by their operating bandwidth. Alternatively, metasurfaces[4–8] have shown to alter the properties of light with the added advantage of being even more ultrathin and lightweight[9]. In addition, by using artificial resonators, the constituent subwavelength nanoantennas, relatively broadband operation has been demonstrated[10,11]. However, current fabrication approaches are complicated by the fact that the feedback loop from design-to-fabrication-to-application is slow and non-trivial[6,7]. This limits quick testing and improvement on an initial design without having to begin anew each time in the cleanroom.

In this paper, we show how arrays of Au pillar-supported bowtie nanoantennas (pBNAs) can be fabricated once in a cleanroom and subsequently used as a template that enables table-top fabrication of multiple, planar optical components using laser-scanning optical microscopy[12,13]. This specialized template can be used to short-circuit the design iteration steps, obviating the need for in-depth knowledge of the phase-modifying behaviour of the constituent nanoparticles[12–14]. Thus, we demonstrate the table-top fabrication of a diffraction grating, Fresnel zone plate (FZP), and a holographic mode converter for generating orbital angular momentum—all using the same template. To achieve this streamlining in fabrication, we sacrifice the subwavelength sculpting of the optical wavefront offered by metasurfaces with one that is diffraction-limited, which is sufficient for many basic applications. We show that the enhanced local heating from plasmonics can enable facile table-top plasmon-assisted etching (PAE) of metal. We also demonstrate that PAE can be used to tune the radial extent of near-field trapping forces of nanotweezers[15,16], and offers a promising route to readily engineering novel nanoantenna arrays that are heterogeneous in both space and material composition[17–20].

Results

Understanding plasmon-assisted etching.
For DIY optics that may be based on the use of planar optical components, metasurfaces are arguably the emerging technology, whereby almost any desired planar optical component can be fabricated. This is especially attractive for applications where both small form factors and nearly negligible mass are sought while maintaining exquisite control of the optical field. Indeed, previous experiments have demonstrated the capability of metasurfaces in fabricating various planar optical components, such as lenses[21–23], blazed gratings[8,24,25], holographic plates[26], polarizers and wave plates[27,28]. In general, metasurfaces provide subwavelength control of the field, by way of judiciously placed nanoantennas, at the expense of an overall complex fabrication process that is slow to adapt to desired changes in end functionality or corrections to errors—a feature that hampers potential application to DIY optics. For example, designing metasurface-based flat lenses of different focal lengths requires first computing for each lens the required phase relationships for the nanostructures and then fabricating each in a cleanroom. PAE provides a complementary approach to fabricating planar optical components by eliminating the need to go back to the cleanroom and rather instead using a one-time fabricated nanoantenna template[12]. A flow diagram comparing our PAE approach to a metasurface-based method for fabricating planar optical components is depicted in Fig. 1. Here, we observe that metasurface fabrication described in Fig. 1a begins at the design stage, whereby a particular arrangement of nanoantennas must be computed for a target functionality, for example, focusing light to a specific distance. The design is then taken for fabrication in the cleanroom and subsequently subject to various characterization experiments. Based on these experiments the last stage could be the realization of the desired planar optical component. However, if characterization reveals errors in the component, or if a particular parameter needs to be tuned, then the entire process has to begin anew from the basic design stage. Now let us evaluate the PAE process described by Fig. 1b. The first step is to establish a template, which for the work carried out here is based on the use of Au pBNAs. The pBNAs template is then fabricated in the cleanroom. Next, the template is taken to a laser-scanning optical microscope, whereby spatially directed pulsed laser illumination is used to debond the Au nanoantennas from their silica pillars in a desired pattern. The fabricated structure can then be characterized and tested for errors. If there are errors or a need to change the parameters of the fabricated component, then the process goes back to the table-top fabrication stage. This is a significant difference compared with the metasurface approach. As a result, PAE offers a more intuitive, fast, and reconfigurable fabrication process with the tradeoff of diffraction-limited shaping of the optical wavefront.

Figure 2 provides an opportunity to take a closer look at the PAE process. A microscope stage is scanned for fixed focused laser illumination of the pBNA template, whereby the Au nanoantennas debond from silica only for the illuminated regions. Note that the beam could have been similarly scanned, but stage scanning was chosen for convenience. We use this approach to etch the initials 'UIUC', as observed in the dark-field image shown in Fig. 2a. Scanning electron micrographs (SEMs) of the etched structure are shown in Fig. 2b for both etched and unetched regions of the pBNA chip. From these images it is clear that this process cleanly debonds the metal, leaving the silica pillars unaffected. To determine the effect of input optical power and scan velocity of the focused laser beam on the PAE process, we independently control these parameters for a fixed pBNA array area ($10 \times 10\ \mu m^2$) and subsequently examine the percentage of metal completely removed in this region. Figure 2c summarizes the results, where the colour used corresponds to the percent efficiency of the process. The white dashed line delineates the threshold at which the PAE efficiency is $>90\%$. We observe that the PAE process has a stronger dependence on average input power than scanning velocity. What is not revealed in the plot is that we find debonding of the metal for some of the pBNAs for

a Planar optical components
via metamaterials

b Planar optical components
via plasmon-assisted etching

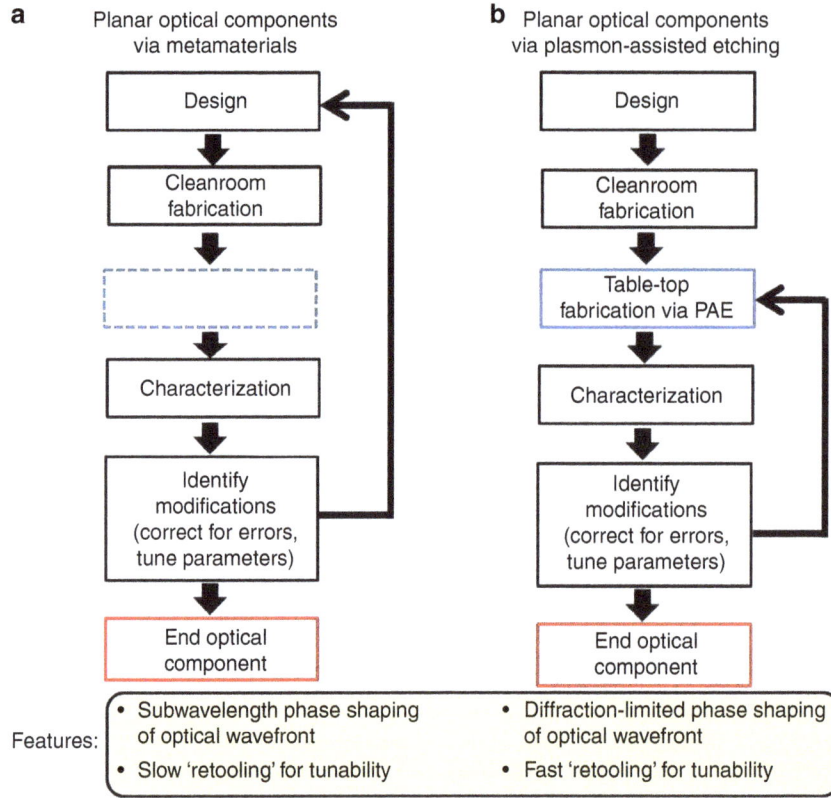

Figure 1 | Fabrication process of planar optical components. A flow diagram comparing the fabrication process of planar optical components using a (**a**) metasurface approach and (**b**) a PAE approach.

average input powers as low as 10 mW. However, due to common inhomogeneities resulting from the electron-beam lithography process, for example, subtle variations in nanoantenna gap size and radius of curvature, an average input power of 65 mW is required to achieve at least 90% PAE efficiency for most of the scanning velocities used.

A straightforward argument can be used to understand the PAE process. To begin, the pBNA structure is immersed in water and illuminated by a focused pulsed laser beam spectrally centred at a wavelength $\lambda = 780$ nm. The excitation source is a 100-fs pulsed, 80-MHz repetition rate, Ti:sapphire laser focused by a 0.6-numerical aperture (NA) microscope objective. Upon optical illumination the metallic nanoantenna structures begin to generate heat via optical absorption, and the corresponding heat power can be estimated through[29]

$$Q = \int_{\lambda_{min}}^{\lambda_{max}} \sigma_{abs}(\lambda) \langle I(\lambda) \rangle \, d\lambda, \quad (1)$$

where $\sigma_{abs}(\lambda)$ is the spectral absorption cross-section of the metal layer of the illuminated pBNAs and $I(\lambda)$ is the incident average intensity. The thermal conductivity ratio between the Au and the Ti ($k_{Au}/k_{Ti} \approx 14$) is much smaller than that between the Au and the surrounding water ($k_{Au}/k_{water} \approx 512$), and the Ti adhesion layer is firmly adhered to the gold bowties. In addition, the gold bowties have a significantly larger volume ($\sim 10\times$) and exhibit larger optical absorption than their Ti adhesion layers. Thus, most of the heat generated is in the gold bowties, and the temperature increase is assumed to be uniform throughout the metal layer. For pulsed illumination, the temperature increase in the bowties can further be estimated through[29]

$$\Delta T = \frac{\sigma_{abs} \langle I_0 \rangle}{V \rho_{Au} c_{Au} f}, \quad (2)$$

where V is the bowtie volume ($0.0011\ \mu m^3$), ρ_{Au} is the density of gold ($19,320\ kg\ m^{-3}$), c_{Au} is the heat capacity of gold ($129\ J\ kg^{-1}\ K^{-1}$) and f is the pulse repetition rate. This results in an absorption cross-section of $0.065\ \mu m^2$ for arrays of 525-nm spacing[30]. For input average powers near 90 mW, the metallic bowtie temperature can easily approach the melting point of bulk Au ($\sim 1,064\ ^\circ C$), where surface melting near highly curved regions already happens[31–33]. When illuminated on resonance, the pBNA structure has stronger absorption and thus larger temperature increment than that illuminated off resonance as described by our simulation results in Supplementary Fig. 1. As a result of the heat generated from this optical absorption, both the metal nanoantennas and the SiO_2 pillar at the interface undergo thermal expansion albeit with different thermal expansion coefficients. This effect leads to the generation of strain in the metal thin film, which later relaxes after the complete separation between the metal thin film and SiO_2 pillar.

In addition to optical illumination, the ambient water alone plays an important role in the debonding process. Previous studies have shown that water can contribute to facile debonding of a metal film from a SiO_2 substrate due to the strong polar interaction with the strained Si–O–Si crack-tip bonds[34]. In the context of the present work, Ti–O–Si bonds are believed to form during the e-beam deposition of the Ti adhesion layer. Once driven by external forces, the Ti–O–Si crack-tip bond reacts with water molecules to form Ti–O–H and Si–O–H bonds on each side of the separated interfaces and this process reduces the critical energy release rate required for delamination of the metal layer from the SiO_2 pillar. This mechanism has been referred to as water-assisted subcritical debonding, and has been used in applications such as the peel-and-stick process[34].

The delamination of the Au layer from the pBNA structure can be understood through an energy framework based on the

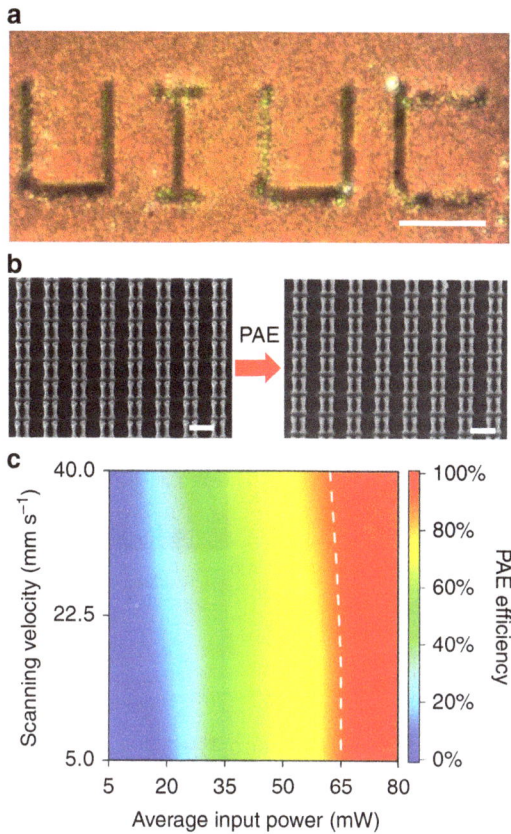

Figure 2 | PAE process. (**a**) Dark-field image of the letters 'UIUC' etched in the pBNA chip. Shaded and orange regions are the etched and unetched areas, respectively. Scale bar, 20 μm. (**b**) Example SEM images of an unetched (left) and etched (right) region. Scale bar, 500 nm. (**c**) Plot of the PAE efficiency as a function of input power and laser beam scanning velocity. A dashed white line indicates a PAE efficiency threshold of 90%.

where the properties of the thin film are expressed in terms of Young's modulus E_f, Poisson ratio v_f, strain ε_m and thickness h_f. The strain in the thin film is equal to $(\alpha_f - \alpha_s)\Delta T$, where α_f and α_s are the linear thermal expansion coefficients for the thin film and pillar substrate respectively, and ΔT is the temperature increase. This expression assumes that the film is much thinner than the substrate, and that the delamination front is straight, which is clearly not perfectly valid here. Nevertheless, we arrive at an estimate of the energy release rate of $\sim 0.9\,\mathrm{J\,m^{-2}}$ for input average powers near 90 mW (refs 31–33).

We may then conclude that while spontaneous delamination in water is not likely, the application of the thermal strain due to the laser heating provides a significant driving force that is on the order of magnitude of the work of adhesion. Factors such as the three-dimensional nature of the strain, possible variations in the interface quality, and uneven heating, all may provide conditions that could lead to delamination under the laser heating conditions. Furthermore, the likely reduction of the strain energy density near the edges of the pillar, where the thermal mismatch strain is partially relaxed by the free surface, may be offset by the reduction of the critical strain energy release rate due to the chemical effects of the water environment, as noted above.

Using PAE to fabricate basic planar, optical components. As shown in Fig. 3a, PAE is used to fabricate a diffraction grating with a period T of 10 μm and duty cycle of 50%. The yellow regions of the grating are the etched areas, displaying the colour of the glass substrate, while the unetched areas exhibit a green hue due to the gold antennas. To estimate the performance of the grating, we first employ FDTD simulations to numerically solve for the normalized reflected intensity as a function of input wavelength λ and diffraction angle θ_r, when the grating is illuminated by normally incident light for either the x- (along the long bowtie axis) or y- (orthogonal to the long bowtie axis) polarization direction, as shown in Fig. 3b, c and Supplementary Figs 2 and 3, respectively. It is found that because of the plasmonic response of the structure, the grating effect emerges in the wavelength range of ~ 600–800 nm for x-polarization and ~ 500–620 nm for y-polarization, whereby $\sim 60\%$ and 35% of the incident light are reflected at resonance, respectively. Within these bands of wavelengths, light is reflected back periodically at the surface of the pBNA chip making the component work as an amplitude grating. Outside the active wavelength regions, the diffraction grating behaves as a normal silica glass showing no diffraction. Thus, this type of structure can be used to route selected wavelengths, while leaving light at other wavelengths unaltered, particularly for applications related to ultra-compact optical systems where frequency demultiplexing is important. Further details on the characterization and simulation results are provided in Supplementary Figs 4–6. The experimentally measured diffraction patterns and associated cross-sectional intensity distributions are shown in Fig. 3d–g for laser wavelengths of 543, 660, 685 and 785 nm. At 660 nm, 65% of the light is concentrated into the first-diffracted order for the x-polarization, thereby behaving more like a blazed grating. In contrast, at 785 nm, most of the energy remains in the zeroth-order. Moreover, as shown in Supplementary Fig. 4, an increasing displacement of the first-order with respect to the wavelength is observed. The behaviour of our pBNA-based grating can be attributed to the wavelength selectivity of the plasmonic response. In this case, the dispersion relation is modified by the spectral envelope of the pBNAs. With incident wavelength approaching resonance, reflection of the unetched area is increased while the reflection from the etched area is kept the same. This effect leads to an increased diffraction efficiency

Griffith criterion[35]. In this framework, strain energy reduction in the Au layer provides a configurational force that promotes delamination. It is possible to derive a precise debonding criterion for some simple geometries; here, however, we can only estimate the relative tendencies. We assume that there is no significant energy dissipation due to plasticity in the Au layer, and that the geometry can be considered as a planar thin film with a straight delamination front and no external loads. We then consider the question of whether relief of strain energy per unit area of delamination of the Au layer, or the energy release rate, is sufficient to overcome the work of adhesion in the film/substrate interface. The work of adhesion, Γ_0, is given by[36]

$$\Gamma_0 = \gamma_f + \gamma_s - \gamma_{fs}, \qquad (3)$$

where γ_f and γ_s are the characteristic surface energy densities for the thin film and pillar substrate materials, respectively, and γ_{fs} is the interfacial characteristic free energy. Using values of the surface energy densities and interfacial free energies found in the literature[37–42], we estimate Γ_0 to be $1.8\,\mathrm{J\,m^{-2}}$ in the presence of water. This value suggests that spontaneous debonding of the thin film is highly unlikely without the heating due to laser illumination.

The work of adhesion is then compared with the strain energy density per unit area of the interface, or the energy release rate, which is given by[36]

$$G = \frac{(1 + v_f)E_f}{2(1 - v_f)}\varepsilon_m^2 h_f, \qquad (4)$$

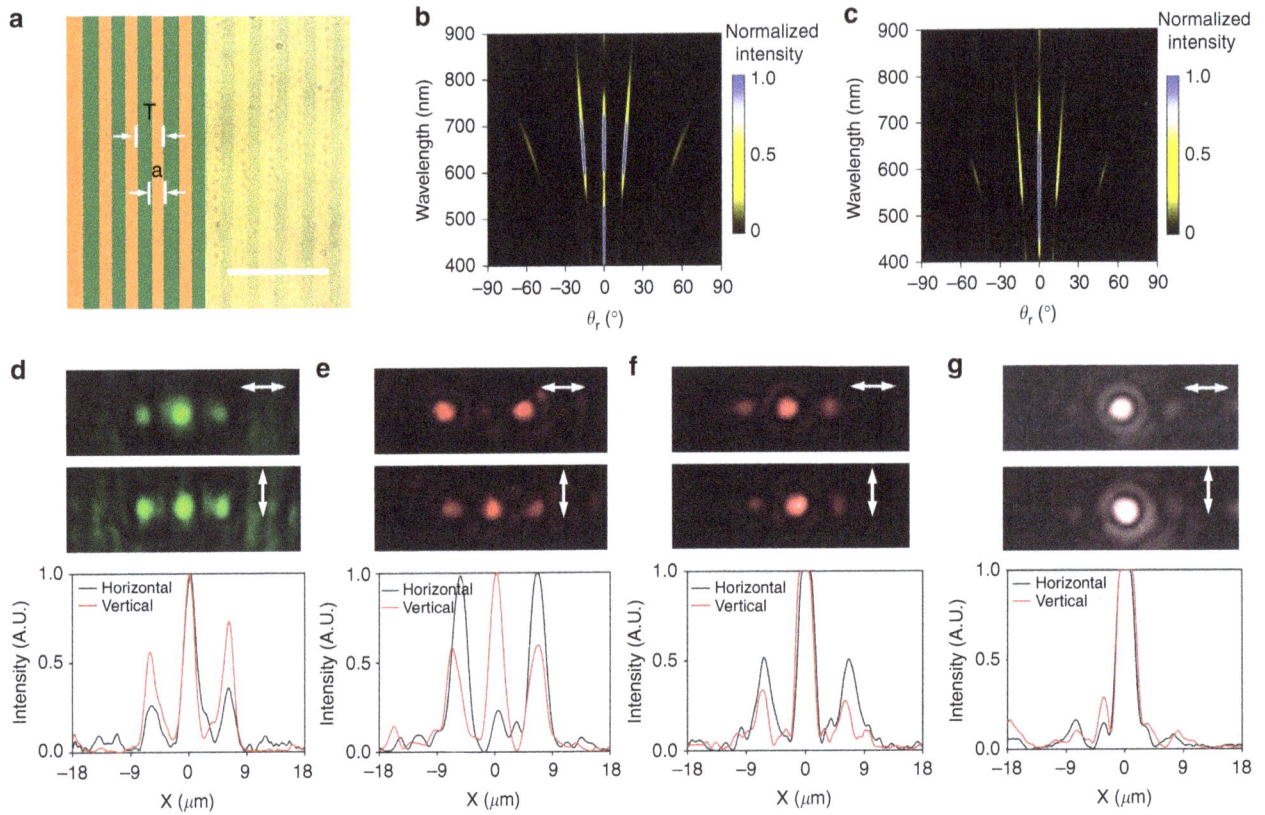

Figure 3 | PAE-fabricated planar diffraction grating. (a) The schematic representation of the grating structure on the left with relevant parameters noted is overlaid with the bright-field image of a fabricated diffraction grating on the right. Scale bar, 20 μm. In **b,c** we have the simulated normalized reflected intensity, for normally incident horizontal and vertical input polarization, respectively, as a function of input wavelength λ and the diffraction angle θ$_r$. **(d–g)** Experimentally obtained intensity distributions and the corresponding cross-sectional intensity profiles for illumination wavelengths of 543, 660, 685 and 785 nm, respectively. For each case, the polarization state of the incident beam is indicated in the top right corner by the arrows.

towards resonance, as shown qualitatively in Fig. 3d–g and quantitatively in Supplementary Fig. 5. Note that the steering angle of the grating increases for longer wavelengths, as this quantity is determined by the diffraction equation $T\sin\theta = n\lambda$, where T is the grating period. In terms of the diffraction efficiency, we observe that for the vertical polarization the strongest is at 543 nm and it decreases as the wavelength increases, whereas for the horizontal polarization the strongest is ~660 nm; the overall shape of the measured efficiency spectrum qualitatively agrees with the predicted spectrum, as shown in Supplementary Fig. 5. On the basis of our experiments, it is clear that although there is a phase contribution due to the plasmon resonance from the pBNAs, our fabricated components behave as 'amplitude-mostly' elements.

In addition, it is also possible to use PAE to fabricate a FZP, as shown in Fig. 4a, with the bright-field image of the actual fabricated pattern overlaid with the schematic representation. The 80×80-μm^2 area of the pBNA chip is divided into 15 alternate concentric circles of etched and unetched regions. The width of each Fresnel zone is governed by the equation: $r_n = \sqrt{n\lambda f + n^2\lambda^2/4}$, where n is an integer, λ is the wavelength of the light which the FZP is designed for and f is the designed focal length of the FZP. We set the focal length to 150 μm at an optical wavelength of 660 nm, the spacings and widths of the Fresnel zones are calculated with different values of n ($n = 1, 2, 3,...,15$). In our particular case, we have 15 alternate zones on a chip. The stepwise edge of rings for the high order n is attributed to the coarse step size in the movement of galvo mirrors. Figure 4b shows the measured contrast for each zone in

comparison with the theoretical value. The radius of the central zone is 10 μm, and the lens radius is about 40 μm. To demonstrate the lensing effect, we measure the cross-sectional intensity distribution in the focal plane with a plane wave broadband source illumination of the PAE-fabricated FZP, and the result is shown in Fig. 4c. A simulated intensity profile with a monochromatic visible light (660 nm) focused by a conventional lens (focal length of 150 μm) is shown in Fig. 4d for comparison. Because of the fact that a broadband light source is used in the experiment whereas a monochromatic source is used in simulation, the lateral width of the experimentally measured focal spot is larger as it is a net combination of many focal points produced by different wavelengths. The detailed description of the optical system used for characterization is explained in Supplementary Fig. 7.

We next use PAE to fabricate a fork dislocation grating via the optical set-up in Supplementary Fig. 8 to produce an optical vortex, as shown in Fig. 5a where the inset represents the schematic. Passage of a plane wave through this holographic structure results in a beam that carries orbital angular momentum (OAM)[41–43]. Optical vortices have been widely studied and play an important role in optical communications and particle trapping[44–47]. Generally, a spatial light modulator or a special organized liquid crystal display encoded with the computer-generated hologram of a 'fork' is used to impart OAM. Our PAE-fabricated fork grating has a period of 10 μm, a 50% duty cycle and a topological charge $l = 1$. Weakly focused light is used to illuminate the fork dislocation grating, resulting at the focal plane with zero and ± 1 diffraction orders as shown in Fig. 5b. As expected, the donut-shaped beam is generated in the ± 1

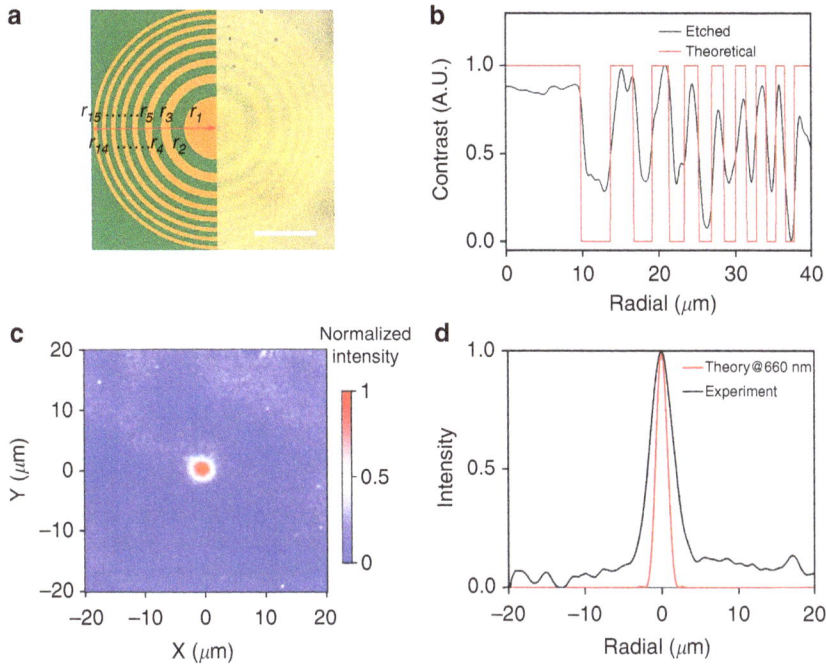

Figure 4 | PAE-fabricated planar Fresnel zone plate. (**a**) Bright-field image of a fabricated Fresnel zone plate with overlaid schematic of the theoretical design. Scale bar, 10 μm. (**b**) Comparison of theoretical (red) and experimental (black) contrast along the radial direction. (**c**) Experimentally obtained image of the focused intensity for illumination by a broadband source illumination. (**d**) Comparison of the theoretical (red) and experimental (black) focal-field intensity for laser illumination with wavelength of 660 nm.

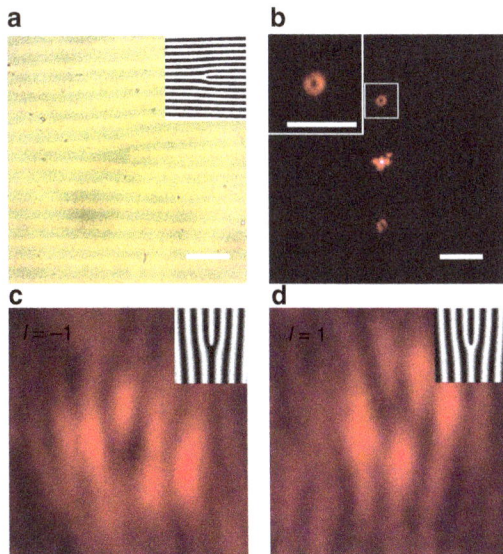

Figure 5 | PAE-fabricated holographic pitch-fork pattern. (**a**) Bright-field image of a fabricated pattern. (**b**) Experimentally obtained optical vortices generated at the focal plane. Two donut beams of ±1 order are shown on either side of the central, zeroth-order beam. The inset shows a zoomed-in view of the +1 vortex. (**c**) Intensity distribution obtained from interfering a plane wave with the $l = -1$ beam, and (**d**) the $l = +1$ beam. The insets show the calculated fork grating design. Scale bar, 20 μm.

diffraction orders due to a phase singularity at the centre. Note that both ±1 diffraction orders carry the same topological charge but opposite in sign. To extract the topological charge information for the diffracted order, the donut-shaped focal spot is interfered with a plane wave. The resulting patterns are shown in Fig. 5c,d for $l = -1$ and 1, respectively, where the forklet in the interference pattern indicates the helical phase

embedded in the diffraction orders. The experimental results agree with the simulated interference pattern, and the estimated diffraction efficiency for the fork dislocation grating is ~20%.

Application to nanotweezers. In addition to fabricating flat optical components, PAE is also a useful approach to locally shape the trapping landscape of the nanoantenna array. Plasmonic optical trapping has become a popular application of nanoantennas. The enhanced electromagnetic-field confinement offered by nanoantennas enables efficient trapping of micro and nano-objects using low-input optical power densities[48–50]. We have shown previously that plasmon-induced heating effect can result in an alteration of the plasmon resonance of the pBNAs by photothermally changing the morphology of the Au nanoparticles. We showed that this effect could be used to tune the local potential energy landscape of the pBNAs[51,52]. In our current work, PAE provides a method to selectively etch out the gold nanoantennas and, thus, form inactive trapping regions. PAE results in zero net optical trapping force at the etched areas leaving unetched areas unaffected. Consequently the trapping effect is more robust in the PAE-fabricated channels since a deeper potential well is created compared with that done by plasmon-assisted heating[12]. Furthermore, optofluidic channels etched by PAE can be made in real time and subsequent optical trapping can be performed in the same aqueous solution[12,53].

To demonstrate the effect of our approach on plasmonic trapping, we apply PAE to create predefined trapping areas using ~35.4 mW μm^{-2} of intensity at the focal plane. As a result, gold nanoantennas with 35-nm gap size are removed from the silica pillars in the exposed area and preserved at the unexposed area. These unexposed gold nanoantennas provide a large trapping force at resonance of ~0.02 pN. Specifically, we fabricate three kinds of predefined trapping patterns: a grating pattern of several line-shaped channels, a pattern of two adjacent crescent-shaped channels with an ~5-μm radius and a 2.5-μm-wide isolation belt, and a pattern of a circular channel of two different radii. For trapping, a water-based colloidal suspension of 1-μm-diameter SiO$_2$ particles is injected into the water solution. Each fabricated pattern is illuminated with an approximately collimated, 25-μm-diameter excitation beam obtained by focusing a 660-nm, horizontally polarized CW laser beam using a 0.6-NA objective. It is observed that particles are trapped merely in the predefined channels for all patterns as shown in Fig. 6. For the pattern of line-shaped channels, three particles in a chain are confined in a narrow channel as observed in Fig. 6a. Despite the activation of the next predefined channel which is 5 μm away in distance, all particles remain in a chain only within the single channel, proving the existence of a sharp potential gradient at the edge of the channel. Next, by translating the sample stage vertically, and hence the pBNA chip, the particles move downward, in the opposite direction, as shown in Fig. 6a (Supplementary Movie 1). When the crescent-shaped channels are illuminated, a cluster of particles is dragged towards the trap by convection and redistributed in the shape of an isolated crescent (Fig. 6b and Supplementary Movie 2). Once all the particles become stabilized, the separation between two clusters is clearly observed. Moreover, as depicted in Fig. 6c (Supplementary Movie 3) and Fig. 6d (Supplementary Movie 4), the predefined trapping area can be reduced in size, so that fewer number of particles are allowed to be trapped until eventually single-particle trapping is achieved.

A route to doubly heterogeneous nanoantenna arrays. The results in Fig. 7 demonstrate another great advantage of the pBNA platform and PAE—the flexibility in creating doubly heterogeneous nanoantenna arrays. The illumination system discussed above is focused at the plane of nanoantenna arrays to scan the left half region of the 80×80-μm^2 pBNA chip. After applying PAE, we deposit a 50-nm layer of Ti onto the entire pBNA template through e-beam evaporation and thus successfully fabricate doubly heterogeneous nanoantenna arrays, where the left half region that is etched consists of nanoantenna arrays with a 50-nm Ti layer on SiO$_2$ pillars, while the right half region that is unetched consists of nanoantenna arrays with a 50-nm Ti layer stacked on a 50-nm Au layer that sits on SiO$_2$ pillars. We investigate the optical response of our etched and unetched regions, as shown in Fig. 7a,b for simulation and experiment, respectively. The optical response of unetched and etched areas are assessed by measuring the spectral reflectance $R = 1$-$R_{raw}/\max(R_{raw})$, where R_{raw} is the raw reflectance obtained by focusing a white-light source onto modified regions. We observe that the reflectance of the pBNAs with the single Ti layer exhibits a dip ~ 550 nm, while that of the metal-stacking pBNAs exhibits a redshifted-dip ~ 590 nm; note that both are blue-shifted compared with the original gold pBNAs before PAE. In the SEMs shown in Fig. 7c, the left two columns of the pBNA structure represent the scanned etched area, where the dark regions on top of the silica pillars indicate the 50-nm single layer of Ti. The right two columns of the pBNA structure represent the unetched area where the metal-stacked pBNAs are successfully fabricated with a 50-nm Ti layer deposited on top of a 50-nm Au layer. The physical appearance at the boundary between the etched and unetched areas of the pBNA arrays is clearly distinguishable under SEM. However, as seen in Fig. 7c, the shape of the second layer of Ti cannot precisely replicate that of the first layer of Au, as Ti accumulates on the side wall of the Au layer as well. The uneven height of the second layer of Ti and the change in the radius of curvature of the nanoantennas attribute to the slight discrepancy in the reflectance curves we observe between simulated and experimental results. Nonetheless, in this case, PAE provides an extra degree of

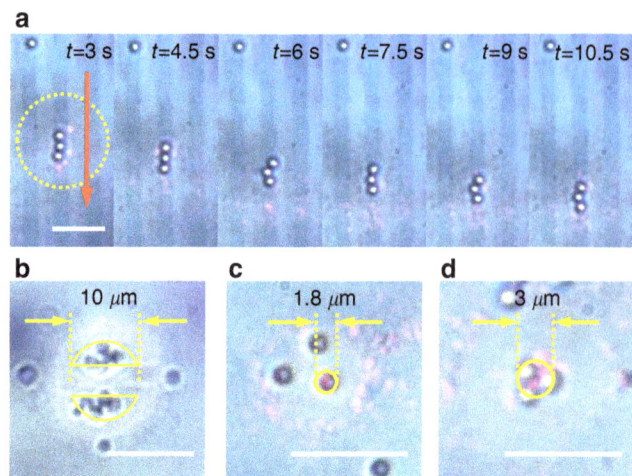

Figure 6 | Using PAE to shape the trapping landscape. (a) Selected frames from a video demonstrating the guiding of 1-μm diameter SiO$_2$ particles in a predefined grating-like channel etched into the pBNA chip. The yellow circle and red arrow indicates, respectively, the optically illuminated region and direction of motion of the approximately collimated beam. Scale bar, 15 μm. **(b)** Passive separation of microspheres into two crescent-shaped regions. **(c)** Demonstration of microspheres conforming to a predefined circular trapping region of 1.8-μm diameter, and **(d)** 3-μm diameter. Scale bars in **b–d**, 10 μm.

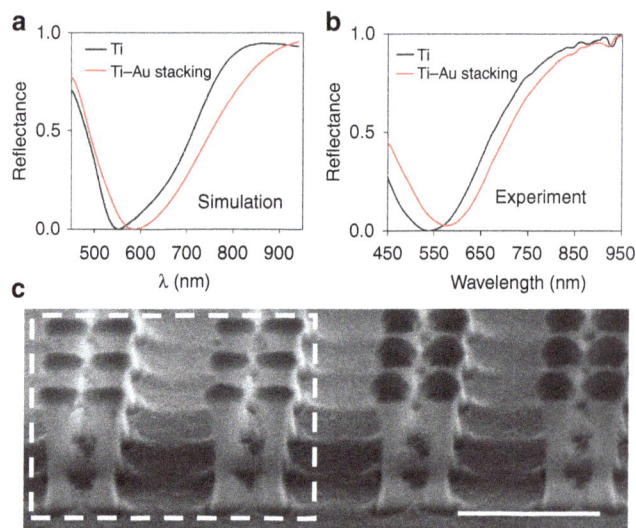

Figure 7 | Metal stacking. (a) Simulated and **(b)** experimental reflection spectra for Ti pBNAs (black) and stacked Ti-Au pBNAs (red). **(c)** Corresponding SEM images of Ti pBNAs (in dashed white box) and Ti-Au pBNAs. Scale bar, 500 nm.

freedom in manipulating the optical properties of such fabricated planar optical components.

Discussion

As a step towards the realization of DIY optical components, we have demonstrated a novel approach to fabricating a class of planar optical components based on a template structure that consists of two-dimensional arrays of gold pBNAs. The uniqueness in our approach is the use of PAE offered by the nanoantennas such that table-top debonding of the gold from the silica pillars can be realized via laser-scanning optical microscopy. Therefore, by simplifying the steps from design-to-fabrication-to-application, PAE could introduce DIY planar optical components to a broad community of researchers who may not necessarily be specialist in nanophotonics, yet alone metasurfaces design. This would especially be true when large-scale, nanomanufacturing technologies[54,55] are utilized in lieu of electron-beam lithography. As part of this study, we have shown the feasibility of utilizing the pBNA template and PAE to fabricate various kinds of optical components including a diffraction grating, FZP, and a holographic mode converter. In addition, we experimentally demonstrated how the pBNA template can be used to spatially tailor the optical potential energy landscape and thus enable preferential trapping and sorting of particles, which offers the possibility for fabricating optofluidic channels 'without walls.' Moreover, the heterogeneity such as material composition and geometry within the pBNA template can also be controllably modified, thereby enabling a promising approach to readily tuning the optical response, such as the dispersive characteristics, of two-dimensional nanoantenna-based surfaces locally.

Methods

Fabrication. The surface bounded bowtie nanoantennas (BNAs) are patterned by electron-beam lithography on 5-nm ITO-coated glass substrates. For patterning the BNAs, a 100-nm-thick PMMA electron-beam resists layer is dispensed on the substrate and baked at 200 °C for 2 min. After exposure with $100\ \mu C\ cm^{-2}$, the resist is developed in IPA:MIBK 3:1 for 45 s, rinsed with isopropyl alcohol for 30 s and dried under a stream of high-purity nitrogen. Using electron-beam evaporation technique, a 5-nm-thick Ti adhesion layer, a 50-nm-thick Au layer followed by an 8-nm Ni layer is deposited, respectively. After deposition, excess metal is removed by soaking the sample in acetone for 45 min. Finally, the pBNA structure fabrication is completed by performing reactive ion etching for 21 min with 70-s.c.c.m. CF_4, 35-mtorr pressure and 90-W power. The fabricated pBNAs have 35-nm gaps with a 525-nm array spacing and pillars with a height of 500 nm. Note that on a single substrate we have 128 sample regions (pBNA chips). Fabrication of new optical elements is achieved simply by translating under the microscope to one of these new regions.

Numerical simulations. Far-field reflectance spectra are calculated using the commercial FDTD software Lumerical FDTD-Solutions. The reflectance and transmittance of nanoantennas arrays are obtained by numerically solving Maxwell's equations under normal incidence plane wave source with polarizations parallel and perpendicular to the long antenna axis. In all simulations, the gold nanoantennas arrays supported by 500-nm SiO_2 pillars are placed on a SiO_2 substrate and situated at least one wavelength away from the edges of simulation box. Periodic boundary conditions and perfect matched layers are applied in the x–y plane and z direction (along the light propagation direction)[56]. The dielectric function of gold is taken from Johnson and Christy[57]. To resolve the nanostructure, the antennas and pillars are discretized into $3 \times 3 \times 3$-nm meshes. Its reflectance and transmittance are calculated by integrating the power flux through a power monitor situated in air 1 μm above and below the sample plane, then normalizing it with respect to the source power.

Microscope-based PAE. The beam used for PAE is derived from a Ti:sapphire laser (Spectra Physics Mai Tai) and polarized along the horizontal axis of the pBNA structure. Before coupling to the microscope, the laser beam is reflected by a pair of galvo mirrors which are operated for beam steering by Labview (National Instruments Corporation), a data acquisition and instrument control platform[58]. The galvo driver is connected to a DAQ board (NI USB-6221) with the position of the mirrors controlled by the output voltage. A 0.6-NA, collar-adjustable microscope objective (Olympus LUCPlanFLN × 40) is used to focus the incident laser beam onto the plane of the pBNA structure, which is placed on the sample stage of a standard microscope. A white-light source (Ocean Optics HL-2000) with an approximate bandwidth over 400–1,000 nm is used to measure the reflectance of the doubly heterogeneous nanoantenna arrays.

References

1. Deek, F. P. & McHugh, J. A. M. *Source: Technology and Policy*, (Cambridge University Press, New York, 2008).
2. Pearce, J. M. Building research equipment with free, open-source hardware. *Science* **337**, 1303–1304 (2012).
3. Grey, F. Creativity unleashed. *Nat. Nanotechnol.* **10**, 480 (2015).
4. Yu, N. F. *et al.* Light propagation with phase discontinuities: generalized laws of reflection and refraction. *Science* **334**, 333–337 (2011).
5. Kildishev, A. V., Boltasseva, A. & Shalaev, V. M. Planar photonics with metasurfaces. *Science* **339**, 1232009 (2013).
6. Zheng, G. *et al.* Metasurface holograms reaching 80% efficiency. *Nat. Nanotechnol.* **10**, 308–312 (2015).
7. Sun, S. L. *et al.* High-efficiency broadband anomalous reflection by gradient meta-surfaces. *Nano Lett.* **12**, 6223–6229 (2012).
8. Lin, D. M., Fan, P. Y., Hasman, E. & Brongersma, M. L. Dielectric gradient metasurface optical elements. *Science* **345**, 298–302 (2014).
9. Kang, M., Feng, T. H., Wang, H. T. & Li, J. S. Wave front engineering from an array of thin aperture antennas. *Opt. Express* **20**, 15882–15890 (2012).
10. Li, Z. Y., Palacios, E., Butun, S. & Aydin, K. Visible-frequency metasurfaces for broadband anomalous reflection and high-efficiency spectrum splitting. *Nano Lett.* **15**, 1615–1621 (2015).
11. Aieta, F., Kats, M. A., Genevet, P. & Capasso, F. Multiwavelength achromatic metasurfaces by dispersive phase compensation. *Science* **347**, 1342–1346 (2015).
12. Roxworthy, B. J., Bhuiya, A. M., Inavalli, V.V.G.K., Chen, H. & Toussaint, K. C. Multifunctional plasmonic film for recording near-field optical intensity. *Nano Lett.* **14**, 4687–4693 (2014).
13. Roxworthy, B. J., Bhuiya, A. M., Yu, X., Chow, E. K. C. & Toussaint, K. C. Reconfigurable nanoantennas using electron-beam manipulation. *Nat. Commun.* **5**, 4427 (2014).
14. Yu, N. F. & Capasso, F. Flat optics with designer metasurfaces. *Nat. Mater.* **13**, 139–150 (2014).
15. Juan, M. L., Righini, M. & Quidant, R. Plasmon nano-optical tweezers. *Nat. Photon.* **5**, 349–356 (2011).
16. Roxworthy, B. J. *et al.* Application of plasmonic bowtie nanoantenna arrays for optical trapping, stacking, and sorting. *Nano Lett.* **12**, 796–801 (2012).
17. Shegai, T., Johansson, P., Langhammer, C. & Kall, M. Directional scattering and hydrogen sensing by bimetallic Pd-Au nanoantennas. *Nano Lett.* **12**, 2464–2469 (2012).
18. Shegai, T. *et al.* A bimetallic nanoantenna for directional colour routing. *Nat. Commun.* **2**, 481 (2011).
19. Appavoo, K. & Haglund, R. F. Polarization selective phase-change nanomodulator. *Sci. Rep.* **4**, 6771 (2014).
20. Pakizeh, T., Abrishamian, M. S., Granpayeh, N., Dmitriev, A. & Kall, M. Magnetic-field enhancement in gold nanosandwiches. *Opt. Express* **14**, 8240–8246 (2006).
21. Aieta, F. *et al.* Aberration-free ultrathin flat lenses and axicons at telecom wavelengths based on plasmonic metasurfaces. *Nano Lett.* **12**, 4932–4936 (2012).
22. Pors, A., Nielsen, M. G., Eriksen, R. L. & Bozhevolnyi, S. I. Broadband focusing flat mirrors based on plasmonic gradient metasurfaces. *Nano Lett.* **13**, 829–834 (2013).
23. Vo, S. *et al.* Sub-wavelength grating lenses with a twist. *IEEE Photon. Technol. Lett.* **26**, 1375–1378 (2014).
24. Aieta, F., Kats, M. A. & Capasso, F. Reflection and refraction of light from metasurfaces with phase discontinuities. *Nano Lett.* **6**, 1720–1706 (2012).
25. Ni, X., Emani, N. K., Kildishev, A. V., Boltasseva, A. & Shalaev, V. M. Broadband light bending with plasmonic nanoantennas. *Science* **335**, 427 (2012).
26. Chen, W. T. *et al.* High-efficiency broadband meta-hologram with polarization-controlled dual images. *Nano Lett.* **14**, 225–230 (2014).
27. Yu, N. & Capasso, F. Flat optics: controlling wavefronts with optical antenna metasurfaces. *IEEE J. Sel. Top. Quantum Electron.* **19**, 4700423 (2013).
28. Yang, B., Ye, W. M., Yuan, X. D., Zhu, Z. H. & Zeng, C. Design of ultrathin plasmonic quarter-wave plate based on period coupling. *Opt. Lett.* **38**, 679–681 (2013).
29. Baffou, G. & Rigneault, H. Femtosecond-pulsed optical heating of gold nanoparticles. *Phys. Rev. B Condens. Matter Mater. Phys.* **84**, 1–13 (2011).
30. Roxworthy, B. J. & Toussaint, K. C. Femtosecond-pulsed plasmonic nanotweezers. *Sci. Rep.* **2**, 1–6 (2012).
31. Petrova, H. *et al.* On the temperature stability of gold nanorods: comparison between thermal and ultrafast laser-induced heating. *Phys. Chem. Chem. Phys.* **8**, 814–821 (2006).
32. Kofman, R. *et al.* Surface melting enhanced by curvature effects. *Surf. Sci.* **303**, 231–246 (1994).
33. Inasawa, S., Sugiyama, M. & Yamaguchi, Y. Laser-induced shape transformation of gold nanoparticles below the melting point: the effect of surface melting. *J. Phys. Chem. B* **109**, 3104–3111 (2005).
34. Lee, C. H. *et al.* Peel-and-stick: mechanism study for efficient fabrication of flexible/transparent thin-film electronics. *Sci. Rep.* **3**, 2917 (2013).
35. Griffith, A. A. The Phenomena of Rupture and Flow in Solids. *Philos. Trans. R. Soc. A Math. Phys. Eng. Sci.* **221**, 163–198 (1921).
36. Freund, L. B. & Suresh, S. *Thin Film Materials: Stress, Defect Formation and Surface Evolution* (Cambridge University Press, 2003).
37. Williams, R. & Goodman, A. M. Wetting of thin layers of SiO2 by water. *Appl. Phys. Lett.* **25**, 531–532 (1974).

38. Vitos, L., Ruban, A. V., Skriver, H. L. & Kollár, J. The surface energy of metals. *Surf. Sci.* **411**, 186–202 (1998).

39. Kinloch, A. J. *Adhesion and Adhesives: Science and Technology (Springer, Cambridge,* 1987).

40. Skriver, H. L. & Rosengaard, N. Surface energy and work function of elemental metals. *Phys. Rev. B Condens. Matter Mater. Phys* **46**, 7157–7168 (1992).

41. Thomas, R. R. Wettability of polished silicon oxide surfaces. *J. Electrochem. Soc.* **143**, 643 (1996).

42. Wojciechowski, K. F. Surface energy of metals: theory and experiment. *Surf. Sci.* **437**, 285–288 (1999).

43. Yao, A. M. & Padgett, M. J. Orbital angular momentum: origins, behavior and applications. *Adv. Opt. Photon.* **3**, 161–204 (2011).

44. Curtis, J. E. & Grier, D. G. Structure of optical vortices. *Phys. Rev. Lett.* **90**, 133901 (2003).

45. Petrov, D. V., Canal, F. & Torner, L. A simple method to generate optical beams with a screw phase dislocation. *Opt. Commun.* **143**, 265–267 (1997).

46. Cai, X. L. *et al.* Integrated compact optical vortex beam emitters. *Science* **338**, 363–366 (2012).

47. Leach, J. *et al.* Quantum correlations in optical angle-orbital angular momentum variables. *Science* **329**, 662–665 (2010).

48. Wang, K., Schonbrun, E., Steinvurzel, P. & Crozier, K. B. Trapping and rotating nanoparticles using a plasmonic nano-tweezer with an integrated heat sink. *Nat. Commun.* **2**, 469 (2011).

49. Zhang, W. H., Huang, L. N., Santschi, C. & Martin, O. J. F. Trapping and sensing 10 nm metal nanoparticles using plasmonic dipole antennas. *Nano Lett.* **10**, 1006–1011 (2010).

50. Donner, J. S., Baffou, G., McCloskey, D. & Quidant, R. Plasmon-assisted optofluidics. *ACS Nano* **5**, 5457–5462 (2011).

51. Behrens, S. H., Plewa, J. & Grier, D. G. Measuring a colloidal particle's interaction with a flat surface under nonequilibrium conditions - total internal reflection microscopy with absolute position information. *Eur. Phys. J. E. Soft Matter* **10**, 115–121 (2003).

52. Burns, M. M., Fournier, J. M. & Golovchenko, J. A. Optical matter - crystallization and binding in intense optical-fields. *Science* **249**, 749–754 (1990).

53. Chen, H., Ding, Q., Roxworthy, B. J., Bhuiya, A. M. & Toussaint, K. C. Optical trapping with pillar bowtie nanoantennas. *Proc. SPIE* **9164**, 91641M–91641M (2014).

54. Ahn, S. H. & Guo, L. J. Large-area roll-to-roll and roll-to-plate nanoimprint lithography: a step toward high-throughput application of continuous nanoimprinting. *ACS Nano* **3**, 2304–2310 (2009).

55. Kwak, M. K., Ok, J. G., Lee, J. Y. & Guo, L. J. Continuous phase-shift lithography with a roll-type mask and application to transparent conductor fabrication. *Nanotechnology* **23**, 344008 (2012).

56. Chen, H., Bhuiya, A. M., Liu, R. Y., Wasserman, D. M. & Toussaint, K. C. Design, fabrication, and characterization of near-IR gold bowtie nanoantenna arrays. *J. Phys. Chem. C* **118**, 20553–20558 (2014).

57. Johnson, P. B. & Christy, R. W. Optical constants of noble metals. *Phys. Rev. B* **6**, 4370–4379 (1972).

58. Chen, H., Bhuiya, A. M., Ding, Q. & Toussaint, K. C. Plasmon-assisted audio recording. *Sci. Rep.* **5**, 9125 (2015).

Acknowledgements

This work was supported by University of Illinois at Urbana-Champaign. We thank S.P. Vanka for very useful conversations.

Author contributions

H.C., A.B. and Q.D. performed the experiments and analysed data. H.C. and K.C.T. designed the experiments and all authors co-wrote the paper. A.B. and Q.D. fabricated the samples.

Additional information

4

Arbitrary cross-section SEM-cathodoluminescence imaging of growth sectors and local carrier concentrations within micro-sampled semiconductor nanorods

Kentaro Watanabe[1,2], Takahiro Nagata[1], Seungjun Oh[1], Yutaka Wakayama[1], Takashi Sekiguchi[1], János Volk[3] & Yoshiaki Nakamura[2]

Future one-dimensional electronics require single-crystalline semiconductor free-standing nanorods grown with uniform electrical properties. However, this is currently unrealistic as each crystallographic plane of a nanorod grows at unique incorporation rates of environmental dopants, which forms axial and lateral growth sectors with different carrier concentrations. Here we propose a series of techniques that micro-sample a free-standing nanorod of interest, fabricate its arbitrary cross-sections by controlling focused ion beam incidence orientation, and visualize its internal carrier concentration map. ZnO nanorods are grown by selective area homoepitaxy in precursor aqueous solution, each of which has a $(0001){:}+c$ top-plane and six $\{1\text{-}100\}{:}m$ side-planes. Near-band-edge cathodoluminescence nanospectroscopy evaluates carrier concentration map within a nanorod at high spatial resolution (60 nm) and high sensitivity. It also visualizes $+c$ and m growth sectors at arbitrary nanorod cross-section and history of local transient growth events within each growth sector. Our technique paves the way for well-defined bottom-up nanoelectronics.

[1] International Center for Materials Nanoarchitectonics, National Institute for Materials Science, 1-1 Namiki, Ibaraki 305-0044, Japan. [2] Graduate School of Engineering Science, Osaka University, 1-3 Machikaneyama-cho, Osaka 560-8531, Japan. [3] MTA EK Institute of Technical Physics and Materials Science, Konkoly Thege M. ut 29-33, Budapest 1121, Hungary. Correspondence and requests for materials should be addressed to K.W. (email: watanabe@ee.es.osaka-u.ac.jp).

Bottom-up growths (for example, wet chemical growth[1,2] and molecular beam epitaxy[3,4]) of semiconductor nanocrystals are paid attention because of the limitation of semiconductor microfabrication. Especially, a free-standing nanowire or nanorod has several exotic properties applicable to unique one-dimensional electronics; high flexibility (large fracture strain) for strain-engineered ultrafast nanowire field-effect transistors[5] and for efficient nanopiezotronic devices[6], large surface-to-volume ratio for sensitive solar cells[7] and gas sensors[8], and wave-guides for Fabry-Perot nanolasers[9–11]. A nanowire with uniform electric properties besides uniform diameter is essential for one-dimensional electronics with high device designability and reproducibility, which should be driven by well-defined current density.

Up to now, electrical properties of a bottom-up nanocrystal and its device properties are reported assuming electrical uniformities. However, this is idealistic because each crystallographic plane surface has an atomic arrangement with unique chemical activity (for example, etching rate[12,13], host crystal growth rate[14,15] and incorporation rates of point-defects), as is reported on plane-dependent donor concentration in ZnO bulk crystal[13,16] and amphoteric Si-doping in GaAs substrates[17,18]. Thus, a nanocrystal grown in multiple crystallographic orientations has corresponding growth sectors with different electric properties, which is critical for above electronic applications. Further, growth sectors within a nanocrystal are not fully described macroscopically by pure crystallographic planes at constant growth rates[19–23]. (For example, a free-standing nanowire with uniform diameter is also idealistic[24].) In reality, a growth form of a nanocrystal may evolve with growth duration depending on the local and temporal growth environment on each crystallographic plane, where spontaneous surface roughening or new plane formation may take place. Thus, it is difficult to foresee the internal spatial distributions of growth sectors. We demand a novel experimental technique which reveals local electrical properties within a nanocrystal, especially those due to the growth sectors, at high spatial resolution and high sensitivity.

A recent report on nanowire Hall effect measurement reveals carrier transport properties of an entire free-standing nanowire between micro-fabricated electrical contacts[25]. However, there is no adequate technique that can probe such electrical non-uniformities within a free-standing nanocrystal at sufficient sensitivity. Atom probe tomography[26,27] and cross-sectional scanning transmission electron microscopy[28] may visualize dopant species at atomic resolutions. However, none of them has sufficient sensitivities to probe local concentration and investigate electrical activity of point-defects within nominally undoped semiconductors. Cross-sectional scanning tunnelling microscopy[29] may visualize electrical activities of dopants at sub-surfaces at atomic resolution. However, atomically smooth specimen cleavage is required to obtain noticeable contrasts of impurity atoms or vacancies, which is not available for a free-standing nanocrystals of interest up to now.

Contrary, cathodoluminescence (CL) nanospectroscopy is promising for its high sensitivity ($\sim 10^{15}\,cm^{-3}$) of shallow level point-defects and nanometric resolution achievable by low-energy electron beam (e-beam) probe[30–34]. Cross-sectional CL imaging of growth sector interface is reported on hydrothermally grown ZnO bulk crystals with $(0001):+c$, $\{1–100\}:m$ and $\{1–101\}:+p$ planes[35] and on hydride vapour phase epitaxy grown bulk GaN $+c$ films containing hexagonal micro-pits with $+p$ planes[36,37]. Plane-dependent CL behaviours of ZnO nanorod are reported recently[38]. However, cross-sectional CL technique is not applied to any semiconductor free-standing nanocrystal up to now.

ZnO single-crystalline free-standing nanorod array is chosen to demonstrate our technique, as it is suitable for low-cost optoelectronics and piezoelectronics. Wurzite ZnO is a piezoelectric wide-gap semiconductor with direct bandgap ($E_g = 3.37\,eV$), a large exciton binding energy ($E_b = 60\,meV$)[39] and a large piezoelectric coefficient along its polar c-axis ($d_{cc} = 12.4\,pm\,V^{-1}$)[40], which can be synthesized at low temperature and at atmospheric pressure in precursor aqueous solution: by simple chemical synthesis[41–54] or by electrodeposition[55–58].

Here we demonstrate a series of techniques to micro-sample a free-standing nanocrystal of interest and to visualize its internal local carrier concentration, especially those due to growth sectors, from arbitrary cross-sectional orientations. Circular growth windows are fabricated by e-beam lithography in a polymethyl methacrylate (PMMA) e-beam-resist film spin-coated on a single-crystalline ZnO (0001) substrate. ZnO single-crystalline free-standing nanorod arrays are then homoepitaxially grown at a time in precursor aqueous solution from circular growth window arrays with different diameters (D_w; Supplementary Fig. 1). Scanning electron microscopy (SEM) observation shows that any nanorod grows from each circular growth window to have a $+c$ top-plane and six m side-planes (Supplementary Figs 2 and 3). A nanorod is micro-sampled and its cross-section is fabricated by controlling crystallographic orientation of focused ion beam (FIB) incidence. Cross-section CL imaging of the nanorod visualizes $+c$ and m growth sectors within an entire nanorod at high spatial resolution ($<60\,nm$). CL nanospectroscopy on $+c$ and m planes reveals their local carrier concentration differences and its quantitative accuracy is confirmed by 'differential' I–V measurements of an individual ZnO free-standing nanorod (Supplementary Fig. 4).

Results

Room temperature CL nanospectroscopy of ZnO nanorods.
Local CL properties are studied for ZnO free-standing nanorods grown homoepitaxially at arrayed circular windows. ZnO nanorod arrays of $D_w = 100$ and $500\,nm$ are investigated by SEM observation and near-band-edge (NBE)/visible (Visible) CL imaging at $3.0\,keV$ from bird's-eye view angles (Fig. 1a,b). Further, local CL spectroscopy is performed at $4.5\,keV$ on a nanorod of $D_w = 500\,nm$ at spot-CL mode. Eight spot-CL spectra at corresponding e-beam spot positions 1–8 are shown in Fig. 1c, which are correlated by the same colour: position 1 at top $+c$ plane, positions 2–7 at side m plane from top to the bottom and position 8 on the ZnO $+c$ substrate covered with transparent PMMA film. Spectral band of NBE and Visible CL imaging are also indicated by the rectangular grey shadows. Visible CL emission band at $2.1\,eV$ exhibits much more uniform spatial distribution and spectral shape within the nanorod (Supplementary Note 2). Contrary, NBE CL images show that $+c$ top-planes are darkest and m side-planes gradually become brighter towards the nanorod bottom by one order of magnitude, regardless of D_w. Spot-CL spectroscopy also shows that each NBE CL peak consists of intrinsic emission ($3.28\,eV$) and red-shifted emission ($3.19\,eV$). Intrinsic NBE CL emission is dominant at position 1, however, red-shifted NBE CL emission intensity increases to be dominant and then saturated as the spot position goes from 2 to 7. Considering that SEM e-beam probes CL from ZnO surface-to-depth of Kanaya–Okayama electron range ($192\,nm$ for $4.5\,keV$ electron)[59], red-shifted emission is unique to m plane and m polar growth thickness seems to be increasing from 0 at the nanorod top-plane to $>200\,nm$ at the nanorod bottom. Spatial distribution of NBE CL emission efficiency within a nanorod is illustrated in Fig. 1d, which is further investigated in

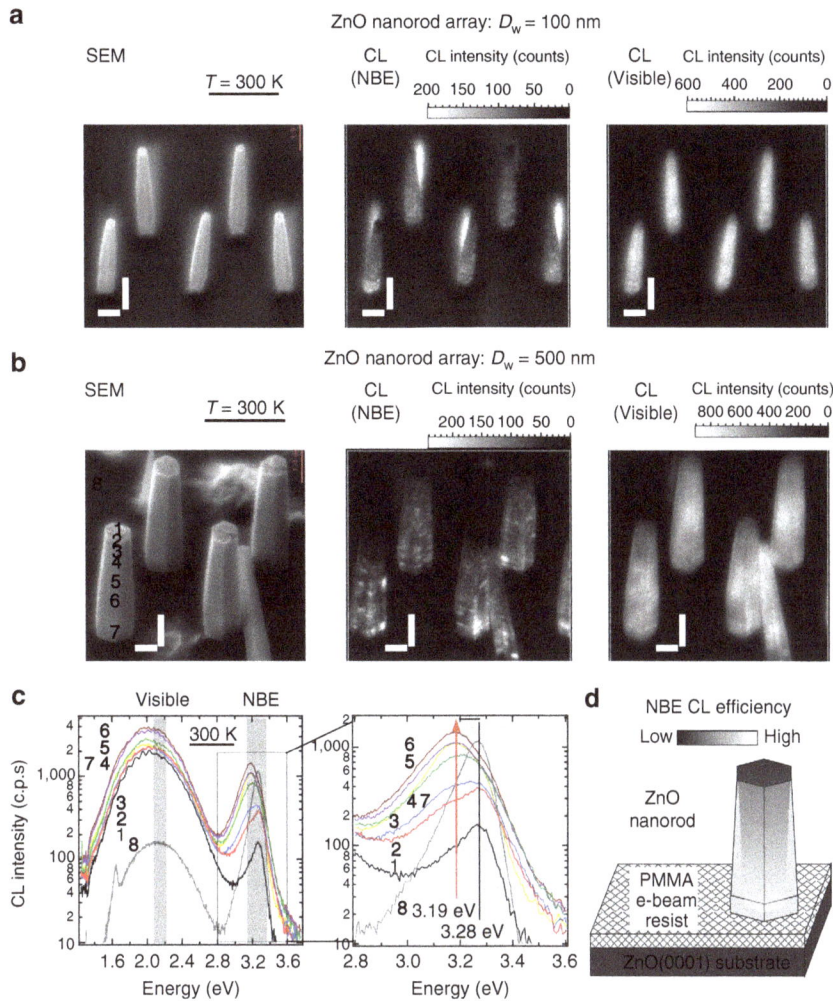

Figure 1 | Bird's-eye view NBE/Visible CL imaging of ZnO free-standing nanorods. SEM and monochromatic (NBE/Visible) CL images of ZnO free-standing nanorod arrays, probed by 3.0 keV e-beam. Each nanorod is homoepitaxially grown from individual growth window with a variety of diameter: (**a**) small diameter growth windows ($D_w = 100$ nm) and (**b**) large diameter growth windows ($D_w = 500$ nm). Nanorods misorientation in $D_w = 100$ nm array takes place during the SEM observation due to electrical charge-up. A bright shadow behind each nanorod array in NBE CL image are signal from ZnO substrate, which is excited due to PMMA thinning by the concentrated e-beam dose. (**c**) Spot-CL spectroscopy at positions 1–8 in **b**, probed by 4.5 keV e-beam. Spectral bands for NBE/Visible CL imaging are highlighted with transparent grey zones. (**d**) A schematic of ZnO free-standing nanorod. Local NBE CL emission efficiency within the nanorod is illustrated by grey scale. The specimen is tilted by 45 degree from ZnO (0001) plane normal. All horizontal and vertical scale bars indicate lengths of 0.5 μm and 1.0 μm, respectively.

Figs 3 and 4. Note that each nanorod exhibits some locally bright NBE CL emission spots (CL spots) on m side-planes. As spectrum 6 on a CL spot shows the same spectral shape as those at other spot positions on m plane, the CL spots have the same properties of m plane.

Low-temperature CL spectroscopy at nanorod top/side plane. To reveal the origin of red-shifted NBE CL emission of m plane, temperature-dependent CL spectroscopy is performed on $+c$ top-plane and m side-plane of a nanorod ($D_w = 500$ nm). A bird's-eye view SEM image of a ZnO free-standing nanorod indicates each e-beam scan area by corresponding solid box superimposed (Fig. 2a). Temperature-dependent 3.0 keV NBE CL spectra are obtained in Area-CL mode at higher wavelength resolution: at 300 K (top), at 80 K (middle) and at 10 K (bottom; Fig. 2b). Each CL peak at 10 K is assigned as follows: 3.37 eV to radiative recombination of free excitons (FX), 3.361 eV to neutral donor-bound excitons (D^0X) observed in naturally n-type ZnO, 3.324 eV to its two electron satellite (TES-D^0X) and two lower energy peaks with 0.072 eV interspacing to nth LO phonon

replica of D^0X peak (D^0X-nLO; $n = 1$ and 2). The energy difference between D^0X (1 s state) and TES-D^0X (2s and 2p states) peaks of 37 meV equals the three-fourth of donor-binding energy $E_D = 49$ meV in the hydrogenic effective mass approach[60]. At elevated temperatures, D^0X peaks quench and FX peaks emerge at 80 K and FX peak locates at 3.280 eV on $+c$ top-plane and at 3.200 eV on m side-plane at 300 K (Fig. 2b). The detailed temperature dependence between 80 and 300 K also reveals the emergence of redshifted FX emission peak above 220 K (Fig. 2c). Such a 80-meV redshift larger than E_D is not attributed to the donor-binding energy itself. Rather, it is attributed to the band-gap shrinkage or band-tailing of a heavily donor-doped semiconductor observable at elevated temperature, both of which originates in donor ionization and its local perturbation of conduction and valence band-edges.

Local carrier concentration evaluated by CL nanospectroscopy. Here, room temperature FX peak energy in Fig. 2b is attributed to the local residual carrier concentration n (cm^{-3}). Room temperature band-edge emission energy of a heavily donor-doped

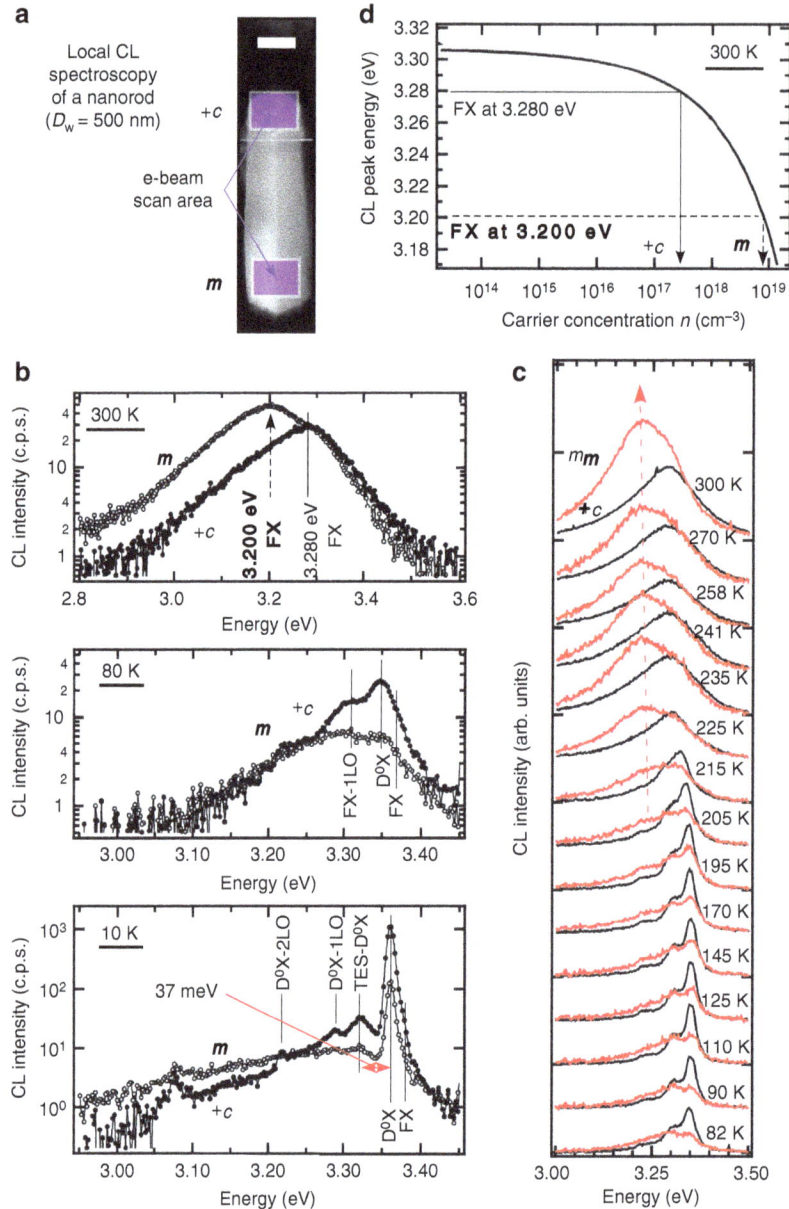

Figure 2 | Temperature-dependent high-resolution CL spectroscopy of a nanorod +c top-plane and m side-plane. (**a**) Bird's-eye view SEM image of a ZnO nanorod investigated by Area-CL spectroscopy. E-beam scan areas on +c top-plane and m side-plane are indicated by transparent purple boxes. A horizontal scale bar indicates length of 0.5 μm. (**b**) Area-CL spectra on +c and m planes at 300, 80 and 10 K, probed by 3.0 keV e-beam. CL peaks are assigned to free exciton (FX), neutral donor-bound exciton (D^0X), its nth LO phonon replica (D^0X-nLO) and its two electron satellite (TES-D^0X). (**c**) Detailed temperature dependence of Area-CL spectra between 80 and 300 K. Emergence of the redshifted FX peak is clearly observed. (**d**) Correlation between FX peak energy E_{FX} (eV) and carrier electron concentration n (cm^{-3}). Residual carrier concentrations n of +c and m growth sectors are evaluated using their correlation: $E_{FX}(n) = 3.307$-$8.39 \times 10^{-15} n^{2/3}$-$3.64 \times 10^{-8} n^{1/3}$ (ref. 61).

semiconductor redshifts due to its band-gap shrinkage or band-tailing. Giles *et al.* investigated n-type ZnO bulk crystals with different carrier concentrations n (cm^{-3}) by photolumine-scence spectroscopy and reported their FX emission energy: $E_{FX}(n)$ (eV) $= 3.307$-$8.39 \times 10^{-15} \cdot n^{2/3}$-$3.64 \times 10^{-8} \cdot n^{1/3}$ (solid curve in Fig. 2d)[61]. Based on this work, we obtain $n_{+c} = 2.8 \times 10^{17}$ cm^{-3} at +c top-plane and $n_m = 8.2 \times 10^{18}$ cm^{-3} at m side-planes. Carrier concentration gap at 300 K is evidenced qualitatively by FX emission intensity gap between dark +c top-plane and bright m side-plane, which is intensified by residual carrier concentration.

Also, room temperature NBE CL emission spectroscopy of a 'ZnO nanocolumn', the nanorod at earlier growth stage, is performed at Spot-CL mode (Supplementary Fig. 3). This

nanocolumn also exhibits 30 meV redshift of FX emission at its m side-plane. NBE CL emission energy of ZnO nanocolumn is also attributed to the local carrier concentration: 3.28 eV emission to $n_{+c} = 2 \times 10^{17}$ cm^{-3} in the axial +c growth sector and 3.25 eV emission to $n_m = 2 \times 10^{18}$ cm^{-3} in the lateral m growth sector. The growth duration-dependent energy redshift also supports our idea that the energy redshift of NBE CL emission is originated from the difference of donor incorporation rates, rather than a certain donor-binding energy.

A quantitative accuracy of n values given by CL nanospectro-scopy is evaluated in comparison with a net carrier concentration estimated from 'differential' I–V measurements of a ZnO free-standing nanorod (Supplementary Fig. 4). Here, σ_e is net electrical conductivity in the nanorod. Electron mobility

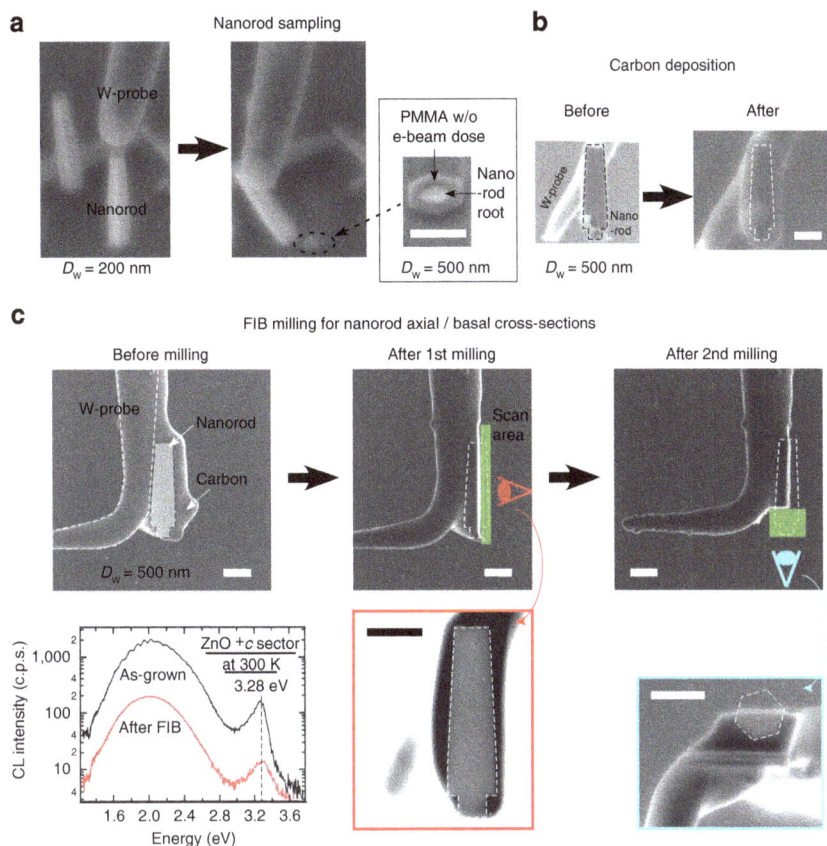

Figure 3 | Nanorod micro-sampling and cross-section fabrication by controlling FIB incidence orientation. (**a**) A nanorod ($D_w = 200$ nm) micro-sampling observed by SEM. The inset SEM image shows the circular nanorod root left after the removal of a nanorod ($D_w = 500$ nm). (**b**) SEM images of a micro-sampled nanorod before and after the amorphous carbon deposition. Nanorod position is indicated by dashed lines. (**c**) A series of FIB milling process observed by scanning ion microscope imaging. Nanorod positions and FIB milling scan areas at each step are indicated by dashed line and half-transparent box, respectively. Axial and basal cross-sections fabricated by first and second FIB milling, respectively, are observed consecutively by SEM. The inset at the bottom left shows Spot-CL spectra of the nanorod $+c$ top-plane before and after the FIB milling, probed by 4.5 keV e-beam. All scale bars indicate lengths of 1 μm.

$\mu_e = 1 \times 10^1$ cm^2 V^{-1} s^{-1} in the nanorod is assumed as aqueous solution synthesized ZnO films with Hall mobility of $\mu_e = 12.5$ cm^2 V^{-1} s^{-1} at $n = 6.66 \times 10^{17}$ cm^{-3} are reported[62]. Considering slightly tapered nanorod shape, 'differential' I–V measurements of an individual ZnO free-standing nanorod yields net electrical conductivity, $\sigma_e = 0.7$ Ω^{-1} cm^{-1}. Calculated net residual carrier concentration $n = \sigma_e/(e\mu_e) = 4 \times 10^{17}$ cm^{-3} falls between n values of $+c$ and m growth sectors, which suggests that n evaluation by CL nanospectroscopy is quantitative.

Arbitral cross-section of micro-sampled nanorod made by FIB. Next we show a sequential process to fabricate cross-sections of a micro-sampled free-standing nanorod in arbitral crystallographic orientation. First, a nanorod is micro-sampled using a tungsten (W) probe attached to a nanomanipulator (Fig. 3a). Note that inset SEM image shows a circular root of a nanorod remained attached to the substrate. This root surrounded by PMMA film with a hexagonal white area without e-beam dose, which is originally covered by the nanorod. This micro-sampling reveals circular cylindrical root of the hexagonal nanorod, which suggests that the nanorod growth is limited by PMMA film and that m plane growth thickness is evaluated from the diameter gap at the nanorod bottom. The micro-sampled nanorod on the W-probe is then embedded in a carbon film by carbon deposition from different orientations (Fig. 3b).

The W-probe with the nanorod is then transferred into FIB system to fabricate nanorod cross-sections at arbitrary crystallographic orientations by two-step FIB milling (Fig. 3c). Sequential scanning ion microscope images show first FIB milling in axial direction and second FIB milling in basal direction. SEM image of each cross-section observed from eye mark direction is also displayed, which highlights each nanorod cross-section with a bright contrast. Note that the SEM image of second basal cross-section shows that first axial cross-section is fabricated exactly on the nanorod axis, which evidences the high position controllability of this FIB milling technique.

Impact of Ga ion beam milling on surface CL properties is studied by 4.5 keV CL spectra at the nanorod top-plane before and after FIB milling (Fig. 3c inset). After the two-step FIB milling, the CL intensity has dropped to one-tenth due to the residual damaged layer of cross-section surface. Nevertheless, CL spectral shape is retained to investigate local CL properties of the nanorod.

Growth sectors visualized at arbitrary nanorod cross-section. Here we show cross-sectional CL observation of a micro-sampled nanorod to visualize its internal $+c/m$ growth sectors. Figure 4a,b shows SEM, NBE CL and Visible CL images of first axial cross-section and second basal cross-section at 3.0 kV, respectively. P1–P3 in Fig. 4c are NBE CL line profiles along Z axis ($//c$) in Fig. 4a. CL probing regions of P1–P3 are indicated by red areas at nanorod core (centre), nanorod shell (side-plane intersection) and nanorod shell (side-plane centre), respectively, in SEM image in Fig. 4b. Fig. 4d is Panchromatic, NBE and

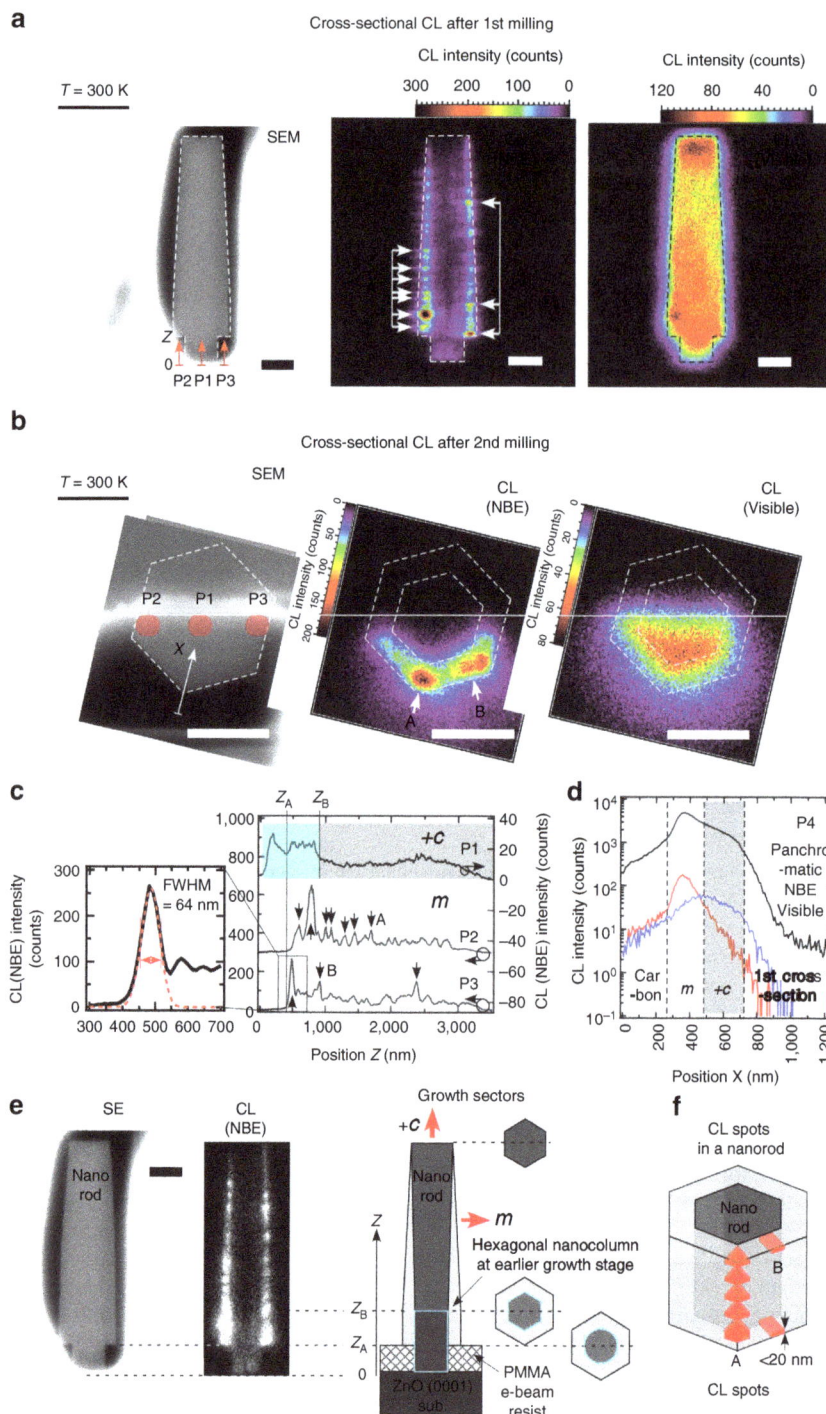

Figure 4 | High-resolution cross-sectional CL imaging of a micro-sampled nanorod. SEM and CL (NBE/Visible) images of first axial cross-section (**a**) and those of second basal cross-section (**b**), both of which are probed by 3.0 keV e-beam. Original nanorod volume is indicated by dashed line for eye-guide. (**c**) NBE CL profiles P1–P3 along coordinates Z at P1–P3 in **a**. CL spatial resolution is evaluated from FWHM of profile P3 around the sharpest CL spot. (**d**) CL (Panchromatic/NBE/Visible) profiles along coordinate X. (**e**) Schematic representation of $+c$ and m growth sectors within the nanorod cross-section, revealed from comparison between SEM and NBE CL images. Nanorod bottom ($Z = Z_A$) and slight NBE CL intensity drop in $+c$ growth sector ($Z = Z_B$) are also indicated in **c,e**. CL profiles P1–P4 and a hexagonal nanocolumn at earlier growth stage are indicated by aqua zones and schematically by an aqua box in a $+c$ growth sector. All scale bars indicate lengths of 0.5 μm. (**f**) Schematic illustration of spatial distribution and shape of CL spots. Observed CL spots in **a-c** are indicated by solid arrows and categorized into either spot A or spot B by their shape and location.

Visible CL profiles along X axis in Fig. 4b. Figure 4b,d shows CL quenches near the first axial cross-section because of the damaged surfaces. Two NBE CL images show the two regions with distinct intensity difference: dark hexagonal column core and bright hexagonal shell, whereas two Visible CL images also show them

with opposite and weaker contrasts. Note that the bright shell region in NBE CL image corresponds to the diameter gap at the nanorod bottom in SEM image. The diameter gap is formed at an earlier growth stage. A single hexagonal nanocolumn with its diameter equivalent to $D_w = 0.5$ μm is formed by the coalescence of multiple nuclei within each growth window. Subsequently, the lateral growth of the nanocolumn is allowed above PMMA film surface to form a diameter gap (Supplementary Fig. 2b). Thus, we assign the dark core and bright shell in NBE CL images to $+c$ and m growth sectors of the nanorod, respectively, where each sector has different donor concentration: $n_{+c} = 2.8 \times 10^{17}$ cm^{-3} in axial $+c$ growth sector and $n_m = 8.2 \times 10^{18}$ cm^{-3} in lateral m growth sectors, as investigated in Fig. 2. Figure 4e summarizes schematic spatial distribution of $+c$ and m growth sectors together with corresponding SEM and highlighted NBE CL images.

Also, CL spots are observed in NBE CL images, as indicated by solid arrows and labelled as spot A and B in Fig. 4a,b and as their locations and shapes are illustrated in Fig. 4f. They extend laterally across m growth sector and CL spots locate more at the m region intersection (spot A) than at their centres (spot B), comparing P2 with P3. A bright CL spot at nanorod bottom is chosen to evaluate the spatial resolution of CL imaging. P3 is locally fitted well with Gaussian curve $[I(Z) = I_0 \cdot \exp\{-(Z-Z_0)^2/2\sigma^2\}]$ of full-width at half-maximum (FWHM) $= 2\sqrt{(2\ln2)} \cdot \sigma = 64$ nm. As this FWHM is comparable to Kanaya–Okayama range[59] of 3.0 keV electron (97 nm), the FWHM is attributed to the lateral resolution of CL imaging.

Note that cross-sectional NBE CL image exhibits two growth turning points, $(Z_A, Z_B) = (402$ nm, 871 nm$)$, where the axial $+c$ growth domain contains a slightly brighter region $(0 < Z < Z_B;$ Fig. 4c,e). As Z_A matches with PMMA film thickness $(0.3\,\mu m)$, the turning point at $Z = Z_A$ is attributed to PMMA film surface. Sequential SEM observation of ZnO nanorods at earlier growth stages reveals that the Z_B–Z_A matches with the typical height $(0.5\,\mu m)$ of each hexagonal nanocolumn above PMMA film (Supplementary Fig. 2b). The turning point at $Z = Z_B$ is then attributed to the formation of single hexagonal nanocolumn (nanorod at earlier stage). CL profiles P1–P4 and a hexagonal nanocolumn at earlier growth stage are indicated by aqua zones and schematically by an aqua box in a $+c$ growth sector (Fig. 4c–e). Spot-CL spectroscopy of a nanocolumn reveals NBE CL emission redshift on its side-plane, which evidences that the top-plane and side-planes of the nanocolumn already have distinct difference in their donor incorporation rates (Supplementary Fig. 3b). As nuclei within a growth window also experience axial and lateral growths, the slightly bright region within this nominal $+c$ growth sector in Fig. 4e is accountable by the lateral m growth sectors of nuclei exposed to the cross-section. Thus, cross-sectional CL imaging of an individual nanorod reveals its growth history from minor contrasts within each growth sector as well as its local carrier concentrations from major contrasts between growth sectors.

Discussion

The origin of unintentional donors is remained to be discussed. Here we tentatively attribute the residual carriers to interstitial hydrogen donors incorporated from growth environment for the following five reasons: (i) wide-scan X-ray photoemission spectroscopy of reference ZnO homoepitaxial thin film (Supplementary Fig. 5e) does not detect any other possible ZnO donor element than hydrogen, such as Al or Ga, and thus its concentration is typically below 0.1 atomic %; (ii) narrow-scan X-ray photoemission spectroscopy (Supplementary Fig. 5f,g) and Raman spectroscopy (Supplementary Fig. 5h) of the ZnO thin film evidence the presence of hydrogen atoms at bond-centred interstitial sites in the form of [-OH] groups[63–67]; (iii) Raman peak observed at 330 cm^{-1} (Supplementary Fig. 5h) is energetically equivalent to the 37 meV gap in 10 K CL peaks in Fig. 2b and thus can be attributed to $1s \rightarrow 2p$ transitions of

interstitial hydrogen donor[66–68], although it can also be ascribed to second-order vibration mode $(E_2^{high}$–$E_2^{low})$ of intrinsic ZnO at 333 cm^{-1} (ref. 69); (iv) D^0X peak in Fig. 2b matches I_4 peak (3.363 eV) associated with H donor[60]; (v) the $E_D = 49$ meV calculated from CL spectrum at 10 K agrees with $E_D = 46.1$ meV of hydrogen donor reported[60]. Interstitial hydrogen donor in our ZnO nanorod may be originated from $Zn(OH)_2$ precursors and incorporated as [-OH] group at oxygen site (Supplementary Note 1). This model is supported by the report on enhanced hydrogen donor incorporation into ZnO (0001) films grown in aqueous solution at higher [OH$^-$] concentration $(n \sim 10^{17}$ cm^{-3} at pH $= 8$ and $n = 1.79 \times 10^{19}$ cm^{-3} at pH $= 10.9)$[62].

Also, we discuss origins of CL spots appeared in m growth sectors. The minimum thickness of CL spot origin is at least half an order of magnitude smaller $(< 20$ nm$)$. CL spots extends from the interface of growth sectors along the nanorod basal plane. NBE CL images also show that each CL spot extends laterally, populate preferentially around nanorod corners, and populate more on nanorods of $D_w = 500$ nm than those of $D_w = 100$ nm. Inspired by the above indirect observations, we tentatively consider that CL spots origin might be related with the growth mode of the nanorod side-planes. However, further studies are required to demonstrate this idea.

In summary, we demonstrate that cross-sectional CL technique evaluates local carrier concentration quantitatively at high spatial resolution and at high sensitivity. It also visualizes internal growth sectors of an entire semiconductor nanorod from arbitrary crystallographic orientations, and even reveals nanorod growth history. Our model also gives suggestions how to improve nanorods: (i) nanorod side-plane growth should be minimized as it is the origin of non-uniform diameter and local high carrier concentration; (ii) for aqueous solution growth, amine additives might be promising to enhance uniformities of nanorod diameter and its electrical properties, as they adsorb selectively on nanorod non-polar side-planes and suppress nanorod lateral growth[41,42]. Above findings are quite general and are valid for various luminescent semiconductor nanocrystals, regardless of semiconductor species and growth methods.

Methods

Selective area homoepitaxy in precursor aqueous solution. In the precursor aqueous solution, ZnO hexagonal nanorods are grown at a time from size-controlled circular holes patterned in PMMA film on an atomically flat ZnO $+c$ substrate (selective area homoepitaxy)[53,54]. The c-axis polarities of these nanorods are expected to be matched considering the recent report on aqueous solution grown ZnO nanowire[70].

First, Zn-terminated ZnO $+c$ substrate was ultrasonically rinsed in acetone, ethanol and deionized water and then annealed at 1,000 °C for 8 h in oxygen atmosphere at 1 atm to make atomically flat surfaces. Then, 300-nm-thick PMMA photoresist is formed on the substrate by spin-coating (Supplementary Fig. 1a). Two hundred circular growth windows per each diameters $(D_w = 100, 150, 200, 300, 400, 500$ nm$)$ are opened in the PMMA film by e-beam lithography to form a 2×100 trigonal lattice with $2.0\,\mu m + D_w$ interspacing (Supplementary Fig. 1b). Buffered aqueous solution $(0.2\,l)$ of equimolar $(8 \times 10^{-3}$ M$)$ zinc nitrate hexahydrate $(>99.9\%,$ Wako$)$ and hexamethylenetetramine $(>99.0\%,$ Wako$)$ without any chemical additive is prepared at room temperature: $T_{aq} = 25$ °C. The specimen substrate is mounted upside-down in the solution and sealed in the polytetrafluoroethylene (PTFE) container. Then, the container is introduced into the multi-purpose oven set at $T_{set} = 85.0$ °C and heated for 3.5 h. Therein, ZnO deposition starts at $T_{aq} > 60$ °C and ZnO nanorods array grow stationary at $T_{aq} = 79$ °C (Supplementary Note 1). After the heating, the container is cooled naturally down to room temperature and the specimen substrate is removed from the container, rinsed by the deionized water and finally dried by nitrogen gas blow. Contrary to our previous publications[30,53,54], PMMA film is not removed from ZnO substrate after the growth.

SEM-CL observation of a ZnO nanorod. SEM-CL observation is conducted in Nanoprobe-CL system[30], which is based on a Schottky SEM (Hitachi High-Technologies SU6600). This system equips a SEM specimen cooling stage (Thermal Block Company), which is modified to mount a triaxial piezoelectric nanomanipulator (Kleindiek Nanotechnik MM3A-EM) and triaxial coaxial cables for nano-manipulation and electrical nano-probing. Low-temperature measurements down to 10 K are available by flowing He or N_2 cooling gas beneath the specimen. SEM-CL nanospectroscopy is performed using e-beam of 3.0 keV and 2.35 nA or that of 4.5 keV and 2.75 nA. E-beam excites CL from ZnO surface to the depth of Kanaya–Okayama range, 97 nm at 3.0 keV and 192 nm at 4.5 keV, respectively[59]. CL is collected by ellipsoidal mirror, dispersed in the spectrometer (Horiba iHR320) and detected by multi-channel charge-coupled device detector (Andor Tech. DU420A-BU2) for CL spectroscopy or by photomultiplier tube (Hamamatsu R943-02) for monochromatic CL imaging. Panchromatic CL analysis is also available using the optical path bypassing the spectrometer. The 300-nm-thick PMMA film on the ZnO substrate is transparent for ZnO CL emission range but plays as an e-beam stopping layer to suppress strong CL from high-quality ZnO substrate. The 3.0-keV CL imaging were performed as short as possible, as e-beam with energy higher than 3.0 keV penetrate PMMA film to excite NBE emission of ZnO substrate and the e-beam dose decompose the PMMA film gradually. CL spectroscopy is performed either in Spot-CL mode at moderate energy resolution (32 meV) and in Area-CL mode at high energy resolution (4.3 meV), where focused e-beam is either spotted or scanned on a nanorod of interest.

Nanorod micro-sampling and cross-section fabrication by FIB. ZnO nanorods are micro-sampled and their cross-sections are made at desired crystallographic orientations by two-step FIB technique. First, specimens were installed in Nanoprobe-CL system[30]. Single ZnO free-standing nanorod is picked up by electrochemically etched W-probe attached to the nanomanipulator. The probe is then transferred to the carbon coater where carbon film is deposited from various directions to embed the nanorod in a thick amorphous carbon layer. Here, the carbon medium is expected to play several roles: (i) surface protective layer and heat sink from Ga ion beam bombardment; (ii) non-luminescent embedding medium for CL imaging; (iii) embedding medium with appropriate hardness for smooth cross-sections. Each nanorod cross-section is fabricated by two-step FIB milling, coarse milling at 30 keV followed by fine milling at 10 keV, to minimize thickness of damaged layer of the cross-section surface and to improve the S/N ratio of CL imaging.

References

1. Peng, X.-G., Wickham, J. & Alivisatos, A. P. Kinetics of II-VI and III-V colloidal semiconductor nanocrystal growth: 'Focusing' of size distributions. *J. Am. Chem. Soc.* **120**, 5343–5344 (1998).
2. Peng, X.-G. *et al.* Shape control of CdSe nanocrystals. *Nature* **404**, 59–61 (2000).
3. Nakamura, Y., Watanabe, K., Fukuzawa, Y. & Ichikawa, M. Observation of the quantum-confinement effect in individual Ge nanocrystals on oxidized Si substrates using scanning tunneling spectroscopy. *Appl. Phys. Lett.* **87**, 133119 (2005).
4. Nakamura, Y., Amari, S., Naruse, N., Mera, Y. & Ichikawa, M. Self-assembled epitaxial growth of high density beta-FeSi2 nanodots on Si (001) and their spatially resolved optical absorption properties. *Cryst. Growth Des.* **8**, 3019–3023 (2008).
5. Li, Y. *et al.* Dopant-free GaN/AlN/AlGaN radial Nanowire. *Nano Lett.* **6**, 1468–1473 (2006).
6. Wang, Z.-L. Nanopiezotronics. *Adv. Mater.* **19**, 889–892 (2007).
7. Tian, B. *et al.* Coaxial silicon nanowires as solar cells and nanoelectronic power sources. *Nature* **449**, 885–890 (2007).
8. Cui, Y., Wei, Q.-Q., Park, H.-K. & Lieber, C. M. Nanowire nanosensors for highly sensitive and selective detection of biological and chemical species. *Science* **293**, 1289–1292 (2001).
9. Huang, M. H. *et al.* Room-temperature ultraviolet Nanowire nanolasers. *Science* **292**, 1897–1899 (2001).
10. Duan, X., Huang, Y., Agarwal, R. & Lieber, C. M. Single-nanowire electrically driven lasers. *Nature* **421**, 241–245 (2003).
11. Chu, S. *et al.* Electrically pumped waveguide lasing from ZnO nanowires. *Nat. Nanotechnol* **6**, 506–510 (2011).
12. Mariano, A. N. & Hanneman, R. E. Crystallographic polarity of ZnO crystals. *J. Appl. Phys.* **34**, 384–388 (1963).
13. Heiland, G. & Kunstmann, P. Polar surfaces of zinc oxide crystals. *Surf. Sci* **13**, 72–84 (1969).
14. Sekiguchi, T., Miyashita, S., Obara, K., Shishido, T. & Sakagami, N. Hydrothermal growth of ZnO single crystals and their optical characterization. *J. Cryst. Growth* **214/215**, 72–76 (2000).
15. He, Y. *et al.* Crystal-plane dependence of critical concentration for nucleation on hydrothermal ZnO nanowires. *J. Phys. Chem. C* **117**, 1197–1203 (2013).
16. Sakagami, N. *et al.* Variation of electrical properties on growth sectors of ZnO single crystals. *J. Cryst. Growth* **229**, 98–103 (2001).
17. Ballingall, J. M. & Wood, C. E. C. Crystal orientation dependence of silicon auto compensation in molecular beam epitaxial gallium arsenide. *Appl. Phys. Lett.* **41**, 947–949 (1982).
18. Wang, W. I., Mendez, E. E., Kuan, T. S. & Esaki, L. Crystal orientation dependence of silicon doping in molecular beam epitaxial AlGaAs/GaAs heterostructures. *Appl. Phys. Lett.* **47**, 826–828 (1985).
19. Wulff, G. Zur frage der geschwindigkeit des wachstums und der auflosung der krystallflachen. *Z. Kristallogr.* **34**, 449–530 (1901).
20. Chernov, A. A. The kinetics of the growth forms of crystals. *Sov. Phys. Crystallogr* **7**, 728–730 (1963).
21. Borgström, L. H. Die geometrische bedingung fur die entstehung von kombinationen. *Z. Kristallogr.* **62**, 1 (1925).
22. Alexandru, H. V. A macroscopic model for the habit of crystals grown from solutions. *J. Cryst. Growth* **5**, 115–124 (1969).
23. Singh, M., Verma, P., Tung, H.-H., Bordawekar, S. & Ramkrishna, D. Screening crystal morphologies from crystal structure. *Cryst. Growth Des.* **13**, 1390–1396 (2013).
24. Dubrovskii, V. G., Timofeeva, M. A., Tchernycheva, M. & Bolshakov, A. D. Lateral growth and shape of semiconductor nanowires. *Semiconductors* **47**, 50–57 (2013).
25. Storm, K. *et al.* Spatially resolved Hall effect measurement in a single semiconductor nanowire. *Nat. Nanotechnol* **7**, 718–722 (2012).
26. Chen, W. *et al.* Boron distribution in the core of Si nanowire grown by chemical vapor deposition. *J. Appl. Phys.* **111**, 094909 (2012).
27. Chen, W. *et al.* Incorporation and redistribution of impurities into silicon nanowires during metal-particle-assisted growth. *Nat. Commun* **5**, 4134 (2014).
28. van Benthem, K. *et al.* Three-dimensional imaging of individual hafnium atoms inside a semiconductor device. *Appl. Phys. Lett.* **87**, 034104 (2005).
29. Çelebi, C. C. *et al.* Surface induced asymmetry of acceptor wave functions. *Phys. Rev. Lett.* **104**, 086404 (2010).
30. Watanabe, K. *et al.* Band-gap deformation potential and elasticity limit of semiconductor free-standing nanorods characterized *in situ* by scanning electron microscope-cathodoluminescence nanospectroscopy. *ACS Nano* **9**, 2989–3001 (2015).
31. Watanabe, K., Nakamura, Y. & Ichikawa, M. Conductive optical-fibre STM probe for local excitation and collection of cathodoluminescence at semiconductor surfaces. *Opt. Exp* **21**, 19261–19268 (2013).
32. Watanabe, K. *et al.* Scanning tunneling microscope-cathodoluminescence measurement of the GaAs/AlGaAs heterostructure. *J. Vac. Sci. Technol. B* **27**, 1874–1880 (2009).
33. Watanabe, K., Nakamura, Y. & Ichikawa, M. Spatial resolution of imaging contaminations on the GaAs surface by scanning tunneling microscope-cathodoluminescence spectroscopy. *Appl. Surf. Sci.* **254**, 7737–7741 (2008).
34. Watanabe, K., Nakamura, Y. & Ichikawa, M. Measurements of local optical properties of Si-doped GaAs (110) surfaces using modulation scanning tunneling microscope cathodoluminescence spectroscopy. *J. Vac. Sci. Technol. B* **26**, 195–200 (2008).
35. Mass, J. *et al.* Cathodoluminescence characterization of hydrothermal ZnO crystals. *Superlattices Microstruct.* **38**, 223–230 (2005).
36. Lee, W. *et al.* Cathodoluminescence study of nonuniformity in hydride vapor phase epitaxy-grown thick GaN films. *J. Electron Microscopy* **61**, 25–30 (2012).
37. Lee, W. *et al.* Cross sectional CL study of the growth and annihilation of pit type defects in HVPE grown (0001) thick GaN. *J. Cryst. Growth* **351**, 83–87 (2012).
38. Lee, W.-W., Kim, S.-B., Yi, J., Nichols, W. T. & Park, W.-I. Surface polarity-dependent cathodoluminescence in hydrothermally grown ZnO hexagonal rods. *J. Phys. Chem. C* **116**, 456–460 (2012).
39. *Numerical Data and Fundamental Relationships in Science and Technology* Vol. 17 of Landolt-Bornstein new series (Springer, 1982).
40. Christman, J. A., Woolcott, Jr. R. R., Kingon, A. I. & Nemanich, R. J. Piezoelectric measurements with atomic force microscopy. *Appl. Phys. Lett.* **73**, 3851–3853 (1998).
41. Law, M., Greene, L. E., Johnson, J. C., Saykally, R. & Yang, P. Nanowire dye-sensitized solar cells. *Nat. Mater.* **4**, 455–459 (2005).
42. Zhou, Y., Wu, W., Hu, G., Wu, H. & Cui, S. Hydrothermal synthesis of ZnO nanorods arrays with the addition of polyethyleneimine. *Mater. Res. Bull.* **43**, 2113–2118 (2008).
43. Vayssieres, L., Keis, K., Lindquist, S.-E. & Hagfeldt, A. Purpose-built anisotropic metal oxide materials: 3D high oriented microrod array of ZnO. *J. Phys. Chem B* **105**, 3350–3352 (2001).
44. Vayssieres, L. Growth of arrayed nanorods and nanowires of ZnO from aqueous solutions. *Adv. Mater.* **15**, 464–466 (2001).
45. Greece, L. E. *et al.* Low-temperature wafer-scale production of ZnO Nanowire arrays. *Angew. Chem. Int. Ed.* **42**, 3031–3034 (2003).
46. Tian, Z. R., Voigt, J. A., Liu, J., Mckenzie, B. & Mcdermott, M. J. Biomimetic arrays of oriented helical ZnO nanorods and columns. *J. Am. Chem. Soc.* **124**, 12954–12955 (2002).

47. Govender, K., Boyle, D. S., Kenway, P. B. & O'Brien, P. Understanding the factors that govern the deposition and morphology of thin films of ZnO from aqueous solution. *J. Mater. Chem.* **14**, 2575–2591 (2004).

48. Sun, Y., Riley, D. J. & Ashfold, M. N. R. Mechanism of ZnO nanotube growth by hydrothermal methods on ZnO film-coated Si substrates. *J. Phys. Chem. B* **110**, 15186–15192 (2006).

49. Ashfold, M. N. R., Doherty, R. P., Ndifor-Angwafor, N. G., Riley, D. J. & Sun, Y. The kinetics of the hydrothermal growth of ZnO nanostructures. *Thin Solid Films* **515**, 8679–8683 (2007).

50. Xu, S., Lao, C., Weintraub, B. & Wang, Z.-L. Density-controlled growth of aligned ZnO nanowire arrays by seedless chemical approach on smooth surfaces. *J. Mater. Res.* **23**, 2072–2077 (2008).

51. Xu, S. *et al.* Optimizing and improving the growth quality of ZnO nanowire arrays guided by statistical design of experiments. *ACS Nano* **3**, 1803–1812 (2009).

52. Xu, S. & Wang, Z.-L. One-dimensional ZnO nanostructures: solution growth and functional properties. *Nano Res* **4**, 1013–1098 (2011).

53. Volk, J. *et al.* Highly uniform epitaxial ZnO nanorods arrays for nanopiezotronics. *Nanoscale Res. Lett.* **4**, 699–704 (2009).

54. Erdérlyi, R. *et al.* Investigations into the impact of the template layer on ZnO Nanowire arrays made using low temperature wet chemical growth. *Cryst. Growth Des.* **11**, 2515–2519 (2011).

55. Pauporté, T. h., Lincot, D., Viana, B. & Pellé, F. Toward laser emission of epitaxial nanorods arrays of ZnO grown by electrodeposition. *Appl. Phys. Lett.* **89**, 233112 (2006).

56. Belghiti, H. E., Pauporté, T. & Lincot, D. Mechanistic study of ZnO nanorod array electrodeposition. *Phys. Stat. Solid.* **205**, 2360–2364 (2008).

57. Könenkamp, R. *et al.* Thin film semiconductor deposition on free-standing ZnO columns. *Appl.Phys. Lett.* **77**, 2575–2577 (2000).

58. Weintraub, B., Deng, Y. & Wang, Z.-L. Position-controlled Seedless growth of ZnO nanorods arrays on a polymer substrate via wet chemical synthesis. *J. Phys. Chem. C* **111**, 10162–10165 (2007).

59. Kanaya, K. & Okayama, S. Penetration and energy-loss theory of electrons in solid targets. *J. Phys. D* **5**, 43–58 (1972).

60. Meyer, B. K. *et al.* Bound exciton and donor-acceptor pair recombinations in ZnO. *Phys. Stat. Sol. (b)* **241**, 231–260 (2004).

61. Giles, N. C. *et al.* Effects of phonon coupling and free carriers on band-edge emission at room temperature in n-type ZnO crystals. *Appl. Phys. Lett.* **89**, 251906 (2006).

62. Zhang, Y. B., Goh, G. K., Ooi, K. F. & Tripathy, S. Hydrogen-related n-type conductivity in hydrothermally grown epitaxial ZnO films. *J. Appl. Phys.* **108**, 083716 (2010).

63. Reynolds, J. G. *et al.* Shallow acceptor complex in p-type ZnO. *Appl. Phys. Lett.* **102**, 152114 (2013).

64. Lavrov, E. V., Weber, J., Börrnert, F., Van der Walle, C. G. & Helbig, R. Hydrogen-related defects in ZnO studied by infrared absorption spectroscopy. *Phys. Rev. B* **66**, 165205 (2002).

65. Lavrov, E. V., Börrnert, F. & Weber, J. Dominant hydrogen-oxygen complex in hydrothermally grown ZnO. *Phys. Rev. B* **71**, 035205 (2005).

66. Lavrov, E. V., Herklotz, F. & Weber, J. Identification of two hydrogen donors in ZnO. *Phys. Rev. B* **79**, 165210 (2009).

67. Koch, S. G., Lavrov, E. V. & Weber, J. Interplay between interstitial and substitutional hydrogen donors in ZnO. *Phys. Rev. B* **89**, 235203 (2014).

68. Janotti, A. & Van der Walle, C. G. Fundamentals of zinc oxide as a semiconductor. *Rep. Prog. Phys* **72**, 126501 (2009).

69. Cuscó, R. *et al.* Temperature dependence of Raman scattering in ZnO. *Phys. Rev. B* **75**, 165202 (2007).

70. Consonni, V. *et al.* Selective area growth of well-ordered ZnO nanowire arrays with controllable polarity. *ACS Nano* **8**, 4761–4770 (2014).

Acknowledgements

This work was supported by JSPS KAKENHI (grant numbers 23760022 and 26790046) and JSPS-HAS Bilateral Joint Research Projects.

Author contributions

K.W. conceived and designed this work. K.W. fabricated the specimen with the assistance of J.V., S.O., T.N. and Y.W. K.W. developed Nanoprobe-CL system and performed micro-sampling, FIB milling, cross-sectional CL nanospectroscopy, and 'differential' *I–V* measurements of a single nanorod. K.W., T.N., and Y.N. performed XPS, XRD, and Raman spectroscopy of referential ZnO thin-films, respectively. T.S. joined some discussions. K.W. wrote the manuscript with the assistance of J.V., T.N. and Y.N. All authors have given approval to the manuscript.

Additional information

5

Frequency comb transferred by surface plasmon resonance

Xiao Tao Geng[1,2], Byung Jae Chun[3], Ji Hoon Seo[4], Kwanyong Seo[4], Hana Yoon[5], Dong-Eon Kim[1,2], Young-Jin Kim[3] & Seungchul Kim[1,2]

Frequency combs, millions of narrow-linewidth optical modes referenced to an atomic clock, have shown remarkable potential in time/frequency metrology, atomic/molecular spectroscopy and precision LIDARs. Applications have extended to coherent nonlinear Raman spectroscopy of molecules and quantum metrology for entangled atomic qubits. Frequency combs will create novel possibilities in nano-photonics and plasmonics; however, its interrelation with surface plasmons is unexplored despite the important role that plasmonics plays in nonlinear spectroscopy and quantum optics through the manipulation of light on a subwavelength scale. Here, we demonstrate that a frequency comb can be transformed to a plasmonic comb in plasmonic nanostructures and reverted to the original frequency comb without noticeable degradation of $<6.51 \times 10^{-19}$ in absolute position, 2.92×10^{-19} in stability and $1\,Hz$ in linewidth. The results indicate that the superior performance of a well-defined frequency comb can be applied to nanoplasmonic spectroscopy, quantum metrology and subwavelength photonic circuits.

[1] Max Planck Center for Attosecond Science, Max Planck POSTECH/KOREA Res. Initiative, Pohang, Gyeongbuk 376-73, South Korea. [2] Department of Physics, Center for Attosecond Science and Technology (CASTECH), POSTECH, Pohang, Gyeongbuk 376-73, South Korea. [3] School of Mechanical and Aerospace Engineering, Nanyang Technological University (NTU), 50 Nanyang Avenue, Singapore 639798, Singapore. [4] Department of Energy Engineering, Ulsan National Institute of Science and Technology (UNIST), Ulsan 689-798, South Korea. [5] Energy Storage Department, Korea Institute of Energy Research (KIER), Daejeon 305-343, South Korea. Correspondence and requests for materials should be addressed to Y.-J.K. (email: yj.kim@ntu.edu.sg) or to S.K. (email: inter99@postech.ac.kr).

The frequency comb of mode-locked femtosecond lasers has led to remarkable advances in high-resolution spectroscopy[1,2], broadband calibration of astronomical spectrographs[3,4], time/frequency transfer over long distances[5,6], absolute laser ranging[7–10] and inter-comparison of atomic clocks[11,12]. It provides millions of well-defined optical modes over a broad spectral bandwidth with high-level phase coherence referenced to an atomic clock. Recently, the potential of frequency comb has expanded to microscopic applications; high inter-mode coherence within a short pulse duration enabled manipulating atomic qubits[13], operating quantum logic gates and performing high-speed molecular detection by coherent Raman spectroscopy through harnessing inter-mode beat frequencies between two frequency combs at different repetition rates[14].

Coupling surface plasmons (SPs)[15,16], collective charge oscillations produced by the resonant interaction of light and free electrons on the interface of metallic and dielectric materials, to frequency comb creates numerous advantages. First, SP can allow for the frequency comb to access nanoscopic volumes that surpass the diffraction limit[17]. Second, the field enhancement by localized SP enables the highly sensitive detection of weak signals, even from a single molecule (for example, surface-enhanced Raman scattering)[18]. Third, next-generation photonic devices and circuits can be implemented within a small subwavelength volume by all-optical control of light properties (amplitude, phase and polarization state) in plasmonic nanostructures within ultrafast time scales[19–22]. However, the superior performance of the frequency comb, such as absolute frequency uncertainty, high-frequency stability and narrow linewidth, could deteriorate during the photon-plasmon conversion process. For exploring novel combination of frequency comb and SP resonance, it is prerequisite to verify that frequency comb maintains its performance under plasmonic resonance; however, there have been no studies to date.

In the following, we report that frequency comb successfully maintains core performances in photon-plasmon conversion by exploiting plasmonic extraordinary transmission through a subwavelength plasmonic hole array. This implies that the original frequency comb can be transformed into a form of plasmonic comb on metallic nanostructures and reverted to an original frequency comb without noticeable degradation in absolute frequency position, stability and linewidth. The superior performance of well-defined frequency combs can therefore be applied to various nanoplasmonic spectroscopy, coherent quantum metrology and subwavelength photonic circuits.

Results

Frequency comb transferred by SP resonance. Figure 1 shows the experimental apparatus to characterize the conservation of frequency comb for the conversion from photon to SP. The frequency comb is split into reference and measurement beams; one part of the beam transmits through an acousto-optic modulator (AOM) for a frequency shift of 40 MHz to construct a reference frequency comb and the other part of the beam passes through the plasmonic sample. The frequency comb structure in SP resonance was generated by the exploitation of a metallic nanohole array used for extraordinary optical transmission (EOT) that converted photon into SP. The small diameter of each hole prevents light passing through the sample based on classical optics. However, the SP-mediated tunnelling effect of nanohole array drastically enhances optical transmittance[23]. These intriguing optical phenomena have been studied widely for high-resolution chemical sensing, ultrafast optical modulation, wavelength-tunable optical filtering and subwavelength lithography[24,25]. The physical origin of EOT has been attributed to resonant SP

polaritons (SPPs)[26]. The appropriate geometrical and material parameters of nanohole array excite the SPP mode that allows the transmission of light that contains plasmonic information inside an EOT sample. The resonant nature of the SP changes the transmitted spectral distribution, depending on sample design, input polarization and incident angle. Plasmonic EOT can also induce wavelength-dependent changes in optical frequency and phase in addition to wavelength-dependent transmittance. The optical frequency of a single frequency comb mode transmitted through the plasmonic sample via SP resonance (f_{MEA}) can be expressed as

$$f_{MEA} = nf_r + f_{ceo} + \Delta f_{sp} \tag{1}$$

where f_r is the pulse repetition frequency, f_{ceo} the carrier-envelope offset frequency, and Δf_{sp} the frequency and phase change generated by SP resonance. Meanwhile, the optical frequency of the single mode passing through the reference path (f_{REF}) can be expressed as

$$f_{REF} = nf_r + f_{ceo} + f_{AOM} \tag{2}$$

where f_{AOM} denotes the intentional frequency shift by AOM. The detection of the heterodyne beat-frequency generated by the interference between the reference and measurement beams enables the measurement of optical frequency difference, ($f_{REF} - f_{MEA}$) at a radio-frequency (RF) regime using a fast avalanche photodiode. This resultant frequency difference can be simplified to $f_{AOM} - \Delta f_{sp}$, where f_{AOM} works as the high-frequency carrier to isolate Δf_{sp} from the relatively strong low frequency noise components.

Plasmonic extraordinary transmission. For transmitting frequency combs through the subwavelength holes by SP resonance, there are three important geometric parameters: hole diameter (d), hole pitch (l) and Au film thickness (t; Fig. 2a). For maximum optical transmission at a wavelength of 840 nm, three parameters were optimized by solving Maxwell's equations using finite-difference time-domain (FDTD) method. Figure 2b,c show the calculated plasmonic field distribution through the optimized sample. The electric field around the hole was significantly enhanced by SP in the periodic apertures, delivering the optical energy through the hole. Figure 2a shows the scanning electron microscope image of the fabricated nanohole array; all dimensions were matched with optimized design parameters within a geometric error of <5%. Figure 2d shows that the transmitted optical spectrum coincided with the numerical FDTD results and validated the numerical analysis. Minor deviations between the two spectrums are expected by focusing geometry onto the plasmonic sample. The plasmonic resonance conditions are dissimilar in given transverse electric-transverse magnetic polarization if the angle of incidence is not surface normal. As a result, plasmonic sample shows different transmission spectra for transverse electric-transverse magnetic polarization at the incidence angle of 45° (Fig. 2e); therefore, optical transmission of our sample is dominated by the plasmonic EOT, not classical diffraction theory.

Frequency comb structure after plasmonic transmission. The transmitted frequency combs through the plasmonic sample results in an interference with the reference frequency comb to verify the frequency comb structure after the photon-plasmon mode conversion by the EOT (Fig. 3a). For comparative analysis, interference signals were obtained at three different wavelength regimes with optical band-pass filters, representing on-resonance (840 nm) and off-resonance (800 and 900 nm) positions.

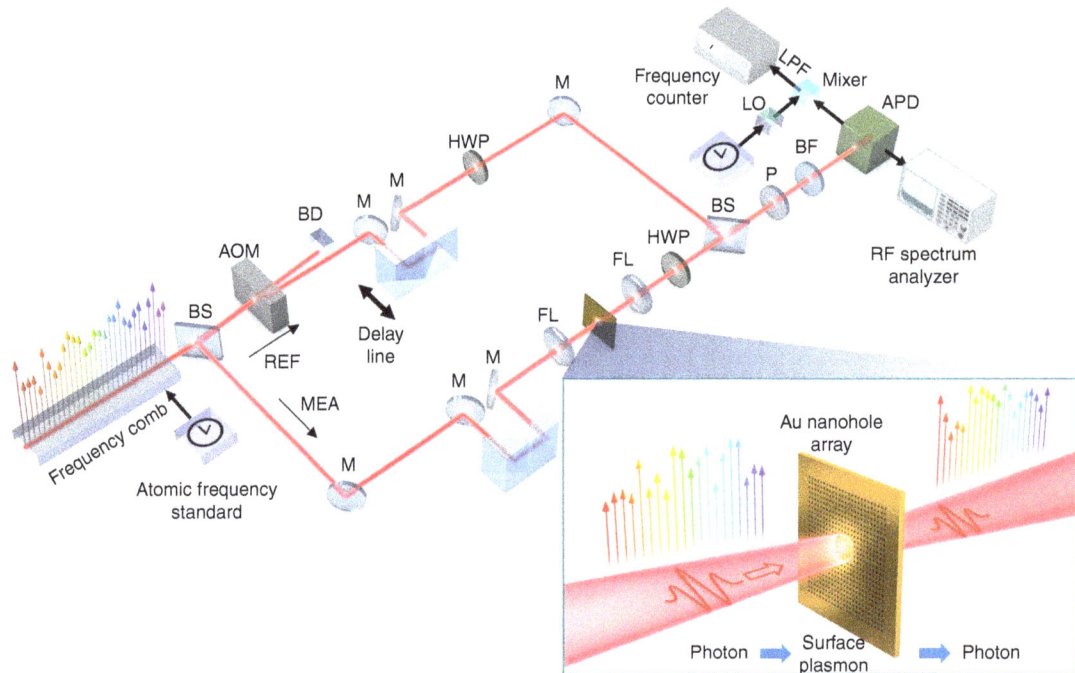

Figure 1 | Generation and characterization of plasmonic frequency comb. Part of the frequency comb experiences plasmonic mode conversion by passing through the plasmonic sample. The sample consists of a subwavelength nanohole array on an Au thin-film, enabling the conversion from photon to SP (from SP to photon). The other part of the frequency comb is used as a reference beam to compare with the frequency comb passed through the plasmonic sample. The frequency combs at two different paths are combined and monitored by APD. The characteristics of the frequency comb at measurement path are analysed by an RF spectrum analyser and a frequency counter. APD, avalanche photo-detector; BD, beam dumper; BF, band-pass filter; BS, beam splitter; FL, focusing lens; HWP, half-wave plate; LO, local oscillator; LPF, low-pass filter; M, mirror; MEA, measurement beam path; P, polarizer; REF, reference beam path.

The coherence of a large number of frequency comb modes can be deteriorated by temporal and spectral plasmonic dispersion, phase noise and frequency noise during the propagation through the plasmonic EOT sample. The frequency comb fundamentally suffers from phase and frequency noises when passing through the optical medium (for example, ambient air and optical fibre) exposed to environmental variations, such as vibration, temperature variation and humidity change. Therefore, it has been an important task to monitor and compensate the temporal and spectral dispersion, phase noise and frequency noise generated in the medium, as reported through long optical-fibre[6] and through ambient air[27]. SPs also suffer from the dispersion and phase change by the medium and environmental disturbances, which have not been investigated with the frequency comb for their quantitative or qualitative analysis. Propagating SPs through the EOT sample experience phase delay depending on their wavelengths and spatial locations before and after tunnelling through each subwavelength hole; this phase delay can be additionally induced by the plasmonic dynamic damping, imperfect sample geometry, surface roughness of the metal film or air refractive index change around the sample. Therefore, the total summation of the electromagnetic waves at the output side of each hole may contain temporal and spectral dispersion, phase distortion and frequency change.

Most noise sources of the frequency comb can be categorized into intra-cavity and extra-cavity sources; intra-cavity noise sources (including cavity length change, cavity loss fluctuations and pump noise) cause frequency noise whereas extra-cavity noise sources (induced by path-length fluctuation, shot noise from the limited power or noise generated during super-continuum generation) result in time-varying phase noise floor[5]. In this investigation, plasmonic mode conversion by the EOT was considered as an extra-cavity noise source that provided wavelength-dependent power attenuation, phase shift and frequency noise, similar to the supercontinuum generation process. Noise contributions should be observed at $f_{AOM} - \Delta f_{SP}$ in the form of linewidth broadening, frequency shift, signal-to-noise (S/N) ratio reduction, increased phase noise or a higher Allan deviation if the plasmonic frequency comb suffers from phase or frequency noise during the plasmonic mode conversion.

Linewidth broadening and S/N ratio reduction in plasmonic mode conversion process was initially evaluated by measuring RF beat linewidth of $f_{AOM} - \Delta f_{SP}$ at three different wavelength regimes (Fig. 3b). With different resolution bandwidths (RBWs), there was no substantial degradation in the linewidth at 840 nm before and after the installation of the plasmonic sample in the beam path. The high-level S/N ratio of ~ 60 dB beat signal indicates that the plasmonic EOT provide no significant phase noise to the frequency comb.

Phase noise and frequency stability was measured for the quantitative analysis of frequency-dependent noise contributions. Figure 4a shows the phase noise spectrum obtained by monitoring one of high harmonics of the beat frequencies at ~ 1.2 GHz with and without the plasmonic sample; this confirms that there was no noticeable frequency noise inclusion. For high-precision frequency position measurement, the beat frequency between reference and measurement frequency comb was measured by a frequency counter for 3,000 s, resulting in 0.24 mHz frequency difference with a s.d. of 61 mHz (Fig. 4b). This corresponds to 6.51×10^{-19}, which proves that plasmonic mode conversion provides no substantial degradation in the frequency accuracy of the frequency comb. The stability of the beat signal was measured to be 4.08×10^{-18} without the plasmonic sample, 4.37×10^{-18} with the plasmonic sample at

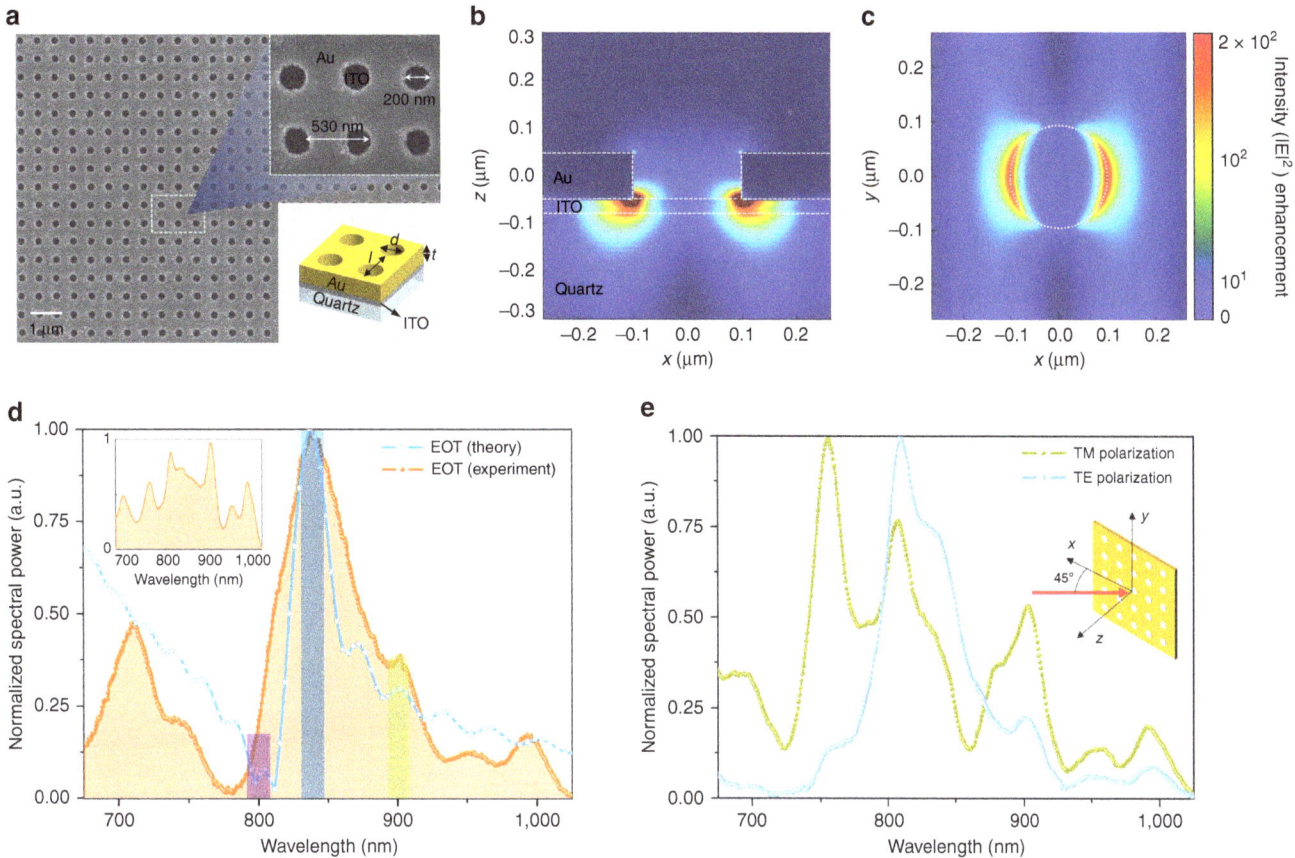

Figure 2 | Numerical simulation and characterization of fabricated plasmonic sample. (**a**) Scanning electron microscope image of the fabricated subwavelength nanohole array for plasmonic EOT. The fabricated nanohole has the diameter (*d*) of 200 nm, pitch (*l*) of 530 nm and thickness (*t*) of 100 nm on 25-nm-thick ITO-coated quartz substrate. (**b**) Calculated intensity distribution of an plasmonic sample taken from the side. (**c**) Calculated intensity distribution at the interface between Au and ITO layer. (**d**) Theoretical (blue line) and experimental (orange line) spectrum of transmitted frequency combs through the plasmonic sample. Purple, blue and green bars represent the selected spectral components (800, 840 and 900 nm) to characterize frequency comb, respectively. Inset (top left) shows the original spectrum of the frequency comb. (**e**) Polarization dependent transmission spectrum through the plasmonic sample at an incident angle of 45°.

resonance wavelength of 840 nm for an averaging time of 100 s, respectively (Fig. 4c). At the off-resonance wavelength, the stability of beat signal was 4.59×10^{-18}, signifying almost no difference between on- and off-plasmonic resonance stabilities. All the experiments pointed that plasmonic mode conversion causes no substantial degradation to the frequency comb in terms of linewidth, frequency position, S/N ratio and frequency stability.

Discussions

All hundreds of thousands optical modes in the frequency comb were firstly converted from photonic to plasmonic mode at the input side of the plasmonic EOT sample and then reverted to photonic mode at the other output side of the sample. It is known to be practically difficult to directly measure the optical frequency of the plasmonic mode so the characteristics of the plasmonic comb were measured here in the far field. Because the plasmonic and photonic modes are assumed to be mutually coherent, if there is any change in the frequency comb characteristics during the plasmonic propagation (in plasmonic mode) through the sample, it should be monitored at the output side in the far field (in photonic mode). Therefore, the beat-frequency detection using the transmitted photonic mode in the far-field regime enabled us to compare the qualities of the plasmonic comb with the original frequency comb, which cannot be implemented in the

near-field regime. As the result of the comparison, there were no noticeable degradation in linewidth, frequency shift, S/N ratio, phase noise and Allan deviation. This implies that SP, the collective electrons, can be regarded as information carrier as precise as the optical frequency comb.

The frequency comb passing through the plasmonic EOT sample experiences the different physical process with the light reflection at a metallic mirror. Although both of the SP resonance and the surface reflection are governed by free-electron oscillation in conduction band of metals, the SP resonance additionally requires the specific momentum matching between incident photon and SP, whose relationship is determined by the plasmonic dispersion relation. Therefore, it is natural to maintain the coherence during the light reflection at metal surface (governed by frequency conservation), which is not the case in plasmonic structures (governed by frequency and momentum matching). Once the incident photon (in photonic mode) is converted into SP, it will propagate through the metal as the form of SPPs (in plasmonic mode). This plasmonic propagation causes temporal and spectral dispersions, phase variations and frequency changes, which may degrade the inherently high coherence of the optical frequency comb.

Plasmonic EOT is governed by not only the hole geometry[28] but also hole pitch. Therefore, the incidence angle tuning of the input beam can provide the change in plasmonic coupling mode without dimensional changes, which can possibly cause some

a

b

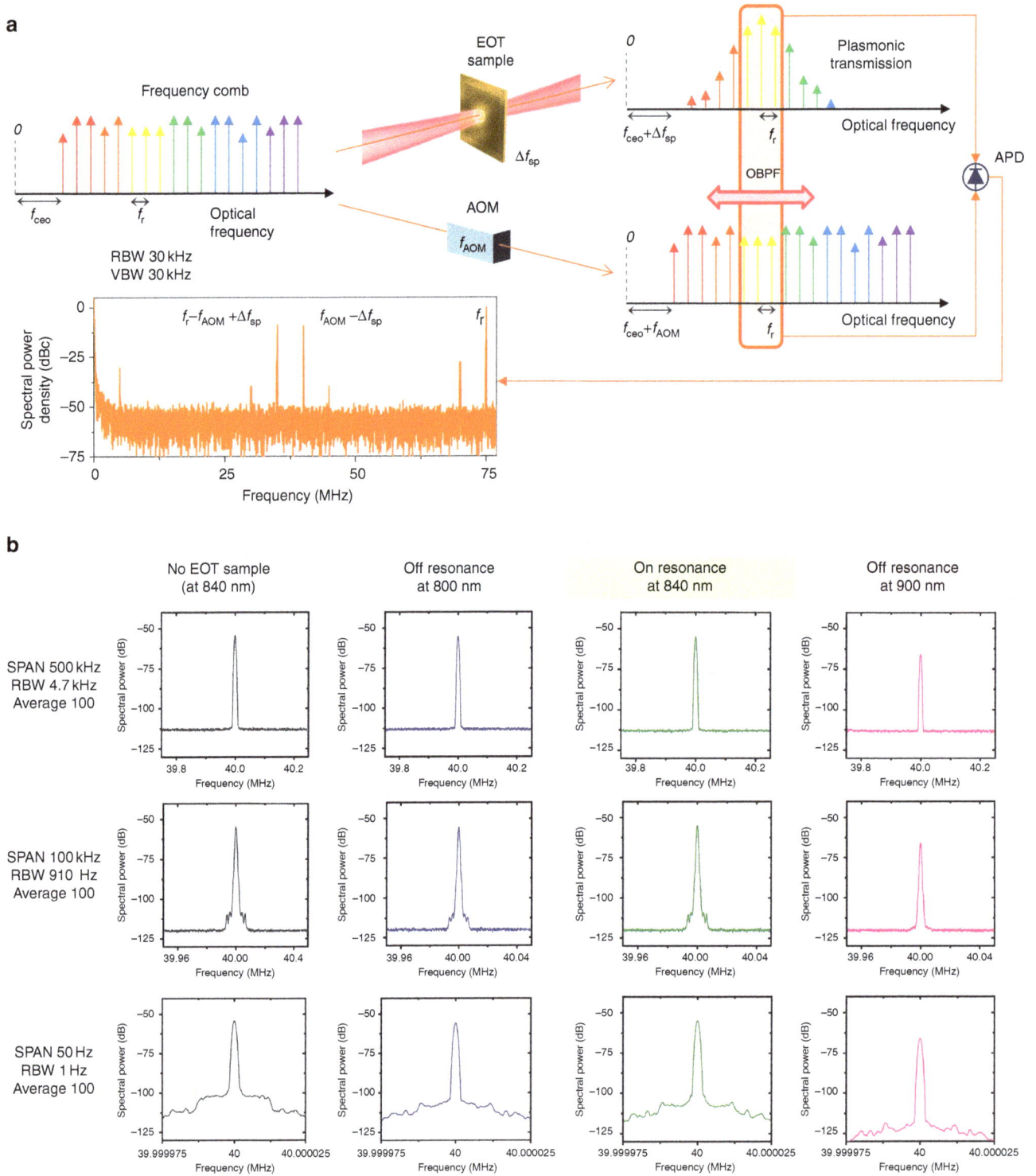

Figure 3 | Evaluation of the plasmonic frequency comb by EOT. (**a**) Generation of RF beats by the interference between frequency-shifted (40 MHz) reference combs and plasmonic EOT combs. The beat spectra of plasmonically transmitted frequency comb and the reference comb are measured at three wavelengths: one at a strong plasmonic resonance position (a 840-nm centre wavelength with a 10-nm bandwidth), two at off-resonance positions (a 800-nm centre wavelength with a 40-nm bandwidth and 900 nm with a 10-nm bandwidth) using three optical band-pass filters. These are compared with a beat spectrum at a 840-nm wavelength, acquired without the plasmonic sample. (**b**) Linewidth measurement of RF beats with different span, RBWs and VBWs. There was no noticeable linewidth degradation by the plasmonic transduction (<1 Hz, limited by RBW of the instrument). APD, avalanche photo-detector; OBPF, optical band-pass filter; VBWs, video bandwidths.

degradation in the frequency characteristics of the frequency comb by providing different plasmonic field distribution and enhancement. To test this, the beat spectrum was monitored while the sample was rotated by up to 45° (for transverse magnetic wave) as shown in Figure 2e. For the given condition, all frequency characteristics were maintained in the same level with

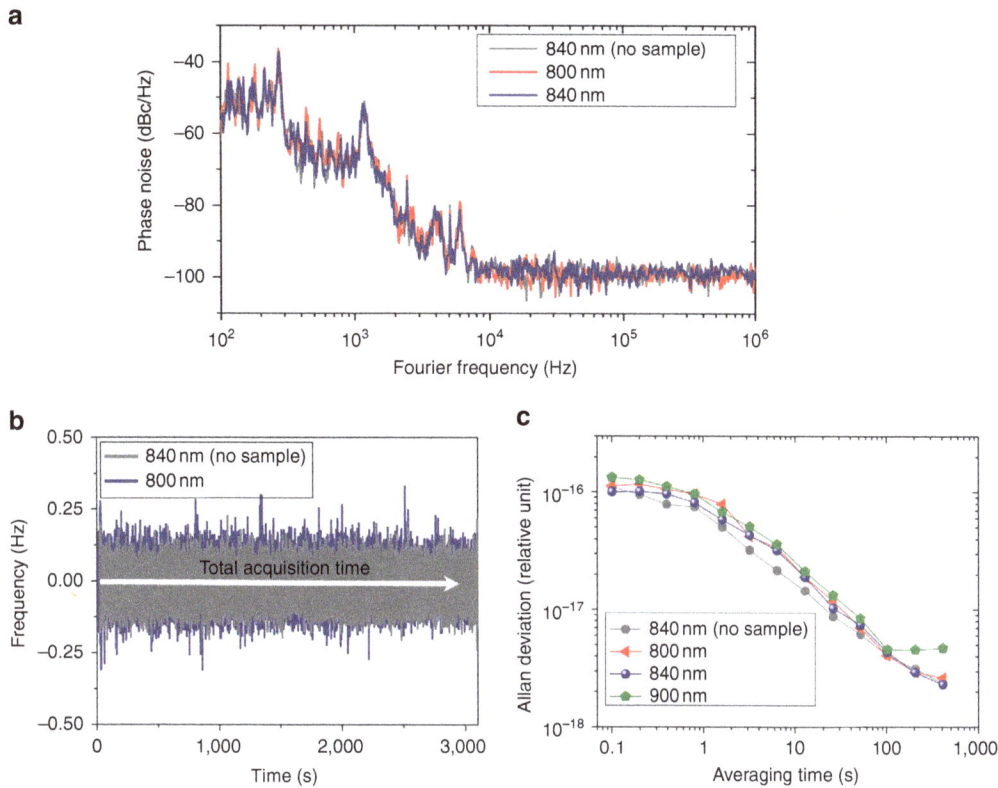

Figure 4 | Plasmonic frequency comb: phase noise and frequency stability. (**a**) Phase noise spectra at a 1.2-GHz RF carrier at on- and off-resonance wavelengths. (**b**) Time trace of the beat frequency with and without the plasmonic sample over 3,000 s. (**c**) Allan deviations of frequency stability with varying average time at the positions of on-resonance and off-resonance wavelengths.

normal incidence case, which shows that no performance degradation exist depending on plasmonic coupling or geometrical parameters of the sample.

The linewidth broadening by plasmonic EOT was evaluated to be <1 Hz, which is limited by RBW of the instrument in use (Fig. 3b). A single RF beat-frequency corresponds to the superposition of small RF beat contributions of >10^4 frequency comb modes, which proves that there is no significant wavelength-dependent frequency or phase noise during the plasmonic EOT. There was minor increase in spectral power in the pedestal peaks at 12, 17 and 21 Hz when the frequency comb passed through the plasmonic sample; this is expected to be caused by the vibrational and thermal noises at the plasmonic sample. The beat frequency, $f_{AOM} - \Delta f_{SP}$, was found to be exactly the same as the driving frequency of the AOM in all measured spectra shown in Fig. 3b, which implies that the absolute frequency position is well maintained in the plasmonic mode. The ambient temperature and vibration on EOT sample were not intentionally controlled so as to evaluate the performance in normal laboratory environment conditions. Our results show that the frequency comb structures are well maintained under environmental disturbances, for example, temperature variation, mechanical vibration and air fluctuation. This will enable us to develop high-sensitivity frequency-comb-referenced SP sensors working in harsh environments. The phase noise spectra in Fig. 4a also shows a number of minor peaks at 0.2, 1.5, 300 and 600 kHz other than the low-frequency spectral peaks at 12, 17 and 21 Hz observed in Fig. 3b. At higher frequency than 10 kHz, there is a flat noise floor without other spectral peaks or broad pedestals. In S/N ratio measurement, the S/N ratio theoretically could reach 68~75 dB in a 100 kHz RBW because there are 10^4~10^5 frequency comb modes in the pass-band of the optical

filter transmittance. The experimental S/N ratio with the plasmonic sample on-resonance position was ~60 dB; this minor deviation could come from imperfect intensity balancing, polarization matching and spatial beam mode-matching. The S/N ratio at 900 nm was 54 dB, relatively lower than that at 800 nm because the quantum efficiency of the avalanche photodetector at 900 nm is ~20% lower than that at 800 nm and the filter bandwidth at 900 nm is 25% of that at 800 nm.

In this article, we have studied SP resonance effects on frequency comb structure in the plasmonic EOT of light through a subwavelength metallic nanohole array. The frequency comb was transduced to plasmonic mode in the sample and reverted to photonic mode without significant changes in linewidth, frequency shift, S/N ratio, phase noise and Allan deviation. The linewidth broadening was <1 Hz (instrument limited), frequency inaccuracy was 6.51×10^{-19}, S/N ratio was higher than 60 dB, Allan deviation increased by 2.92×10^{-19} at 100 s averaging time. This outstanding frequency comb performance in plasmonic nanostructures enables a highly sensitive, high accurate and broadband measurement with direct traceability to standards. This inclusion of frequency comb has the potential to accelerate progresses in various plasmonic applications such as bio-chemical spectroscopy or sensing, quantum optics and sub-diffraction-limit biomedical-imaging. With the aid of SP, frequency-comb-referenced high-speed coherent anti-stokes Raman spectroscopy[14] can be implemented in much smaller nanoscopic volume being requested for single-molecule detection, for example, surface-enhanced coherent anti-stokes Raman spectroscopy[29]. A large number of optical modes in a frequency comb as the time and frequency standard can be coupled at the same time with SP for broadband quantum metrology for entangled atomic qubits or information carrier in subwavelength scale[13,30].

Localized field enhancement of SP will enable highly efficient nonlinear optics[31] coupled with high precision of frequency comb, which is prerequisite for novel sub-diffraction-limit nonlinear biomedical imaging and spectroscopy.

Methods

Frequency comb. A Ti:sapphire femtosecond laser delivers 4.8 fs pulses at a repetition rate of 75 MHz over a broad spectral bandwidth from 1.03 to 2.06 eV (Venteon UB, Venteon). For establishing a frequency comb, the pulse repetition frequency (f_r) and carrier-envelope offset frequency (f_{ceo}) were precisely locked to a reference Rb atomic clock (FS725, Stanford Research Systems) with the aid of a f–$2f$ interferometer and phase-locked control loops (AVR32, TEM-Messtechnik & XPS800-E, Menlosystems). One part of the beam was diverted to and transmitted through an AOM for the frequency shift of 40 MHz to construct a reference frequency comb. If the plasmonic frequency comb suffers from the phase or frequency noise during the plasmonic mode conversion, the noise contributions should be observed at $f_{AOM} - \Delta f_{SP}$ in the forms of linewidth broadening, frequency shift, S/N ratio reduction, increased phase noise or higher Allan deviation. The frequency comb excited the plasmonic sample with the whole-broadband spectrum in a loose focusing geometry with an aspheric lens of 100 mm focal length. The focused peak intensity at the plasmonic sample was set to be $<0.1\,MW\,cm^{-2}$ not to exceed the thermal damage threshold ($\sim 1\,TW\,cm^{-2}$ for Au). Input polarization state was set linear and its direction is parallel to the x axis of periodic holes on plasmonic sample as denoted in Fig. 2e.

Plasmonic EOT sample: design and development. We exploited FDTD solution (XFDTD8.3, Lumerical) to solve Maxwell's equation for plasmonic near-field distribution and transmitted spectrum. Through a series of iterative computations, the optimal geometric parameters were determined as $d = 200\,nm$, $l = 530\,nm$ and $t = 100\,nm$. The thickness, t was designed to be much thicker than the Au skin depth ($\sim 20\,nm$) here to block the direct transmission through the Au film. The designed nanohole array was fabricated using electron-beam lithography (Raith 150) onto 25-nm-thick ITO-coated quartz substrate.

Evaluation of plasmonic frequency comb. For comparative analysis, interference signals were obtained at three different wavelength regimes – one at plasmonic on-resonance (840 nm) and the others at off-resonance positions (800 and 900 nm) – using optical band-pass filters. The resulting interference beat signal was obtained by high-speed avalanche photodiode and analysed using a high-resolution RF spectrum analyzer (N9020A, Agilent) and a RF frequency counter (53230A, Keysight Technologies). An exemplary RF spectrum is shown in Fig. 3b; the repetition rate (f_r) is located at 75 MHz, the beat frequency ($f_{AOM} - \Delta f_{sp}$) between the frequency-shifted reference frequency comb and the plasmonic frequency comb is at ~ 40 MHz, and the beat frequency ($f_r - f_{AOM} + \Delta f_{sp}$) between the other nearby reference frequency comb modes and the plasmonic frequency comb is at ~ 35 MHz. Other minor spurious peaks are due to the imperfect sinusoidal modulation of AOM; their positions match with beat frequencies between the f_{AOM}-harmonics and the reference frequency comb of $2f_{AOM} - f_r$, $2f_r - 3f_{AOM}$, $3f_{AOM} - f_r$ and $2(f_r - f_{AOM})$.

References

1. Hänsch, T. W. Nobel Lecture: passion for precision. *Rev. Mod. Phys.* **78**, 1297–1309 (2006).
2. Hall, J. L. Nobel Lecture: defining and measuring optical frequencies. *Rev. Mod. Phys.* **78**, 1279–1295 (2006).
3. Steinmetz, T. *et al.* Laser frequency combs for astronomical observations. *Science* **321**, 1335–1337 (2008).
4. Wilken, T. *et al.* A spectrograph for exoplanet observations at the centimetre-per-second level. *Nature* **485**, 611–614 (2012).
5. Predehl, K. *et al.* A 920-kilometer optical fiber link for frequency metrology at the 19th decimal place. *Science* **336**, 441–444 (2012).
6. Giorgetta, F. R. *et al.* Optical two-way time and frequency transfer over free space. *Nat. Photon.* **7**, 434–438 (2013).
7. Newbury, N. R. Searching for applications with a fine-tooth comb. *Nat. Photon.* **5**, 186–188 (2011).
8. Minoshima, K. & Matsumoto, H. High-accuracy measurement of 240-m distance in an optical tunnel by use of a compact femtosecond laser. *Appl. Opt.* **39**, 5512–5517 (2000).
9. Coddington, I., Swann, W. C., Nenadovic, L. & Newbury, N. R. Rapid and precise absolute distance measurements at long range. *Nat. Photon.* **3**, 351–356 (2009).
10. Lee, J., Kim, Y.-J., Lee, K., Lee, S. & Kim, S.-W. Time-of-flight measurement with femtosecond light pulses. *Nat. Photon.* **4**, 716–720 (2010).
11. Rosenband, T. *et al.* Frequency ratio of Al + and Hg + single-ion optical clocks; metrology at the 17th decimal place. *Science* **319**, 1808–1812 (2008).
12. Reinhardt, S. *et al.* Test of relativistic time dilation with fast optical atomic clocks at different velocities. *Nat. Phys.* **3**, 861–864 (2007).
13. Hayes, D. *et al.* Entanglement of atomic qubits using an optical frequency comb. *Phys. Rev. Lett.* **104**, 140501 (2010).
14. Ideguchi, T. *et al.* Coherent Raman spectro-imaging with laser frequency combs. *Nature* **502**, 355–358 (2013).
15. Stockman, M. Nanoplasmonics: past, present, and glimpse into future. *Opt. Express* **19**, 22029–22106 (2011).
16. Kauranen, M. & Zayats, A. V. Nonlinear plasmonics. *Nat. Photon.* **6**, 737–748 (2012).
17. Willets, K. A. & Van Duyne, R. P. Localized surface plasmon resonance spectroscopy and sensing. *Annu. Rev. Phys. Chem.* **58**, 267–297 (2007).
18. Zhang, Y. *et al.* Coherent anti-Stokes Raman scattering with single-molecule sensitivity using a plasmonic Fano resonance. *Nat. Commun.* **5**, 4424 (2014).
19. Tame, M. S. *et al.* Quantum plasmonics. *Nat. Phys.* **9**, 329–340 (2013).
20. Melikyan, A. *et al.* High-speed plasmonic phase modulators. *Nat. Photon.* **8**, 229–233 (2014).
21. Haffner, C. *et al.* All-plasmonic Mach–Zehnder modulator enabling optical high-speed communication at the microscale. *Nat. Photon.* **9**, 525–528 (2015).
22. Dennis, B. S. *et al.* Compact nanomechanical plasmonic phase modulators. *Nat. Photon.* **9**, 267–273 (2015).
23. Ebbesen, T. W., Lezec, H. J., Chaemi, H. F., Thio, T. & Wolff, P. A. Extraordinary optical transmission through sub-wavelength hole arrays. *Nature* **391**, 667–669 (1998).
24. Barnes, W. L., Dereux, A. & Ebbesen, T. W. Surface plasmon subwavelength optics. *Nature* **424**, 824–830 (2003).
25. Wurtz, G. A. & Zayats, A. V. Nonlinear surface plasmon polaritonic crystals. *Laser Photon. Rev.* **2**, 125–135 (2008).
26. Garcia-Vidal, F. J., Martin-Moreno, L., Ebbesen, T. W. & Kuipers, L. Light passing through subwavelength apertures. *Rev. Mod. Phys.* **82**, 729–787 (2010).
27. Newbury, N. R. & Swann, W. C. Low-noise fiber-laser frequency combs. *J. Opt. Soc. Am. B* **24**, 1756–1770 (2007).
28. Yue, W. *et al.* Enhanced extraordinary optical transmission (EOT) through arrays of bridged nanohole pairs and their sensing applications. *Nanoscale* **6**, 7917–7923 (2014).
29. Steuwe, C., Kaminski, C. F., Baumberg, J. J. & Mahajan, S. Surface enhanced coherent anti-Stokes Raman scattering on nanostructured gold surfaces. *Nano Lett.* **11**, 5339–5343 (2011).
30. Altewischer, E., Van Exter, M. P. & Woerdman, J. P. Plasmon-assisted transmission of entangled photons. *Nature* **418**, 304–306 (2002).
31. Almeida, E. & Prior, Y. Rational design of metallic nanocavities for resonantly enhanced four-wave mixing. *Sci. Rep.* **5**, 10033 (2015).

Acknowledgements

This work was supported by the Basic Science Research Program (NRF-2013R1A1A2004932; NRF-2014R1A1A1004885), by Global Research Laboratory Program (Grant No. 2009-00439), by the Leading Foreign Research Institute Recruitment Program (Grant No. 2010-00471), by the Max Planck POSTECH/KOREA Research Initiative Program (Grant No. 2011-0031558), by the NRF Grant (No. 2010-0021735), by the Leading Foreign Research Institute Recruitment Program (Grant No. 2012K1A4A3053565) through the NRF funded by the MEST. This work was also supported by a Grant (14CTAP-C077584-01) from Infrastructure and Transportation Technology Promotion Research Program funded by Ministry of Land, Infrastructure and Transport of Korean government. This work was also supported by Singapore National Research Foundation (NRF-NRFF2015-02) and Singapore Ministry of Education under its Tier 1 Grant (RG85/15).

Author contributions

The project was planned and overseen by Y.-J.K., D.E.K. and S.K. Plasmonic sample was prepared and characterized by J.H.S., H.Y. and K.S. Frequency comb experiments were performed by X.T.G., B.J.C., Y.-J.K and S.K. All authors contributed to the manuscript preparation.

Additional information

Competing financial interests: The authors declare no competing financial interests.

6

Computational multiqubit tunnelling in programmable quantum annealers

Sergio Boixo[1], Vadim N. Smelyanskiy[1,2], Alireza Shabani[1], Sergei V. Isakov[1], Mark Dykman[3], Vasil S. Denchev[1], Mohammad H. Amin[4,5], Anatoly Yu. Smirnov[4], Masoud Mohseni[1] & Hartmut Neven[1]

Quantum tunnelling is a phenomenon in which a quantum state traverses energy barriers higher than the energy of the state itself. Quantum tunnelling has been hypothesized as an advantageous physical resource for optimization in quantum annealing. However, computational multiqubit tunnelling has not yet been observed, and a theory of co-tunnelling under high- and low-frequency noises is lacking. Here we show that 8-qubit tunnelling plays a computational role in a currently available programmable quantum annealer. We devise a probe for tunnelling, a computational primitive where classical paths are trapped in a false minimum. In support of the design of quantum annealers we develop a nonperturbative theory of open quantum dynamics under realistic noise characteristics. This theory accurately predicts the rate of many-body dissipative quantum tunnelling subject to the polaron effect. Furthermore, we experimentally demonstrate that quantum tunnelling outperforms thermal hopping along classical paths for problems with up to 200 qubits containing the computational primitive.

[1]Google, Venice, California 90291, USA. [2]NASA Ames Research Center, Moffett Field, California 94035, USA. [3]Department of Physics and Astronomy, Michigan State University, East Lansing, Michigan 48824, USA. [4]D-Wave Systems Inc., Burnaby, British Columbia, Canada V5C 6G9. [5]Department of Physics, Simon Fraser University, Burnaby, British Columbia, Canada V5A 1S6. Correspondence and requests for materials should be addressed to S.B. (email: boixo@google.com) or to V.N.S. (smelyan@google.com).

Quantum annealing[1–5] is a technique inspired by classical simulated annealing[6] that aims to take advantage of quantum tunnelling. In classical cooling optimization algorithms such as simulated annealing, the initial temperature must be high to overcome tall energy barriers. As the algorithm progresses, the temperature is gradually lowered to distinguish between local minima with small energy differences. This causes the stochastic process to freeze once the thermal energy is lower than the height of the barriers surrounding the state. In contrast, quantum tunnelling transitions are still present even at zero temperature. Therefore, for some energy landscapes, one might expect that quantum dynamical evolutions can converge to the global minimum faster than the corresponding classical cooling process.

The goal of quantum annealing is to find low-energy states of a 'problem Hamiltonian'

$$H_{\mathrm{P}} = -\sum_{\mu} h_{\mu}\sigma_{\mu}^{z} - \sum_{\mu\nu} J_{\mu\nu}\sigma_{\mu}^{z}\sigma_{\nu}^{z}, \qquad (1)$$

where the Pauli matrices σ_{μ}^{z} correspond to spin variables with values $\{\pm 1\}$. The local fields $\{h_{\mu}\}$ and couplings $\{J_{\mu\nu}\}$ define the problem instance. Quantum annealing is characterized by evolution under the Hamiltonian

$$H_{0}(s) = A(s)H_{\mathrm{D}} + B(s)H_{\mathrm{P}}, \qquad (2)$$

where $H_{\mathrm{D}} = -\sum_{\mu}\sigma_{\mu}^{x}$. The annealing parameter s slowly increases from 0 to 1 throughout the annealing time t_{qa}. Initially, $A(0) \gg B(0)$. With increasing s, $A(s)$ monotonically decreases to 0 for $s = 1$, whereas $B(s)$ increases.

The performance of chips designed to implement quantum annealing using superconducting electronics has been studied in a number of recent lines of work[7–24]. Given a quantum annealer operating at finite temperature, noise and dissipation strengths, does it utilize tunnelling or thermal activation for computation? Here we address precisely this question. We introduce a 16-qubit probe for tunnelling, a computational primitive where classical paths are trapped in a false minimum. We present a nonperturbative theory of multiqubit tunnelling, which takes into account both high- and low-frequency noises. To distinguish between tunnelling and thermal activation, we study the thermal dependence of the probability of success for the computational primitive. Thermal activation shows an increasing probability of success with increasing temperature, as expected. Multiqubit tunnelling, on the other hand, shows a decreasing probability of success with increasing temperature, both in theory and experiment. Finally, we study a generalization of the computational primitive to a larger number of qubits that contain the same 'motif' multiple times. Quantum tunnelling outperforms thermal hoping for these problems under similar parameters.

Results

Quantum tunnelling probe. We now describe a probe for computational tunnelling: a non-convex optimization problem consisting of just one global minimum and one false (local) minimum. For concreteness, we use the problem Hamiltonian depicted in Fig. 1, implementable in a D-Wave Two quantum annealer (a description of the flux qubits used in D-Wave Two is given in the Supplementary Note 1). The problem Hamiltonian consists of two cells, left and right, each with $n = 8$ qubits. The local fields $0 < h_{\mathrm{L}} < 0.5$ and $h_{\mathrm{R}} = -1$ are equal for all the spins within each cell, and all the couplings $J = 1$ are ferromagnetic. The spins within each cell tend to move together as clusters because of symmetry and the strong intracell ferromagnetic coupling energy. We choose $|h_{\mathrm{R}}| > |h_{\mathrm{L}}|$ so that in the low-energy states of H_{P} the right cluster is pointing along its own local field as seen in Fig. 1. The difference in energy of the states with opposite polarization in the left cluster is $n(J - 2h_{\mathrm{L}})$. Choosing $h_{\mathrm{L}} < J/2 = 0.5$, the global minimum corresponds to both clusters having the same orientation, while in the false minimum they have opposite orientations.

We can gain an intuitive understanding of the effective energy landscape if we represent each qubit by a mean field spin vector in the xz plane. Denote by θ_{μ} the angle of the spin vector for qubit μ with the x quantization axis. We assume that all the qubits in the left (right) cluster have the same angle θ_{L} (θ_{R}). This assumption is based on symmetry and the strong intracluster ferromagnetic energy. The resulting energy potential can also be derived using more formal methods, such as the Villain representation (see Supplementary Note 2 and ref. 25). Figure 2 plots the effective energy potential for the left cluster as a function of θ_{L} with $h_{\mathrm{L}} = 0.44$. At the beginning of the annealing process $B(s)/A(s) \ll 1$, and we have $\langle\sigma_{\mu}^{z}\rangle \simeq h_{\mu}B(s)/A(s)$ (the coupling terms are quadratic in the z polarizations $\langle\sigma_{\mu}^{z}\rangle$ and therefore negligible at this point). As h_{L} and h_{R} have opposite signs, so will the z-projections of spins in the two clusters early in the evolution. To escape this path classically all spins in the left cluster must flip sign, which requires traversing an energy barrier. The barrier peak corresponds to zero total z-polarization of the left cluster. Therefore, the barrier grows with the ferromagnetic energy of the cluster $(n/2)^2 J$. The barrier height is much greater than the residual energy, which grows with $n(J - 2h_{\mathrm{L}})$.

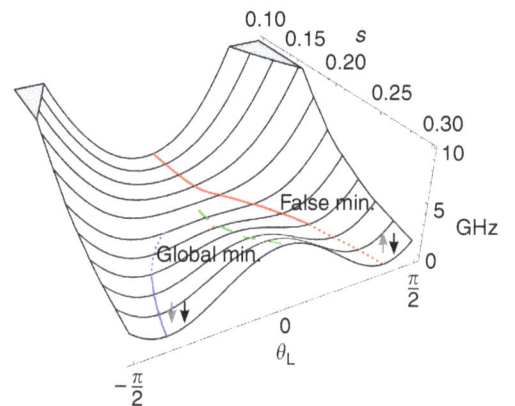

Figure 2 | Effective energy potential using $h_{\mathrm{L}} = 0.44$. The mean field potential is plotted versus annealing parameter s and tilt angle θ_{L} of each spin vector in the left cluster. The red line corresponds to a path that starts in the initial global minimum and follows the instantaneous local energy minimum. A second local minimum (dashed blue line) forms at the bifurcation point $s = 0.18$. The global minimum is found in this second path after $s = 0.24$ (dashed to continuous blue line). To reach this global minimum the system state has to traverse the energy barrier between them (dashed green line), either by thermal activation or by quantum tunnelling.

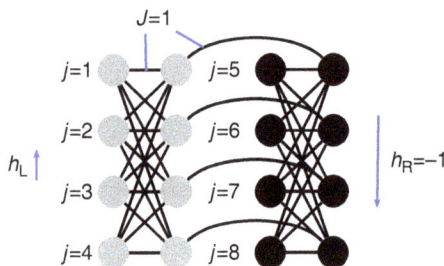

Figure 1 | Graph of the tunnelling probe Hamiltonian. The 16 qubits are coupled ferromagnetically with $J = 1$ (lines). The applied fields are $0 < h_{\mathrm{L}} < J/2$ ($h_{\mathrm{R}} = -1$) for the left (right) qubit cell. The symmetry and strong intracell ferromagnetic coupling makes each 8-qubit cluster evolve together.

Quantum mechanically, if $h_L < J/2$ the system evolution goes through an 'avoided-crossing' where the two lowest eigenstates $E_1(s)$ and $E_0(s)$ approach closely to, and then repel from, each other (see inset in Fig. 3). Higher-energy states remain well separated during the evolution. This level repulsion occurs because of the collective tunnelling of qubits in the left cluster between the opposite z polarizations. At the point where the gap $\hbar\Omega_{10}(s) = E_1(s) - E_0(s)$ reaches its minimum, the corresponding adiabatic eigenstates are formed by the symmetric and antisymmetric superpositions of the cluster orientations. The size of the minimum gap varies with h_L as seen in Fig. 3. The position of the avoided-crossing can be estimated to occur at the point where $|\langle\sigma_\mu\rangle| \sim 2h_L/J$ and moves towards $s = 1$ as h_L approaches $J/2$. Note that for $h_L = J/2$ the residual energy $n(J - 2h_L)$ vanishes and the final ground space is degenerate. There is no avoided-crossing when $h_L > J/2$ (Supplementary Fig. 1).

Characterization of noise and dissipation. Under realistic conditions, a quantum annealer can be strongly influenced by coupling to the environment. We introduce a detailed phenomenological open quantum system model based on single-qubit measurable noise parameters. We shall assume that each qubit is coupled to its own environment with an independent noise source. In the concrete case studied here, this is consistent with experimental data[9], and the coupling of the environment to each flux qubit is proportional to a σ^z qubit operator (flux fluctuations).

In the analysis of the transitions between the states we start from the initial (gapped) stage when the instantaneous energy gap $\hbar\Omega_{10}(s)$ between the two lowest eigenstates $|\psi_0(s)\rangle$, $|\psi_1(s)\rangle$ is sufficiently large compared with the linewidth $\hbar W$. Then, the coupling to the environment can be treated as a perturbation and the transition rate between these states is given by Fermi's golden rule $\Gamma_{1\rightarrow 0}(s) \approx a(s)S(\Omega_{10}(s))/\hbar^2$, where $S(\omega)$ is the noise spectral density (see Methods). Here

$$a(s) = \sum_{\mu=1}^{2n}\left|\langle\psi_0(s)|\sigma_\mu^z|\psi_1(s)\rangle\right|^2 \qquad (3)$$

is a sum of (squared) transition matrix elements between the two eigenstates.

In the minimum gap region, the (squared) matrix element $a(s)$ for the transition rate is large, and the system is thermalized (Fig. 4). More precisely, we have $\Gamma_{1\rightarrow 0} \gg 1/t_{qa}$, where the inverse of the annealing time $1/t_{qa}$ is an approximation for the annealing rate. The ground-state population is given by the Boltzmann distribution at the experimental temperature.

After the avoided-crossing region (at $s = 0.255$) we observe a steep exponential fall-off of the matrix element $a(s)$ with s, eventually causing multiqubit freezing (Fig. 4). Multiqubit freezing is quite distinct from single-qubit freezing. Single-qubit tunnelling[10] decays slowly as the magnitude of the transverse field $A(s)$ decreases. The multiqubit transition rate, however, decays exponentially fast (see inset of Fig. 4). This is due to the increasing effective barrier width (Fig. 2), which results in an exponential decrease in quantum tunnelling and a slowdown of the transition rate $\Gamma_{1\rightarrow 0}$. Formally, the barrier width corresponds to the Hamming distance

$$h(s) = \sum_{\mu=1}^{2n}\left|\langle\psi_0|\sigma_\mu^z|\psi_0\rangle - \langle\psi_1|\sigma_\mu^z|\psi_1\rangle\right|^2/4 \qquad (4)$$

between the opposite z orientations of the left cluster in the two lowest-energy eigenstates. The exponential sensitivity of multiqubit tunnelling to the width or Hamming distance $h(s)$ is the cause of the exponential decay of the matrix element $a(s)$, and of the multiqubit freezing.

We distinguish a slowdown phase (roughly, $0.1 < t_{qa}\Gamma_{1\rightarrow 0} < 10$) and a frozen phase ($t_{qa}\Gamma_{1\rightarrow 0} < 0.1$). In the frozen phase, there are no dynamics. Part of the system population remains trapped in the excited state $|\psi_1(s)\rangle$ corresponding to the false minimum of the effective potential until the end of the quantum annealing process (Fig. 4).

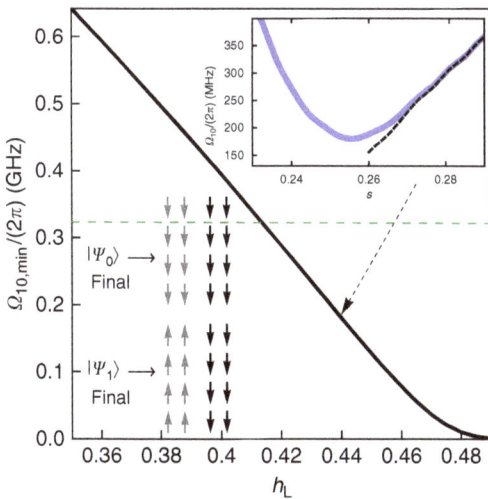

Figure 3 | Quantum energy gap. Inset shows the gap $\hbar\Omega_{10} = E_1(s) - E_0(s)$ versus s, using $h_L = 0.44$. The dashed line is the gap in the diabatic (pointer) basis. In the main plot, the minimum gap decreases with h_L. The horizontal green dashed line (324 MHz) corresponds to 15.5 mK, the lowest temperature in our experiments. The lower inset shows the spin configurations of the two lowest eigenstates at the end of the annealing.

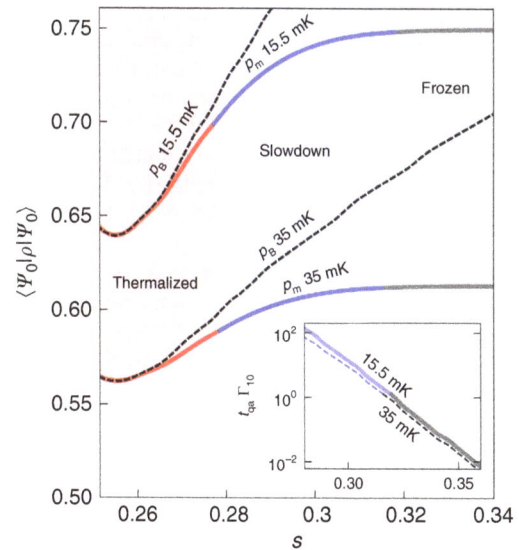

Figure 4 | Multiqubit freezing. Solid lines correspond to the modelled population p_m of the lowest-energy eigenstate along the quantum annealing process using $h_L = 0.44$ at 15.5 mK (top line) and 35 mK (bottom line). Dashed lines correspond to the thermal equilibrium population p_B. In the thermalization phase (red) the transition rate is fast and the population remains close to thermal equilibrium. As the multiqubit energy barrier increases, the transition rates are exponentially reduced with s, as shown in the inset. We define the slowdown regime (blue) as $t_{qa}\Gamma_{1\rightarrow 0} < 10$ and the frozen regime (grey) as $t_{qa}\Gamma_{1\rightarrow 0} < 0.1$. Comparing the data at 15.5 and 35 mK, we see a small change in the transition rate relative to the larger change in the thermal equilibrium ground-state population. Therefore, the probability of success is lower at higher temperature.

When the energy gap is similar to (or smaller than) the noise linewidth W, the environment cannot be treated as a perturbation. We develop a multiqubit nonperturbative analysis in the spirit of the Non-interacting Blip Approximation (NIBA)[26] that covers all quantum annealing (QA) stages. In the slowdown phase, when the Hamming distance approaches its maximum value, $h \sim n$, the instantaneous decay rate of the first excited state takes the form

$$\Gamma_{1 \to 0} = \int_{-\infty}^{\infty} d\tau \, e^{i\Omega_{10}\tau - h\left(i\epsilon_p\tau + (W\tau)^2/2\right)} \left[\frac{\pi\tau_c}{i\beta} \csc h \frac{(\tau - i\tau_c)}{\beta/\pi}\right]^{\frac{h\eta}{2\pi}} D(\tau), \quad (5)$$

where W and ϵ_p are the linewidth and reconfiguration energy from the low-frequency noise, η and $1/\tau_c$ are the high-frequency Ohmic noise coupling and cutoff and β is the inverse temperature $\hbar/k_B T$. The dependence on the annealing parameter s is implicit. The factor $D(\tau)$ is related to the tunnelling permeability of the potential barrier in Fig. 2 (similar to the coefficient a). It has the expression

$$D(\tau, s) = \left[\left(\epsilon_p(s)c_+(s) - ic_-(s)\partial_\tau\zeta(\tau, s)\right)^2 + a(s)\partial_{\tau\tau}\zeta(\tau, s)\right]e^{-i\epsilon_p(s)d(s)\tau}$$

(6)

where

$$Z_\mu^{\gamma\gamma'}(s) = \left\langle \psi_\gamma(s)\left|\sigma_\mu^z\right|\psi_{\gamma'}(s)\right\rangle, \quad \gamma, \gamma' = 0, 1$$

$$c_\pm(s) = \frac{1}{4}\sum_{\mu=1}^{2n} Z_\mu^{10}(s)\left(Z_\mu^{11}(s) \pm Z_\mu^{00}(s)\right),$$

$$d(s) = \frac{1}{4}\sum_{\mu=1}^{2n} \left[\left(Z_\mu^{11}(s)\right)^2 - \left(Z_\mu^{00}(s)\right)^2\right]$$

(7)

$$\zeta(\tau, s) = i\epsilon_p(s)\tau + \frac{(W(s)\tau)^2}{2}$$

Equation (5) describes collective tunnelling of the left qubit cluster assisted by the environment. The crucial difference from the single-qubit theory[27,28] is that the parameters of the environment in the transition rate are rescaled by the barrier width or Hamming distance $h(s)$. The effective low-frequency noise linewidth is $h^{1/2}(s)W(s)$, the reconfiguration energy is $h(s)\epsilon_p(s)$ and the Ohmic coefficient is $h(s)\eta(s)$. This is important at the late stages of quantum annealing when $h \sim n \gg 1$.

Comparison of NIBA with data. We observe a very close correspondence between the results of the analysis with the NIBA Quantum Master Equation for the dressed cluster states and the D-Wave Two data displayed in Fig. 5, which shows the dependence of the ground-state population on h_L. We emphasize that for NIBA (and the standard Redfield equation with Ohmic $S_{oh}(\omega)$, see Methods) we do not have any parameter fitting: the parameters are obtained from experiments (see Methods). The success probability of quantum annealing is (roughly) determined by the thermal equilibrium ground-state population during the slowdown phase (Fig. 4). When the temperature grows, the ground-state population decreases appreciably, while the transition rate changes little. Consequently, quantum mechanically, the probability of success decreases with increasing temperature, as seen in Fig. 6, for sufficiently big gaps. Figure 6 shows the dependence of the ground-state population with temperature.

For h_L closer to the degeneracy value $h_L = J/2$, the minimum gap Ω_{10}^{min} becomes smaller, as seen in Fig. 3. Where $\Omega_{10} \ll W$, the adiabatic basis of the instantaneous multiqubit states $\{|\psi_0(s)\rangle, |\psi_1(s)\rangle\}$ loses its physical significance. Because the coupling to the bath is relatively strong here, the system quickly approaches the states corresponding to predominantly opposite cluster orientations, similar to diabatic states (see inset of Fig. 3).

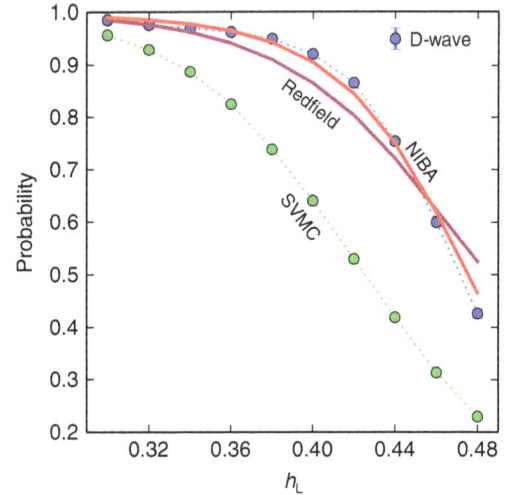

Figure 5 | Probability of success versus h_L. We plot the probability of measuring the final ground state for different values of h_L. The physical temperature for D-Wave is 15.(5) mK, and the annealing time is 20 μs. Both Redfield and NIBA use only measured parameters (no fitting). SVMC uses an algorithmic temperature equal to the physical temperature and 128,000 sweeps, as explained in the text. Error bars (s.e.m.) are smaller than markers.

Transitions between these states, also called pointer states[29], occur at a much slower rate as a consequence of the polaronic effect. As a result, for sufficiently small mininum gaps the multiqubit freezing starts before the avoided-crossing and the success probability increases with temperature[12]. This is captured by the multiqubit NIBA equation, but not by the standard Redfield Quantum Master Equation.

Spin vector Monte Carlo. We want to distinguish quantum tunnelling from thermal activation along classical paths of product states (which preclude multiqubit tunnelling). To give a more precise description of the classical paths of product states, let each qubit be represented by a mean field spin vector in the xz plane and denote by θ_μ the angle of the spin vector for qubit μ with the x quantization axis, as before. A classical path (red line in Fig. 2) that follows the local minimum of the effective energy potential gets trapped in a false minimum and fails to solve the corresponding optimization problem, as explained above. In the absence of quantum tunnelling, the global minimum could be reached through thermal excitations for over-the-barrier escape from the false minimum. This thermal activation results in an increasing probability of success with rising temperature.

This intuition is supported by spin vector Monte Carlo (SVMC), a numerical algorithm consisting of thermal Metropolis updates of the spin vectors. SVMC was introduced recently in ref. 20 and studied in related lines of work[21,23]. The dynamics are constrained to spin–vector product states, with one spin vector per qubit. For a given Hamiltonian $H_0(s)$, we denote the corresponding energy by $E_s(\theta_1,...,\theta_{n_q})$, where n_q is the number of qubits. The evolution consists of a sequence of sweeps along the Hamiltonian path $\{H_0(s)\}$. In each sweep, a Monte Carlo update is proposed for each qubit in two steps. First, a new angle θ'_μ is drawn from the uniform distribution in $[0, 2\pi]$. Second, the spin vector for qubit μ is updated $\theta_\mu \leftarrow \theta'_\mu$ according to the Metropolis rule for the energy difference

$$D = E_s\left(\ldots, \theta'_\mu, \ldots\right) - E_s\left(\ldots, \theta_\mu, \ldots\right).$$

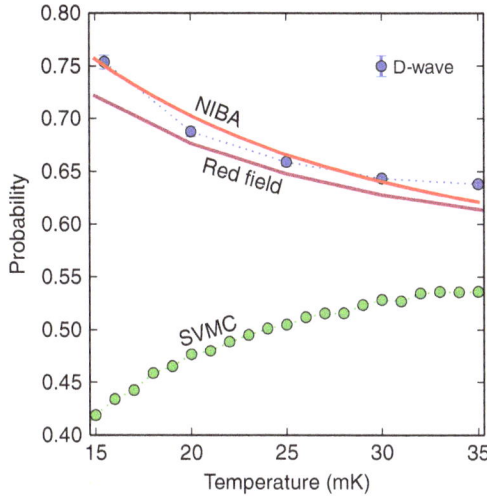

Figure 6 | Probability of success versus temperature for $h_L = 0.44$. The decreasing probability with increasing temperature is only matched with theories based on quantum tunnelling. This is the opposite tendency to thermal activation (SVMC). The annealing time is 20 μs. Both Redfield and NIBA use only measured parameters (no fitting). SVMC uses an algorithmic temperature equal to the physical temperature and 128,000 sweeps. In this temperature range the lowest two states (the double-well potential) account for all the probability (0.9998 for D-Wave, 0.99998 for SVMC). Error bars (s.e.m.) are smaller than markers.

That is, the move is always accepted if D is negative, and with probability given by the Boltzmann factor $\exp(-D/k_B T)$ if D is positive.

The initial state is chosen to be the global minimum of the transverse field. For low T and sufficient sweeps, the evolution proceeds along the false minima path of Fig. 2. This numerical method allows us to study thermal hopping between the classical paths. To check this correspondence, we studied the height of the energy barrier obtained from Kramers' theory applied to SVMC. For the effective potential at a fixed value of s, we initialized the spin vector state at a local minima, and watch for Kramers events. A Kramers event corresponds to the arrival at the other minima under thermal activation. According to Kramers' theory, the dependence on temperature for the expected number of sweeps necessary for a Kramers event follows the formula $\exp(\Delta U/T)$, where ΔU is the height of the energy barrier. We extract the energy barrier by fitting the curve of the average number of sweeps for different T. We find that this matches almost exactly the energy barrier height in Fig. 2 for different values of s (Fig. 7).

A disadvantage of SVMC as outlined above and introduced in ref. 20 is that there is no natural choice to relate the number of sweeps to the physical evolution time. As in other lines of work, we will choose the number of sweeps to obtain a good correlation with the probability of success of the D-Wave chip for a benchmark of random Ising models with binary couplings $J_{\mu\nu} \in \{1, -1\}$ (refs 15,18,20,23). This will allow us to phenomenologically correlate the number of sweeps to physical time. We set the algorithmic temperature of SVMC to be the same as the physical temperature because we are interested in the dependence of the success probability with temperature. There are no important differences for the correlation with other temperature choices. The correlation with the random Ising benchmark for 128,000 sweeps (Fig. 8) is 0.92, and the residual probabilities $P_{SVMC} - P_{D-Wave}$ have a mean of 0.05 and a s.d. of 0.12. This is consistent with the best values found over a wide range of parameters. We therefore use 128,000 sweeps at 15 mK as our reference parameters for SVMC.

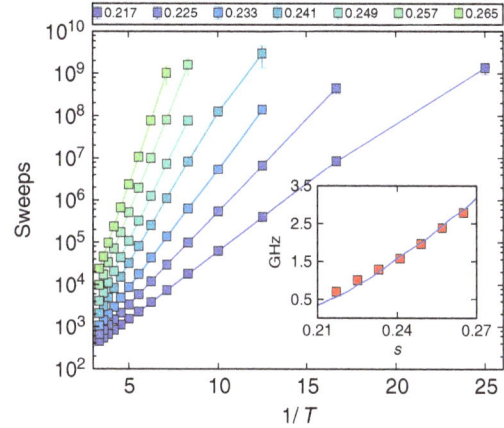

Figure 7 | Analysis of the activation energy for Kramer's scape for SVMC. The main figure shows, in a semilog scale, the average number of sweeps as a function of temperature. We plot lines for different points in the annealing schedule, from $s = 0.217$ (dark blue, see legend) to $s = 0.265$ (green). Error bars correspond to the s.e.m. The embedded figure shows the activation energy (red dots) and the semiclassical energy barrier (blue). There is a good correspondence between SVMC and the effective energy potential.

Figure 6 confirms the thermal activation in SVMC. This is opposite to both open quantum system theory and experiments with the D-Wave chip, which show a reduction in the probability of success with rising temperature, as explained above. Furthermore, Fig. 5 shows that the probability of success for SVMC is lower than the probability of success for D-Wave and open system quantum models.

Larger problems. A generalization of the 16-qubit problem to a larger number of qubits is achieved by studying problems that contain the same 'motif' (Fig. 1) multiple times within the connectivity graph (Fig. 9). The success probabilities for up to 200 qubits are shown in Fig. 10. We fit the average success probability as $P(n_q) \propto \exp(-\alpha n_q)$, where n_q is the number of qubits. The fitting exponent α for the D-Wave Two data is $(1.1 \pm 0.05) \times 10^{-2}$, while the fitting exponent for the SVMC numerics is $(2.8 \pm 0.17) \times 10^{-2}$. We conclude that, for instance with multiqubit quantum tunnelling, the D-Wave Two processor returns the solution that minimizes the energy with consistently higher probability than physically plausible models of the hardware that only employ product states and do not allow for multiqubit tunnelling transitions.

Discussion

The role of multiqubit tunnelling as a computational resource is an open problem of active research. Nevertheless, it is instructive to consider some plausible estimates for the case of minimization of an Ising problem that contains pairwise interactions between all qubits. A way to think of multiqubit tunnelling as a computational resource is to regard it as a form of large neighbourhood search. Collective tunnelling transitions involving K qubits explore a K variable neighbourhood, and there is a combinatorial number of such neighbourhoods. Using standard resources, the cost of exhaustively searching on a Hamming ball of binary strings of radius K is $\sum_{i=1}^{K} \binom{n}{i}$, which is bounded from below by $\binom{n}{K}$, where n is the total number of qubits. Therefore, for $K \ll n$, the cost is $\sim n^K/K! \sim \exp(K \log(n))$. As one can see, the exponent in K is very steep ($\log n$). On the other hand, for $K \sim n/2$ the exponent is that of the exhaustive search in the entire n-bit string space (2^n). We now compare this with the tunnelling rate across a barrier of

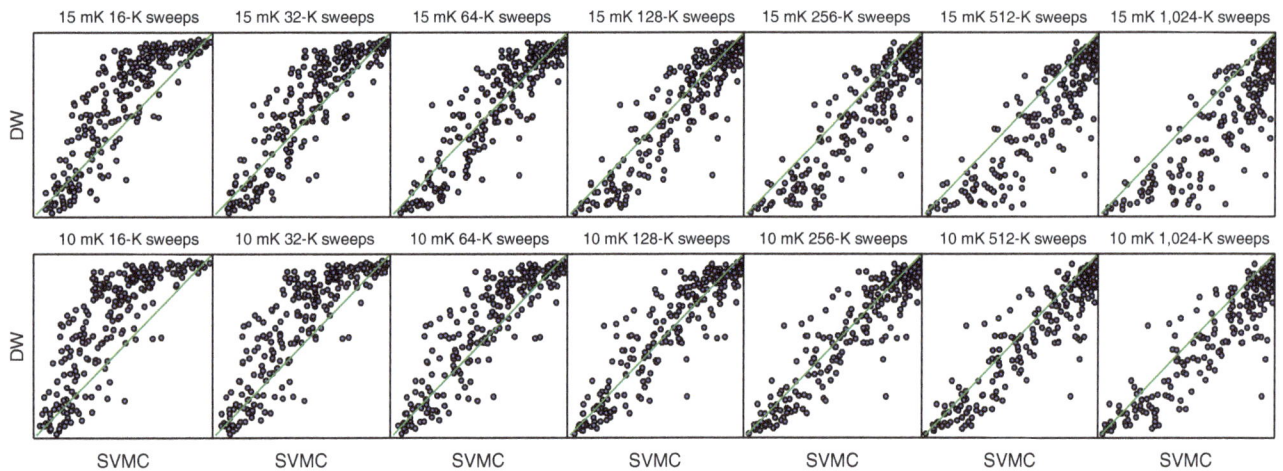

Figure 8 | Scatter plots showing the correlation of D-Wave Two data and SVMC. Correlations for the random Ising benchmark for different algorithmic temperatures (in mK) and number of sweeps. We will use the parameters $T = 15$ mK and sweeps $= 128{,}000$ for SVMC in the rest of the paper.

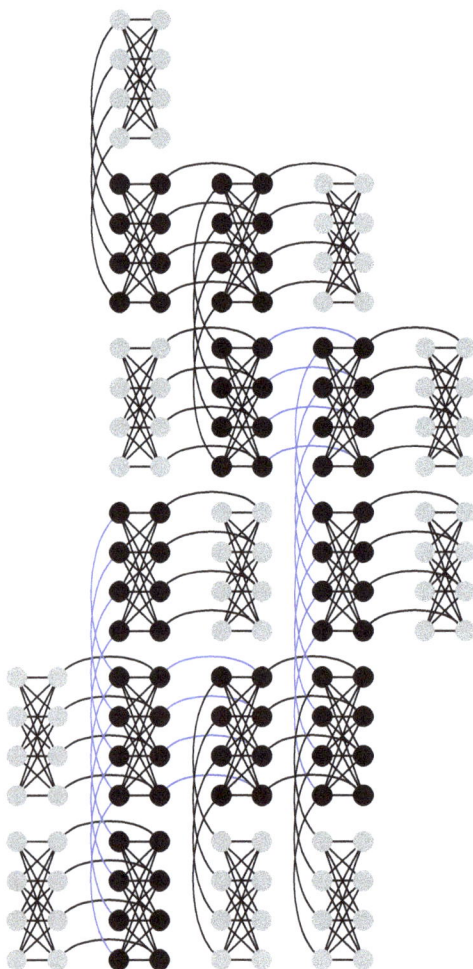

Figure 9 | Larger problems that contain the tunnelling probe 'motif' as subproblems. As in Fig. 1, the black (grey) cluster has a strong $h_R = -1$ (weak $h_L = 0.44$) local field. The black clusters are connected in a glassy manner to make the problem less regular: all connections between any two neighbouring black clusters are set randomly to either -1 or $+1$. The -1 connections are depicted in blue.

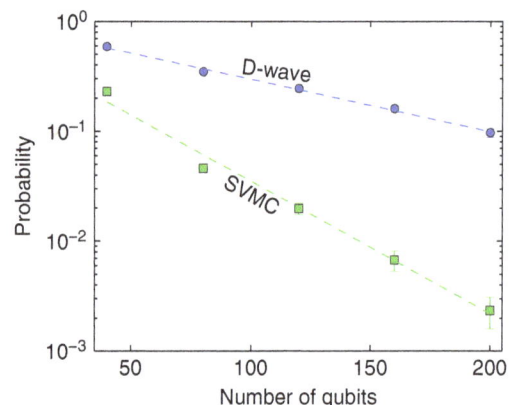

Figure 10 | Success probability for a glass of clusters as a function of the number of qubits. We fit the mean probability of success $P(n_q) \propto \exp(-\alpha n_q)$ as a function of the number of qubits n_q (dashed lines). The fitting exponent α for the D-Wave Two data is $(1.1 \pm 0.05) \times 10^{-2}$, while the fitting exponent for the SVMC numerics is $(2.8 \pm 0.17) \times 10^{-2}$. The error estimates for the exponents (s.e.m.) are obtained by bootstrapping.

width K. It is given by $\exp(-cK)$ for some constant c. If K is smaller than n then, as n increases, $\exp(cK) \ll \exp(K \log(n))$ and tunnelling can be faster than large neighbourhood search. On the other hand, in many problems similar to the two-cluster problem the barrier width is K in $O(n)$. In these cases the tunnelling rate can still be $\exp(cn) \ll 2^n$ (refs 30–35). Therefore, again tunnelling can still provide a dramatically faster search option.

We find that the current-generation D-Wave Two annealer enables tunnelling transitions involving at least 8 qubits. It will be an important future task to determine the maximal K attainable by current technology and how large it can be made in next generations. The multiqubit polaronic quantum master equation presented here lays down the theory to answer this question. It guides the design of next-generation architectures and helps to understand for which computational problems quantum-enhanced optimization may offer an advantage. The larger the K the easier it should be to translate the quantum resource 'K-qubit tunnelling' into a possible computational speedup. We want to emphasize that this paper does not claim to have established a quantum speedup. To this end, one would have to demonstrate that no known classical algorithm finds the optimal solution as fast as the quantum process. To establish such an advantage it will be important to study to what degree collective tunnelling can be emulated in classical algorithms such as Quantum Monte Carlo or by employing cluster update methods. However, the collective tunnelling phenomena demonstrated here present an important step towards what we would like to call a physical speedup: a speedup relative to a hypothetical version of the hardware operated under the laws of classical physics.

Methods

Experimental properties of the noise. The properties of the noise are determined by the noise spectral density $S(\omega)$, which is characterized by single-qubit macroscopic resonant tunnelling (MRT) experiments in a broad range of biases (0.4 MHz–4 GHz) and temperatures (21–38 mK) for tunnelling amplitudes of a single flux qubit below 1 MHz. The MRT data collected are surprisingly well described[28,36] by a phenomenological 'hybrid' thermal noise model $S(\omega) = S_{lf}(\omega) + S_{oh}(\omega)$. Here $S_{oh}(\omega) = \hbar^2 \eta \omega e^{-\omega \tau_c}/\left(1 - e^{-\hbar\omega/k_B T}\right)$ denotes the high-frequency part and has Ohmic form with dimensionless coupling η and cutoff frequency $1/\tau_c$ (assumed to be very large). The low-frequency part S_{lf} is of the $1/f$ type[37], and in current D-Wave chips this noise is coupled to the flux qubit relatively strongly. Its effect can be described with only two parameters: the width W and the Stokes shift ϵ_p of the MRT line[27]. The experimental shift value is related to the width by the fluctuation-dissipation theorem $\left(\epsilon_p = \hbar W^2/2k_B T\right)$ and represents the reorganization energy of the environment. The values of the noise parameters measured at the end of the annealing ($s = 1$) for the D-Wave Two chip are $W/(2\pi) = 0.40(1)$ GHz and $\eta = 0.24(3)$.

References

1. Finnila, A. B., Gomez, M. A., Sebenik, C., Stenson, C. & Doll, J. D. Quantum annealing: a new method for minimizing multidimensional functions. *Chem. Phys. Lett.* **219**, 343–348 (1994).
2. Kadowaki, T. & Nishimori, H. Quantum annealing in the transverse Ising model. *Phys. Rev. E* **58**, 5355–5363 (1998).
3. Farhi, E., Goldstone, J. & Gutmann, S. Quantum adiabatic evolution algorithms versus simulated annealing. Preprint at http://arxiv.org/abs/quant-ph/0201031(2002).
4. Brooke, J., Bitko, D., Rosenbaum, T. F. & Aeppli, G. Quantum annealing of a disordered spin system. *Science* **284**, 779–781 (1999).
5. Santoro, G. E., Martoňák, R., Tosatti, E. & Car, R. Theory of quantum annealing of an Ising spin glass. *Science* **295**, 2427–2430 (2002).
6. Ray, P., Chakrabarti, B. K. & Chakrabarti, A. Sherrington-Kirkpatrick model in a transverse field: absence of replica symmetry breaking due to quantum fluctuations. *Phys. Rev. B* **39**, 11828–11832 (1989).
7. Mooij, J. et al. Josephson persistent-current qubit. *Science* **285**, 1036–1039 (1999).
8. Harris, R. et al. Experimental investigation of an eight-qubit unit cell in a superconducting optimization processor. *Phys. Rev. B* **82**, 024511 (2010).
9. Lanting, T. et al. Cotunneling in pairs of coupled flux qubits. *Phys. Rev. B* **82**, 060512 (2010).
10. Johnson, M. et al. Quantum annealing with manufactured spins. *Nature* **473**, 194–198 (2011).
11. Boixo, S., Albash, T., Spedalieri, F. M., Chancellor, N. & Lidar, D. A. Experimental signature of programmable quantum annealing. *Nat. Commun.* **4**, 2067 (2013).
12. Dickson, N. et al. Thermally assisted quantum annealing of a 16-qubit problem. *Nat. Commun.* **4**, 1903 (2013).
13. McGeoch, C. C. & Wang, C. in *Proceedings of the ACM International Conference on Computing Frontiers* 23 (ACM, 2013).
14. Dash, S. A note on qubo instances defined on chimera graphs. Preprint at http://arxiv.org/abs/1306.1202 (2013).
15. Boixo, S. et al. Evidence for quantum annealing with more than one hundred qubits. *Nat. Phys.* **10**, 218–224 (2014).
16. Lanting, T. et al. Entanglement in a quantum annealing processor. *Phys. Rev. X* **4**, 021041 (2014).
17. Santra, S., Quiroz, G., Ver Steeg, G. & Lidar, D. A. MAX 2-SAT with up to 108 qubits. *New J. Phys.* **16**, 045006 (2014).
18. Rønnow, T. F. et al. Defining and detecting quantum speedup. *Science* **345**, 420–424 (2014).
19. Vinci, W. et al. Hearing the shape of the Ising model with a programmable superconducting-flux annealer. *Sci. Rep.* **4**, 5703 (2014).
20. Shin, S. W., Smith, G., Smolin, J. A. & Vazirani, U. How "quantum" is the D-Wave machine? Preprint at http://arxiv.org/abs/1401.7087 (2014).
21. Albash, T., Vinci, W., Mishra, A., Warburton, P. A. & Lidar, D. A. Consistency tests of classical and quantum models for a quantum annealer. *Phys. Rev. A* **91**, 042314 (2015).
22. Venturelli, D. et al. Quantum optimization of fully connected spin glasses. *Phys. Rev. X* **5**, 031040 (2015).
23. Albash, T., Rønnow, T., Troyer, M. & Lidar, D. Reexamining classical and quantum models for the D-Wave One processor: the role of excited states and ground state degeneracy. *Eur. Phys J. Special Top.* **224**, 111–129 (2015).
24. Martin-Mayor, V. & Hen, I. Unraveling quantum annealers using classical hardness. *Sci. Rep.* **5**, 15324 (2015).
25. Boulatov, A. & Smelyanskiy, V. N. Quantum adiabatic algorithm and large spin tunnelling. *Phys. Rev. A* **68**, 062321 (2003).
26. Leggett, A. J. et al. Dynamics of the dissipative two-state system. *Rev. Mod. Phys.* **59**, 1–85 (1987).
27. Amin, M. H. S. & Averin, D. V. Macroscopic resonant tunneling in the presence of low frequency noise. *Phys. Rev. Lett.* **100**, 197001 (2008).
28. Lanting, T. et al. Probing high-frequency noise with macroscopic resonant tunnelling. *Phys. Rev. B* **83**, 180502 (2011).
29. Zurek, W. H. Pointer basis of quantum apparatus: into what mixture does the wave packet collapse? *Phys. Rev. D* **24**, 1516–1525 (1981).
30. Farhi, E., Goldstone, J., Gutmann, S. & Sipser, M. Quantum computation by adiabatic evolution. Preprint at http://arxiv.org/abs/quant-ph/0001106(2000).
31. Roland, J. & Cerf, N. J. Quantum search by local adiabatic evolution. *Phys. Rev. A* **65**, 042308 (2002).
32. Reichardt, B. W. in *Proceedings of the Thirty-Sixth Annual ACM Symposium on Theory of Computing* 502–510 (New York, NY, USA, 2004).
33. Somma, R. D., Boixo, S., Barnum, H. & Knill, E. Quantum simulations of classical annealing processes. *Phys. Rev. Lett.* **101**, 130504 (2008).
34. Somma, R. D. & Boixo, S. Spectral gap amplification. *SIAM J. Comput.* **42**, 593–610 (2013).
35. Kechedzhi, K. & Smelyanskiy, V. N. Open system quantum annealing in mean field models with exponential degeneracy. Preprint at http://arxiv.org/abs/1505.05878(2015).
36. Harris, R. et al. Probing noise in flux qubits via macroscopic resonant tunnelling. *Phys. Rev. Lett.* **101**, 117003 (2008).
37. Sendelbach, S. et al. Decoherence of a superconducting qubit due to bias noise. *Phys. Rev. B* **67**, 094510 (2003).

Acknowledgements
We thank John Martinis, Edward Farhi and Anthony Leggett for useful discussions and reviewing the manuscript. We also thank Ryan Babbush and Bryan O'Gorman for reviewing the manuscript, and Damian Steiger, Daniel Lidar and Tameem Albash for comments about the temperature experiment. The work of V.N.S. was supported in part by the Office of the Director of National Intelligence (ODNI), Intelligence Advanced Research Projects Activity (IARPA) via IAA 145483 and by the AFRL Information Directorate under grant F4HBKC4162G001.

Author contributions
S.B., V.N.S. and H.N. designed the project. H.N., V.N.S. and S.B. proposed frustration patterns for tunnelling. V.N.S., M.D., M.H.A., A.Y.S. and S.B. completed the open quantum system study. S.B., A.S., S.V.I. and V.D. performed numerical studies. All authors contributed to several tasks, such as analysis of the results and discussions of the draft.

Additional information

7

Semiconductor-inspired design principles for superconducting quantum computing

Yun-Pil Shim[1,2] & Charles Tahan[1]

Superconducting circuits offer tremendous design flexibility in the quantum regime culminating most recently in the demonstration of few qubit systems supposedly approaching the threshold for fault-tolerant quantum information processing. Competition in the solid-state comes from semiconductor qubits, where nature has bestowed some very useful properties which can be utilized for spin qubit-based quantum computing. Here we begin to explore how selective design principles deduced from spin-based systems could be used to advance superconducting qubit science. We take an initial step along this path proposing an encoded qubit approach realizable with state-of-the-art tunable Josephson junction qubits. Our results show that this design philosophy holds promise, enables microwave-free control, and offers a pathway to future qubit designs with new capabilities such as with higher fidelity or, perhaps, operation at higher temperature. The approach is also especially suited to qubits on the basis of variable super-semi junctions.

[1] Laboratory for Physical Sciences, College Park, Maryland 20740, USA. [2] Department of Physics, University of Maryland, College Park, Maryland 20742, USA. Correspondence and requests for materials should be addressed to Y.-P.S. (email: ypshim@lps.umd.edu) or to C.T. (email: charlie@tahan.com).

Spin qubits[1] are based on the fundamental and intrinsic properties of semiconductor systems, such as electron spins trapped in the potential of a quantum dot[2] or a chemical impurity[3]. Spins can be naturally protected from charge noise due to weak spin–orbit coupling. In fact, the tiny matrix element between spin qubit states can allow spin qubits to operate at temperatures above the Zeeman splitting[4,5]. While a benefit to qubit coherence, this property of spins also leads to relatively slow single-qubit gates via, for example, a microwave pulse. It turns out that nature provides a solution: a very fast and robust two-qubit gate via the exchange interaction. This has led to "encoded" qubit schemes where the qubit is embedded logically in two to four physical spin qubits[6–8]. The fact that electrons are real particles can be used for fast initialization and readout techniques. Exchange-only qubits[7,9] allow all electrical implementation of qubit-gate operations, and enable universal quantum computation (QC) while providing some immunity to global field and timing fluctuations via a decoherence free subsystem, at the cost of more physical qubits and extra operations per encoded gate.

This work investigates how superconducting Josephson junction quantum circuits[10], whose properties can be engineered, can be improved by mimicking some of best properties of spin qubit systems. We propose a first step: an encoded superconducting qubit approach, which does not require microwave control, and thus divorces qubit frequency from control electronics. In analogy to the exchange-only qubit in semiconductor spin qubit systems, encoded qubits enable microwave-free control of the qubit states via fast DC-like voltage or flux pulses. In contrast to the exchange-only qubits, logical gate operations of this encoded superconducting (SC) qubit can be done with minimal overhead (zero overhead in physical two-qubit gates) in terms of control operations, a surprising result. We describe how to initialize the encoded qubit and implement single- and two-qubit logical gates using only z-control pulse sequences (via tunable frequency qubits). In the process we also lay out possible opportunities for future research on the basis of other insights from spin-based QC. To encourage implementation, we give an explicit protocol on the basis of qubits in operation today.

Small systems of superconducting qubits based on variations of the transmon qubit[11,12] have already demonstrated gates with fidelities approaching 99.99% along with rudimentary quantum algorithms including error correction cycles[13–19]. Note that because these architectures rely on single-qubit gates via microwaves, the future design space is constrained by the availability and convenience of microwave generators.

An alternative approach to combining the best properties of semiconductor and superconducting quantum systems is to take advantage of true superconducting-semiconductor systems. The appearance of superconductivity in conventional semiconductors[20,21] such as silicon[22–25] or germanium[26,27] could potentially allow for a new type of fully epitaxial super-semi devices[28,29]. And epitaxial super-semi Josephson juction devices based on the proximity effect have already led to new superconducting circuits[30,31]. Epitaxial super-semi systems may improve noise properties, but perhaps more importantly they enable gate-tunable Josephson junctions, which we can also take advantage of in our proposal introduced below.

Results
From encoded spins to tunable qubits.
Spins in quantum dots, say in silicon, are typically assumed to have equivalent g-factors, so that in a magnetic field the frequency of each qubit is the same. Thus, to achieve universal QC (an ability to do arbitrary rotations around the Bloch sphere plus a two-qubit entangling gate), one

needs at minimum three spins. In this case, an encoded one-qubit gate requires around 3 pulses and a CNOT gate requires roughly 20 pulses[7], a hefty overhead. Two-qubit encodings are possible, but require the complication of a magnetic field gradient (via for example a micromagnet). Superconducting qubits, on the other hand, can be man-made such that the qubit frequency is tunable. This allows arbitrary one-qubit rotations with just two-physical qubits, in theory.

In this work, we consider a qubit encoded in a system of two capacitively coupled-SC qubits. We take tunable transmons[16,32] like xmons[13] or gatemons[30] as our prototypical SC qubits (see Fig. 1a) and suggest one possible implementation following the capacitively-coupled xmon architecture of Martinis et al.[14] to encourage near-term realization. Although we explicitly chose the xmon geometry to be more specific about our proposed protocol, the general idea can easily be applied to other types of SC qubits, such as traditional transmons or capacitively-shunted flux qubits[33,34], which we will discuss later. A transmon qubit[11] is

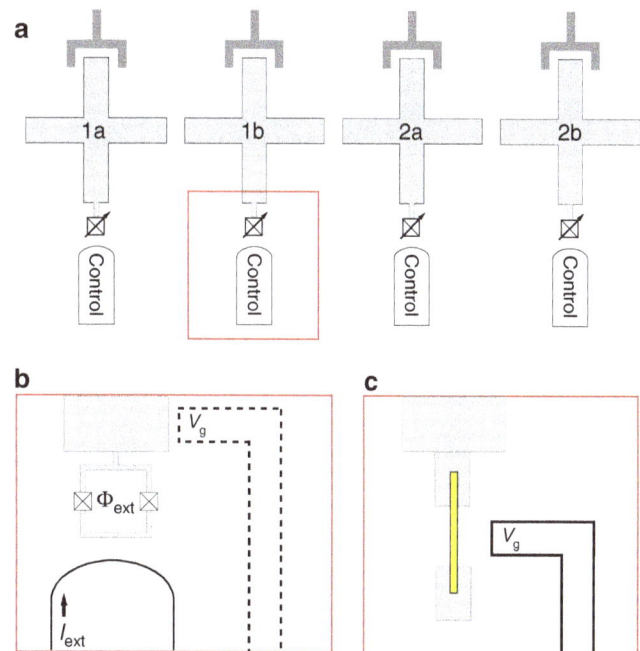

Figure 1 | Encoded superconducting qubits and tunable Josephson junctions. (a) Schematic diagram of a possible encoded superconducting qubit scheme as described in the text. An encoded qubit consists of two tunable physical SC qubits (for example, tunable transmons such as the xmon depicted here), with the encoded qubit states $|0\rangle_Q = |01\rangle$ and $|1\rangle_Q = |10\rangle$. In this picture, two encoded qubits are shown (for example, physical qubits 1a and 1b form an encoded qubit) and more encoded qubits can be introduced in a straightforward manner. Each SC qubit has a z-control line which tunes the Josephson energy E_J, and there are no additional microwave xy-control lines. All manipulation of the qubit states are done by the z-control pulses. Each transmon is capacitively-coupled to neighbouring transmons and also coupled to, as an example, a transmission line resonator for readout. (b) Double JJs in a loop act as a tunable JJ, controlled by an externally applied magnetic flux. In the SQUID tunable approach, one of the transmons in each encoded qubit needs a separate voltage control to tune the gate charge number n_g which may be needed to initialize the encoded qubit state. (c) electrostatically tunable JJ on the basis of a proximitized superconducting-semiconductor nanowire connecting two superconductors[30] used for gatemons. The nanowire is coated with SC and a portion is lifted off to form a semiconductor nanowire weak link. The JJ energy E_J is tuned by a side-gate voltage V_G, which can also serve as a capacitive tuning for initialization.

described by the charge qubit Hamiltonian:

$$H_X = 4E_C(\hat{n} - n_g)^2 - E_J \cos \hat{\theta}, \qquad (1)$$

where $E_C = e^2/2C_\Sigma$ is the electron charging energy for total capacitance C_Σ and E_J is the Josephson energy. $\hat{n}$ and $\hat{\theta}$ are the number and phase operators, respectively, and n_g is the gate charge number that can be tuned by a capacitively-coupled external voltage. The qubit frequency $f_Q = \varepsilon/h$, where ε is the energy difference between the first excited state and the ground state, and $f_Q \simeq \sqrt{8E_C E_J}/h$ in the transmon regime, $E_J \gg E_C$. The Josephson energy of a JJ is determined by the material properties and geometry of the JJ, but a double JJ can be considered as a tunable JJ[35] where an externally applied magnetic flux through the double JJ loop can tune the effective coupling energy $E_J = E_{J0} \cos(\pi\Phi_{ext}/\Phi_0)$ (see Fig. 1b). Φ_{ext} is the external magnetic flux and Φ_0 is the SC flux quantum. Individual transmon qubits are typically controlled by tuning the qubit frequency with tunable E_J for z-control and by applying microwaves for x control.

Recently, there has been progress in an alternative approach for a tunable JJ using a superconductor proximitized semiconductor weak-link junction[30,31]. In ref. 30, an InAs nanowire was used to connect two superconductors (Al). The nanowire was epitaxially coated with Al and a small portion of the wire was etched off to form a semiconductor nanowire bridging two SCs (Fig. 1c). A side-gate voltage was used to tune the carrier density under the exposed portion of the wire and thus the Josephson energy of this SNS JJ. The gatemon, a tunable transmon based on this gate-tunable JJ, has several advantages. It requires only a single JJ that can be quickly tuned by a electrostatic voltage. It removes the need for external flux and hence reduces dissipation by a resistive control line and allows the device to operate in a magnetic field. The epitaxial growth of the nanowire JJ and its clean material properties[36,37] demonstrate the potential of a bottom-up approach for SC quantum devices[28,29].

Our encoded qubit is defined in a two-transmon system. The Hamiltonian for two transmons with the capacitive xx coupling is:

$$\begin{aligned} H_{2X} &= \sum_{k=a,b} \left[4E_C^{(k)}\left(\hat{N}_k - n_g^{(k)}\right)^2 - E_J^{(k)} \cos \hat{\theta}_k \right] \\ &\quad + E_{cc}\left(\hat{N}_a - n_g^{(a)}\right)\left(\hat{N}_b - n_g^{(b)}\right) \\ &= \varepsilon_a \tilde{\sigma}_a^z + \varepsilon_b \tilde{\sigma}_b^z + \varepsilon' \tilde{\sigma}_a^x \otimes \tilde{\sigma}_b^x, \end{aligned} \qquad (2)$$

where E_{cc} is the capacitive coupling energy and $\tilde{\sigma}_k^i$ ($i = x, y, z$) is the Pauli operator for k-th transmon in a reduced subspace of transmon qubit states. ε_k is the qubit energy of the k-th transmon, and $\varepsilon' = E_{cc}\alpha_a\alpha_b$ with $\alpha_k = \langle 1|\hat{N}_k|0\rangle$ where $|0\rangle$ and $|1\rangle$ are the two lowest energy states of individual transmons. In transmon qubit systems the capacitive coupling is usually turned on (off) by tuning the qubit frequencies to on (off) resonance. The capacitive xx coupling conserves the parity $\tilde{\sigma}_a^z \otimes \tilde{\sigma}_b^z$ of the two-transmon system and the Hamiltonian (equation (2)) is block-diagonal in the basis of $\{|00\rangle, |01\rangle, |10\rangle, |11\rangle\}$. We define our encoded qubit in the subspace of $\langle\{|01\rangle, |10\rangle\}\rangle$, since the other subspace $\langle\{|00\rangle, |11\rangle\}\rangle$ has states with a very large energy difference (much larger than the capacitive coupling), effectively turning off the capacitive coupling all the time.

In the encoded qubit basis $\{|0\rangle_Q, |1\rangle_Q\}$ where

$$|0\rangle_Q = |01\rangle, \quad |1\rangle_Q = |10\rangle, \qquad (3)$$

the single qubit Hamiltonian is

$$H_Q = \begin{pmatrix} -\varepsilon_a + \varepsilon_b & \varepsilon' \\ \varepsilon' & \varepsilon_a - \varepsilon_b \end{pmatrix} = \frac{\varepsilon_a + \varepsilon_b}{2}\mathbb{1} + \Delta\varepsilon\hat{\sigma}^z + \varepsilon'\hat{\sigma}^x, \qquad (4)$$

where $\Delta\varepsilon = (\varepsilon_b - \varepsilon_a)/2$ and $\hat{\sigma}^i$ ($i = x, y, z$) is the Pauli operator for

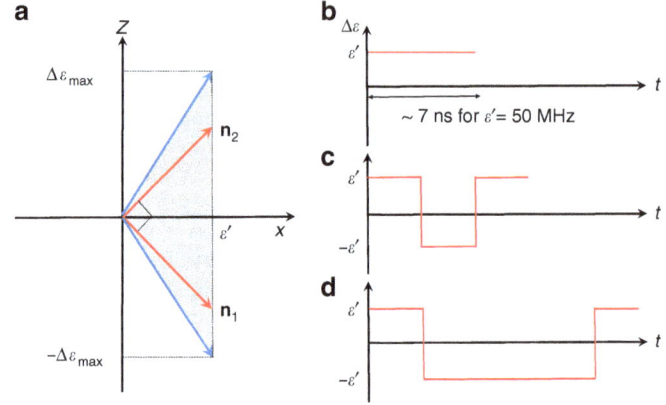

Figure 2 | Encoded single qubit operations. (a) Possible rotation axes in xz plane. The shaded grey region depicts the range of the direction of the possible rotation axis. The two red directions indicate a set of two orthogonal rotation axes, which can be used to implement any arbitrary single qubit gates in three steps. **(b–d)** Schematically shows an implementation of some logical gates in terms of rotations around $\hat{n}_1 = \mathbf{n}_1/|\mathbf{n}_1|$ and $\hat{n}_2 = \mathbf{n}_2/|\mathbf{n}_2|$. **(b)** Pulse shape for Hadamard gate. **(c)** Pulse shape for X gate. **(d)** Pulse shape for Z gate.

the encoded qubit. The qubit energies ε_a and ε_b can be controlled by the tunable JJ of each tunable transmon or gatemon, enabling logical gate operations with only fast DC-like voltage or flux pulses. In the following we will describe the logical gate operations, initialization and measurement schemes for this encoded qubit architecture.

Single-qubit operations. The Hamiltonian for an encoded qubit is given by equation (4). For a fixed capacitive coupling between SC qubits, ε' is fixed, and the single-qubit operations can be implemented by pulsing the qubit energy ε through the z-control of individual transmons, in at most three rotations. Since the tunable range of $\Delta\varepsilon$ (order of GHz) is much greater than ε' (tens or hundreds of MHz), the rotation axis can be in almost any direction in the right half of the xz plane (see Fig. 2a), and most logical single-qubit gates can be implemented in two rotations[38]. In general, all single-qubit-gate operations can be implemented as a three-step Euler angle rotations around two orthogonal rotation axes (for example, see the two red axes in Fig. 2a).

We now provide implementations for a few representative single-qubit gates. The Hadamard gate, $H = ((1,1),(1,-1))/\sqrt{2}$, is a single-qubit gate that is almost ubiquitous in quantum circuits. Figure 2b shows an implementation of H gate as a single rotation $H = iR(\hat{n}_2, \pi)$ around $\hat{n}_2 = (1, 0, 1)/\sqrt{2}$. It can be achieved by tuning $\delta\varepsilon = \varepsilon'$. Here $R(\hat{n}, \phi)$ is a rotation by angle ϕ around $\hat{n}$ axis. Pauli X gate can be realized as a single rotation by tuning the two xmons on resonance ($\Delta\varepsilon = 0$), or three-step rotations such as $X = iR(\hat{n}_2, \pi/2)R(\hat{n}_1, \pi/2)R(\hat{n}_2, \pi/2)$, where $\hat{n}_1 = (1, 0, -1)/\sqrt{2}$ and $\hat{n}_2 = (1, 0, 1)/\sqrt{2}$, as was shown in Fig. 2c. Z gate requires three-step rotations: $Z = -iR(\hat{n}_2, \pi/2)R(\hat{n}_1, 3\pi/2)R(\hat{n}_2, \pi/2)$. The above examples are for ideal systems with precise control over the system parameters. In real systems with fluctuating parameters, recently developed dynamical error-cancelling pulse sequences[39,40] could be useful for gate operations with higher fidelity.

Given that single-qubit gates in transmon systems through z-control have already demonstrated fidelities better than 0.999 (ref. 14), we expect the logical single-qubit gates (which require at most three rotation steps through z-control of transmons) will be

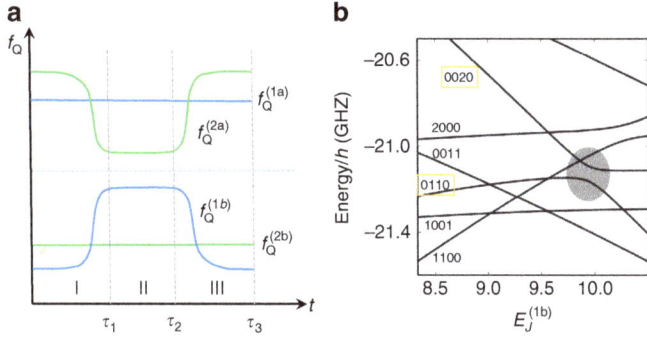

Figure 3 | Two-qubit gate operation. (a) Pulse scheme for two-qubit gate operations. y axis (f_Q) is the qubit frequencies of each transmon in Fig. 1a. $f_k^{(0)}$ is the idle qubit frequency of k-th transmon. The two blue curves (transmons 1a and 1b) form an encoded qubit, and the two green curves (transmons 2a and 2b) form the other encoded qubit. The transmons 1b and 2a are brought close to resonance while still far from being resonant with other transmons (transmon 1a and 2b), then are brought back to respective idle frequencies. **(b)** Energy spectrum for the process. The system is brought to the shaded area where (0110) and (0020) states are mixed. (0110) state accumulates non-trivial phase during this process, which leads to a CPHASE gate between transmon 1b and transmon 2a. This provides a non-trivial encoded two-qubit gate necessary for universal QC.

able to reach a fidelity better than $F_1 \geq F_z^3 = 0.999^3 = 0.997$ using currently available experimental techniques.

Two-qubit operations. For a scalable qubit architecture, we need to plan for the transmon qubit frequencies such that unnecessary resonances are avoided, especially if the two-qubit interaction cannot be completely shut off via, for example, a tunable coupler[41]. An encoded qubit has two transmons with idle frequency difference much larger than the capacitive coupling, so we can effectively turn the coupling off. In the two-encoded qubit system (four transmon system), we set the idle frequencies of next-nearest neighbour transmons to be different by more than the direct capacitive coupling between them, which is order of MHz[15]. We also set the encoded qubit frequencies $\Delta\varepsilon$ of the neighbouring encoded qubits to be different so we can mitigate some unintended resonances. For the calculations in this section, we set the four transmon qubit idle frequencies $f_Q^{(k)}$ as 5.6, 4.6, 5.9, 4.8 GHz for $k = $ 1a, 1b, 2a, 2b, respectively (see Fig. 1a. In this section and the following, we set $E_C^{(k)}/h = 375$ MHz and $E_{cc} = 30$ MHz for all transmons. Transmon qubit frequencies are controlled by tuning $E_J^{(k)}$.

Two-qubit operations can be implemented by adopting the adiabatic two-qubit CPHASE operations[14,42] between two-transmon qubits. By tuning the qubit frequencies of two-transmon qubits such that (11) and (02) states become resonant and then bringing them back to their idle frequencies, a unitary gate equivalent to the CPHASE gate between two qubits up to single-qubit unitary gates can be achieved[43]. This scheme has already been used in experiments and achieved reported fidelity better than 0.99 (ref. 14). In a similar manner, we can implement the CPHASE gate between two encoded qubits up to single-qubit unitary gates. Figure 3a shows schematically the pulse sequence of the transmon qubit frequencies, changing the qubit frequencies of transmon 1b and transmon 2a in Fig. 1a. First, we bring the transmons 1b and 2a closer during time τ_1 such that (0110) and (0020) states are on resonance in step (I). Then, in step (II), they stay there for a time period $\tau_{12} = \tau_2 - \tau_1$, and finally we bring them back to initial point at time $\tau_3 = \tau_2 + \tau_1$

in step (III). The (0110) state gets mixed with (0020) due to the capacitive coupling during the pulse sequence with strength $\sqrt{2}\varepsilon'$. During this process the (0110) state obtains some non-trivial phase due to the interaction with (0020) while the other qubit states, (0101), (1001) and (1010), obtain only trivial phases since they do not get close to any other states that can mix. This process results in a unitary operation in the encoded qubit space, up to a global phase,

$$U = \begin{pmatrix} e^{i\phi_2} & 0 & 0 & 0 \\ 0 & e^{i(\phi_2 + \phi_3 + \delta\phi)} & 0 & 0 \\ 0 & 0 & 1 & 0 \\ 0 & 0 & 0 & e^{i\phi_3} \end{pmatrix}. \quad (5)$$

This is equivalent to the CPHASE gate $(1, 1, 1, e^{i\delta\phi})^T$ up to single-qubit operations.

$$\begin{aligned} \text{CPHASE} &= \left[\begin{pmatrix} 0 & 1 \\ e^{-i\phi_2} & 0 \end{pmatrix} \otimes \begin{pmatrix} 1 & 0 \\ 0 & 1 \end{pmatrix} \right] \\ &\times U \times \left[\begin{pmatrix} 0 & 1 \\ 1 & 0 \end{pmatrix} \otimes \begin{pmatrix} 1 & 0 \\ 0 & e^{-i\phi_3} \end{pmatrix} \right]. \end{aligned} \quad (6)$$

Note that, unlike refs 14,42, we tune both transmons 1b and 2a instead of tuning only one of them. If we only tuned transmon 2a to bring the (0110) state close to the (0020) state, then transmon 2a and transmon 2b would be close to resonance. Because the transmon–transmon interaction through capacitive coupling can be turned on and off by bringing the transmons on and off resonance, this will result in a complicated, unintended operation as well as leakage. So it is necessary to tune transmons 1b and 2a simultaneously so that transmons 1a and 2b do not come into play during the process. The resonance between next-nearest neighbours can also lead to some small anti-crossing, but these resonances only occur during the fast ramping up and down steps and thus can be negligible. This scheme is preferable to directly using the xx coupling between transmons 1b and 2a, since xx coupling drives the system outside of the encoded qubit space and hence leads to leakage, requiring a rather long sequence of pulse gates to implement a two-qubit logical operation[7,9]. The physical CPHASE gate has been successfully implemented for xmon qubits with gate time of ~ 40 ns (ref. 14), which can be directly applied for logical two-qubit gate here, too.

Figure 4 shows simulated numerical results of this physical two-qubit interaction between transmons 1b and 2a. We use an error function shape ramping up and down, similar to ref. 44,

$$\begin{aligned} &E_J^{(1b)}(t) \\ &= \begin{cases} E_{J0}^{(1b)} + \frac{E_{res}^{(1b)} - E_{J0}^{(1b)}}{2}\left(1 + \text{erf}\left(\frac{t - \tau_1/2}{\sqrt{2}\sigma}\right)\right) & \text{(I)} \\ E_{res}^{(1b)} & \text{(II)} \\ E_{res}^{(1b)} - \frac{E_{res}^{(1b)} - E_{J0}^{(1b)}}{2}\left(1 + \text{erf}\left(\frac{t - \tau_1/2 - \tau_2}{\sqrt{2}\sigma}\right)\right) & \text{(III)} \end{cases} \end{aligned} \quad (7)$$

and $E_J^{(2a)}(t) = E_{J0}^{(1b)} + E_{J0}^{(2a)} - E_J^{(1b)}(t)$. E_{J0} is the idle value and E_{res} is for resonant (11) and (02) states. To find optimal solutions of this form, we change τ_1 and choose $\sigma = \tau_1/4\sqrt{2}$. $\tau_{12} = \tau_2 - \tau_1$ is calculated analytically using a perturbative expression such that the whole process will result in the U with desired $\delta\phi$. Figure 4a shows τ_{12} and the total time τ_3 needed to implement a CZ gate ($\delta\phi = \pi$).

Due to the mixing with higher energy states which are out of the encoded qubit space, leakage error could pose a problem. We can compute the leakage error as follows. The full unitary operation matrix U can be written in a block-form

$$U = \begin{pmatrix} U_{AA} & U_{AB} \\ U_{BA} & U_{BB} \end{pmatrix}, \quad (8)$$

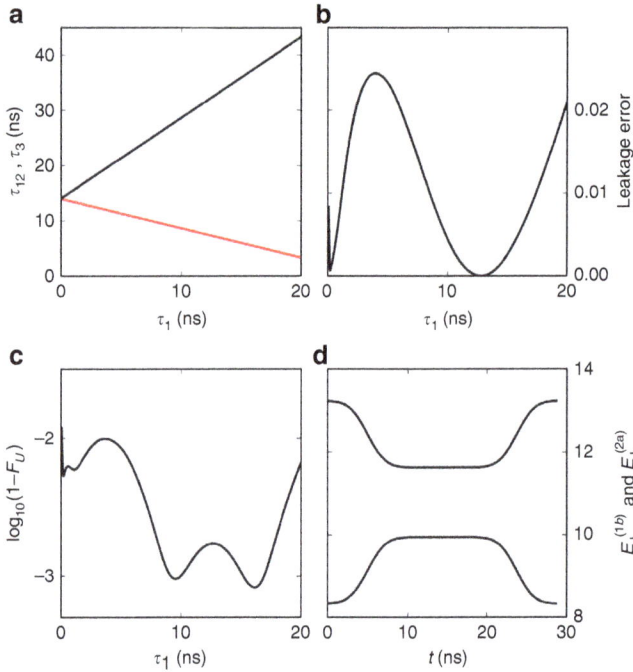

a

b

c

d

Figure 4 | Fidelity of adiabatic CZ interaction operation. (**a**) Operation time. The red curve is the staying time $\tau_{12} = \tau_2 - \tau_1$ and the black curve is the total time τ_3. (**b**) Leakage error during the process. (**c**) Gate fidelity in terms of Makhlin invariants. This gives a measure of how close the unitary gate is to the CZ gate up to single qubit operations. (**d**) Pulse shape for $\tau_1 = 10$ ns. We used error function to model a smooth pulse shapes for $E_J^{(1b)}$ and $E_J^{(2a)}$.

where A is the encoded qubit subspace and B is the complementary subspace. For any qubit state $|\psi_A\rangle$ in the encoded qubit space, the leaked portion is $U_{BA}|\psi_A\rangle$ and $\|U_{BA}|\psi_A\rangle\|^2 = \langle\psi_A|U_{BA}^\dagger U_{BA}|\psi_A\rangle \equiv \langle\psi_A|W_{AA}|\psi_A\rangle$. $W_{AA} = U_{BA}^\dagger U_{BA}$ is positive definite and the leakage error E_{leak} can be defined as $\max\langle\psi_A|W_{AA}|\psi_A\rangle = \max_\lambda\{w_\lambda\}$ where w_λ are the eigenvalues of W_{AA}. The leakage error (Fig. 4b) can be a few percent, but if we choose optimal τ_1, it can be significantly reduced, well below 1%. Note too that leakage can be dealt with algorithmically[45,46]; such circuit-based leakage reduction algorithms will likely be required in any quantum computing implementation.

Figure 4c shows the fidelity of this two-qubit unitary gate U from numerical simulation of the procedure. The fidelity of the unitary gate was defined as:

$$F_U = 1 - [f_1(U_{CZ}) - f_1(U)]^2 - [f_2(U_{CZ}) - f_2(U)]^2 \quad (9)$$

where f_1 and f_2 are the two Makhlin invariants[47] for two-qubit gates. Makhlin invariants are identical for different two-qubit unitary gates if they are equivalent up to single-qubit operations. We find that fidelity better than 99% is achievable for $\tau_1 \simeq 10$ ns, which also leads to very small leakage. Figure 4d shows the pulse shape of $E_J^{(1b)}$ and $E_J^{(2a)}$ for $\tau_1 = 10$ ns. The total time duration for the whole process is about 30 ns. In real devices, the fidelity can be lower due to other sources of noise, but here we use only a simple form for the pulse shapes which are not fully optimized as in refs 44,48, so there is some room for improvement. We also considered Gaussian shape pulses and obtained similar results.

We can estimate the realistic fidelity of the encoded CPHASE gate constructed here from the fidelities of the z-control pulses and the adiabatic process. Since any single-qubit logical gate involves at most three rotations (that is, three pulse steps), the encoded CPHASE gate requires at most 12 pulse steps. Assuming

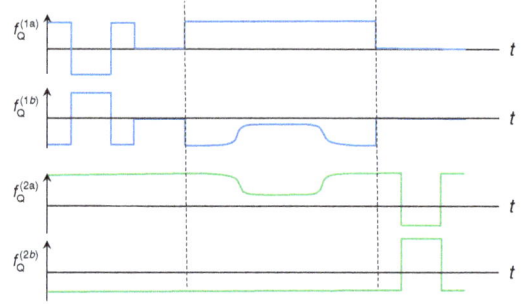

Figure 5 | Pulse sequence for encoded CPHASE gate. This schematically shows a sequence for CPHASE gate in equation (6). Single qubit phase gate is implemented with three-step Euler rotations, and Pauli X and identity gates are implemented as a single rotation. The two vertical dashed lines separate single qubit gates and two-qubit adiabatic operation.

the z-pulse fidelity of 0.999 and a fidelity of the adiabatic gate U in equation (6) between two transmons of about 0.99, the fidelity of the total process can be estimated to be better than $F_2 > 0.999^{12} \times 0.99 \simeq 0.978$. Better optimization or different sequences may improve the fidelity. Of critical comparison, the already demonstrated physical CPHASE gate fidelity of 0.99 (ref. 14) also includes a single-adiabatic operation and single-qubit corrective operations, so the encoded CPHASE gate should be achievable with a similar fidelity. The encoded CNOT gate can be implemented with CPHASE gate and single-qubit gates, and we can expect similar fidelity for CNOT gate.

Figure 5 schematically depicts a sequence of DC pulses for the logical CPHASE gate, using the expression in equation (6). The first three pulses in encoded qubit 1 implement a phase gate and the next resonant pulse realizes a Pauli X gate. The second encoded qubit is pulsed to qubit frequencies such that the encoded qubit 2 rotates by $2n\pi$ to implement the identity operation. Then the two-qubit adiabatic gate between transmons 1b and 2a is applied. After that, an X gate is applied to encoded qubit 1 as a single-resonant pulse step and a phase gate is applied to encoded qubit 2 in three rotations. This particular implementation of CPHASE contains only 9 single-qubit operations, better than the general 12 single-qubit gates we discussed above.

Our choice of encoded qubit is for the sake of simplicity and straightforward incorporation of physical qubit operations into logical gate operations. We also considered an alternative choice, $(|01\rangle \pm |10\rangle)/\sqrt{2}$ in the same subspace, which more closely resembles choice for encoded spin qubits. With this encoded qubit, the constant capacitive coupling leads to a constant energy gap between encoded qubit states and the z-control of each physical qubit allows tunable $\hat{\sigma}^x$ operation. Single-qubit logical gates can be implemented in a similar way, and the adiabatic two-qubit operation will need additional single-qubit unitary gates to transform to the CPHASE gate due to the basis change of the encoded qubit.

The capacitive coupling between transmons is typically constant and determined solely by the geometry of the SC islands. This coupling is effectively turned on and off by the qubit frequency differences. With more complicated control circuits as in the gmon architecture[41,49], the capacitive coupling can also be tunable and completely turned off, giving a very large on/off ratio. The tunable capacitive coupling removes the need to detune each transmon to avoid unwanted resonances, hence significantly simplifying the qubit frequency controls during the CPHASE operation. This also allows rotating the encoded qubit around any axis in the full xz plane, reducing the necessary rotation steps to two for any single-qubit logical gates[38].

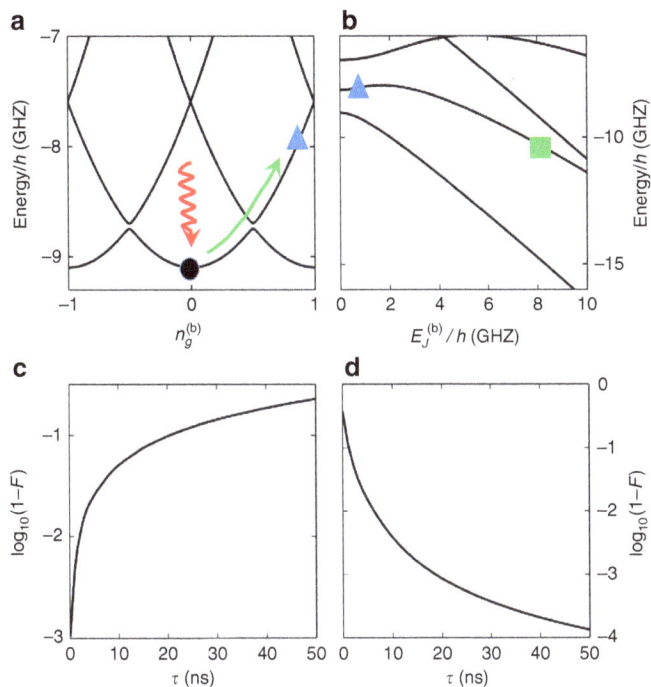

Figure 6 | Initialization scheme for the encoded qubit. (**a**) The energy spectrum of an encoded qubit with the second transmon in the charge qubit regime. After thermalizing the qubit into the ground state (black dot), $n_g^{(b)}$ is tuned from 0 to some value between 0.5 and 1. If this change is done fast enough, the qubit is in the first excited state (blue triangle). Then the qubit is adiabatically moved into the transmon regime (green square) by increasing $E_J^{(b)}$ as shown in (**b**). (**c,d**) Fidelity of these processes as a function of total time duration.

Initialization. In spin systems the encoded qubit can be initialized fast and with high fidelity by loading pairs of electrons in the singlet state directly from the Fermi sea provided by the leads supplying the quantum dots, then adiabatically separating the singlet into two dots[6]. Electrons' fermionic and particle nature enables this—a quantum property that may be emulated with engineered many-body photonic systems (for example, refs 50,51)), but which is in no way practical in the near-term. One could also engineer a two-qubit system where the ground state is a singlet, for example, by making the coupling between the two qubits much greater than the qubit splittings (and, for example, waiting for relaxation to the ground state). Here, although, one would want to quickly move out of this regime to do gates at an implementable speed in addition to turning off as much as possible qubit–qubit couplings, which would be very challenging. Here, we provide an alternative initialization scheme that only requires fast DC pulses.

The ground state of the two-transmon system is $|00\rangle$, which is not in the encoded qubit subspace defined by equation (3). To initialize the system into $|0\rangle_Q = |01\rangle$ without microwave control, we propose using a process analogous to the Landau–Zener (LZ) tunnelling[52,53]. For this procedure, we need tunability of the gate charge $n_g^{(b)}$ of the second transmon, which can be provided by connecting a capacitor with a voltage control to the transmon (see Fig. 1b) or by using the side-gate for gatemons. The initialization procedure is as follows. First, we tune the transmon qubit into the charge qubit regime where $E_J^{(b)}$ is much smaller than $E_C^{(b)}$ by tuning Φ_{ext} (or V_g for a variable super-semi JJ) with $n_g^{(b)} \simeq 0$. Then, via thermalization (by waiting the relaxation time or by coupling to a dissipative reservoir) the qubit reaches the ground state (black dot in Fig. 6a). (The thermalization could

instead be done before tuning to the charge qubit regime.) In this charge qubit regime, the two lowest energy states anti-cross at the sweet spot $n_g^{(b)} = 0.5$. By changing the gate charge $n_g^{(b)}$ from 0 to a value larger than 0.5, we can induce the LZ tunnelling to prepare the charge qubit in the first excited state (blue triangle). Then, we can tune $E_J^{(b)}$ back to the operating transmon regime ($E_J^{(b)} \gg E_C^{(b)}$) (green square in Fig. 6b). If we tune $E_J^{(b)}$ exactly to be zero, then there is a crossing instead of anti-crossing, and the fidelity will be much better. But some finite value will be allowable as long as we can change $n_g^{(b)}$ fast enough.

Figure 6c shows the calculated fidelity of the LZ tunnelling in the charge qubit regime of Fig. 6a as a function of the total time τ taken to change $n_g^{(b)}$. Here fidelity is defined as $F = |\langle \Psi_{target} | \Psi_{final} \rangle|$. We have used system parameters easily available in real systems, $E_C^{(a)}/h = E_C^{(b)}/h = 375$ MHz, $E_J^{(a)}/h = 12$ GHz, $E_J^{(b)}/h = 50$ MHz, $E_{cc} = 30$ MHz. $n_g^{(b)}$ was changed from 0 to 0.8. As is the case for typical LZ tunnelings, the fidelity is better with faster change of the parameter. We expect to see fidelity better than 99% for a LZ process of a few nanoseconds. Tuning back to the transmon regime is essentially an adiabatic process, and the fidelity increases with slower change (Fig. 6d). We changed $E_J^{(b)}/h$ from 50 MHz to 8.33 GHz, and the fidelity is better than 99% for a process of a few tens of nanoseconds. So this initialization process will take ~ 20 ns to prepare the logical qubit state with fidelity of $\sim 99\%$. The effect of charge and quasiparticle noise during this process is a concern that should be investigated experimentally, but charge qubits have been shown to have T_1 times up to 0.2 ms (ref. 54). Variants of the flux qubit are especially stable to quasiparticle and charge noise fluctuations even at small qubit splittings[34].

Measurement of the qubit states. Since an encoded qubit is in a state

$$|\Psi\rangle = \alpha|0\rangle_Q + \beta|1\rangle_Q = \alpha|01\rangle + \beta|10\rangle, \qquad (10)$$

the encoded qubit can be measured by measuring either of the physical qubits using a standard method, such as dispersive measurement[55–58] (which can be multiplexed). The choice of our encoded qubit in equation (3) allows us to translate the single-qubit state into the encoded qubit states. With a choice of a singlet-triplet-like encoded qubit states, $(|01\rangle \pm |10\rangle)/\sqrt{2}$, the encoded qubit state can also be measured after some single-qubit gates are applied to turn them into the encoded qubit states as above, or they could be measured directly by dispersive measurement since these states correspond to different resonator frequencies[59,60].

Unlike the spin system where measurement of a singlet can be done electrostatically using a projective measurement[6], the dispersive measurement of SC qubits using a transmission line resonator still requires a microwave carrier, which is fine as a proof of concept. We would prefer a measurement approach that takes full advantage of our encoded qubit architecture, with qubit energy completely separated from microwave source. One possibility is to convert the encoded qubit to another quantum system (or measurement qubit) that is long-lived classically, but can be read out digitally or with fast base-band pulses (in other words a latched readout), for example[61]. A compromise option is to do dispersive measurement but still utilize lower bandwidth lines: we can either tune E_J directly or swap the qubit with another one with a different frequency such that it can be readily measured.

Discussion

We proposed a scheme for a "dual rail" superconducting quantum computer where each qubit consists of two tunable physical qubits. Encoded two-qubit operations are found to require only a single physical two-qubit gate and single-qubit pulses. Since physical two-qubit gates are typically much more costly in time and fidelity, this means that the overhead of encoded operations as proposed here is not significant, especially compared to spin qubits.

In this encoded qubit architecture all qubit manipulations are achieved solely by the z-control pulse sequences of individual qubits. This removes the requirement of microwave xy-control lines necessary in conventional transmon or similar qubit devices, simplifying classical control circuitry significantly. In addition, the encoded approach may allow lower requirements for available bandwidth per line, the potential for less crosstalk, and a reduction in needed timing accuracy as the encoded qubit states are nearly degenerate. Removing the need for microwave control frees the choice of qubit frequency from the cost and availability of microwave electronics. One is then able to design physical qubits with higher (or much lower) frequency that might enable higher temperature qubit operation (which may benefit from work already underway to enable high magnetic field compatible circuits for Majorana experiments[62] in higher-T_c materials) or qubits made from degenerate quantum circuits as in symmetry protected approaches[63-65], of which there is a natural connection to how spin qubits are inherently protected.

Encoding a qubit in a two-dimensional subspace in a larger Hilbert space introduces leakage error. For our encoded qubits, the relaxation process of individual transmons will lead to leakage out of the encoded qubit space. For a single-gate operation such as CNOT of duration τ, the leakage error due to the T_1 process would be $1 - \exp(-\tau/T_1) \simeq 0.04$ % for $\tau = 40$ ns and $T_1 = 100\,\mu$s, which would slightly reduce the error threshold for quantum error correction[66]. While a single-gate operation of a few tens of ns does not lead to significant leakage errors, a long sequence of gate operations in a large system can be a problem. Particularly, a single-logical qubit for fault-tolerant quantum computing such as the surface code will consist of many encoded qubits and a logical operation will be a sequence of operations on those encoded qubits. Therefore, leakage reduction units[67] will likely be essential. For example, a full-leakage reduction unit on the basis of one-bit teleportation[66] would require an ancilla qubit for each encoded qubit and additional CNOT operations, and measurements after each logical CNOT operation. Qubits especially designed for large relaxation times, such as variants of fluxonium[68], may be particularly promising for our approach (for example, a T_1 time of 1 ms would lead to a leakage error per CNOT of 4×10^{-5}) and would reduce the overhead for leakage mitigation dramatically.

The recently demonstrated capacitively-shunted flux qubits[33,34] (or "fluxmon") may also provide a promising alternative. They have comparable coherence times and a larger anharmonicity than transmons. Qubit–qubit coupling through mutual inductance would also provide transversal xx coupling like the capacitive coupling between transmon qubits, so the formalism used in this work should be applicable as well. They also offer benefits for initialization as they can be tuned to the flux qubit regime down to very small qubit splitting while being protected to T_1 processes that flux qubits offer, and readout can also be done by using a DC SQUID[69,70] without a transmission line.

In the next phase of this design philosophy one can consider how to mimic other beneficial properties of spin qubits: very weak coupling between qubit states to charge noise and phonons, a fast and selective two-qubit gate via a Pauli exclusion like mechanism or an interaction that mimics it, very large ON/OFF ratios and fast initialization via some as yet unknown method.

References

1. Awschalom, D. D., Bassett, L. C., Dzurak, A. S., Hu, E. L. & Petta, J. R. Quantum spintronics: engineering and manipulating atom-like spins in semiconductors. *Science* **339**, 1174–1179 (2013).
2. Loss, D. & DiVincenzo, D. P. Quantum computation with quantum dots. *Phys. Rev. A* **57**, 120–126 (1998).
3. Kane, B. E. A silicon-based nuclear spin quantum computer. *Nature* **393**, 133–137 (1998).
4. Tyryshkin, A. M. et al. Electron spin coherence exceeding seconds in high-purity silicon. *Nat. Mater.* **11**, 143–147 (2012).
5. Saeedi, K. et al. Room-temperature quantum bit storage exceeding 39 minutes using ionized donors in silicon-28. *Science* **342**, 830–833 (2013).
6. Petta, J. R. et al. Coherent manipulation of coupled electron spins in semiconductor quantum dots. *Science* **309**, 2180–2184 (2005).
7. DiVincenzo, D. P., Bacon, D., Kempe, J., Burkard, G. & Whaley, K. B. Universal quantum computation with the exchange interaction. *Nature* **408**, 339–342 (2000).
8. Bacon, D., Kempe, J., Lidar, D. A. & Whaley, K. B. Universal fault-tolerant quantum computation on decoherence-free subspaces. *Phys. Rev. Lett.* **85**, 1758–1761 (2000).
9. Fong, B. H. & Wandzura, S. M. Universal quantum computation and leakage reduction in the 3-qubit decoherence free subsystem. *Quant. Inf. Comput.* **11**, 1003–1018 (2011).
10. Devoret, M. H. & Schoelkopf, R. J. Superconducting circuits for quantum information: an outlook. *Science* **339**, 1169–1174 (2013).
11. Koch, J. et al. Charge-insensitive qubit design derived from the Cooper pair box. *Phys. Rev. A* **76**, 042319 (2007).
12. Schreier, J. A. et al. Suppressing charge noise decoherence in superconducting charge qubits. *Phys. Rev. B* **77**, 180502 (R) (2008).
13. Barends, R. et al. Coherent Josephson qubit suitable for scalable quantum integrated circuits. *Phys. Rev. Lett.* **111**, 080502 (2013).
14. Barends, R. et al. Superconducting quantum circuits at the surface code threshold for fault tolerance. *Nature* **508**, 500–503 (2014).
15. Kelly, J. et al. State preservation by repetitive error detection in a superconducting quantum circuit. *Nature* **519**, 66–69 (2015).
16. Reed, M. D. et al. Realization of three-qubit quantum error correction with superconducting circuits. *Nature* **482**, 382–385 (2012).
17. Sun, L. et al. Tracking photon jumps with repeated quantum non-demolition parity measurements. *Nature* **511**, 444–448 (2014).
18. Chow, J. M. et al. Implementing a strand of a scalable fault-tolerant quantum computing fabric. *Nat. Commun.* **4**, 4015 (2014).
19. Córcoles, A. D. et al. Demonstration of a quantum error detection code using a square lattice of four superconducting qubits. *Nat. Commun.* **6**, 6979 (2015).
20. Blase, X., Bustarret, E., Chapelier, C., Klein, T. & Marcenat, C. superconducting group-IV semiconductors. *Nat. Mater.* **8**, 375–382 (2009).
21. Bustarret, E. Superconductivity in doped semiconductors. *Physica C* **514**, 36–45 (2015).
22. Bustarret, E. et al. Superconductivity in doped cubic silicon. *Nature* **444**, 465–468 (2006).
23. Marcenat, C. et al. Low-temperature transition to a superconducting phase in boron-doped silicon films grown on (001)-oriented silicon wafers. *Phys. Rev. B* **81**, 020501 (R) (2010).
24. Dahlem, F. et al. Subkelvin tunneling spectroscopy showing Bardeen–Cooper–Schrieffer superconductivity in heavily boron-doped silicon epilayers. *Phys. Rev. B* **82**, 140505 (R) (2010).
25. Grockowiak, A. et al. Superconducting properties of laser annealed implanted Si:B epilayers. *Supercond. Sci. Technol.* **26**, 045009 (2013).
26. Herrmannsdörfer, T. et al. Superconducting state in a gallium-doped germanium layer at low temperatures. *Phys. Rev. Lett.* **102**, 217003 (2009).
27. Skrotzki, R. et al. The impact of heavy Ga doping on superconductivity in germanium. *Low Temp. Phys.* **37**, 877–883 (2011).
28. Shim, Y.-P. & Tahan, C. Bottom-up superconducting and Josephson junction device inside a group-IV semiconductor. *Nat. Commun.* **5**, 4225 (2014).
29. Shim, Y.-P. & Tahan, C. Superconducting-semiconductor quantum devices: from qubits to particle detectors. *IEEE J. Sel. Top. Quant. Electron.* **21**, 9100209 (2015).
30. Larsen, T. W. et al. Semiconductor-nanowire-based superconducting qubit. *Phys. Rev. Lett.* **115**, 127001 (2015).
31. deLange, G. et al. Realization of microwave quantum circuits using hybrid superconducting-semiconducting nanowire Josephson elements. *Phys. Rev. Lett.* **115**, 127002 (2015).
32. Ristè, D. et al. Detecting bit-flip errors in a logical qubit using stabilizer measurements. *Nat. Commun.* **6**, 6983 (2015).
33. You, J. Q., Hu, X., Ashhab, S. & Nori, F. Low-decoherence flux qubit. *Phys. Rev. B* **75**, 140515 (2007).
34. Yan, F. et al. The flux qubit revisited. Preprint at http://arxiv.org/abs/1508.06299 (2015).

35. Makhlin, Y., Schön, G. & Shnirman, A. Josephson-junction qubits with controlled couplings. *Nature* **398**, 305–307 (1999).
36. Chang, W. *et al.* Hard gap in epitaxial semiconductorsuperconductor nanowires. *Nat. Nanotechonol.* **10**, 232–236 (2015).
37. Krogstrup, P. *et al.* Epitaxy of semiconductorsuperconductor nanowires. *Nat. Mater.* **14**, 400–406 (2015).
38. Shim, Y.-P., Fei, J., Oh, S., Hu, X. & Friesen, M. Single-qubit gates in two steps with rotation axes in a single plane. Preprint at http://arxiv.org/abs/1303.0297 (2013).
39. Wang, X. *et al.* Composite pulses for robust universal control of singlettriplet qubits. *Nat. Commun.* **3**, 997 (2012).
40. Wang, X., Bishop, L. S., Barnes, E., Kestner, J. P. & Sarma, S. D. Robust quantum gates for singlet-triplet spin qubits using composite pulses. *Phys. Rev. A* **89**, 022310 (2014).
41. Chen, Y. *et al.* Qubit architecture with high coherence and fast tunable coupling. *Phys. Rev. Lett.* **113**, 220502 (2014).
42. DiCarlo, L. *et al.* Demonstration of two-qubit algorithms with a superconducting quantum processor. *Nature* **460**, 240–244 (2009).
43. Strauch, F. W. *et al.* Quantum logic gates for coupled superconducting phase qubits. *Phys. Rev. Lett.* **91**, 167005 (2003).
44. Ghosh, J. *et al.* High-fidelity controlled-σ^z gate for resonator-based superconducting quantum computers. *Phys. Rev. A* **87**, 022309 (2013).
45. Wu, L.-A., Byrd, M. S. & Lidar, D. A. Efficient universal leakage elimination for physical and encoded qubits. *Phys. Rev. Lett.* **89**, 127901 (2002).
46. Fowler, A. G. Coping with qubit leakage in topological codes. *Phys. Rev. A* **88**, 042308 (2013).
47. Makhlin, Y. Nonlocal properties of two-qubit gates and mixed states, and the optimization of quantum computations. *Quant. Inf. Proc* **1**, 243–252 (2002).
48. Martinis, J. M. & Geller, M. R. Fast adiabatic qubit gates using only σ_z control. *Phys. Rev. A* **90**, 022307 (2014).
49. Geller, M. R. *et al.* Tunable coupler for superconducting Xmon qubits: perturbative nonlinear model. *Phys. Rev. A* **92**, 012320 (2015).
50. Greentree, A. D., Tahan, C., Cole, J. H. & Hollenberg, L. C. L. Quantum phase transitions of light. *Nat. Phys.* **2**, 856–861 (2006).
51. Hartmann, M., Brandão, F. & Plenio, M. Quantum many-body phenomena in coupled cavity arrays. *Laser Photon. Rev.* **2**, 527–556 (2008).
52. Landau, L. Zur Theorie der Energieubertragung. II. *Phys. Z. Sowjetunion* **2**, 46–51 (1932).
53. Zener, C. Non-Adiabatic Crossing of Energy Levels. *Proc. R. Soc. Lond. A* **137**, 696–702 (1932).
54. Kim, Z. *et al.* Decoupling a Cooper-pair box to enhance the lifetime to 0.2 ms. *Phys. Rev. Lett.* **106**, 120501 (2011).
55. Blais, A., Huang, R.-S., Wallraff, A., Girvin, S. M. & Schoelkopf, R. J. Cavity quantum electrodynamics for superconducting electrical circuits: an architecture for quantum computation. *Phys. Rev. A* **69**, 062320 (2003).
56. Wallraff, A. *et al.* Strong coupling of a single photon to a superconducting qubit using circuit quantum electrodynamics. *Nature* **431**, 162–167 (2004).
57. Schuster, D. I. *et al.* ac stark shift and dephasing of a superconducting qubit strongly coupled to a cavity field. *Phys. Rev. Lett.* **94**, 123602 (2005).
58. Wallraff, A. *et al.* Approaching unit visibility for control of a superconducting qubit with dispersive readout. *Phys. Rev. Lett.* **95**, 060501 (2005).
59. Gambetta, J. M., Houck, A. A. & Blais, A. Superconducting qubit with purcell protection and tunable coupling. *Phys. Rev. Lett.* **106**, 030502 (2011).
60. Srinivasan, S. J., Hoffman, A. J., Gambetta, J. M. & Houck, A. A. Tunable coupling in circuit quantum electrodynamics using a superconducting charge qubit with a V-shaped energy level diagram. *Phys. Rev. Lett.* **106**, 083601 (2011).
61. Berkley, A. J. *et al.* A scalable readout system for a superconducting adiabatic quantum optimization system. *Supercond. Sci. Technol.* **23**, 105014 (2010).
62. van Woerkom, D. J., Geresdi, A. & Kouwenhoven, L. P. One minute parity lifetime of a NbTiN Cooper-pair transistor. *Nat. Phys.* **11**, 547550 (2015).
63. Douçot, B. & Ioffe, L. B. Physical implementation of protected qubits. *Rep. Prog. Phys.* **75**, 072001 (2012).
64. Kitaev, A. Y. Fault-tolerant quantum computation by anyons. *Ann. Phys.* **303**, 2–30 (2003).
65. Brooks, P., Kitaev, A. & Preskill, J. Protected gates for superconducting qubits. *Phys. Rev. A* **87**, 052306 (2013).
66. Suchara, M., Cross, A. W. & Gambetta, J. M. Leakage suppression in the toric code. *Quant. Inf. Comp.* **15**, 997–1016 (2015).
67. Aliferis, P. & Terhal, B. M. Fault-tolerant quantum computation for local leakage faults. *Quant. Inf. Comp.* **7**, 139–156 (2007).
68. Manucharyan, V. E., Koch, J., Glazman, L. I. & Devoret, M. H. Fluxonium: single Cooper-pair circuit free of charge offsets. *Science* **326**, 113–116 (2009).
69. Bylander, J. *et al.* Noise spectroscopy through dynamical decoupling with a superconducting flux qubit. *Nat. Phys.* **7**, 565–570 (2011).
70. Jin, X. Y. *et al.* Z-gate operation on a superconducting flux qubit via its readout SQUID. *Phys. Rev. Applied* **3**, 034004 (2015).

Acknowledgements

We thank C.M. Marcus, A. Mizel, W.D. Oliver, B. Palmer and K.D. Petersson for useful discussions.

Author contributions

All the authors contributed to the planning of the project, interpretation of the results, discussions and writing of the manuscript. Y.-P.S performed the theoretical and numerical calculations.

Additional information

Competing financial interests: The authors declare no competing financial interests.

8

Super-crystals in composite ferroelectrics

D. Pierangeli[1], M. Ferraro[1], F. Di Mei[1,2], G. Di Domenico[1,2], C.E.M. de Oliveira[3], A.J. Agranat[3] & E. DelRe[1]

As atoms and molecules condense to form solids, a crystalline state can emerge with its highly ordered geometry and subnanometric lattice constant. In some physical systems, such as ferroelectric perovskites, a perfect crystalline structure forms even when the condensing substances are non-stoichiometric. The resulting solids have compositional disorder and complex macroscopic properties, such as giant susceptibilities and non-ergodicity. Here, we observe the spontaneous formation of a cubic structure in composite ferroelectric potassium–lithium–tantalate–niobate with micrometric lattice constant, 10^4 times larger than that of the underlying perovskite lattice. The 3D effect is observed in specifically designed samples in which the substitutional mixture varies periodically along one specific crystal axis. Laser propagation indicates a coherent polarization super-crystal that produces an optical X-ray diffractometry, an ordered mesoscopic state of matter with important implications for critical phenomena and applications in miniaturized 3D optical technologies.

[1] Dipartimento di Fisica, Università di Roma 'La Sapienza', Rome 00185, Italy. [2] Center for Life Nano Science@Sapienza, Istituto Italiano di Tecnologia, Rome 00161, Italy. [3] Department of Applied Physics, Hebrew University of Jerusalem, Jerusalem 91904, Israel. Correspondence and requests for materials should be addressed to E.D. (email: eugenio.delre@uniroma1.it).

Textbook models of global symmetry-breaking include a low-symmetry low-temperature state with a fixed infinitely extended coherence. In contrast, the spontaneous polarization observed as spatial inversion symmetry is broken during a paraelectric–ferroelectric phase transition generally leads to a disordered mosaic of polar domains that permeate the finite samples[1]. Coherent and ordered ferroelectric states with remarkable properties of both fundamental and technological interest[2-5] can emerge when ferroelectricity is influenced by external factors, such as system dimensionality[6], strain gradients[7-9], electrostatic coupling[10,11] and magnetic interaction[12,13].

Here we report the spontaneous formation of an extended coherent three-dimensional (3D) superlattice in the nominal ferroelectric phase of specifically grown potassium–lithium–tantalate–niobate (KLTN) crystals[14-17]. Visible-light propagation reveals a polarization super-crystal with a micrometric lattice constant, a counterintuitive mesoscopic phase that naturally mimics standard solid-state structures but on scales that are thousands of times larger. The phenomenon is achieved using compositionally disordered ferroelectrics[18-27]. At one given temperature, these have the interesting property of manifesting a single perovskite phase whose dielectric properties depend on the specific composition[28-30]. For example, a compositional gradient along the pull axis leads to a position-dependent Curie point $T_C(\mathbf{r})$, so that for a given value of crystal temperature T a phase separation occurs, where regions with $T > T_C$ are paraelectric and those with $T < T_C$ have a spontaneous polarization[31]. Specifically tailored growth schemes are even able to achieve an oscillating T_C along a given direction, say the x axis[32,33]. Under these conditions, we can expect that, at a given T in proximity of the average (macroscopic) T_C, the sample will be in a hybrid state with alternating regions with and without spontaneous polarization. Crossing the Curie point, under conditions in which perovskite polar domains pervade the volume forming 90° configurations to minimize the free energy associated with polarization charge[34], this oscillation can form a full 3D periodic structure.

Results

Observation of a compositionally induced super-crystal. To investigate the matter, we make use of top-seeded ferroelectric crystals with an oscillating composition along the growth axis achieved using an off-centre growth technique in the furnace[33,35]. We obtain a zero-cut 2.4 mm by 2.0 mm by 1.7 mm, along the x,y,z directions, respectively, optical-quality KLTN sample with a periodically oscillating niobium composition of period $\Lambda = 5.5\,\mu m$ along the x axis, with an average composition $K_{1-\alpha}Li_\alpha Ta_{1-\beta}Nb_\beta O_3$, where $\alpha = 0.04$ and $\beta = 0.38$ (see Methods). When the crystal is allowed to relax at $T = T_C - 2\,K$, that is, in proximity of the spatially averaged room-temperature Curie point $T_C = 294\,K$, laser light propagating through the sample suffers relevant scattering with strongly anisotropic features (Fig. 1a–d). Typical results are reported in Fig. 1b–d, and they appear as an optical analogue of X-ray diffraction in low-temperature solids. This optical diffractometry provides basic evidence of a 3D superlattice at micrometric scales. Probing the principal crystal directions reveals several diffraction orders that map the entire reciprocal space. The large-scale super-crystal, which permeates the whole sample, overlaps—along the x direction—with the built-in compositional oscillating seed (see Methods). The superlattice extends in full three dimensions, with the same periodicity $\Lambda = 5.5\,\mu m$ of the x-oriented compositional oscillation, also along the orthogonal y and z directions. In particular, Fig. 1d indicates

that in the plane perpendicular to the built-in dielectric microstructure Γ vector, that is, where spatial symmetry should be unaffected by the microstructure in composition, the ferroelectric phase transition leads to a spontaneous pattern of transverse scale Λ. The corresponding elementary structure on micrometric spatial scales is reported in Fig. 1e; it can be represented as a face-centred cubic structure in which the occupation of one of the three faces ($z - y$ face) is missing[36]. The structure, which is, to our knowledge, not observed at atomic scales, can be reduced to a simple cubic structure with a threefold basis and lattice parameter $a = \Lambda$.

As the crystal is brought below the average Curie point, it manifests a metastable (supercooled) and a stable (cold) phase, as analysed in Fig. 2 both in the reciprocal (Fourier) and direct (real) space. In the nominal paraelectric phase, at $T = T_C + 2\,K$ (Fig. 2a), we observe the first Bragg diffraction orders (± 1) consistent with the presence of the seed microstructure, a one-dimensional (1D) transverse sinusoidal modulation acting as a diffraction grating; the distance from the central zero order fulfills the Bragg condition, that is, scattered light forms an angle $\theta_B = \lambda/2n_0\Lambda \simeq 7°$ with the incident wavevector $\mathbf{k}$. Crossing the ferroelectric phase-transition temperature T_C (see Methods), we detect a supercooled metastable state that has an apparently analoguous diffraction effect (Fig. 2b) that is dynamically superseded by the stable and coherent cold superlattice phase (Fig. 2c), in which spatial correlations are extended to the whole crystal volume. In real space, transmission microscopy (see Methods) shows unscattered optical propagation through the paraelectric sample at $T = T_C + 2\,K$ (Fig. 2d), which turns into critical opalescence and scattering from oblique random domains at the structural phase transition (Fig. 2e,f), and into unscattered transmission in the metastable ferroelectric phase at $T = T_C - 2\,K$ (Fig. 2g). After dipolar relaxation has taken place, the cold super-crystal appears in this case as a periodic intensity distribution on micrometric scales, as shown in Fig. 2h.

Spontaneous polarization underlying the ferroelectric superlattice. To further analyse these supercooled and cold phases, we inspect the supercooled 1D phase (Fig. 2b) that is accessible through linear (unbiased) and electro-optic (biased) polarization-resolved Bragg diffraction measurements. In particular, referring to the set-up illustrated in Fig. 3a, we measure the diffraction efficiency $\eta = P_B/(P_B + P_0)$, where P_B and P_0 are, respectively, the diffracted and non-diffracted powers, in the first Bragg resonance condition, that is, with the incident wavevector $\mathbf{k}$ forming the angle θ_B with respect to the z axis. The diffraction efficiency η is reported in Fig. 3b for different input light polarization and temperature across the average Curie point. Diffraction strongly depends both on the nominal crystal phase and on the polarization of the incident wave: a large increase in η is found for light polarized in the x,z plane (H-polarized). For $T > T_C$, the dependence on light polarization is consistent with what expected in standard periodically index-modulated media (wave-coupled theory), that is, a weak temperature dependence and a maximum η for light polarized normal to the grating vector (V-polarized). In this case, the difference in η_H ($\triangle$) and η_V ($\square$) can be related to the different Fresnel coefficients governing interlayer reflections and is congruently $\eta_V > \eta_H$ by an amount that decreases for larger θ_B (refs 37,38). Consistently, the (H + V)-polarized curve ($\bigcirc$), that is, when the input linear polarization is at 45° with respect to the H and V polarizations, falls between these two curves. Standard behaviour is violated for $T < T_C$, where a large enhancement in η_H rapidly leads to a regime with $\eta_V < \eta_H$.

The physical underpinnings of the super-crystal can be grasped considering the simple model illustrated in Fig. 3c. Here we

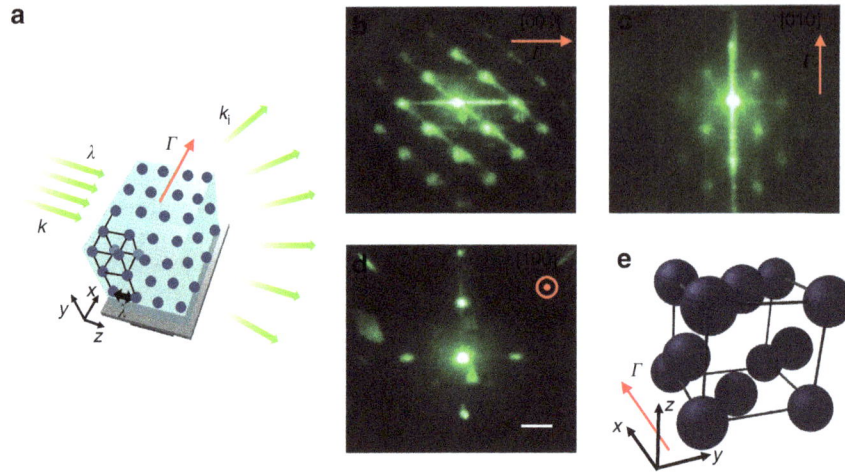

Figure 1 | Super-crystal in the ferroelectric phase. (**a**) Sketch of visible-light diffraction from micrometric structures through a transparent crystal and (**b-d**) 3D superlattice probed at $T = T_C - 2\,K$ along the principal symmetry direction of the crystal, respectively, with the incident wavevector **k** parallel to (**b**) the z direction, (**c**) y direction and (**d**) x direction. Crystallographic analysis reveals the elementary cubic structure of lattice constant Λ shown in **e**. Scale bar, 1.2 cm.

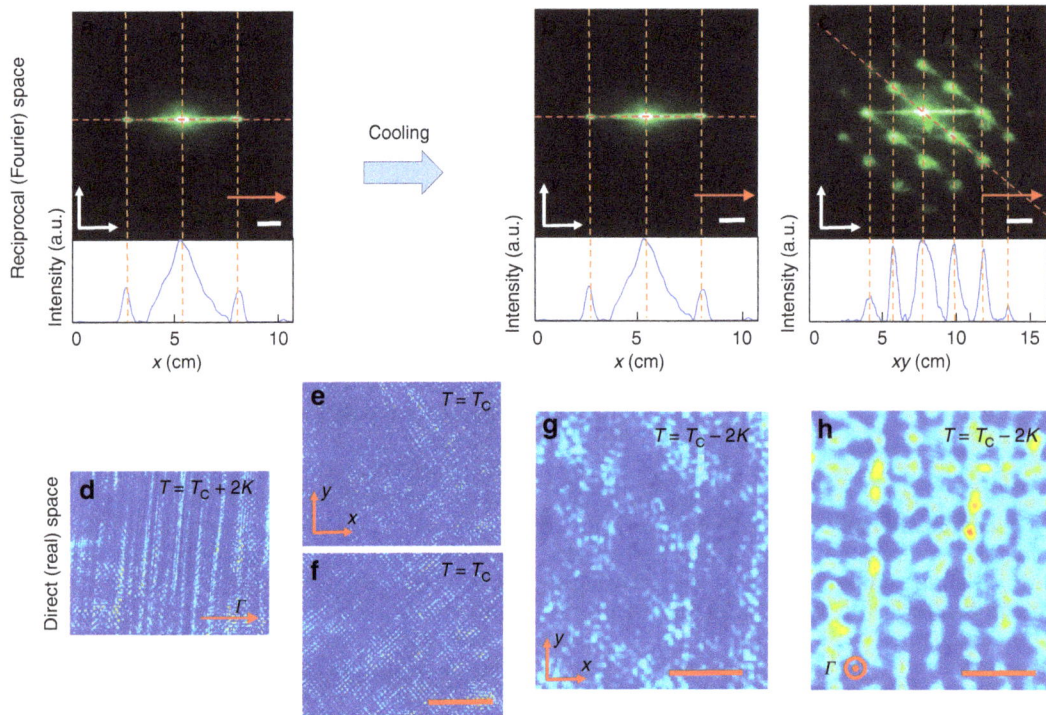

Figure 2 | Light diffraction above and below T_c. (**a**) Reciprocal space probed at $T = T_C + 2\,K$ (hot paraelectric phase), showing the first diffraction orders due to the one-dimensional sinusoidal compositional modulation. Cooling below the critical point results at $T = T_C - 2\,K$ (super-crystal ferroelectric phase) in (**b**) a supercooled (metastable) 1D superlattice with the same diffraction orders that relaxes at the steady state into (**c**) the cold (stable) super-crystals. In both **b**,**c** the direction of incident light is othogonal to Γ, as in **a**. (**d-h**) Corresponding transmission microscopy images revealing (**d**) unscattered optical propagation, (**e**,**f**) scattering at the phase transition, (**g**) unscattered optical propagation in the metastable superlattice and (**h**) periodic intensity distribution underlining the 3D superlattice. Metastable and stable (equilibrium) phases are inspected, respectively, at times $t \approx 1\,min$ and $t \approx 1\,h$ after the structural transition at $T = T_C$. Bottom profiles in **a-c** are extracted along the red dotted line. Scale bars (**a-c**), 1.2 cm, (**d-f**), 100 μm and (**g**,**h**), 10 μm.

consider the metastable 1D superlattice (Fig. 2b) before tensorial effects cause the full 3D superlattice relaxation (Fig. 2c). Specifically, for a given T, regions with a local value of T_C such that $T < T_C$ (dark shading) will manifest a finite spontaneous polarization $P_S \neq 0$, whereas region with $T > T_C$ (light shading) will have a $P_S \simeq 0$. Optical measurements are sensitive to the square of the crystal polarization $\langle \mathbf{P} \cdot \mathbf{P} \rangle \simeq P_S^2$ through the

resulting index pattern modulated via the quadratic elecro-optic response $\delta n(P) = -(1/2)n^3 g P^2$, where n is the unperturbed refraction index and g is the corresponding perovskite elecro-optic coefficient[25,39]. Enhanced Bragg-scattering of light polarized parallel to the seed direction Γ (H in Fig. 3b—super-crystal) indicates that $P_S(x)$ is parallel to the seed direction (x axis), where the elecro-optic coefficient g has its maximum value

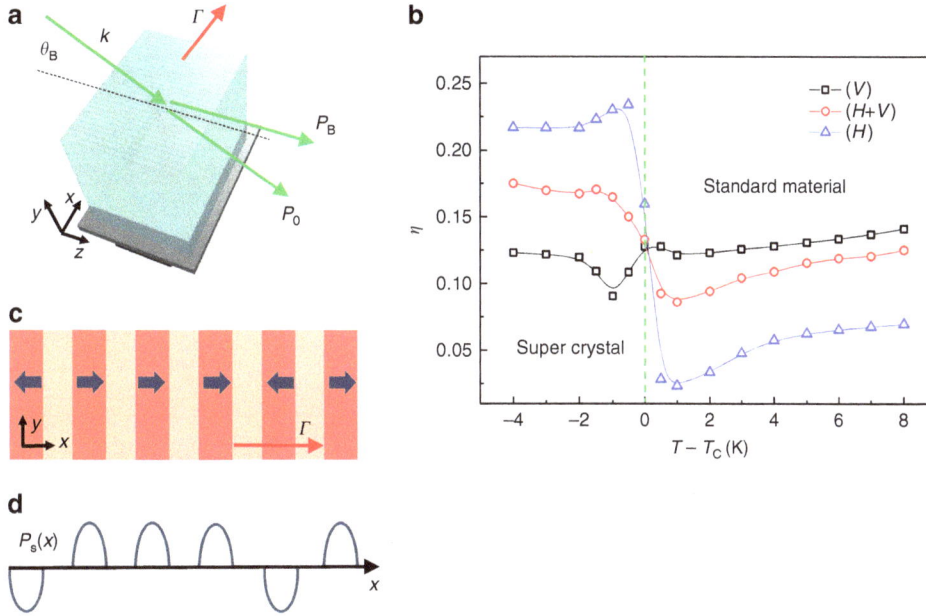

Figure 3 | Diffractive behaviour of the 1D supercooled superlattice. (a) Sketch of the experimental geometry and **(b)** detected diffraction efficiency (dots) as a function of temperature in the proximity of ferroelectric transition for different wave polarizations. An anomaly appears crossing T_C for H-polarized light signalling the emergence of the super-crystal. Lines are interpolations serving as guidelines. **(c)** Scheme of the periodically ordered ferroelectric state along the x direction undelying the super-crystal phase for $T < T_C$ and giving the spontaneous polarization $P_S(x)$ sketched in **d**.

$g = 0.16\,\mathrm{m^4\,C^{-2}}$. The resonant response at θ_B and the absence of higher harmonics (Fig. 2b) indicate that this $P_S(x)^2$ distribution is sinusoidal with wavevector Γ. Hence, although in general it may be that macroscopically $\langle \mathbf{P} \rangle \simeq 0$, it turns out that $\langle \mathbf{P}^2 \rangle \simeq P_S^2 \neq \langle \mathbf{P} \rangle^2 \neq 0$ on the micrometric scales, in analogy with optical response in crystals affected by polar nanoregions[25,27,40]. Optical diffraction efficiency reported in Fig. 3b then occurs considering $\eta = \sin^2\left(\frac{\pi d(\delta n)}{\lambda \cos\theta_B}\right)$, with resonant enhanced diffraction for $T < T_C$ caused by $\delta n = \delta n_0 + \delta n(P)$, where $\delta n_0 \sim 10^{-4}$ is the polarization-independent index change due to the periodic composition variation (Sellmeier's index change).

Electro-optical diffraction analysis. To validate this picture, we perform electro-optic diffractometry experiments, in which a macroscopic polarization activating the nonlinear periodic response is induced via an external static field E applied along x. Results are reported in Fig. 4; in particular, in Fig. 4a the polarization and field dependence of η are shown at $T = T_C + 2\,\mathrm{K}$. We observe a nearly field-independent behaviour for V-polarized light, which arises from its low electro-optic coupling (bias field and light polarization are orthogonal, $g = -0.02\,\mathrm{m^4\,C^{-2}}$); differently, η_H increases with the field showing a 'discontinuity' at the critical field $E_C = (1.4 \pm 0.1)\,\mathrm{kV\,cm^{-1}}$. The strong similarity between this enhancement and those observed under unbiased conditions at T_C (Fig. 3b) indicates that E_C coincides with the coercive field, and the discontinuity corresponds to the field-induced phase transition[16,26,35]. In fact, in Fig. 4b we repeat this experiment, enhancing the experimental field sensitivity and acquiring data also for decreasing field amplitudes. The result is a partial hysteretic loop for the diffraction efficiency that demontrates the field-induced transition and underlines that, both in the linear and nonlinear (electro-optic) case, the effect of the seeded ferroelectric ordering is to provide a periodic spontaneous polarization along x. We also note a slight asymmetry with respect to positive/negative fields; this is associated with a residual fixed space–charge field that may play an important role in the spontaneous polarization alignment

process and hence in leading to a residual $\langle \mathbf{P} \rangle \neq 0$. The existence of a periodic spontaneous polarization distribution in the superlattice (Fig. 3c) is confirmed in Fig. 4c, where electro-optic Bragg diffraction below T_C is reported. An oscillating full-hysteretic behaviour is observed as a function of the external field, consistently with the prediction $\eta(E) = \sin^2\left(\frac{\pi d(\delta n(E))}{\lambda \cos\theta_B}\right)$ with $\delta n(E) = \delta n_0 + (1/2)n^3 g\left(P_S^2 + 2\varepsilon_0 \chi \langle P_S \rangle E + \varepsilon_0^2 \chi^2 E^2\right)$. The increase in η due to the superlattice polarization allows us to explore its full sinusoidal behaviour, which usually requires extremely large fields in the paraelectric phase and reduces to a parabolic behaviour (Fig. 4d)[41]. From this parabolic behaviour detected at $T = T_C + 5\,\mathrm{K}$ we estimate that the resulting amplitude in the point-dependent Curie temperature due to the compositional modulation is $\Delta T_C \simeq 2\,\mathrm{K}$ (ref. 32). Agreement with the periodic polarization model is further stressed by deviations emerging in $\eta(E)$, especially for low and negative increasing fields, where the dependence on $\langle P_S \rangle$ makes observations weakly dependent on the specific experimental realization.

Discussion

An interesting point arising from the experimental results and analysis is how the periodically ordered polarization state along the x direction leads to the super-crystal. Since we pass spontaneously from a metastable to a stable mesoscopic phase, polar-domain dynamics in the presence of the fixed spatial scale Λ play a key role. In fact, we note that the 1D superlattice sketched in Fig. 3c involves the appearence of charge density and associated strains between polar planes, so that the ferroelectric crystal naturally tends to relax into a more stable configuration. In standard perovskites, equilibrium configurations are mainly those involving a 180° and 90° orientation between adjacent polar domains, as schematically shown in Fig. 5a. To explain the 3D polar state and its periodical features underlying the super-crystal, we consider the 90° configuration, which is characterized by 45° domain walls that we observe in a disordered configuration during the ferroelectric phase transition at T_C (Fig. 2f). Owing to the periodic constraint along the x axis, this arrangement has the

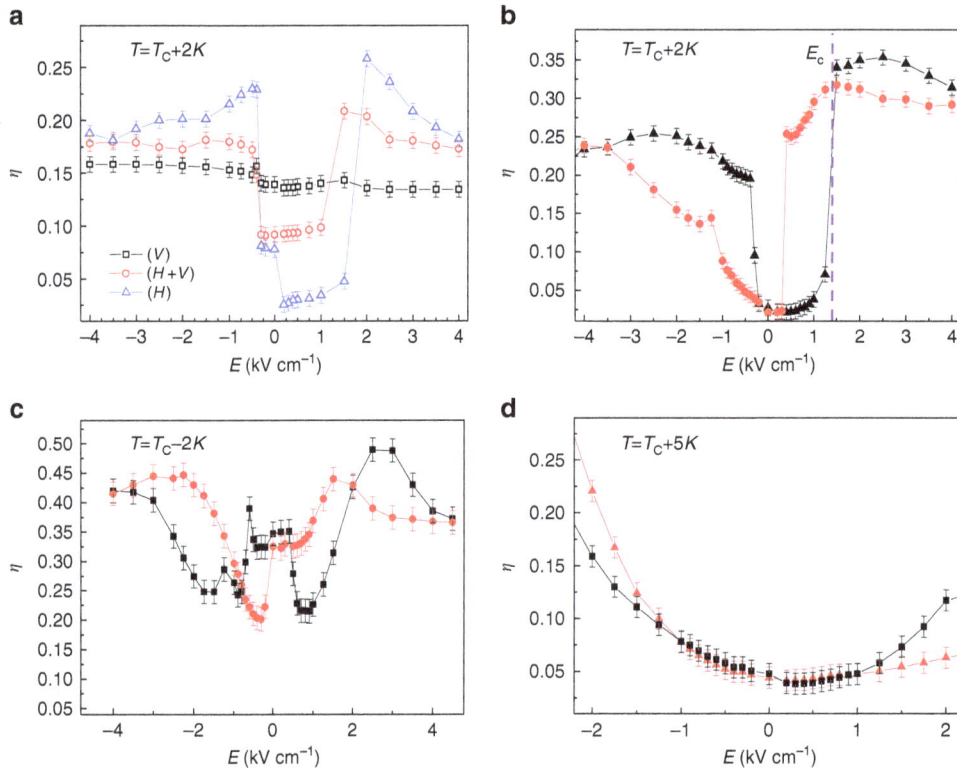

Figure 4 | Electro-optic Bragg diffraction in the critical region. (**a**) Diffraction efficiency as a function of the external applied field for different light polarizations at $T = T_C + 2$ K; (**b**) hysteresis loop at the same temperature and (**c**) at $T = T_C - 2$ K for H-polarization. (**d**) Expected [32] weak-histeretic paraelectric (parabolic) behaviour at $T = T_C + 5$ K. In **b–d**, black and red dots indicate data obtained, respectively, increasing and decreasing the bias fields. Lines are interpolations serving as guidelines. Error bars are given by the statistics on five experimental realizations.

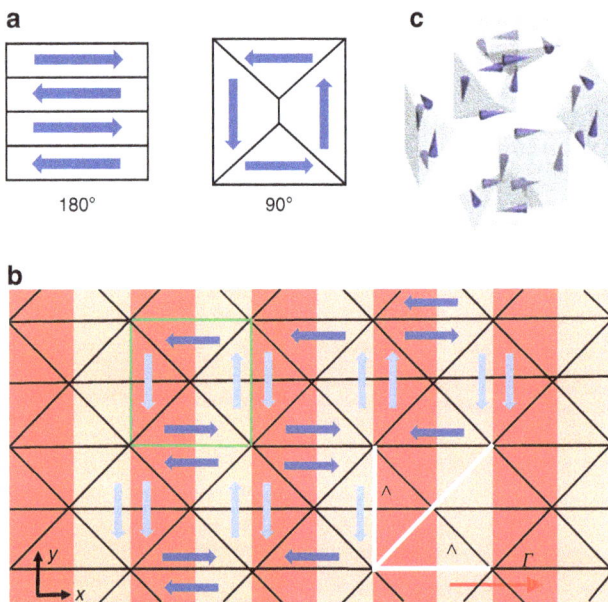

Figure 5 | Polar-domain configuration underlying the 3D superlattice.
(**a**) Typical 180° and 90° domain configurations in perovskite ferroelectrics. (**b**) Planar domain arrangement scheme in the stable super-crystal phase obtained with elementary blocks of 90° configurations (green cell). In this periodically ordered ferroelectric state, the compositional modulation (as for Fig. 3c), other domain walls ruling optical diffractometry (black lines), and periods along x, y and xy axes (white bars) are highlighted. Vertical polarizations have a lighter colour to stress their weak optical response in our KLTN sample. (**c**) Extension of the single unit cell (green cell in **b**) in three dimensions.

unique property of reproducing our observations, minimizing energy associated to internal charge density and transferring the built-in 1D order to the whole volume with the same spatial scale Λ. We illustrate the domain pattern in Fig. 5b for the $x - y$ plane, whereas in Fig. 5c the elementary cell is shown in the 3D case, where it maintains its stability features in terms of charge density energy. In particular, in Fig. 5b, domain walls resulting in the diffraction orders of Fig. 1b are marked, as well as the 45° correlation period, which agrees with optical observations of the reciprocal space. We further stress that vertical domains (light blue in Fig. 5b) are optically analoguous to paraelectric regions; moreover, 180° rotations in the polarization direction in each polar region has no effect on the optical response. In view of the symmetry of this arrangement, the observed diffraction aniso-tropy (Fig. 1d) is then associated to the absence of grating planes in the $y - z$ face.

Further insight on the 3D domain structure requires numerical simulations based on Monte Carlo methods[42,43] and phase-field models[44–47]; they may confirm our picture and reveal new aspects for ferroelectricity, such as polar dynamics, spontaneous long-range ordering and the role of polar strains in composite ferroelectrics with built-in compositional microstructures. In fact, the effect of the composition profile is here crucial in triggering the spontaneous formation of the macroscopic coherent structure, as it sets the typical domain size along the x direction and so rules the whole dynamic towards the equilibrium state. We expect that a different amplitude and period of the modulation may affect the formation, stability, time- and temperature dynamics of the super-crystal; indeed, the parameters of the compositional gradient may be important in determining the interaction between polar regions. Advanced growth techniques[32] can open future perpectives in this direction, as well as towards composite ferroelectrics with different compositional shapes of fundamental and applicative interest.

To conclude, we have reported the formation of a mesoscopic polarization super-crystal in a nanodisordered sample of KLTN. The large-scale coherent state is triggered by a periodically modulated change in composition. Our results show how ferroelectricity can be arranged into new phases, so that in proximity of an average critical temperature a structural order can emerge with a micrometric lattice constant so as to cause light to suffer diffraction as occurs for X-rays in standard crystals. The effect not only opens new avenues in the optical exploration of critical properties and large-scale structures in disordered systems, but also suggests methods to predict and engineer new states of matter. It can also have an impact on the development of innovative technologies, such as nonvolatile electronic and optical structured memories[2–4], microstructured piezo devices and spatially resolved miniaturized electro-optic devices[27,41,48].

Methods

Growth and properties of the microstructured KLTN sample.
We consider a compositionally disordered perovskite of KLTN, $K_{1-x}Li_xTa_{1-\beta}Nb_\beta O_3$ with $\alpha = 0.04$ and $\beta = 0.38$, grown through the top-seeded solution method by extracting a zero-cut 2.4 mm by 2.0 mm by 1.7 mm, along the x, y, z directions, respectively, optical-quality specimen. It shows, through low-frequency dielectric spectroscopy measurements, the spatial-averaged Curie point, which signals the transition from the high-temperature symmetric paraelectric phase to the low-temperature ferroelectric phase, at the room temperature $T_C = 294$ K. A 1D seed microstructure is embedded into the sample as it is grown through the off-centre growth technique so as to manifest a sinusoidal variation in the low-frequency dielectric constant, and thus in the critical temperature T_C, along the growth axis (x direction)[33,35]. This dielectric volume microstructure causes an index of refraction oscillation of period $\Lambda = 5.5\,\mu m$, which is able to diffract light linearly and electro-optically[49]. Details on the technique employed in the sample growth can be found in ref. 33. We note here that the composition amplitude of the periodic microstructure can be estimated from $\Delta\beta/\Delta T$, where $\Delta\beta$ is the amplitude variation in niobium composition and ΔT is the change in the growth temperature incurred by the off-centre rotation. At the growth temperature of $\sim 1{,}470$ K, the ratio $\Delta\beta/\Delta T \approx 0.35$ mol K^{-1} has been extracted from the phase diagram of KTN. The temperature variation incurred by the off-centre rotation was measured to be 3 K, from which we obtain $\Delta\beta \approx 1.05\%$ mol.

Optical diffraction experiments.
The macroscopic linear and electro-optic diffractive properties of the crystal have been investigated launching low-power (mW) plane waves at $\lambda = 532$ nm that propagate normally and parallelly to the grating vector Γ ($\Gamma = 2\pi/\Lambda$), which is along the x direction (Fig. 3a). Light diffracted by the medium is detected using a broad-area CCD (charge-coupled device) camera placed at $d = 0.2$ m from the crystal output facet or collected into Si power meters. In real-space measurements (Fig. 2d–h), the output crystal facet is imaged on the CCD camera and a cross-polarizer set-up[25,27] has been used to highlight contrast due to polarization inhomogeneities. The time needed to obtain a fully correlated state corresponding to the 3D super-crystal depends on the cooling rate τ and on the details of the thermal environment. Considering, for instance, as a thermal protocol a cooling rate $\tau = 0.05$ K s^{-1} and an environment at $T = T_C + 1$ K (weak thermal gradients), we have found that the metastable 1D lattice state at $T = T_C - 2$ K (Fig. 2b), in which correlations involve mainly in the direction including the Γ vector, lasts ~ 1 h. In this stage, although no macroscopic order occurs in the other directions[50], we observe optimal optical transmission of the sample (Fig. 2g); output light is not affected by scattering related to the existence of random domains and this undelines the presence of a mesoscopic ordering process in which the typical domain size is set. As regards the inspected temperature range, we have found that the super-crystal forms for temperatures till $T = 288$ K, although correlations are weaker at the lower temperatures. This is consistent with the fact that at these temperatures also the regions with a lower local T_C are well below the transition point.

References

1. Rabe, K. M., Ahn, C. H. & Triscone, J. M. *Physics of Ferroelectrics: a Modern Perspective*, Vol. 105 (Springer Science & Business Media, 2007).
2. Choi, K. J. *et al.* Enhancement of ferroelectricity in strained BaTiO3 thin films. *Science* **306,** 1005–1009 (2004).
3. Lee, H. N., Christen, H. M., Chisholm, M. F., Rouleau, C. M. & Lowndes, D. H. Strong polarization enhancement in asymmetric three-component ferroelectric superlattices. *Nature* **433,** 395–399 (2005).
4. Garcia, V. *et al.* Giant tunnel electroresistance for non-destructive readout of ferroelectric states. *Nature* **460,** 81–84 (2005).
5. Kim, W.-H., Son, J. Y., Shin, Y.-H. & Jang, H. M. Imprint control of nonvolatile shape memory with asymmetric ferroelectric multilayers. *Chem. Mater.* **26,** 6911–6914 (2014).
6. Dawber, M., Rabe, K. M. & Scott, J. F. Physics of thin-film ferroelectric oxides. *Rev. Mod. Phys.* **77,** 1083 (2005).
7. Catalan, G. *et al.* Polar domains in lead titanate films under tensile strain. *Phys. Rev. Lett.* **96,** 127602 (2006).
8. Catalan, G. *et al.* Flexoelectric rotation of polarization in ferroelectric thin films. *Nat. Mater.* **10,** 963–967 (2011).
9. Biancoli, A., Fancher, C. M., Jones, J. L. & Damjanovic, D. Breaking of macroscopic centric symmetric in paraelectric phases of ferroelectric materials and implications for flexoelectricity. *Nat. Mater.* **14,** 224–229 (2014).
10. Bousquet, E. *et al.* Improper ferroelectricity in perovskite oxide artificial superlattices. *Nature* **452,** 732–736 (2008).
11. Callori, S. J. *et al.* Ferroelectric PbTiO3/SrRuO3 superlattices with broken inversion symmetry. *Phys. Rev. Lett.* **109,** 067601 (2012).
12. Lawes, G. *et al.* Magnetically driven ferroelectric order in Ni3V2O8. *Phys. Rev. Lett.* **95,** 087205 (2005).
13. Baledent, V. *et al.* Evidence for room temperature electric polarization in RMn2O5 multiferroics. *Phys. Rev. Lett.* **114,** 117601 (2015).
14. DelRe, E., Spinozzi, E., Agranat, A. J. & Conti, C. Scale-free optics and diffractionless waves in nanodisordered ferroelectrics. *Nat. Photon.* **5,** 39–42 (2011).
15. DelRe, E. *et al.* Subwavelength anti-diffracting beams propagating over more than 1,000 Rayleigh lengths. *Nat. Photon.* **9,** 228–232 (2015).
16. Pierangeli, D., DiMei, F., Conti, C., Agranat, A. J. & DelRe, E. Spatial rogue waves in photorefractive ferroelectrics. *Phys. Rev. Lett.* **115,** 093901 (2015).
17. Di Mei, F. *et al.* Observation of diffraction cancellation for nonparaxial beams in the scale-free-optics regime. *Phys. Rev. A* **92,** 013835 (2015).
18. Shvartsman, V. V. & Lupascu, D. C. Lead-free relaxor ferroelectrics. *J. Am. Ceram. Soc.* **95,** 1–26 (2012).
19. Kutnjak, Z., Blinc, R. & Petzelt, J. The giant electromechanical response in ferroelectric relaxors as a critical phenomenon. *Nature* **441,** 956–959 (2006).
20. Glinchuk, M. D., Eliseev, E. & Morozovska, A. Superparaelectric phase in the ensemble of noninteracting ferroelectric nanoparticles. *Phys. Rev. B* **78,** 134107 (2008).
21. Pirc, R. & Kutnjak, Z. Electric-field dependent freezing in relaxor ferroelectrics. *Phys. Rev. B* **89,** 184110 (2014).
22. Manley, M. E. *et al.* Phonon localization drives polar nanoregions in a relaxor ferroelectric. *Nat. Commun.* **5,** 3683 (2014).
23. Phelan, D. *et al.* Role of random electric fields in relaxors. *Proc. Natl Acad. Sci. USA* **111,** 1754–1759 (2014).
24. Chang, Y. C., Wang, C., Yin, S., Hoffman, R. C. & Mott, A. G. Giant electro-optic effect in nanodisordered KTN crystals. *Opt. Lett.* **38,** 4574–4577 (2013).
25. Pierangeli, D. *et al.* Observation of an intrinsic nonlinearity in the electro-optic response of freezing relaxors ferroelectrics. *Opt. Mater. Express* **4,** 1487–1493 (2014).
26. Tian, H. *et al.* Double-loop hysteresis in tetragonal KTa0.58Nb0.42O3 correlated to recoverable reorientations of the asymmetric polar domains. *Appl. Phys. Lett.* **106,** 102903 (2015).
27. Tian, H. *et al.* Dynamic response of polar nanoregions under an electric field in a paraelectric KTa0.61Nb0.39O3 single crystal near the para-ferroelectric phase boundary. *Sci. Rep.* **5,** 13751 (2015).
28. Wemple, S. H. & DiDomenico, M. Oxygen-octahedra ferroelectrics. II. electro-optical and nonlinear-optical device applications. *J. Appl. Phys.* **40,** 735–752 (1969).
29. Sakamoto, T., Sasaura, M., Yagi, S., Fujiura, K. & Cho, Y. In-plane distribution of phase transition temperature of KTa1-xNbxO3 measured with single temperature sweep. *Appl. Phys. Express* **1,** 101601 (2008).
30. Li, H., Tian, H., Gong, D., Meng, Q. & Zhou, Z. High dielectric tunability of KTa0.60Nb0.40O3 single crystal. *J. App. Phys.* **114,** 054103 (2013).
31. Tian, H. *et al.* Variable gradient refractive index engineering: design, growth and electro-deflective application of KTa1-xNbxO3. *J. Mater. Chem. C* **3,** 10968–10973 (2015).
32. Agranat, A. J., Kaner, R., Perpelitsa, G. & Garcia, Y. Stable electro-optic striation grating produced by programmed periodic modulation of the growth temperature. *Appl. Phys. Lett.* **90,** 192902 (2007).
33. de Oliveira, C. E. M., Orr, G., Axelrold, N. & Agranat, A. J. Controlled composition modulation in potassium lithium tantalate niobate crystals grown by off centered TSSG method. *J. Cryst. Growth.* **273,** 203–206 (2004).
34. Lines, M. E. & Glass, A. M. *Principles and Applications of Ferroelectrics and Related Materials* (Oxford University Press, 1977).
35. Agranat, A. J., deOliveira, C. E. M. & Orr, G. Dielectric electrooptic gratings in potassium lithium tantalate niobate. *J. Non-Cryst. Sol.* **353,** 4405–4410 (2007).
36. Ramachandran, G. N. *Advanced Methods of Crystallography* (Academic Press, 1964).

37. Weber, M. F., Stover, C. A., Gilbert, L. R., Nevitt, T. J. & Ouderkirk, A. J. Giant birefringent optics in multilayer polymer mirrors. *Science* **287**, 2451–2456 (2000).

38. Kogelnik, H. Coupled wave theory for thick hologram gratings. *Bell Syst. Tech. J.* **48**, 2909–2947 (1969).

39. Bitman, A. *et al.* Electroholographic tunable volume grating in the g44 configuration. *Opt. Lett.* **31**, 2849–2851 (2006).

40. Gumennik, A., Kurzweil-Segev, Y. & Agranat, A. J. Electrooptical effects in glass forming liquids of dipolar nano-clusters embedded in a paraelectric environment. *Opt. Mater. Express* **1**, 332–343 (2011).

41. Wang, L. *et al.* Field-induced enhancement of voltage-controlled diffractive properties in paraelectric iron and manganese co-doped potassium-tantalate-niobate crystal. *Appl. Phys. Express* **7**, 112601 (2014).

42. Potter, Jr B. G., Tikare, V. & Tuttle, B. A. Monte Carlo simulation of ferroelectric domain structure and applied field response in two dimensions. *J. Appl. Phys.* **87**, 4415–4424 (2000).

43. Li, B. L., Liu, X. P., Fang, F., Zhu, J. L. & Liu, J. M. Monte Carlo simulation of ferroelectric domain growth. *Phys. Rev. B* **73**, 014107 (2006).

44. Chen, L. Q. Phase-field method of phase transitions/domain structures in ferroelectric thin films: a review. *J. Am. Ceram. Soc.* **91**, 1835–1844 (2008).

45. Li, Y. L., Hu, S. Y., Liu, Z. K. & Chen, L. Q. Phase-field model of domain structures in ferroelectric thin films. *Appl. Phys. Lett.* **78**, 3878–3880 (2001).

46. Chu, P. *et al.* Kinetics of 90 domain wall motions and high frequency mesoscopic dielectric response in strained ferroelectrics: a phase-field simulation. *Sci. Rep.* **4**, 5007 (2014).

47. Li, Q. *et al.* Giant elastic tunability in strained BiFeO3 near an electrically induced phase transition. *Nat. Commun.* **6**, 8950 (2015).

48. Parravicini, J. *et al.* Volume integrated phase-modulator based on funnel waveguides for reconfigurable miniaturized optical circuits. *Opt. Lett.* **40**, 1386–1389 (2015).

49. Pierangeli, D. *et al.* Continuous solitons in a lattice nonlinearity. *Phys. Rev. Lett.* **90**, 203901 (2015).

50. Kounga, A. B., Granzow, T., Aulbach, E., Hinterstein, M. & Rödel, J. High-temperature poling of ferroelectrics. *J. App. Phys.* **107**, 024116 (2008).

Acknowledgements

The research leading to these results was supported by funding from grants PRIN 2012BFNWZ2 and Sapienza 2014 Projects. A.J.A. acknowledges the support of the Peter Brojde Center for Innovative Engineering.

Author contributions

D.P., M.F. and E.D. conceived and developed experiments and theory and wrote the article; A.J.A. and C.E.M.d.O. designed and fabricated the KLTN samples and participated in the interpretation of results; and F.D.M. and G.D.D. participated in the experiments, data analysis and interpretation of results.

Additional information

Competing financial interests: The authors declare no competing financial interests.

9

Efficient and stable perovskite solar cells prepared in ambient air irrespective of the humidity

Qidong Tai[1,2], Peng You[1], Hongqian Sang[2], Zhike Liu[1], Chenglong Hu[2], Helen L.W. Chan[1] & Feng Yan[1]

Poor stability of organic–inorganic halide perovskite materials in humid condition has hindered the success of perovskite solar cells in real applications since controlled atmosphere is required for device fabrication and operation, and there is a lack of effective solutions to this problem until now. Here we report the use of lead (II) thiocyanate (Pb(SCN)$_2$) precursor in preparing perovskite solar cells in ambient air. High-quality CH$_3$NH$_3$PbI$_{3-x}$(SCN)$_x$ perovskite films can be readily prepared even when the relative humidity exceeds 70%. Under optimized processing conditions, we obtain devices with an average power conversion efficiency of 13.49% and the maximum efficiency over 15%. In comparison with typical CH$_3$NH$_3$PbI$_3$-based devices, these solar cells without encapsulation show greatly improved stability in humid air, which is attributed to the incorporation of thiocyanate ions in the crystal lattice. The findings pave a way for realizing efficient and stable perovskite solar cells in ambient atmosphere.

[1] Department of Applied Physics, The Hong Kong Polytechnic University, Hung Hom, 999077 Kowloon, Hong Kong. [2] Institute for Interdisciplinary Research and Key Laboratory of Optoelectronic Chemical Materials and Devices of Ministry of Education, Jianghan University, 430056 Wuhan, China. Correspondence and requests for materials should be addressed to F.Y. (email: apafyan@polyu.edu.hk).

Organolead halide ($CH_3NH_3PbX_3$, X = Cl, Br or I) perovskite solar cells have stepped into the spotlight of solar cell community for an unprecedented increase of their efficiencies from 3.8% to over 20% in <5 years[1-11]. $CH_3NH_3PbX_3$ can be synthesized by the reaction of PbX_2 and CH_3NH_3X via one- or two-step solution processing methods[3,4], vapour deposition methods[5] or vapour-assisted solution methods[12]. Besides the unique features of strong light absorption, ambipolar charge transport, long carrier lifetime and solution processability, $CH_3NH_3PbX_3$ perovskites have distinguished themselves from other solar cell materials for being compatible with a wide range of device architectures[11,13,14].

Despite the success in obtaining excellent photovoltaic performance, the instability of $CH_3NH_3PbX_3$ to water and ambient moisture is still an open problem[15]. In most cases, perovskite films have to be processed in inert atmosphere and bare devices cannot survive long in air, which hamper the real production and applications of the perovskite solar cells. So far, effective solutions to this issue still lack although some related studies have been carried out[16-18]. For example, water-resistive coating can be used to improve the stability of perovskite solar cells; however, the intrinsic vulnerability of perovskites to moisture remains unchanged[19]. Better moisture stability was observed for a two-dimensional, layered perovskite formed by incorporating phenylethylammonium ($C_6H_5(CH_2)_2NH_3^+$) into $CH_3NH_3PbI_3$ matrix. Such a perovskite film could be fairly fabricated in ambient condition, while its photovoltaic performance was less satisfactory[18]. Similar observations were also reported for mixed halide perovskites $CH_3NH_3PbI_{3-x}Br_x$, when 10–15 mol% I^- was substituted by Br^-. The enhanced stability was attributed to the stronger interaction between Br^- and $CH_3NH_3^+$ (refs 6,16). These studies suggest that ion doping might be a possible way to improve the inherent stability of perovskite films.

It is also perceived that both the photovoltaic performance and stability of perovskite films are closely related to the film quality that is governed by the methods and materials used. To date, PbI_2 and $PbCl_2$ have been ubiquitously used to prepare the state-of-art $CH_3NH_3PbI_3$ perovskite films, and in the latter case, the perovskite is often referred as $CH_3NH_3PbI_{3-x}Cl_x$ although the existence of Cl in the final film is still under debate[20,21]. Recent studies showed that non-halide lead precursors could also be used

to prepare high-quality perovskite films[22-25]. Although no superior device performance has been obtained yet, such precursors would lead to versatile approaches for the preparation of novel perovskite films with different optoelectronic properties and air stability.

Here we report a type of efficient and stable perovskite solar cells by using lead (II) thiocyanate ($Pb(SCN)_2$) precursor. The perovskite layer was formed on mesoporous TiO_2 (mp-TiO_2) scaffold in ambient air by a two-step sequential deposition method, as shown in Fig. 1a. Fine perovskite films can be readily prepared in our lab even when the average relative humidity (RH) exceeds 70%, suggesting excellent moisture stability of the perovskite. Upon optimization, the devices demonstrated average power conversion efficiencies (PCEs) over 13%, together with the currently maximum value higher than 15%. More importantly, these devices without encapsulation showed much better stability in ambient air than typical perovskite solar cells prepared from PbI_2 precursor.

Results

Preparation of perovskite solar cells. As shown in Fig. 1a, dimethylsulfoxide (DMSO) was used as a solvent to prepare $Pb(SCN)_2$ films, and highly dense and uniform film morphology was obtained (Fig. 1b), which should be beneficial for obtaining high-quality perovskite films. In contrast, a quite poor film morphology was found in the $Pb(SCN)_2$ film derived from the commonly used solvent N,N'-dimethylformamide (Supplementary Fig. 1). After reacting with CH_3NH_3I, both the colour and morphology of $Pb(SCN)_2$ films changed strikingly (Fig. 1c). The resulting films exhibited quite similar absorption spectrum and X-ray diffraction pattern to the state-of-art tetragonal $CH_3NH_3PbI_3$ perovskite (Fig. 1d,e; ref. 4). It is notable that both X-ray diffraction and Raman spectroscopy (Supplementary Fig. 2) do not show any peak of $Pb(SCN)_2$ after the reaction, suggesting a complete conversion of $Pb(SCN)_2$ into perovskite in the film. On the other hand, S was clearly found in the final perovskite films of different batches under X-ray photoelectron spectroscopy (XPS; Supplementary Fig. 3). The feature characteristic of S $2p$ core electrons was observed, and the concentration of S was estimated to be 2.7–8.4 mol% with respect to I. We speculate that the observed S atoms are from SCN^-

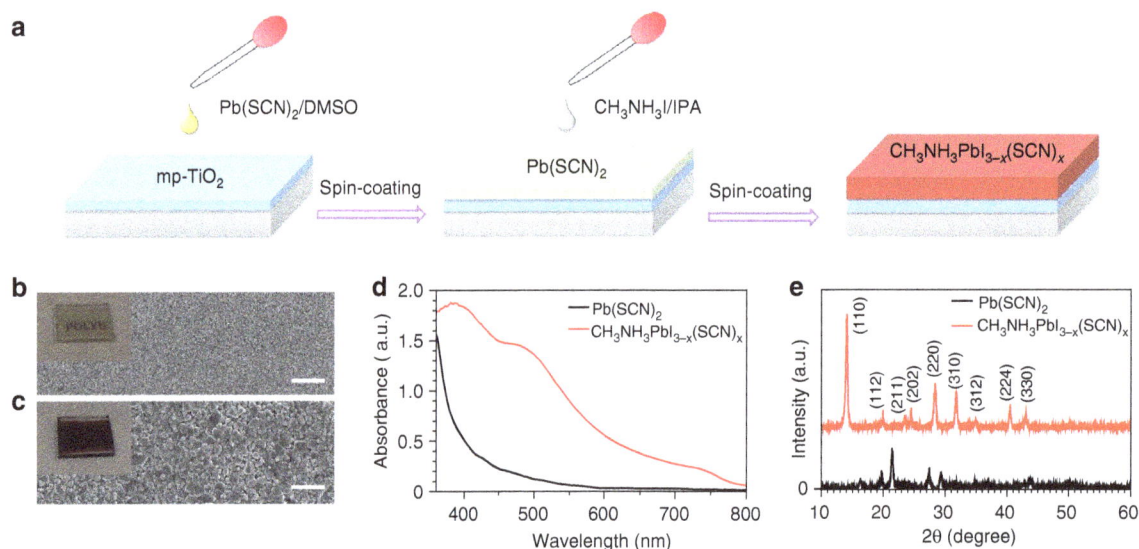

Figure 1 | Fabrication and characterization of $CH_3NH_3PbI_{3-x}(SCN)_x$ perovskite film. (**a**) Schematic illustration of the method used for preparing perovskite films. Digital and SEM images of (**b**) $Pb(SCN)_2$ film and (**c**) the resulting perovskite film ($CH_3NH_3PbI_{3-x}(SCN)_x$). (**d**) Absorption spectra and (**e**) XRD patterns of $Pb(SCN)_2$ film and the resulting perovskite film ($CH_3NH_3PbI_{3-x}(SCN)_x$). Scale bars = 2 μm (**b,c**).

incorporated into $CH_3NH_3PbI_3$ perovskite, although the concentration is rather low. Considering that SCN^- is known as a pseudohalide group with its ionic radius (0.215–0.22 nm) very close to that of I^- (0.22 nm; ref. 26), it is possible for some of SCN^- to substitute I^- in the perovskite lattice. This speculation is also supported by recent studies in which the incorporation of BF_4^- (with a similar ionic radius of 0.218 nm) into $CH_3NH_3PbI_3$ perovskite was theoretically and experimentally confirmed[27,28]. Therefore, the perovskite prepared in this work will be noted as $CH_3NH_3PbI_{3-x}(SCN)_x$ hereafter. When the $CH_3NH_3PbI_{3-x}(SCN)_x$ film was left to degrade in moisture, the final product was mainly identified as PbI_2 (Supplementary Fig. 4), indicating that $Pb(SCN)_2$ follows similar reaction mechanism with $PbCl_2$ and lead acetate ($PbAc_2$), and most of the SCN^- ions have evaporated during the annealing of perovskite film[20,21,25].

Figure 2a–e shows the scanning electron microscope (SEM) images of the perovskite films prepared at different conditions. The spin-coating speed used for the deposition of $Pb(SCN)_2$ has a significant influence on the morphology and thickness of the final perovskite films. Flat polygonal-grained morphologies are observed for perovskite films prepared at the spin-coating speed higher than 3,000 r.p.m. and the grain size increases with decreasing speed. However, according to the cross-sectional views, incomplete coverage on mp-TiO_2 is found for the perovskite films prepared at the spin-coating speeds above 4,000 r.p.m., and 100% surface coverage (upper layer thickness: ~ 140 nm) is only achievable at the intermediate speed of 3,000 r.p.m. On the other hand, very rough structures are observed for perovskite films prepared at 2,000 and 1,500 r.p.m. with the upper layer thicknesses of ~ 240 and ~ 260 nm, respectively. Such a rough surface morphology is mainly due to the high viscosity of DMSO that cannot be spread well at low spin-coating speeds.

Then perovskite solar cells were fabricated with a device configuration of fluorine doped tin oxide (FTO)/compact TiO_2/mp-TiO_2/perovskite/2,2',7,7'-Tetrakis(N,N-di-p-methoxyphenyl amine)-9,9'-spirobifluorene (spiro-MeOTAD)/Au, as shown in Supplementary Fig. 5, and their photovoltaic performance was characterized systematically. The representative current density–voltage (J-V) characteristics, corresponding external quantum efficiencies (EQEs) and photovoltaic parameters, are presented in Fig. 2f,g. Perovskites prepared at the speeds of 5,000 r.p.m. and 4,000 r.p.m. give average open-circuit voltages (V_{OC}) of 0.99 ± 0.03 and 0.98 ± 0.01 V, short circuit current densities (J_{SC}) of 13.6 ± 0.6 and 16.3 ± 0.7 mA cm^{-2}, fill factors (FFs) of 0.575 ± 0.05 and 0.638 ± 0.007, and PCEs of $7.79 \pm 1.05\%$ and $10.24 \pm 0.32\%$, respectively. Improved J_{SC} of 17.7 ± 1.1 mA cm^{-2},

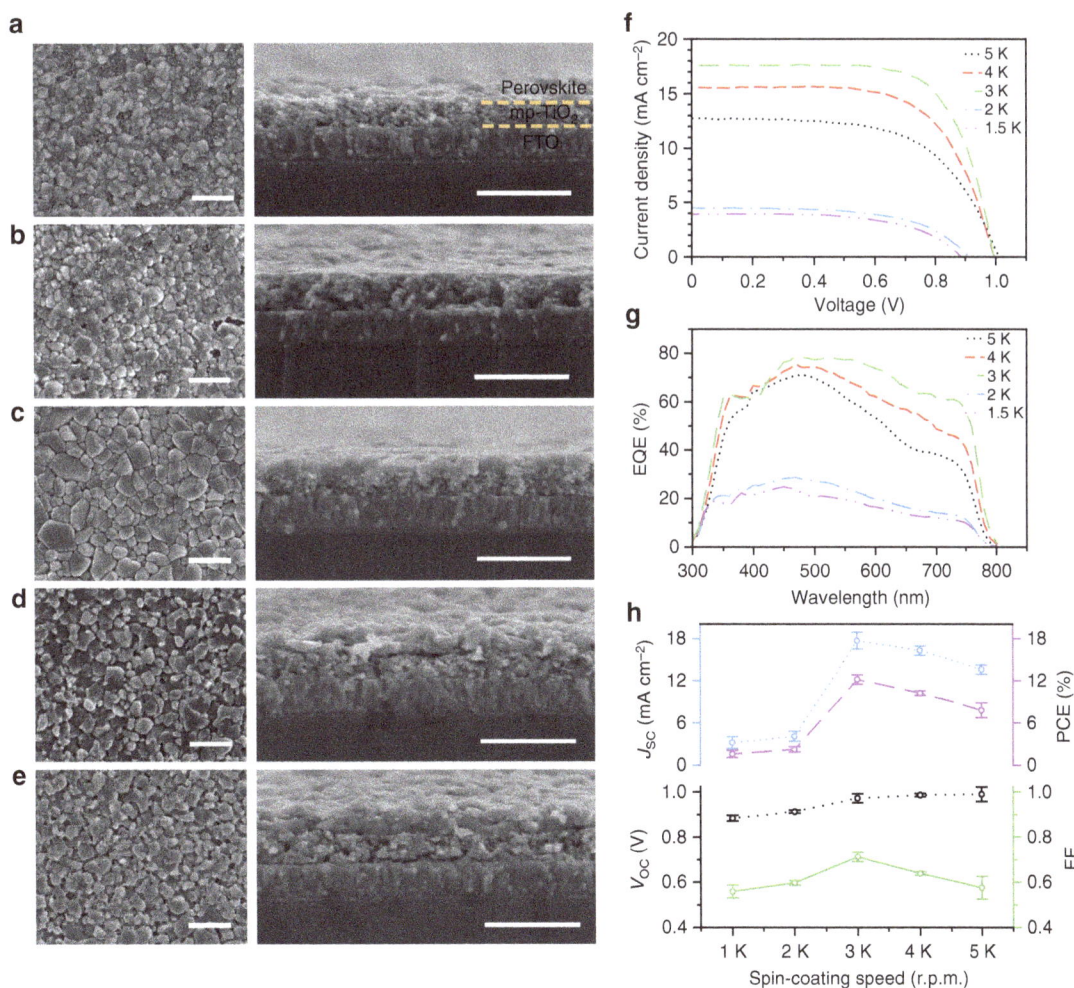

Figure 2 | Morphology and performance of $CH_3NH_3PbI_{3-x}(SCN)_x$ perovskite. SEM top view and cross-sectional view images of $CH_3NH_3PbI_{3-x}(SCN)_x$ films derived from $Pb(SCN)_2$ deposited on mp-TiO_2 films at the spin-coating speeds of (**a**) 5,000 r.p.m., (**b**) 4,000 r.p.m., (**c**) 3,000 r.p.m., (**d**) 2,000 r.p.m. and (**e**) 1,500 r.p.m., respectively. Scale bars $= 1 \mu m$ (**a–e**). The representative (**f**) J-V curves, (**g**) EQEs and (**h**) statistical photovoltaic parameters (J_{SC}, PCE, V_{OC} and FF) of $CH_3NH_3PbI_{3-x}(SCN)_x$ perovskite solar cells prepared at different spin-coating conditions. Error bars represent s.d. calculated from five devices prepared at the same conditions.

FF of 0.712 ± 0.021 and PCE of $12.20 \pm 0.66\%$ are obtained in the case of 3,000 r.p.m. while the change of V_{OC} is negligible. The increase of J_{SC} is attributed to the improved EQE values in the long wavelength region (500–800 nm) for the increase of the film thickness. Improved film coverage at the intermediate spin-coating speed accounts for the increase of FF as the interface carrier recombination is reduced. In contrast, striking drop of the PCE to $\sim 2\%$ is observed for perovskites prepared at 2,000 and 1,500 r.p.m., which is mainly caused by the reduction of J_{sc}. This phenomenon can be understood by the following reasons[29]: (1) the perovskite layer is too thick to have efficient charge separation; (2) the poor contact between crystals in the perovskite film may hamper the charge transport process; and (3) the rough surface of the perovskite film may cause an incomplete coverage of the hole transport layer, resulting in direct carrier recombination at the perovskite/Au interface.

Performance of the $CH_3NH_3PbI_{3-x}(SCN)_x$ perovskite solar cells. By further optimizing the fabrication conditions of mp-TiO_2 layer, higher PCEs can be obtained from devices fabricated in open air regardless of the ambient moisture. As shown in Fig. 3a,b, a device shows the PCE of 15.12% at reverse scan and 14.52% at forward scan, despite the fact that the RH of the ambient air is above 70%. The average PCE of 20 devices prepared at the same conditions is $13.49 \pm 1.01\%$ (Supplementary Fig. 6), suggesting an excellent moisture stability of $CH_3NH_3PbI_{3-x}(SCN)_x$ perovskite. We also noticed that the stabilized PCE of a $CH_3NH_3PbI_{3-x}(SCN)_x$ solar cell characterized in humid air is very close to that obtained from its $J-V$ curves (Supplementary Fig. 7).

To better illuminate the moisture effect, control $CH_3NH_3PbI_3$ solar cells were fabricated in air under the same conditions except using PbI_2 precursor with DMSO solvent. As shown in Fig. 3a, the control device shows the PCEs of 8.78% and 8.02% in the reverse scan and forward scan, respectively. The detailed photovoltaic parameters of the two devices are summarized in Supplementary Table 1. One possible reason for the low efficiency of the control device is the poor morphology of $CH_3NH_3PbI_3$ film caused by the ambient high RH, which is rough and full of pinholes. SEM images show that the uncovered regions in some places are larger than 1 μm scale (Supplementary Fig. 8). Such a poor morphology will cause severe recombination for the direct contact between hole-transporting material (HTM) and TiO_2. In contrast, the $CH_3NH_3PbI_3$ solar cells prepared in controlled atmosphere (N_2-filled glovebox) with the same PbI_2 precursor show higher efficiencies of $\sim 12\%$ due to the much improved film morphology (Supplementary Fig. 9). The above results demonstrate that the preparation of perovskite solar cells is more moisture resistive through $Pb(SCN)_2$ route than through PbI_2 route.

To further explore the origin of the different performance of $CH_3NH_3PbI_{3-x}(SCN)_x$- and $CH_3NH_3PbI_3$-based devices, impedance spectroscopy (IS) measurements were performed and the corresponding IS spectra are presented in Supplementary Fig. 10. As reported in the literature[30–33], the IS response of a perovskite solar cell is composed of three features (arcs in Nyquist plot). The first arc in high-frequency region is possibly related to the charge transport in the TiO_2 layer, HTM and/or at HTM/Au interface[30–34]; this feature may become ambiguous if the charge transport resistance (R_{HTM}) is low. The second arc in the intermediate-frequency region is associated with the charge transport-recombination process in the active film, from which the recombination resistance (R_{rec}) can be obtained and this feature is typically dominant in the IS response; The third arc in the low-frequency region that usually appears under a high applied voltage is correlated to slow dynamic processes in the perovskite film. Here the second arcs in the intermediate-frequency region can be clearly observed in all of the devices, which reflect the carrier recombination processes.

Figure 3c shows the R_{rec} values of the two devices under different applied voltages. Both devices show a similar decrease of R_{rec} with increasing bias voltage due to increased carrier densities[31,35], and a higher R_{rec} value is observed in the $CH_3NH_3PbI_{3-x}(SCN)_x$ solar cell at any bias voltage in

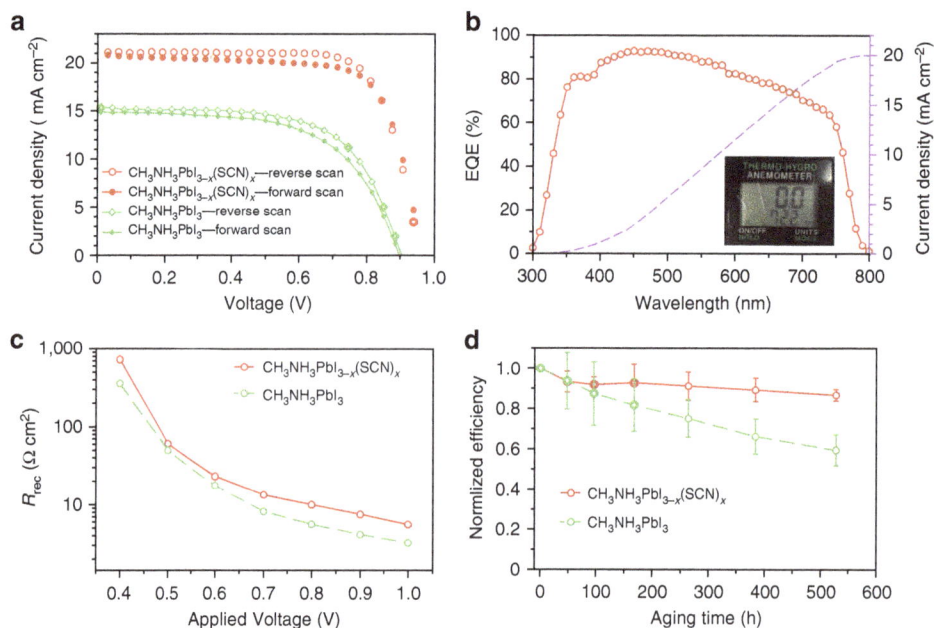

Figure 3 | Characteristics of $CH_3NH_3PbI_{3-x}(SCN)_x$ perovskite solar cells and control devices. (a) $J-V$ curves of $CH_3NH_3PbI_{3-x}(SCN)_x$ ($Pb(SCN)_2$-derived) and $CH_3NH_3PbI_3$ (PbI_2-derived)-based solar cells prepared in ambient air. **(b)** Corresponding EQE of the $CH_3NH_3PbI_{3-x}(SCN)_x$-based device. The inset photo shows the RH of the ambient air of 72.2%. **(c)** The representative recombination resistances (R_{rec}) determined from IS under different applied bias voltages. **(d)** Evolution of the PCEs of $CH_3NH_3PbI_{3-x}(SCN)_x$- and $CH_3NH_3PbI_3$-based solar cells upon ageing in air without encapsulation. Error bars represent s.d. calculated from five devices prepared at the same conditions.

comparison with the control device, thereby indicating a slower recombination rate in the former, which is favourable for device performance. Thus, the observation is consistent with the better performance of the $CH_3NH_3PbI_{3-x}(SCN)_x$ solar cell than the control device. The reduced recombination in the $CH_3NH_3PbI_{3-x}(SCN)_x$ solar cell can be attributed to two possible reasons. One is the pinhole-free morphology of the $CH_3NH_3PbI_{3-x}(SCN)_x$ film, which can avoid direct carrier recombination at HTM/TiO_2 interface. The other is the lower density of trap states in $CH_3NH_3PbI_{3-x}(SCN)_x$ than in $CH_3NH_3PbI_3$ film, as evidenced in the following experiments.

The carrier recombination processes in the perovskite films were characterized by time-resolved photoluminescence measurements (Supplementary Fig. 11). The lifetimes related to the recombination of free carriers in $CH_3NH_3PbI_{3-x}(SCN)_x$ and $CH_3NH_3PbI_3$ films are extracted to be 260.35 and 30.5 ns, respectively. Obviously, the 8.5-times-longer carrier lifetime in the former indicates lower trap density in the $CH_3NH_3PbI_{3-x}(SCN)_x$ film. In addition, it has been theoretically and experimentally proved that lower density of trap states can be found in perovskite film grown from iodide-free precursor than that grown from iodide-containing precursor[22]. Hence, we can conclude that the $CH_3NH_3PbI_{3-x}(SCN)_x$ solar cells have longer carrier lifetime and lower trap density than the control devices, which is a major reason for their better performance.

Stability of perovskite solar cells in humid air. Since excellent moisture resistance has been observed in $CH_3NH_3PbI_{3-x}(SCN)_x$ solar cells during the fabrication process, it is not surprising to find a good long-term ambient stability of them. As shown in Fig. 3d, $CH_3NH_3PbI_{3-x}(SCN)_x$ solar cells without encapsulation maintain 86.7% of the initial average PCE after being stored in open air with the average RH level above 70% for over 500 h. In comparison, the $CH_3NH_3PbI_3$ control devices lose nearly 40% of the initial average PCE (see the evolution of V_{OC}, J_{SC}, and FF in Supplementary Fig.12).

It has been reported that the morphology of a $CH_3NH_3PbI_3$ perovskite film may influence its stability in air[17]. Since the $CH_3NH_3PbI_{3-x}(SCN)_x$ film shows more uniform and compact surface morphology than the $CH_3NH_3PbI_3$ film prepared in air, the better morphology of the former is also a possible reason for the improved ambient stability of the devices. To clarify this issue, $CH_3NH_3PbI_3$ solar cells prepared in controlled atmosphere with the efficiency of ~15% were tested for hundreds of hours in ambient air (Supplementary Fig. 13). In these control devices, the morphology of the $CH_3NH_3PbI_3$ film is similar to that of the $CH_3NH_3PbI_{3-x}(SCN)_x$ film shown in Fig. 2c. However, the PCEs

of the devices decreased rapidly with time, indicating that the morphology of the films plays a minor role on the stability of the devices in air with high RH. These results clearly demonstrate that the intrinsic stability of the perovskite material is the dominating factor for the device stability in our experiments.

Although little J–V hysteresis can be found in a fresh $CH_3NH_3PbI_{3-x}(SCN)_x$ solar cell, it becomes obvious upon ageing of the device in ambient air. The similar change of J–V hysteresis has been observed in the ageing tests of $CH_3NH_3PbI_3$ solar cells in either this work or other reports[36]. Based on IS analysis, we are able to get an insight into the origin of such an ageing-dependent J–V hysteresis as well as the degradation mechanism of perovskite solar cells. Figure 4a displays the J–V curves of a $CH_3NH_3PbI_{3-x}(SCN)_x$ solar cell measured at three representative ageing times, that is, initial stage (0 h), middle stage (168 h), when the hysteresis starts to be obvious, and final stage (528 h). Figure 4b shows the corresponding IS spectra measured in dark with 1 V bias voltage, and the derived R_{HTM} and R_{rec} values are presented in Supplementary Table 2. Since the V_{OC} of the aged device decreased to ~0.9 V, the change of R_{HTM} and R_{rec} values were also measured with a 0.9-V applied voltage (Supplementary Fig. 14), and the trend is similar. Clearly, R_{HTM} increases gradually with ageing time, whereas R_{rec} goes in the opposite direction. As mentioned above, the former indicates an increased charge transport resistance in TiO_2 or Spiro-MeOTAD HTM layer or at the HTM/Au interface, whereas the latter means an increased recombination rate in the device. Both issues may lead to the degradation of the device performance.

Since the TiO_2 layer and Au electrode are very stable in air, the increased R_{HTM} is presumably caused by the hygroscopic nature of Li-bis(trifluoromethanesulfonyl)-imide (Li-TFSI), which is typically used as a p-type dopant to increase the conductivity of Spiro-MeOTAD[37,38]. Meanwhile, the Spiro-MeOTAD oxidation might also account for the increase of R_{HTM} (ref. 39). To clarify this point, a control experiment was conducted, in which the perovskite solar cells were kept in dry air and this would allow us to focus on the influence of air exposure on R_{HTM} without considering the moisture effect. We did not observe much change of the R_{HTM} values, even when the devices were exposed to dry air for 168 h (Supplementary Fig. 15). In contrast, the R_{HTM} values almost doubled in humid air (Fig. 4b), indicating that the Spiro-MeOTAD oxidation is not a reason for the increased R_{HTM}.

The reduction of R_{rec} upon ambient air exposure is related to the moisture-induced defects in the perovskite, which is probably caused or accelerated by the presence of Li-TFSI[19,37,38], and these defects account partially for the ageing-induced J–V hysteresis[36,40]. Besides, negative capacitance is observed in the aged device in the low-frequency region, which can enhance the

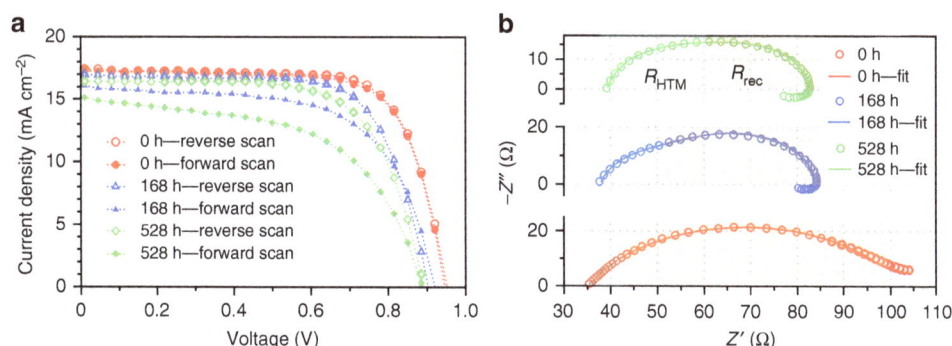

Figure 4 | Ageing tests of a $CH_3NH_3PbI_{3-x}(SCN)_x$ perovskite solar cell. (**a**) J–V curves and (**b**) IS spectra (1 V bias voltage in dark) of a $CH_3NH_3PbI_{3-x}(SCN)_x$ solar cell with characteristic ageing time of 0, 168 and 528 h. The high- and intermediate-frequency responses in the IS are fitted with the equivalent circuit shown in Supplementary Fig. 5, and the corresponding charge transfer resistance in the hole transport material (R_{HTM}) and the recombination resistance (R_{rec}) of the device are highlighted with different colours. The width of each region represents the value of the corresponding resistance.

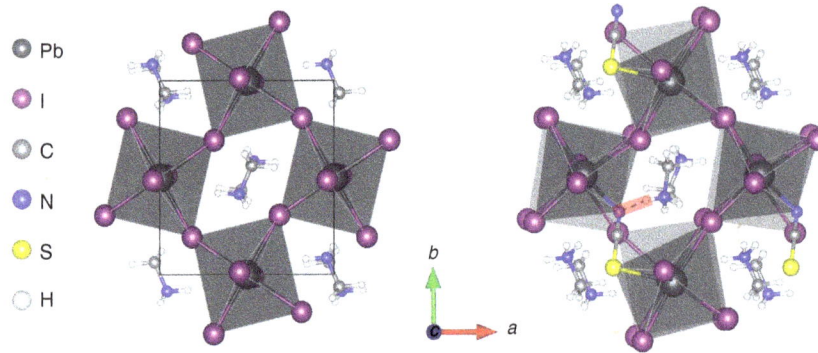

Figure 5 | Theoretical calculation of perovskite materials. Calculated crystal structures of $CH_3NH_3PbI_3$ (left) and $CH_3NH_3PbI_{3-x}(SCN)_x$ based on a chemical formula of $(CH_3NH_3)_4Pb_4I_{11}SCN$(right). The hydrogen bond between SCN^- and $CH_3NH_3^+$ (dash line) is highlighted with red colour.

hysteresis since it is closely related to a slow relaxation process in the device[32]. The physical origin of such a negative capacitance is still unclear, which may be related to the interfacial ion reorganization or dielectric relaxation in a perovskite film[30,32,33]. In our case, either origin should be associated with the air exposure that can induce more traps in the active layers.

Theoretical calculation of $CH_3NH_3PbI_{3-x}(SCN)_x$ perovskite. For a better understanding of the structural and chemical properties of $CH_3NH_3PbI_{3-x}(SCN)_x$, density functional theory calculations were conducted (Fig. 5). Here $x = 0.25$, corresponding to the ratio of SCN^- to I^- numbers in the lattice of $\sim 9\%$, which is close to the maximum molar ratio (8.4%) of SCN^- obtained by the XPS measurement. Our calculations show that the incorporation of SCN^- group into the perovskite lattice is thermodynamically stable. In case of $CH_3NH_3PbI_3$, orthorhombic structure is energetically favoured and the $CH_3NH_3^+$ aligns in parallel along b axis (the orthorhombic phase will change into tetragonal at $T > 162$ K due to the free rotation of $CH_3NH_3^+$ and the influence of temperature is not taken into account in our calculations[41]). The stabilized $CH_3NH_3PbI_{3-x}(SCN)_x$ shows a pseudo-orthorhombic structure, as the SCN^- group tends to align along the same direction as $CH_3NH_3^+$, resulting in a slight tilt of the crystal lattice (the lattice tilt could be greatly alleviated at room temperature owing to the small amount of SCN^- and the free rotation of $CH_3NH_3^+$). The band structure of $CH_3NH_3PbI_{3-x}(SCN)_x$ was calculated (Supplementary Fig. 16), which is similar to that of $CH_3NH_3PbI_3$. Fortunately, no electronic levels (traps) are introduced in the gap, which is critical to carrier recombination process in the perovskite material. This theoretical result is consistent with the observed long carrier lifetime of the material in the photoluminescence measurement. More importantly, our calculations indicate strong ionic interactions between SCN^- and adjacent Pb atoms, and hydrogen bonds can be formed between SCN^- and $CH_3NH_3^+$, which should contribute to improved chemical stability of perovskites. This is very similar to the case for Br-doped perovskite that shows improved air stability for enhanced chemical bonding in perovskite lattice[6,16].

Discussion

According to the above study, the much better moisture resistance of $CH_3NH_3PbI_{3-x}(SCN)_x$ film can be mainly attributed to its good intrinsic stability of the perovskite material. In a very recent study, Jiang *et al.*[42] reported an improved moisture stability of $CH_3NH_3PbI_3$ perovskite by replacing two I^- with SCN^-. Although the methodology, composition and performance of their perovskite solar cells are quite different

from our work, the finding of the better moisture stability of SCN^- containing perovskite is shared.

In conclusion, we report the use of $Pb(SCN)_2$ as a precursor for preparing perovskite solar cells in ambient air with RH higher than 70%, and a $CH_3NH_3PbI_{3-x}(SCN)_x$ formula is adopted for the resulting perovskite based on elemental analysis and theoretical calculations. The $CH_3NH_3PbI_{3-x}(SCN)_x$ perovskite exhibits less trap density and better intrinsic stability than conventional PbI_2-derived $CH_3NH_3PbI_3$ perovskite, and thus better and more stable device performance. The advantages of such perovskite solar cells are demonstrated by the PCEs up to 15% obtained without humidity control and highlighted by the little efficiency loss ($<15\%$) upon long-term (>500 h) ageing in humid air without encapsulation. The slow decrease of the device efficiency with time is attributed to the increased charge transport resistance in Spiro-MeOTAD layer and the increased recombination rate in the perovskite. Ageing-induced J–V hysteresis is also observed during the stability test, which can be attributed to moisture-induced defects and slow dynamic processes (negative capacitance) in perovskite films. Further improvement of the device performance is expected by optimizing the morphology of mp-TiO_2 films and the use of non-hygroscopic HTM. All of the findings will offer useful insights for obtaining efficient and stable perovskite solar cells at ambient conditions.

Methods

Preparation of TiO_2 on electrodes. FTO glass was first patterned and cleaned, and then a 50–80-nm compact TiO_2 layer was formed by spin-coating a 0.2-M titanium isopropoxide solution in ethanol with the addition of 0.02 M HCl at the spin-coating speed of 4,000 r.p.m., followed by sintering at 450 °C for 30 min. A ~ 220-nm mp-TiO_2 layer was then deposited by spin-coating TiO_2 paste in ethanol at 3,000 r.p.m., which contains 4.5 wt% P25 (Degussa) and 5% (weight to TiO_2) ethyl cellulose (22 cp). The paste was ground at 400 r.p.m. for 72 h before use. After that, the mp-TiO_2 was sintered at 500 °C for 30 min, followed by 40-mM $TiCl_4$ treatment at 70 °C for 30 min and sintering at 500 °C for another 30 min.

Preparation of pervoskite solar cells. $Pb(SCN)_2$ powder (Sigma-Aldrich) was dissolved in DMSO at 500 mg ml^{-1} and filtered with 0.45-μm nylon filter to get a clear solution. The solution was then spin-coated on mp-TiO_2 scaffold and heated at 90 °C in air for 1 h to get $Pb(SCN)_2$ film. Then CH_3NH_3I solution (10 mg ml^{-1} in isopropanol) was dropped on top of the $Pb(SCN)_2$ film and kept for 20 s, followed by spin-coating at 3,000 r.p.m. for 20 s. This procedure was repeated three times to guarantee a complete conversion of $Pb(SCN)_2$ into perovskite. The resulting perovskite film was rinsed with pure isopropanol and annealed at 80 °C in air for 20 min. The 2,2′,7,7′-Tetrakis(N,N-di-p-methoxyphenylamine)-9,9′ -spiro-obifluorene (spiro-MeOTAD) layer was prepared by spin-coating a chlorobenzene solution containing 80 mM spiro-MeOTAD, 64 mM tert-butylpyridine and 24 mM Li-TFSI (255 mg ml^{-1} in acetonitrile) at 4,000 r.p.m. for 60 s. Finally, 80-nm-thick Au electrode was deposited via thermal evaporation. The $CH_3NH_3PbI_3$-based control devices were prepared by exactly the same method instead of using PbI_2 precursor.

Characterization. The optical and structural characterizations were performed on Shimadzu UV-2550 spectrophotometer and Rigaku SmartLab X-Ray diffractometer, respectively. The XPS was measured with VG Thermo Escalab 220i-XL. The film morphology was observed under JEOL JSM 6335F SEM. The J–V curves were measured under AM 1.5G-simulated illumination (Oriel 91160) with a power density of $100\,mW\,cm^{-2}$, and the light intensity was calibrated with a standard reference cell (Oriel 91150 V). The scan rate of the measurements is $\pm 0.03\,V\,s^{-1}$. The EQE was measured with an EQE system containing a xenon lamp (Oriel 66902), a monochromator (Newport 66902), a Si detector (Oriel 76175_71580) and a dual-channel power meter (Newport 2931_C). The IS measurements of the devices were carried out in dark by using a HP 4294 impedance analyser in a frequency ranging from 40 Hz to 1 MHz with an oscillating voltage of 30 mV and applied DC bias voltages varying from 0 to 1.0 V.

Calculation details. Density functional theory calculations were performed using VASP package[43]. The algorithm implemented in the code is based primarily on using plane-wave basis set and norm-conserving pseudopotentials. A plane wave with cutoff energy of 500 eV was sufficient to converge the total energy. The semi-local d-electrons of Pb were considered as valence electron in the calculation. The Perdew–Burke–Ernzerhof-generalized gradient approximation method[44] was used as the exchange-correlation density functional. A gamma-centered k-point grid spacing of $0.2\,Å^{-1}$ was used for reciprocal space integration in structure optimization and spacing of $0.1\,Å^{-1}$ in electronic structure calculations. Geometry relaxation in most calculations was run until the forces on atoms that were allowed to relax were no $>0.01\,eV\,Å^{-1}$. Calculations were performed without taking into account spin–orbital coupling.

References

1. Kojima, A., Teshima, K., Shirai, Y. & Miyasaka, T. Organometal halide perovskites as visible-light sensitizers for photovoltaic cells. *J. Am. Chem. Soc.* **131**, 6050–6051 (2009).
2. Kim, H. S. et al. Lead iodide perovskite sensitized all-solid-state submicron thin film mesoscopic solar cell with efficiency exceeding 9%. *Sci. Rep.* **2**, 591 (2012).
3. Lee, M. M., Teuscher, J., Miyasaka, T., Murakami, T. N. & Snaith, H. J. Efficient hybrid solar cells based on meso-superstructured organometal halide perovskites. *Science* **338**, 643–647 (2012).
4. Burschka, J. et al. Sequential deposition as a route to high-performance perovskite-sensitized solar cells. *Nature* **499**, 316–319 (2013).
5. Liu, M., Johnston, M. B. & Snaith, H. J. Efficient planar heterojunction perovskite solar cells by vapour deposition. *Nature* **501**, 395–398 (2013).
6. Jeon, N. J. et al. Solvent engineering for high-performance inorganic–organic hybrid perovskite solar cells. *Nat. Mater.* **13**, 897–903 (2014).
7. Yang, W. S. et al. High-performance photovoltaic perovskite layers fabricated through intramolecular exchange. *Science* **348**, 1234–1237 (2015).
8. Mei, A. et al. A hole-conductor–free, fully printable mesoscopic perovskite solar cell with high stability. *Science* **345**, 295–298 (2014).
9. Zhou, H. et al. Interface engineering of highly efficient perovskite solar cells. *Science* **345**, 542–546 (2014).
10. Jeon, N. J. et al. Compositional engineering of perovskite materials for high-performance solar cells. *Nature* **517**, 476–480 (2015).
11. You, P., Liu, Z., Tai, Q., Liu, S. & Yan, F. Efficient semitransparent perovskite solar cells with graphene electrodes. *Adv. Mater.* **27**, 3632–3638 (2015).
12. Chen, Q. et al. Planar heterojunction perovskite solar cells via vapor-assisted solution process. *J. Am. Chem. Soc.* **136**, 622–625 (2014).
13. Green, M. A., Ho-Baillie, A. & Snaith, H. J. The emergence of perovskite solar cells. *Nat. Photon.* **8**, 506–514 (2014).
14. Jung, H. S. & Park, N. G. Perovskite solar cells: from materials to devices. *Small* **11**, 10–25 (2015).
15. Gratzel, M. The light and shade of perovskite solar cells. *Nat. Mater.* **13**, 838–842 (2014).
16. Noh, J. H., Im, S. H., Heo, J. H., Mandal, T. N. & Seok, S. I. Chemical management for colorful, efficient, and stable inorganic–organic hybrid nanostructured solar cells. *Nano Lett.* **13**, 1764–1769 (2013).
17. Kim, J. H., Williams, S. T., Cho, N., Chueh, C. C. & Jen, A. K. Y. Enhanced environmental stability of planar heterojunction perovskite solar cells based on blade-coating. *Adv. Energy Mater.* **5**, 1401229 (2015).
18. Smith, I. C., Hoke, E. T., Solis-Ibarra, D., McGehee, M. D. & Karunadasa, H. I. A layered hybrid perovskite solar-cell absorber with enhanced moisture stability. *Angew. Chem. Int. Ed.* **53**, 11232–11235 (2014).
19. Habisreutinger, S. N. et al. Carbon nanotube/polymer composites as a highly stable hole collection layer in perovskite solar cells. *Nano. Lett.* **14**, 5561–5568 (2014).
20. Colella, S. et al. Elusive presence of chloride in mixed halide perovskite solar cells. *J. Phys. Chem. Lett.* **5**, 3532–3538 (2014).
21. Dar, M. I. et al. Investigation regarding the role of chloride in organic–inorganic halide perovskites obtained from chloride containing precursors. *Nano Lett.* **14**, 6991–6996 (2014).
22. Buin, A. et al. Materials processing routes to trap-free halide perovskites. *Nano Lett.* **14**, 6281–6286 (2014).
23. Aldibaja, F. K. et al. Effect of different lead precursors on perovskite solar cell performance and stability. *J. Mater. Chem. A* **3**, 9194–9200 (2015).
24. Wang, F., Yu, H., Xu, H. & Zhao, N. HPbI₃: a new precursor compound for highly efficient solution-processed perovskite solar cells. *Adv. Funct. Mater.* **25**, 1120–1126 (2015).
25. Zhang, W. et al. Ultrasmooth organic–inorganic perovskite thin-film formation and crystallization for efficient planar heterojunction solar cells. *Nat. Commun.* **6**, 6142 (2015).
26. Iwadate, Y., Kawamura, K., Igarashi, K. & Mochinaga, J. Effective ionic radii of nitrite and thiocyanate estimated in terms of the Boettcher equation and the Lorentz-Lorenz equation. *J. Phys. Chem.* **86**, 5205–5208 (1982).
27. Hendon, C. H., Yang, R. X., Burton, L. A. & Walsh, A. Assessment of polyanion (BF_4^- and PF_6^-) substitutions in hybrid halide perovskites. *J. Mater. Chem. A* **3**, 9067–9070 (2015).
28. Nagane, S., Bansode, U., Game, O., Chhatre, S. & Ogale, S. $CH_3NH_3PbI_{(3-x)}(BF_4)_x$: molecular ion substituted hybrid perovskite. *Chem. Commun.* **50**, 9741–9744 (2014).
29. Eperon, G. E., Burlakov, V. M., Docampo, P., Goriely, A. & Snaith, H. J. Morphological control for high performance, solution-processed planar heterojunction perovskite solar cells. *Adv. Funct. Mater.* **24**, 151–157 (2014).
30. Dualeh, A. et al. Impedance spectroscopic analysis of lead iodide perovskite-sensitized solid-state solar cells. *ACS Nano* **8**, 362–373 (2013).
31. Gonzalez-Pedro, V. et al. General working principles of $CH_3NH_3PbX_3$ perovskite solar cells. *Nano Lett.* **14**, 888–893 (2014).
32. Sanchez, R. S. et al. Slow dynamic processes in lead halide perovskite solar cells. characteristic times and hysteresis. *J. Phys. Chem. Lett.* **5**, 2357–2363 (2014).
33. Bisquert, J., Bertoluzzi, L., Mora-Sero, I. & Garcia-Belmonte, G. Theory of impedance and capacitance spectroscopy of solar cells with dielectric relaxation, drift-diffusion transport, and recombination. *J. Phys. Chem. C* **118**, 18983–18991 (2014).
34. Pockett, A. et al. Characterizations of planar lead halide perovskite solar cells by impedance spectroscopy, open-circuit photovoltage decay, and intensity-modulated photovoltage/photocurrent spectroscopy. *J. Phys. Chem.* **119**, 3456–3465 (2015).
35. Kim, H. S. et al. High efficiency solid-state sensitized solar cell-based on submicrometer rutile TiO_2 nanorod and $CH_3NH_3PbI_3$ perovskite sensitizer. *Nano Lett.* **13**, 2412–2417 (2013).
36. Tress, W. et al. Understanding the rate-dependent J-V hysteresis, slow time component, and aging in $CH_3NH_3PbI_3$ perovskite solar cells: the role of a compensated electric field. *Energy Enrion. Sci.* **8**, 995–1004 (2015).
37. Liu, J. et al. A dopant-free hole-transporting material for efficient and stable perovskite solar cells. *Energy Enrion. Sci.* **7**, 2963–2967 (2014).
38. Zheng, L. et al. A hydrophobic hole transporting oligothiophene for planar perovskite solar cells with improved stability. *Chem. Commun.* **50**, 11196–11199 (2014).
39. Jaramillo-Quintero, O. A., Sanchez, R. S., Rincon, M. & Mora-Sero, I. Bright visible-infrared light emitting diodes based on hybrid halide perovskite with Spiro-OMeTAD as a hole-injecting layer. *J. Phys. Chem. Lett.* **6**, 1883–1890 (2015).
40. Kim, H. S. & Park, N. G. Parameters affecting I–V hysteresis of $CH_3NH_3PbI_3$ perovskite solar cells: effects of perovskite crystal size and mesoporous TiO_2 layer. *J. Phys. Chem. Lett.* **5**, 2927–2934 (2014).
41. Geng, W., Zhang, L., Zhang, Y. N., Lau, W. M. & Liu, L. M. First -principle study of lead iodide perovskite tetragonal and orthorhombic phases for photovoltaics. *J. Phys. Chem. C* **118**, 19565–19571 (2014).
42. Jiang, Q. et al. Pseudohalide-induced moisture tolerance in perovskite $CH_3NH_3Pb(SCN)_2I$ thin films. *Angew. Chem. Int. Ed.* **54**, 7617–7620 (2015).
43. Kresse, G. & Furthmüller, J. Efficient iterative schemes for ab initio total-energycalculations using a plane-wave basis set. *Phys. Rev. B.* **54**, 11169–11186 (1996).
44. Perdew, J. P., Burke, K. & Ernzerhof, M. Generalized gradient approximation made simple. *Phys. Rev. Lett.* **77**, 3865–3868 (1996).

Acknowledgements

This work is financially supported by the Research Grants Council of Hong Kong, China (project no. C4030-14G), The Hong Kong Polytechnic University (project nos. 1-ZVAW and 1-ZVCG) and the National Natural Science Foundation of China (grant nos. 21403089, 11547243, 51303066). We acknowledge the support from the Center for High Performance Computing of JiangHan University.

Author contributions

Q.T. and F.Y. conceived the experiments. Q.T. fabricated and characterized the devices. P.Y. and Z.L. assisted some experiments on device preparation and material characterization. H.S. contributed to the theoretical calculations. C.H. contributed to the Raman characterizations. H.L.W.C. provided valuable suggestions and discussions. The manuscript was written by Q.T. and F.Y., and discussed, edited and approved by all of the authors.

Additional information

Competing financial interests: The authors declare no competing financial interests.

10

Ultra-high gain diffusion-driven organic transistor

Fabrizio Torricelli[1,2], Luigi Colalongo[1], Daniele Raiteri[2], Zsolt Miklós Kovács-Vajna[1] & Eugenio Cantatore[2]

Emerging large-area technologies based on organic transistors are enabling the fabrication of low-cost flexible circuits, smart sensors and biomedical devices. High-gain transistors are essential for the development of large-scale circuit integration, high-sensitivity sensors and signal amplification in sensing systems. Unfortunately, organic field-effect transistors show limited gain, usually of the order of tens, because of the large contact resistance and channel-length modulation. Here we show a new organic field-effect transistor architecture with a gain larger than 700. This is the highest gain ever reported for organic field-effect transistors. In the proposed organic field-effect transistor, the charge injection and extraction at the metal–semiconductor contacts are driven by the charge diffusion. The ideal conditions of ohmic contacts with negligible contact resistance and flat current saturation are demonstrated. The approach is general and can be extended to any thin-film technology opening unprecedented opportunities for the development of high-performance flexible electronics.

[1] Department of Information Engineering, University of Brescia, via Branze 38, Brescia 25123, Italy. [2] Department of Electrical Engineering, Eindhoven University of Technology, Groene Loper 19, PO Box 513, Eindhoven 5600MB, The Netherlands. Correspondence and requests for materials should be addressed to F.T. (email: fabrizio.torricelli@unibs.it).

Transistors fabricated with organic, polymeric, amorphous-oxide and carbon-based materials are the basis of emerging technologies for the development of lightweight, large-area and flexible electronics[1-6]. Large-area electronics manufactured at near-to-room temperature on plastic foils aims at enabling new applications where mechanical flexibility, integration in wrapping materials and ultra-low cost are paramount. To fabricate a transistor in flexible technologies, nanometre-thick layers of metals, insulators and semiconductor are stacked together and the semiconductor is directly contacted with the metal electrodes. The overall transistor performance intimately depends on three physical processes: the charge injection from the source electrode to the semiconductor, the charge transport through the semiconductor and the charge extraction at the drain electrode. The impressive development of high-mobility semiconductors[7-9] and short channel-length transistors[10,11] urgently demand high-quality contacts and proper transistor design[12,13]. Unfortunately, the energetic matching between abruptly contacted metal–semiconductor materials is challenging, especially at near-to-room temperature[13]. Electrons and holes must overcome large energy barriers to flow from a material to the other, resulting in a large contact resistance, large device-to-device variations and low transistor amplification[13-18].

The figure-of-merit that determines the intrinsic amplification of a transistor is the gain $= g_m/g_o$, where $g_m = \partial I_D/\partial V_G$ is the transconductance and $g_o = \partial I_D/\partial V_D$ is the output conductance. High-gain transistors are essential for the development of large-scale and robust circuits, high-sensitivity sensors, and adequate signal amplification in sensing systems. Unfortunately, organic field-effect transistors (OFETs) typically show a gain of the order of tens[17,18]. The low gain measured in OFETs is due to the large contact resistance that results in a small g_m and to the channel length modulation that results in a large g_o. Therefore, high-gain OFETs need, at the same time, both high-quality contacts and flat current saturation.

Ohmic contacts with small contact resistance require efficient charge injection and extraction. In organic electronics, the contact optimization is performed on a case-by-case basis, depending on the semiconductor, electrodes and device architecture. Despite ad-hoc approaches[18-25,] such as doping, surface treatments and materials blending enable to reduce the contact resistance, a general and simple method is desirable. In addition, the channel length modulation dependents on the specific OFET architecture and geometries, which determine how the charge carriers are extracted at the drain[17,18].

Here we show a new organic transistor with high-quality contacts and flat current saturation. Thanks to the charge diffusion triggered by the transistor architecture, the charge carriers are efficiently injected and extracted from the contacts to the channel, independently of the energy barrier at the contacts. As a prototype and remarkable example, we fabricate Diffusion-driven Organic Field-Effect Transistors (named DOFETs) on flexible plastic substrates with an industrial thin-film technology. The theoretical and experimental analysis unambiguously show that the diffusion-driven contact, proposed in this work for the first time, is fundamental to dramatically improve the charge injection and extraction in organic thin-film field-effect transistors. The ideal conditions of negligible contact resistance and fully flat current saturation are demonstrated. These conditions maximize together the transconductance and the output resistance of the transistors, resulting in OFETs with exceptionally high gain (> 700).

Results

Structure and electrical characteristics of the transistor. The top-view image and the three-dimensional structure of the diffusion-driven organic transistor are shown in Fig. 1a,b. The

transistors are bottom-gate co-planar where the gate is patterned first by using photolithography. Thereafter, we deposited by spin coating a photoimageable polymer (polyvinylphenol) used as a gate insulator (named insulator 1) followed by gold source (S) and drain (D) electrodes patterned by a lift-off process. A 100-nm-thick film of pentacene is deposited by spin coating and patterned. A thick layer of polyvinylphenol (named insulator 2) is deposited by spin coating and used as insulator and capping layer. Finally, two electrodes named 'control source' (CS) and 'control drain' (CD) are patterned on the top of the insulator 2 in front of the source and drain electrodes. The transistors are fabricated on a plastic polyethylene naphthalate foil (Fig. 1c) and the overall process temperature is lower than 150 °C. Further details on the transistors fabrication and geometries are shown in Supplementary Fig. 1. The measured transfer and output curves are shown in Fig. 1d–f and Supplementary Fig. 2.

Operation of the transistor. The DOFET operates as follow. An appropriate voltage applied to CS and CD, creates a vertical electric field orthogonal to the S/D contact surface. It triggers a charge injection from the upper surface of S/D into the semiconductor (Fig. 2a). In equilibrium ($V_S = V_D = 0$ V, no current flows), the electric-field below CS/CD is counterbalanced by the injected-charges that are accumulated in the semiconductor region below CS/CD. When a source-drain voltage is applied ($|V_{DS}| > 0$), the charge carriers flow from source to drain despite the contact energy barriers and the potential drop at the contacts is negligible (Fig. 2b).

More in detail (Fig. 2c), the charge carriers accumulated in the CS region move to the right-edge of the CS region itself, attracted by the drain potential (Fig. 2c, arrow 2). As a consequence, the vertical electric field at the left-hand-side of the CS region is not shielded anymore and, despite the energy barrier, other charges can be injected by the source electrode (arrow 1). The excess of the charge carriers at the right-hand-side of the CS region are pushed to the bottom channel by the diffusion against the vertical electric field (arrow 3). As shown in Fig. 2d, few nanometres far from the CS region the vertical electric-field changes direction under the influence of the gate potential and the charge carriers are eventually pulled into the transistor channel (arrow 4). As a result, the CS region acts as an ideal source. The key physical mechanism triggered by the transistor architecture is the charge diffusion, which takes place in less than $L_{diff} = 50$ nm (Fig. 2d) when the semiconductor thickness is $t_S = 100$ nm. We also verified that the diffusion length scales accordingly with the semiconductor thickness (that is, $L_{diff} \cong 25$ nm when $t_S = 50$ nm).

The charge carriers injected into the channel drift to the drain (Fig. 2c, arrow 5) under the force of the longitudinal electric field. When the charge carriers reach the right edge of the channel, they are blocked by the energy barrier at the drain contact, and the local concentration increases. The charges are no more counterbalanced by the gate electric field, and they can diffuse to the CD region (arrow 6) in correspondence of the CD region edge. As shown in Fig. 2e, few nanometres far from the channel the vertical electric field changes direction, the charge carriers are pulled into the CD region (arrow 7) and eventually diffuse (arrow 9) to the drain. The CD region acts as an ideal drain.

The idea is that in the DOFETs the charge injection and extraction do not take place directly from the source and drain metal electrodes as in conventional transistors but, instead, the charge carries are injected by the CS region and are extracted by the CD region. The injection and the extraction are driven by the diffusion triggered by the transistor architecture. As a result, when enough charge carriers are accumulated in the CS and CD regions, the charge injection and extraction are independent of

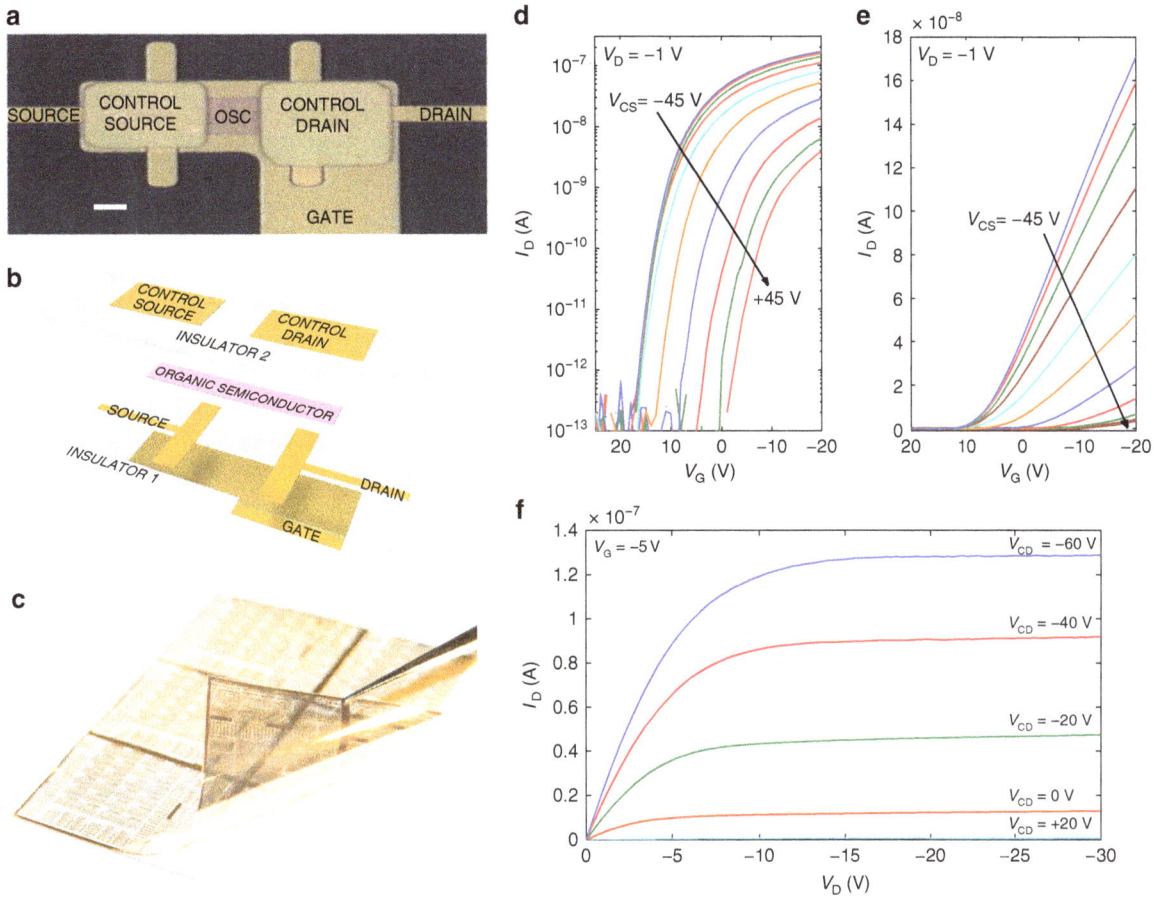

Figure 1 | Transistor architecture and characteristics. (a) Top-view optical image of a diffusion-driven organic field-effect transistor (DOFET) fabricated on plastic foil OSC is the organic semiconductor. Scale bar, 5 μm. **(b)** DOFET components. Photolithographically patterned gold is used for metal electrodes (named gate, source, drain, control source, control drain), the insulators (insulators 1 and 2) are photoimageable polymers (polyvinylphenol), and the organic semiconductor is a solution-processed pentacene. The material thicknesses are detailed in the Supplementary Fig. 1. **(c)** Photograph of the plastic (PEN) foil with the measured transistors detached from the glass substrate. The transistors are fabricated with an industrial thin-film technology with three metal layers. **(d,e)** Measured transfer characteristics at several control source voltages. The V_{CS} step is 10 V, $V_S = 0$ V and $V_{CD} = 0$ V. The DOFET channel width and length are $W = 100$ μm and $L = 12.5$ μm, respectively. **(f)** Measured output characteristics at several control drain voltages.

the applied voltages (viz. V_{CS} and V_{CD}) and the CS and CD regions behave like ideal source and drain for the transistor channel. Therefore, as confirmed by the two-dimensional (2D) numerical simulations shown in Supplementary Fig. 3, the gate electrode is not required to overlap the source and drain electrodes.

The potential at the insulator1–organic interface calculated by means of 2D numerical simulations is shown in Fig. 2b. In the DOFET, the potential drop at the contacts is negligible even if the energy barrier at the metal–semiconductor contacts is 0.5 eV, that is a typical barrier at the metal–organic contacts. In contrast, in a conventional organic transistor (viz. without CS and CD), the charge carriers must overcome the energy barrier flowing from the channel to the S/D electrodes and vice versa. Owing to the energy barrier, the channel is disconnected from the S/D electrodes and more than the half of the drain voltage drops at the contacts (Fig. 2b). The large contact resistance severely limits the transistor performances and this is even worse in case of high-mobility semiconductors and/or short-channel lengths.

Experimental analysis. The effectiveness of the proposed approach is further assessed by means of the experimental results shown in Fig. 3. Figure 3a shows the measured contact resistance

R_P as a function of the gate voltage V_G. The contact resistance of the DOFET biased at $V_{CS} = -40$ V (that corresponds to an electric field $|E_{Y\text{-}VCS}| = 0.28$ MV cm^{-1}) is equal to $R_{P[DOFET]} = 20$ kΩ cm, which is lower than the contact resistance in conventional OFETs with Au-pentacene-doped contacts[20] and, more importantly, $R_{P[DOFET]}$ is independent of V_G. In contrast, the contact resistance of an organic transistor without CS/CD (conventional coplanar transistor) fabricated with the same materials and process is V_G dependent. It is up to 24 times larger than that of the DOFET and, even at large gate voltages ($V_G = -25$ V, that is, $|E_{Y-VG}| = 0.7$ MV cm^{-1}), $R_{P[OFET]} > 5 \times R_{P[DOFET]}$. Analogous results are obtained comparing the DOFET with a conventional staggered OFET (Supplementary Fig. 4).

To give more insight, Fig. 3b shows the R_P-V_{CS} characteristic of two nominally identical DOFETs for several V_G. R_P is controlled by V_{CS} despite the gate voltage. Indeed, at low gate voltage ($V_G = -5$ V, that is, $|E_{Y-VG}| = 0.14$ MV cm^{-1}), V_{CS} modulates R_P by more than four orders of magnitude, and at large $V_G = -20$ V ($|E_{Y-VG}| = 0.57$ MV cm^{-1}), R_P still depends on V_{CS}. Interestingly, when $V_{CS} < -20$ V (that is, $|E_{Y\text{-}VCS}| = 0.14$ MV cm^{-1}) the contact resistance is negligible compared with the channel resistance (Supplementary Fig. 5) and it is independent of both V_G and V_{CS}. As confirmed by the measurements shown in

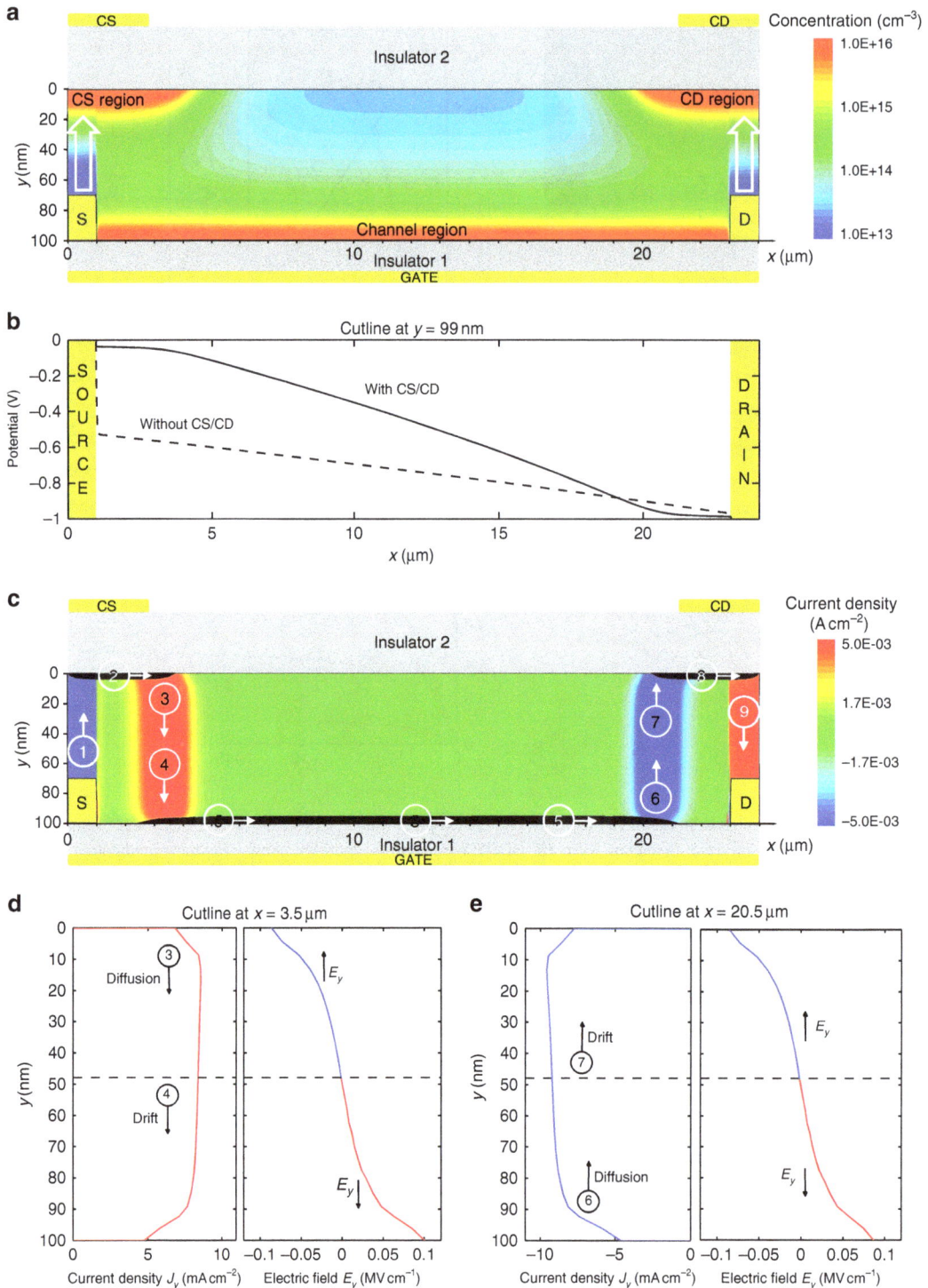

Figure 2 | DOFET operation. Two-dimensional numerical simulations. The applied voltages are $V_G = -5.1\,V$, $V_S = 0\,V$, $V_D = -1\,V$, $V_{CS} = -60\,V$, $V_{CD} = -60\,V$. Geometrical and physical parameters are listed in the Supplementary Fig. 1 and in the Methods section, respectively. (**a**) Charge concentration in the organic semiconductor. The white arrows depict the charge injection from the source and drain electrodes into the semiconductor when the control source and control drain electrodes are biased. The x-to-y scale ratio is 1:200. (**b**) Quasi-Fermi potential at $y = 99\,nm$ with (full line) and without (dashed line) CS/CD. Without CS/CD about half of V_{DS} drops at the source and it is required for the charge injection. (**c**) Current density: x-component J_X (black area) is equal to $1\,A\,cm^{-2}$, and the y-component J_Y is shown with colour scale levels. (**d**) Current density J_Y and electric field E_Y along the y-direction at $x = 3.5\,\mu m$. In the range $y = [0\text{-}47]\,nm$, the current is driven by the diffusion, and in the range $y = [47\text{-}100]\,nm$, the current is driven by the drift. (**e**) J_Y and E_Y along the y-direction at $x = 20.5\,\mu m$. In the range $y = [0\text{-}47]\,nm$, the current is driven by the drift, and in the range $y = [47\text{-}100]\,nm$, the current is driven by the diffusion.

the inset of Fig. 3b, this is the experimental evidence that the current enhancement originates from the improved charge injection at the source. According to the physical insight obtained by means of the 2D numerical simulations, at $V_{CS} < -20\,V$, the accumulated CS-region is an 'infinite' charge reservoir, the charge diffusion efficiently sustain the charge injection required by the

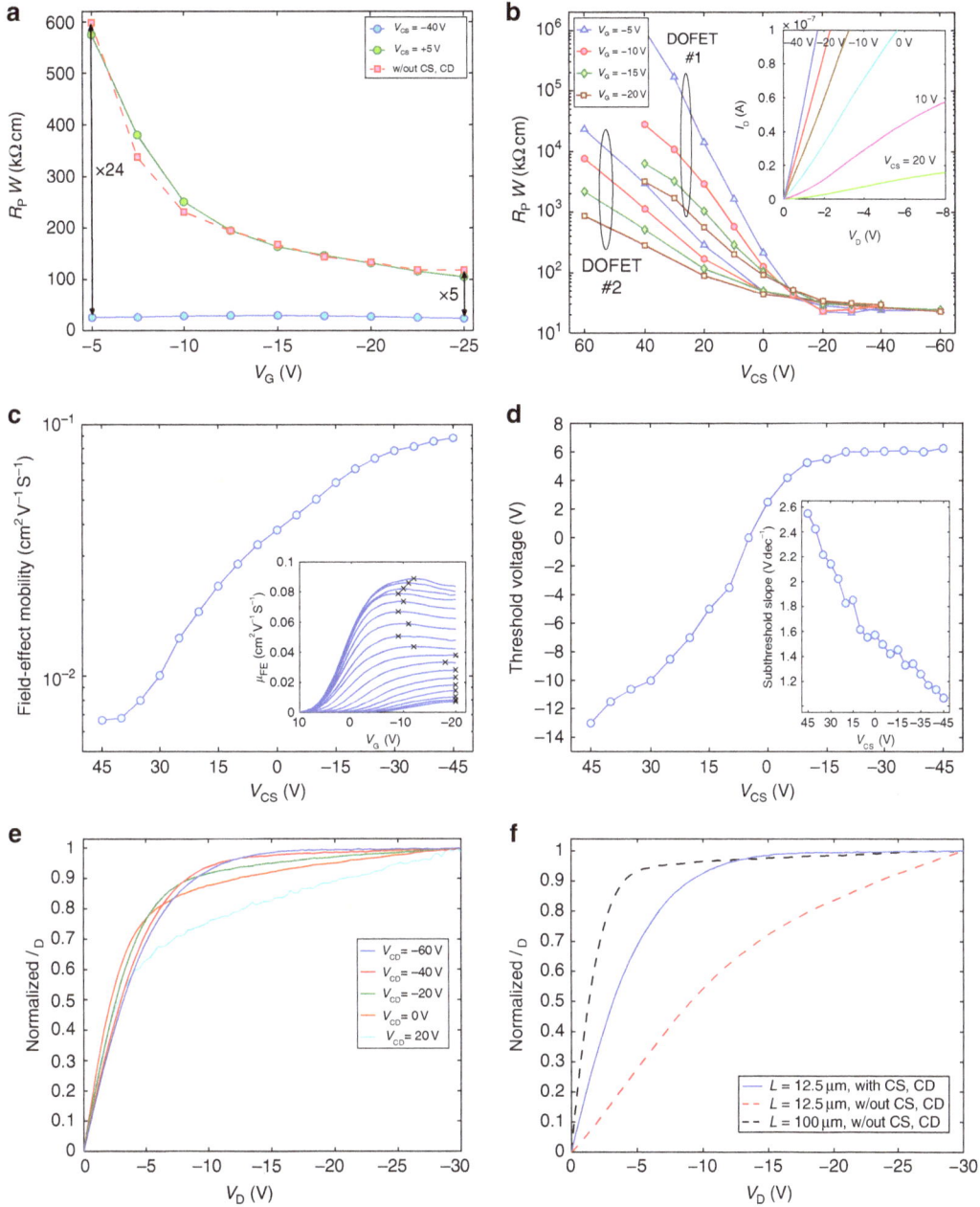

Figure 3 | DOFET measurements and parameters. When it is not specified, the applied voltages are: $V_S = 0$ V, $V_D = -1$ V, $V_{CS} = 0$ V, $V_{CD} = 0$ V, and the transistors geometries are: $W = 100$ μm, $L = 12.5$ μm. (**a**) Width-normalized contact resistance R_P as a function of the gate voltage V_G. R_P is calculated with the method[26]. In the conventional OFET (viz. without CS and CD), R_P decreases with V_G, whereas in the DOFET, R_P is independent of V_G. When the control source is biased at $V_{CS} = +5$ V, the DOFET works as a conventional coplanar OFET. (**b**) R_P vs V_{CS} at various V_G measured on two nominally identical DOFETs. When $V_{CS} < -10$ V, R_P is the same for both the DOFETs and it is independent of both V_G and V_{CS}. Inset: measured output characteristics of a DOFET at several V_{CS}. (**c**) Maximum overall field-effect mobility vs V_{CS}. The inset shows the field-effect mobility as a function of the gate voltage: $\mu_{FE} = (L/W)\,(\partial I_D/\partial V_G)/(C_i\,V_D)$. The × symbol is the maximum value of each curve. (**d**) Threshold voltage (V_{TH}) as a function of V_{CS}. V_{TH} is the intercept to the V_G-axis of the I_D linear fit. Inset: Subthreshold slope as a function of V_{CS}. (**e**) Normalized output characteristics of the DOFET measured at various V_{CD}. I_D is normalized by its maximum value at $V_D = -30$ V. In saturation, the DOFET is an ideal current generator because the current is diffusion driven. The most important short-channel effect due to the channel-length modulation vanishes. The V_{CD} controls the charge extraction at the drain electrode, which has a strong impact on the output conductance ($g_O = \partial I_D/\partial V_D$). (**f**) Normalized output characteristics of a DOFET and two conventional OFETs (viz. without CS and CD).

channel, and the CS-region behaves like an ohmic contact. On the other hand, at $V_{CS} > +5$ V, the diffusion-driven charge injection is turned-off, the contact resistance increases, the drain current lowers and it increases super-linearly with V_D as usually obtained in contact limited transistors[13,15,16,19]. We can conclude that it is possible to control (enhance or reduce) the charge injection at the source contact through nanometre-scale charge diffusion.

Comparing the R_P obtained for two nominally identical DOFETs (Fig. 3b), it results that when the virtual-ohmic source contact is not formed, the transistors show different R_P, whereas as soon as the virtual-ohmic source contact is formed ($V_{CS} < -20$ V), R_P becomes the same for both the DOFETs. According to refs 10,12,13, these measurements suggest that the metal–organic contact is a source of variability. As the DOFET suppresses the contact resistance, the variability due to R_P is reduced as well. This feature is essential for the large-scale integration of flexible circuits. Moreover, the improved charge injection results in a larger overall field-effect mobility (Fig. 3c) as well as in a reduced threshold voltage (Fig. 3d) and steeper subthreshold slope (inset Fig. 3d). Figure 3c shows that the maximum mobility of a DOFET with $L_{[DOFET]} = 12.5\,\mu m$ is close to $0.1\,cm^2\,V^{-1}\,s^{-1}$ and it corresponds to the mobility measured in long-channel OFETs ($L_{[OFET]} = 100\,\mu m$), where the contact resistance is negligible. Figure 3d shows that by means of V_{CS} the DOFET can be turned into a multi-threshold transistor and the improved DOFET ($V_{CS} < -20\,V$) operates in depletion-mode. In unipolar technologies, depletion-mode transistors are essential to design high-performance circuits[27,28] and the electrical control of the threshold voltage is extremely important to improve the circuit robustness[27–29].

When the transistor operates in linear region, the energy barrier at the drain side of the channel is smaller than that at the source side. On the other hand, in saturation ($|V_G| < |V_D|$), a wider energy barrier is present at the drain, independently of the metal/semiconductor properties (Fig. 4a). Therefore, we investigated the impact of the control drain in saturation. The output characteristics (I_D–V_D) of the DOFET measured at various V_{CD} are shown in Fig. 1f. As expected, I_D increases with V_{CD} and, more importantly, at large (negative) V_{CD} the DOFET shows fully flat current saturation. The impact of V_{CD} on the current saturation is readily visible in Fig. 3e where the I_D–V_D characteristics are normalized with respect to the maximum I_D measured at $V_D = -30\,V$. At $V_{CD} < -40\,V$, the detrimental effect of the channel length modulation on the drain current is completely suppressed and the DOFET behaves like an ideal current generator.

This can be explained in the light of the previous analysis. In saturation, the charge carriers drift to the right-edge of the channel (pinch-off region), and diffuse to the CD region (Fig. 4b, arrow 6). Few nanometres far from the channel edge, the vertical electric-field changes direction because of the control drain

voltage and, in turn, the charge carriers are pulled into the CD region (arrow 7). Now, the excess charges are no more in equilibrium with the vertical electric-field and can diffuse to the drain (arrow 9). As the charge-extraction from the accumulated layer (viz. CD region) is diffusion driven, the drain current is independent of the drain voltage as far as V_{CD} is greater than V_D.

Figure 3f shows the comparison between a DOFET with a channel length $L_{[DOFET]} = 12.5\,\mu m$ (full line), and two conventional coplanar OFETs with $L_{[OFET1]} = 12.5\,\mu m$ (red dashed line) and $L_{[OFET2]} = 100\,\mu m$ (black dashed line). Interestingly, the channel length modulation of the DOFET biased at $V_{CD} = -60$ V is completely suppressed: it is even smaller than that of the long-channel OFET2. This is also more evident when the DOFET is compared with a conventional staggered OFET (Supplementary Fig. 6) where the channel length modulation is very large because the source and drain electrodes are placed at the opposite side of the gate. These results confirm that V_{CD} controls channel length modulation and in turn the output resistance of the DOFET. The channel length modulation is one of the most important short-channel effects and it limits the transistor amplification.

Figure 5 shows the maximum gain measured in a DOFET as a function of V_{CD} (full line with symbols). According to Figs 1f and 3e, the gain depends on V_{CD} because it controls both the contact resistance at the drain and the channel length modulation. When $V_{CD} = -60\,V$, the gain is larger than 700. This is the largest gain ever reported for OFETs. It is one order of magnitude larger than the gain usually obtained in OFETs[11,16–18,30–35].

Discussion

The ultra-high gain measured in the DOFET is achieved thanks to the diffusion-driven charge injection and extraction. In particular, when the CS and CD regions are accumulated, they act as ideal contacts for the channel and the diffusion enables the efficient and voltage-independent charge injection and extraction. In the DOFET, the CS and CD regions are at the opposite side of the channel and resemble a staggered OFET with ideal ohmic

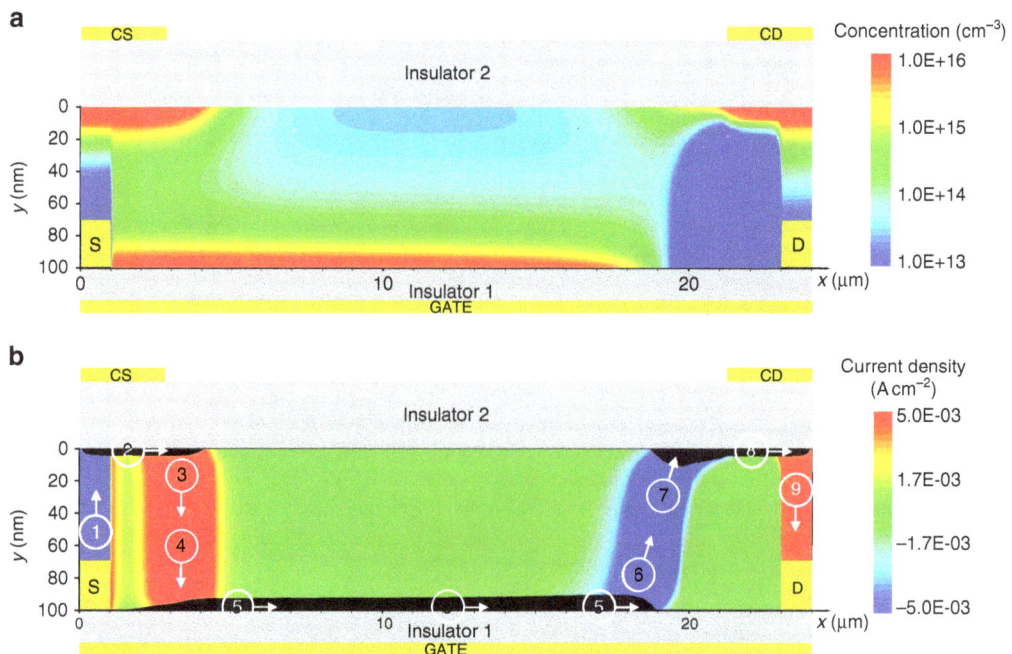

Figure 4 | 2D numerical simulations of a DOFET operating in saturation. The applied voltages are $V_G = -5.1\,V$, $V_S = 0\,V$, $V_D = -10\,V$, $V_{CS} = -60\,V$, $V_{CD} = -60\,V$. (**a**) Charge concentration in the organic semiconductor. (**b**) Current density: x-component J_X (black area) is equal to $1\,A\,cm^{-2}$, and the y-component J_Y is shown with colour scale levels. For the sake of clarity, the positions of control source (CS), control drain (CD) and gate electrodes are shown. Geometrical and physical parameters are listed in the Supplementary Fig. 1.

contacts. It is important to note that in the DOFET this condition is always achieved, thanks to the accumulated CS and CD regions. The charge flow from/to the CS/CD regions and the channel is driven by the charge diffusion, and thus the contact resistance is independent of the gate (Fig. 3a) and drain (inset Fig. 3b) voltages, the saturation current is independent of V_D, and an ultra-high gain is obtained.

As a comparison, the gain measured in the conventional OFET1 ($L_{[OFET1]} = 12.5\,\mu m$, red dashed line) and OFET2 ($L_{[OFET2]} = 100\,\mu m$, black dashed line) are shown in Fig. 5. As expected in both cases, the gain is much lower than that measured in the DOFET at any V_{CD} because in the OFET1 the current is contact limited and the channel modulation is large, whereas in the OFET2 the contact resistance is negligible but the channel length is large and hence g_m is small. In OFETs, the contact resistance can be reduced by means of the contact engineering and optimization[18–25], and the proper choice of the transistor architecture[36,37]. Indeed, staggered OFETs are more tolerant to

the contact resistance with respect to the coplanar OFETs because in the staggered transistors the contact area (of the order of microns) is larger than that of coplanar transistors (of the order of nanometres). On the other hand, in staggered transistors the source and drain electrodes are at the opposite side of the channel and, when operated in saturation, the channel length modulation is larger than that in coplanar OFETs. As an alternative approach, the split-gate OFETs[33,34] are based on a coplanar architecture and lower the contact resistance thanks to the gate bias-assisted charge injection[38]. However, the channel length modulation is not suppressed because the secondary gates are coplanar with the source and drain electrodes, the charge extraction is not diffusion driven and, as a result, the gain is comparable with that typically measured in OFETs (of the order of tens).

In addition to the high-gain, another advantage offered by the DOFET is the possibility to maximize the charge injection/extraction area at the source and drain electrodes, whereas minimizing the overlap between the gate and the electrodes. The 2D numerical simulations in Supplementary Fig. 3 and Fig. 6 show that the gate is not required to overlap the source and drain electrodes because the charge injection/extraction takes place from/to the CS/CD accumulated regions. At the same time, the CS and CD electrodes can be overlapped (without the drawback of extra capacitance) with the source and drain electrodes in order to exploit the full area of the electrodes that is typically in the range 5–10 μm (in our DOFET it is 5 μm). Thanks to the charge diffusion, taking place at the edge of the accumulated CS and CD regions, also the overlap between the gate and the CS and CD electrodes is not required. Moreover, the numerical simulations in Fig. 6 show that the equivalent contact length where the charges are injected/extracted is only $L_C = 0.25\,\mu m$, which is suitable for the megahertz operation[11].

Finally, it is worth noting that the voltages required to form the charge-accumulated CS and CD regions are independent of the DOFET operation. For example, by setting $V_{CS} = V_{CD} = -40\,V$, the DOFET operates as a conventional OFET with ideal ohmic contacts and ultra-high gain. Therefore, the two control electrodes can be connected together and the external circuit design and lines required for the proposed transistor structure is the same of that required for dual-gate transistors. The latter have been successfully used to fabricate an organic microprocessor with 3,381 dual-gate OFETs[39]. Moreover, an alternative approach is to replace the CS and CD electrodes with fixed charges trapped into the insulator 2 (ref. 40). Another very interesting approach

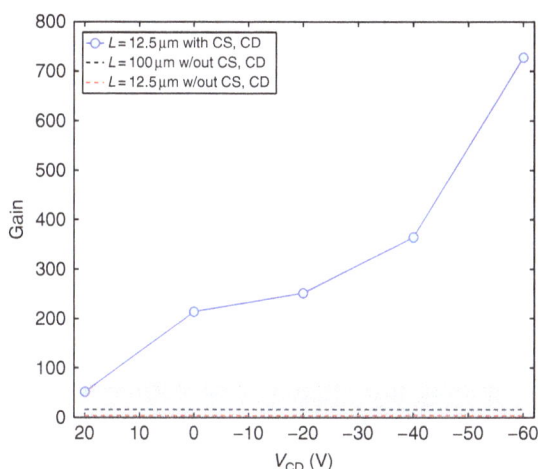

Figure 5 | DOFET gain. Measured gain as a function of V_{CD}. The applied voltages are $V_G = -5\,V$, $V_S = 0\,V$, $V_{CS} = -20\,V$. The transistors width is $W = 100\,\mu m$. The DOFET (full line with symbols) length is $L = 12.5\,\mu m$. The OFET lengths are $L = 12.5\,\mu m$ (red dashed line) and $L = 100\,\mu m$ (black dashed line). The other geometries are the same. The DOFET and OFET are fabricated with the same materials (Supplementary Fig. 1). The grey area shows the gain obtained in OFETs[11,16–18,30–35].

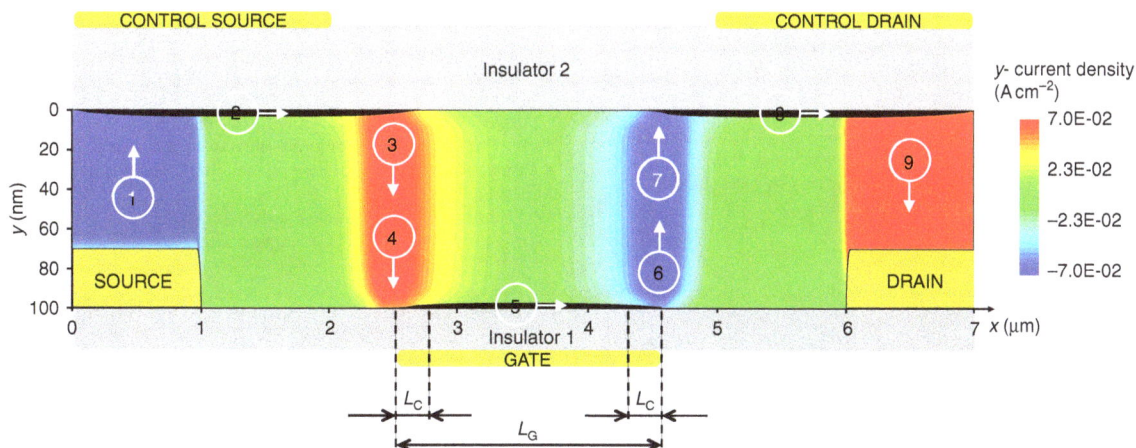

Figure 6 | 2D numerical simulations of a DOFET with minimized capacitances. Current density: x-component J_X (black area) is 10 A cm^{-2}, and the y-component J_Y is shown with colour scale levels. For the sake of clarity, the positions of control source (CS), control drain (CD) and gate electrodes are shown. Geometrical and physical parameters are listed in the Supplementary Fig. 1. The applied voltages are $V_G = -5.1\,V$, $V_S = 0\,V$, $V_D = -1\,V$, $V_{CS} = -60\,V$, $V_{CD} = -60\,V$.

would be the replacement of both the CS and CD electrodes and the insulator 2 with electric dipoles (Supplementary Fig. 7) by local molecular self-assembly functionalization[41,42] of the top surface of the organic semiconductor in front of the source and drain electrodes.

In summary, the DOFET shows that it is possible to dramatically enhance the charge injection and extraction at the metal/semiconductor contacts by means of the nanometre-scale charge diffusion. The enhanced charge injection allowed us to reduce the threshold voltage by more than 15 V, and to increase the field-effect mobility about ten times, approaching the organic semiconductor transport limit also in short-channel transistors. The enhanced charge extraction enables the complete suppression of the channel-length modulation. We show that a short-channel DOFET behaves like an ideal current generator: its channel-length modulation is even smaller than that of an eight times longer organic transistor fabricated in the same technology. These features lead to the fabrication of high performance organic transistors with a unique benefits combination: negligible contact resistance, small device-to-device variability, and exceptionally high gain (> 700).

Thanks to the transistor here proposed we theoretically explain and experimentally demonstrate for the first time that the charge diffusion can play a crucial role in organic transistors. Moreover, the ability to independently enhance or reduce the charge injection, transport and extraction in organic semiconductors makes the DOFET the ideal test-bed to study the fundamental physical processes taking place in organic semiconductors and at the metal–organic interfaces.

The proposed approach is a universal method to obtain high-quality contacts without the need of materials or process optimizations. Moreover, according to the approach proposed in ref. 43, the DOFET combined with ambipolar semiconductors could be used to electrically enhance the charge injection of one charge type and to suppress the other. This feature is very relevant for the low-cost fabrication of high-gain and low-power ambipolar complementary electronics.

The diffusion-driven organic transistor opens up new opportunities for the large-scale integration of flexible electronics, high-sensitivity sensors and ultra-large signal amplification in sensing systems.

Methods

Two-dimensional numerical simulations. The coupled drift–diffusion, Poisson and current continuity equations are solved together[43–45]. The simulation parameters are the following: relative permittivity of semiconductor $\varepsilon_{rs} = 3$, relative permittivity of insulators (1 and 2) $\varepsilon_{ri} = 3.757$, highest occupied molecular orbital (HOMO) energy level $E_{HOMO} = 2.8$ eV, lowest unoccupied molecular orbital (LUMO) energy level $E_{LUMO} = 5.2$ eV, effective density of HOMO states $N_{HOMO} = 10^{21}$ cm^{-3}, effective density of LUMO states $N_{LUMO} = 10^{21}$ cm^{-3}, holes effective mobility $\mu_h = 0.1$ cm^2 V^{-1} s^{-1}, electrons effective mobility $\mu_e = 0.1$ cm^2 V^{-1} s^{-1}, metal electrodes work function $\Phi_m = 4.7$ eV (the hole energy barrier at the source/drain metal-semiconductor is $\Phi_B = 0.5$ eV), Schottky barrier lowering $\Delta\Phi_B = e \, [e \, E/(4 \, \pi \, \varepsilon_0 \, \varepsilon_{rs})]^{\wedge}(1/2)$, where e is the elementary charge, E is the electric field and ε_0 is the vacuum permittivity.

References

1. Kaltenbrunner, M. et al. An ultra-lightweight design for imperceptible plastic electronics. Nature **499**, 458–463 (2013).
2. Nomura, K. et al. Room-temperature fabrication of transparent flexible thin-film transistors using amorphous oxide semiconductors. Nature **432**, 488–492 (2004).
3. Cao, Q. et al. Medium-scale carbon nanotube thin-film integrated circuits on flexible plastic substrates. Nature **454**, 495–500 (2008).
4. Kuribara, K. et al. Organic transistors with high thermal stability for medical applications. Nat. Commun **3**, 1–7 (2012).
5. Takei, K. et al. Nanowire active-matrix circuitry for low-voltage macroscale artificial skin. Nat. Mater. **9**, 821–825 (2010).
6. Gelinck, G. H. et al. Flexible active-matrix displays and shift registers based on solution-processed organic transistors. Nat. Mater. **3**, 106–110 (2005).
7. Minemawari, H. et al. Inkjet printing of single-crystal films. Nature **475**, 364–367 (2011).
8. Yan, H. et al. A high-mobility electron-transporting polymer for printed transistors. Nature **457**, 679–686 (2009).
9. Yuan, Y. et al. Ultra-high mobility transparent organic thin film transistors grown by an off-centre spin-coating method. Nat. Commun **5**, 1–9 (2014).
10. Palfinger, U. et al. Fabrication of n- and p-type organic thin film transistors with minimized gate overlaps by self-aligned nanoimprinting. Adv. Mater. **22**, 5115–5119 (2010).
11. Ante, F. et al. Contact resistance and megahertz operation of aggressively scaled organic transistors. Small **8**, 73–79 (2012).
12. Arias, A. C., MacKenzie, J. D., McCulloch, I., Rivnay, J. & Salleo, A. Materials and applications for large area electronics: solution-based approaches. Chem. Rev. **110**, 2–24 (2010).
13. Natali, D. & Caironi, M. Charge injection in solution-processed organic field-effect transistors: physics, models and characterization methods. Adv. Mater. **24**, 1357–1387 (2012).
14. Li, L. et al. The electrode's effect on the stability of organic transistors and circuits. Adv. Mater. **24**, 3053–3058 (2012).
15. Léonard, F. & Talin, A. A. Electrical contacts to one- and two-dimensional nanomaterials. Nat. Nanotech **6**, 773–783 (2011).
16. Valletta, A. et al. Contact effects in high performance fully printed p-channel organic thin film transistors. Appl. Phys. Lett. **99**, 233309 (2011).
17. Raiteri, D., Cantatore, E. & Van Roermund, A. H. M. Circuit Design on Plastic Foils (Springer International Publishing, 2015).
18. Klauk, H. Organic Electronics II: More Materials and Applications (Wiley-VCH, 2012).
19. Ante, F. et al. Contact doping and ultrathin gate dielectrics for nanoscale organic thin-film transistors. Small **7**, 1186–1191 (2011).
20. Schaur, S., Stadler, P., Meana-Esteban, B., Neugebauer, H. & Sariciftci, N. S. Electrochemical doping for lowering contact barriers in organic field effect transistors. Org. Electron. **13**, 1296–1301 (2012).
21. Gwinner, M. C., Jakubka, F., Gannott, F., Sirringhaus, H. & Zaumseil, J. Enhanced ambipolar charge injection with semiconducting polymer/carbon nanotube thin films for light-emitting transistors. ACS Nano **6**, 539–548 (2012).
22. Asadi, K., Gholamrezaie, F., Smits, E. C. P., Blom, P. W. M. & De Boer, B. Manipulation of charge carrier injection into organic field-effect transistors by self-assembled monolayers of alkanethiols. J. Mater. Chem. **17**, 1947–1953 (2007).
23. Hamadani, B. H., Corley, D. A., Ciszek, J. W., Tour, J. M. & Natelson, D. Controlling charge injection in organic field-effect transistors using self-assembled monolayers. Nano Lett. **6**, 1303–1306 (2006).
24. Liu, Z., Kobayashi, M., Paul, B. C., Bao, Z. & Nishi, Y. Contact engineering for organic semiconductor devices via Fermi level depinning at the metal-organic interface. Phys. Rev. B **82**, 035311 (2010).
25. Chai, Y. et al. Low-resistance electrical contact to carbon nanotubes with graphitic interfacial layer. IEEE Trans. Electron Devices **59**, 12–19 (2012).
26. Torricelli, F., Ghittorelli, M., Colalongo, L. & Kovacs-Vajna, Z. M. Single-transistor method for the extraction of the contact and channel resistances in organic field-effect transistors. Appl. Phys. Lett. **104**, 093303 (2014).
27. Nausieda, I. et al. Dual threshold voltage organic thin-film transistor technology. IEEE Trans. Electron Devices **57**, 3027–3032 (2010).
28. Raiteri, D. et al. A 6 b 10 MS/s current-steering DAC manufactured with amorphous Gallium-Indium-Zinc-Oxide TFTs achieving SFDR > 30 dB up to 300 kHz. IEEE Int. Solid-State Circuits Conf. Digest of Technical Papers **55**, 314–315 (2012).
29. Cosseddu, P., Vogel, J.-O., Fraboni, B., Rabe, J. P., Koch, N. & Bonfiglio, A. Continuous tuning of organic transistor operation from enhancement to depletion mode. Adv. Mater. **21**, 344–348 (2009).
30. Li, Y. et al. Quick fabrication of large-area organic semiconductor single crystal arrays with a rapid annealing self-solution-shearing method. Sci. Rep **5**, 13195 (2015).
31. Abdinia, S. et al. Variation-based design of an AM demodulator in a printed complementary organic technology. Org. Electron. **15**, 904–912 (2014).
32. Lassnig, R. et al. Optimizing pentacene thin-film transistor performance: temperature and surface condition induced layer growth modification. Org. Electron. **26**, 420–428 (2015).
33. Nakayama, K., Hara, K., Tominari, Y., Yamagishi, M. & Takeya, J. Organic single-crystal transistors with secondary gates on source and drain electrodes. Appl. Phys. Lett. **93**, 153302 (2008).
34. Hsu, B. B. Y. et al. Split-gate organic field effect transistors: control over charge injection and transport. Adv. Mater. **22**, 4649–4653 (2010).
35. Rapisarda, M. et al. Analysis of contact effects in fully printed p-channel organic thin film transistors. Org. Electron. **13**, 2017–2027 (2012).
36. Shim, C.-H., Maruoka, F. & Hattori, R. Structural analysis on organic thin-film transistor with device simulation. IEEE Trans. Electron Devices **57**, 195–200 (2010).

37. Street, R. A. & Salleo, A. Contact effects in polymer transistors. *Appl. Phys. Lett.* **81**, 2887–2889 (2002).

38. Brondijk, J. J., Torricelli, F., Smits, E. C. P., Blom, P. W. M. & De Leeuw, D. M. Gate-bias assisted charge injection in organic field-effect transistors. *Org. Electron.* **13**, 1526–1531 (2012).

39. Myny, K. *et al.* An 8-Bit, 40-instructions-per-second organic microprocessor on plastic foil. *IEEE J. Solid-State Circuits* **47**, 284–291 (2012).

40. Huang, C., West, J. E. & Katz, H. E. Organic field-effect transistors and unipolar logic gates on charged electrets from spin-on organosilsesquioxane resins. *Adv. Funct. Mater.* **17**, 142–153 (2007).

41. Calhoun, M. F., Sanchez, J., Olaya, D., Gershenson, M. E. & Podzorov, V. Electronic functionalization of the surface of organic semiconductors with self-assembled monolayers. *Nat. Mater.* **7**, 84–89 (2008).

42. Hirata, I. *et al.* High-resolution spatial control of the threshold voltage of organic transistors by microcontact printing of alkyl and fluoroalkylphosphonic acid self-assembled monolayers. *Org. Electron.* **26**, 239–244 (2015).

43. Torricelli, F. *et al.* Ambipolar organic tri-gate transistor for low-power complementary electronics. *Adv. Mater.* **28**, 284–290 (2016).

44. Mariucci, L. *et al.* Current spreading effects in fully printed p-channel organic thin film transistors with Schottky source-drain contacts. *Org. Electron.* **14**, 86–93 (2013).

45. Brondijk, J. J., Spijkman, M., Torricelli, F., Blom, P. W. M. & de Leeuw, D. M. Charge transport in dual-gate organic field-effect transistors. *Appl. Phys. Lett.* **100**, 023308 (2012).

Acknowledgements

We acknowledge funding from the Dutch Technology Foundation STW, which is the Applied Science Division of the Netherlands Organisation for Scientific Research (NWO), and by the Technology Programme of the Ministry of Economic Affairs. We also thank the Polymer Vision for the transistors fabrication. F.T. and D.R. thank Holst Centre for the patent application submission. F.T. thank Matteo Ghittorelli for the useful discussions, suggestions and for his contribution to Figure 2.

Author contributions

F.T. conceived the DOFET architecture, designed the devices, performed the 2D numerical calculations, measured the transistors, analysed the data, developed the physical analysis and wrote the manuscript. L.C. contributed to the physical analysis and wrote the manuscript. D.R. contributed to the DOFET architecture idea and performed the measurements. Z.M.K.-V. contributed to the performance analysis, supported the manuscript preparation and revised the manuscript. E.C. supervised the project and revised the manuscript. All the authors discussed the results and commented on the manuscript.

Additional information

11

Superlattice-based thin-film thermoelectric modules with high cooling fluxes

Gary Bulman[1], Phil Barletta[1], Jay Lewis[1], Nicholas Baldasaro[1], Michael Manno[2], Avram Bar-Cohen[2] & Bao Yang[2]

In present-day high-performance electronic components, the generated heat loads result in unacceptably high junction temperatures and reduced component lifetimes. Thermoelectric modules can, in principle, enhance heat removal and reduce the temperatures of such electronic devices. However, state-of-the-art bulk thermoelectric modules have a maximum cooling flux q_{max} of only about $10\,\mathrm{W\,cm^{-2}}$, while state-of-the art commercial thin-film modules have a $q_{max} < 100\,\mathrm{W\,cm^{-2}}$. Such flux values are insufficient for thermal management of modern high-power devices. Here we show that cooling fluxes of $258\,\mathrm{W\,cm^{-2}}$ can be achieved in thin-film Bi_2Te_3-based superlattice thermoelectric modules. These devices utilize a p-type Sb_2Te_3/Bi_2Te_3 superlattice and n-type δ-doped $Bi_2Te_{3-x}Se_x$, both of which are grown heteroepitaxially using metalorganic chemical vapour deposition. We anticipate that the demonstration of these high-cooling-flux modules will have far-reaching impacts in diverse applications, such as advanced computer processors, radio-frequency power devices, quantum cascade lasers and DNA micro-arrays.

[1] RTI International, Electronics and Applied Physics Division, Research Triangle Park, North Carolina 27709, USA. [2] Department of Mechanical Engineering, University of Maryland, College Park, Maryland 20742, USA. Correspondence and requests for materials should be addressed to P.B. (email: pbarletta@rti.org) or to B.Y. (email: baoyang@umd.edu).

Thermoelectric modules have long been considered for cooling high-power electronic[1-5] and opto-electronic devices[6], and there have been consistent efforts to improve the cooling flux values of thermoelectric modules in the past decade[7-15]. The maximum cooling flux q_{max} of a thermoelectric module is defined as the maximum cooling flux that the thermoelectric module is capable of providing at a temperature difference across the module (ΔT) of zero and is given by equation (1)[16,17] below.

$$q_{max} = \frac{(S_p - S_n)^2 T_C^2}{4(\rho_p + \rho_n)} \cdot \frac{f}{l} \qquad (1)$$

where S_p and S_n are the Seebeck coefficients of the p- and n-type elements, respectively, ρ_p and ρ_n are the electrical resistivities of the p- and n-type elements, respectively, T_C is the temperature of the cold side of the thermoelectric module, f is the packing fraction and l is the thickness of the thermoelectric elements. Equation (1) shows that q_{max} of the thermoelectric module depends on the thermoelectric material properties, the element thickness and the packing fraction.

One approach for increasing q_{max} is to develop thermoelectric materials with a high thermoelectric figure of merit ZT (refs 1,7-15,18-24), as defined in equation (2).

$$ZT = \frac{S^2}{\rho k} T \qquad (2)$$

where S, ρ, k and T are the Seebeck coefficient, electrical resistivity, thermal conductivity and absolute temperature, respectively. Significant progress has been made in recent years to increase ZT using nanostructured materials such as thin-film superlattices[1,18,25,26], thick films of quantum-dot superlattices[14] and nanocomposites[7-9].

Alternatively, q_{max} of the thermoelectric module can be increased by reducing the element thickness, as q_{max} is inversely proportional to l (refs 1,18,26). However, the reduction of element thickness is limited by two factors: the synthesis method for the thermoelectric elements; and the electrical contact resistance between the thermoelectric elements and the copper trace. Electrical constant resistance has a marked impact on the performance of thin thermoelectric modules since the magnitude of the contact resistance can be comparable to that of thermoelectric element itself. Bulk thermoelectric materials cannot be thinned below a few hundred microns, resulting in the modest maximum cooling flux of approximately $10\,W\,cm^{-2}$ (refs 27,28). Epitaxial semiconductor films can be grown much thinner, resulting in a higher maximum cooling flux for thin-film modules. As a point of reference, commercially available thin-film thermoelectric modules have elements that are approximately 20-μm thick and have maximum cooling fluxes around

$100\,W\,cm^{-2}$ (ref. 29). However, even the cooling flux of state-of-the art thermoelectric modules is insufficient for advanced thermal management application. For instance, heat generation in a silicon microprocessor is non-uniform, with localized heat flux possibly larger than $200\,W\,cm^{-2}$, and GaN-based transistors can produce hotspots with heat fluxes in excess of $1\,kW\,cm^{-2}$ (refs 4,30,31).

In this paper, we present thermoelectric module capable of producing a cooling flux of $258\,W\,cm^{-2}$, more than double that of the current state-of-the-art value. The enhancement in the module-level cooling flux is a result of thin (8.1 μm) Bi_2Te_3-based thin-film superlattice materials with high intrinsic ZT, the δ-doped n-type structure, and reduced electrical contact resistances (2.68×10^{-7} $\Omega\,cm^2$ for n-type and $1.36 \times 10^{-6}\,\Omega\,cm^2$ for p-type).

Results

Material synthesis and module fabrication. Figure 1a,b show the structure of the thermoelectric module. The thermoelectric material at the heart of the module consists of p-type 10/50 Å Bi_2Te_3/Sb_2Te_3 superlattice and n-type δ-doped $Bi_2Te_{3-x}Se_x$, both of which are grown heteroepitaxially using metalorganic chemical vapour deposition[18]. The n-type structure is grown by periodically interrupting the growth of $Bi_2Te_{3-x}Se_x$ and dosing the flow with Te and Se species. This δ-doping process can result in an increase in carrier concentration without a reduction in electron mobility. In the present experiment, 8.1-μm-thick thin films are grown and fabricated into cooling devices.

The ZT values of the Bi_2Te_3/Sb_2Te_3 superlattice materials used in this study have been previously measured by two different methods. One is the direct measurement of ZT by Harman method, which reported ZT > 2 in ref. 18. The second method is the determination of individual thermoelectric properties, such as Seebeck coefficient, electrical resistivity and thermal conductivity, followed by calculation of ZT. The ZT values of representative p- and n-type materials used in the devices in the present study were estimated, using this indirect technique, to be 1.4 and 1.5, respectively. These data have been added in Table 1. These values were determined through measurement of the in-plane electrical resistivity and Seebeck coefficient of representative material along with an estimation of thermal conductivity via an in-couple-property-validation model. Details of this model are given in ref. 32. Both direct measurement and indirect measurement of ZT were conducted at $T = 300\,K$. These estimated ZT values by the indirect method are slightly less than the ZT data measured directly via the Harman method for similar materials in ref. 18.

In the device fabrication process, the top epitaxial surfaces of the p- and n-type materials are first metalized and then bonded to a common metalized AlN die header. Following chemical removal of the substrate, electrical contacts are fabricated on the exposed surface (typically with two n- and two p-type circular contacts designated as the 2N–2P configuration). This die header subassembly is then inverted and bonded to a second AlN header that contains the electrical traces used to power the module.

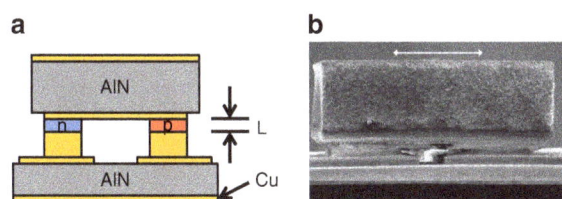

Figure 1 | Cross-sectional views of a Bi_2Te_3-based thin-film thermoelectric module. (a) Illustration of thin-film-based thermoelectric module, showing top and bottom AlN headers, Cu traces and n-type and p-type active elements. L represents the length of the active elements, which is 8 μm in the present work. Figure is not to scale. (b) Scanning electron microscope image of the upper portion of a completed thin-film superlattice thermoelectric module. Scale bar, 250 μm.

Table 1 | ZT data from representative p- and n-type thermoelectric materials.

	Electrical resistivity ($\Omega\,m$)	Seebeck coefficient ($\mu V\,K^{-1}$)	Thermal conductivity ($W\,m^{-1}\,K^{-1}$)	ZT
p-type Bi_2Te_3/Sb_2Te_3	1.02e-5	238	1.2	1.4
n-type δ-doped $Bi_2Te_{3-x}Se_x$	1.37e-5	−276	1.1	1.5

As shown in Fig. 1b, the basic completed module structure has a $600 \times 600\text{-}\mu m^2$ top header area and is $550\,\mu m$ tall.

Electrical contact resistance. One significant barrier to reaching high cooling fluxes with thin-film thermoelectric modules is electrical contact resistance between the metal electrodes and the thermoelectric elements, especially for Bi_2Te_3-based materials with low intrinsic electrical resistivity ($10^{-3}\,\Omega\,cm$)[1,26]. In the present experiment, two metallization methods (plated Au and evaporated Cr/Ni/Au metallization) and two different superlattice structures (standard structures and δ-doped structures) are explored to improve the electrical contact resistance. Au diffusion into the Bi_2Te_3 lattice may very well be happening in the thermoelectric devices with plated Au contacts. The detailed effects of this potential Au diffusion need further study in the future.

Electrical contact resistivity has been measured using the transmission line measurement technique[33], which measures the resistance across an annular gap as a function of gap width, fabricated on a broad area of metallization on the top surface of the thin-film superlattice material. The metalized contacts are formed on the thin-film superlattice surface as shown in Fig. 2, through the use of photoresist masks (gold is the metal contact, white is the thin-film superlattice surface) to form a set of six gaps with different gap widths. A four-wire probe is used to apply a small current across each gap (with two of the probe wires) and measure the voltage drop across the thin-film superlattice gap (with the other two probe wires). The gap resistance R_g (measured ΔV over current) is a function of the gap width, governed by the relationship shown in equation (3).

$$R_g = \frac{R_s}{2\pi}\left[\ln\left(\frac{R_1}{R_1 - d}\right) + L_T\left(\frac{1}{R_1 - d} + \frac{1}{R_1}\right)\right] \quad (3)$$

The gap resistance R_g versus gap width d data is fitted to equation (3), with R_s (sheet resistance) and L_T (transfer length) as fitting parameters. Contact resistivity ρ_c can then be calculated as

$R_s \times L_T^2$ with units of $\Omega\,m^2$. In the present modules, plated Au was used as the contact to the source (top) side of the n-type δ-doped $Bi_2Te_{3-x}Se_x$, while evaporated Cr/Ni/Au was used as the source (top) side of the p-type Sb_2Te_3/Bi_2Te_3. Plated Au was used as the sink (bottom) side contact of both the n- and p-type elements. Electric contact resistivity for thin-film superlattice materials and selected contact metals is shown in Table 2. The R_s, L_T and ρ_c values given in Table 2 are an average of the 6–8 experimental values measured for each material type and metallization scheme. The s.d. among these measurements is provided as well.

Maximum temperature difference ΔT_{max}. To evaluate the maximum temperature difference ΔT_{max}, the thermoelectric module is placed directly on a water-cooled heat sink, maintained at $25\,^\circ C$, and the T_C and T_H values are measured using $25\,\mu m$ diameter thermocouples as a function of current supplied to the module, I. The voltage V is also monitored as a function of electric current under vacuum (pressure, $P < 1\,mTorr$), up to the current producing the maximum ΔT_{max}, which defines the current value of I_{max}.

The measured temperature difference and voltage behaviour of the thermoelectric module are given by[28,34]:

$$\Delta T = \frac{S}{K}T_C I - \frac{R}{2K}I^2 - R_{th}IV - Q_P\left(\frac{1}{K} + R_{th}\right) \quad (4)$$

$$V = \frac{IR + S\Delta T + SR_{th}Q_P}{1 - SR_{th}I}, \quad (5)$$

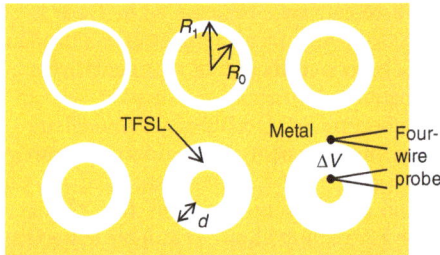

Figure 2 | Example of circular TLM patterns used for contact resistivity determination. In this measurement technique, six gaps of increasing widths ($d = R_1 - R_0$) are patterned onto the sample surface. The gold colour represents the metal contact, while the white is the thin-film superlattice (TFSL) surface. A four-wire probe is used to apply a small current across each gap and measure the voltage drop (ΔV).

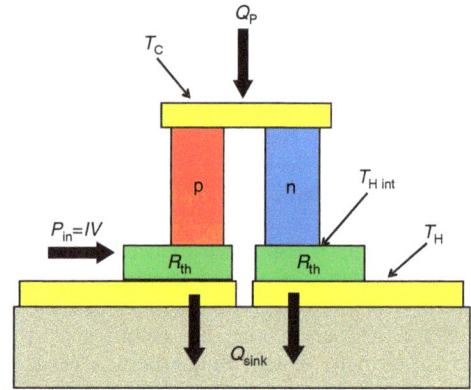

Figure 3 | Thin-film thermoelectric module structure. The red and blue rectangles represent the p- and n-type active TE materials, respectively. Yellow represents the current traces and tan represents the ceramic bottom header of the module. Heat entering the module is given by Q_P, while the heat being pumped out is given by Q_{sink}. T_C, T_H and $T_{H\,int}$ are the cold-side temperature, externally measured hot-side temperature and internal element hot-side temperature, respectively. The green rectangles represent the parasitic thermal resistances R_{th}, which reduce the externally observed ΔT value. Figure is not drawn to scale.

Table 2 | Specific electric contact resistivity, as measured by transmission line model (TLM) technique, for superlattice thermoelectric elements with different structures and metallization.

Sample	Growth information		Contact metal	Contact resistivity			
	Type	Target structure		R_s (Ω per sq)	L_T (μm)	ρ_c ($\Omega\,cm^2$)	ρ_c (s.d.)
A	n	δ-doped n type	Plated Au	1.57	4.20	2.68e-7	6.88e-8
B	p	Bi_2Te_3/Sb_2Te_3	Plated Au	0.93	12.26	1.36e-6	4.08e-7
C	n	δ-doped n type	Evap Cr/Ni/Au	1.94	7.81	1.16e-6	1.73e-7
D	p	Bi_2Te_3/Sb_2Te_3	Evap Cr/Ni/Au	1.15	11.74	1.42e-6	7.95e-7

Evap, evaporated.

In the above equations, S is the module Seebeck coefficient, which is determined by the voltage between the two electrical leads, V, and the temperature difference between the two AIN headers, ΔT. The module electric resistance is denoted by R, and consists of the electric resistance of the n and p elements $R_{element}$, the electric resistance of the metal traces R_{trace}, and the electric contact resistance between the elements and the metal traces $R_{contact}$. The module electric resistance R can be determined by the voltage between the two electrical leads, V, and the electrical current I through the electrical leads. K is the module thermal conductance between the two AIN headers. R_{th} is the parasitic thermal resistance of the module, which is the difference between the module thermal resistance and the thermal resistance of the n and p thermoelectric element pair (Fig. 3). In the ΔT_{max} measurements, there is no heat pumped by the module (cooling power, $Q_P = 0$) and the module parameter values can be determined by fitting equation (4) to the experimental ΔT as a function of I, V, T_C and T_H data, and similarly fitting equation (5) to the voltage data. The inclusion of a parasitic thermal resistance R_{th} is necessary when considering thin-film thermoelectric

modules, because the high-heat fluxes that occur within the module produce an internal element hot-side temperature $T_{H\ int}$ that is different from the externally measured value T_H.

The upper portion of Fig. 4 shows ΔT versus I data for the module with the largest contact diameter (230 μm). The circles indicate the experimental external ΔT values measured with thermocouples, and the curve drawn through these points is the fit of equation (4) to the data, using a parasitic thermal resistance of $R_{th} = (3.08 \pm 1.98)\,\mathrm{K\,W^{-1}}$. An external ΔT_{max} value of 43.54 K is observed at an electric current of 14.8 A, and the upper dashed curve shows the predicted internal ΔT ($T_{H\ int} - T_C$) occurring inside the module, which has a maximum value of 49.3, 5.7 K higher than the external value. The difference in internal and external ΔT is caused by the internal thermal resistance R_{th}, which reduces the external ΔT that can be measured by the thermocouples. The multidimensional fit of equation (4) to experimental $\Delta T_{external}$ versus I data also yields the values of the ratios S/K (0.0228 μV W^{-1}) and R/K (0.349 Ω K W^{-1}).

Figure 4b shows the voltage versus current data for the same module. R can be determined via a one-dimensional least squares fit of equation (4) to this experimental V versus I data, which subsequently allows for the calculation of K and S from the previously determined R/K and S/K values. The resulting values of the total Seebeck coefficient $S = (450 \pm 48)$ μV K^{-1}, thermal conductance $K = (19.7 \pm 2.1)$ mW K^{-1} and electric resistance $R = (6.87 \pm 0.01)$ mΩ are given in the legend of Fig. 4b. Details of this procedure have been described elsewhere[34].

Maximum cooling flux q_{max}. The cooling power for a thermoelectric module in the presence of a parasitic thermal resistance R_{th} is shown in equation (6), where ΔT is the externally measured temperature difference[28,34].

$$Q_P = \frac{1}{1 + KR_{th}}\left[ST_C I - \frac{1}{2}RI^2 - KR_{th}IV - K\Delta T\right] \quad (6)$$

In the case of $R_{th} = 0$, equation (6) reduces to the standard thermoelectric heat-pumping equation[35], where the maximum cooling power Q_P occurs when $I = ST_C/R$ (which is I_{max}). To determine I_{max} in the case of non-zero R_{th}, the voltage dependence in equation (6) is eliminated using equation (5), and after some rearrangement the expression for Q_P becomes:

$$Q_P = \frac{\left(ST_C I - \frac{1}{2}RI^2\right)(1 - SR_{th}I) - KRR_{th}I^2 - K\Delta T}{(1 + KR_{th})(1 - SR_{th}I) + KR_{th}^2 SI} \quad (7)$$

Equation (7) is then differentiated with respect to I and solved numerically for the value of I that yields a zero of the resulting expression. The I_{max} value is then substituted into equation (7), along with the individual fitted module parameter values to obtain the Q_{max} value that is expected. The I_{max} values calculated using equation (5) are found to agree with the experimentally observed values in these modules and are smaller than what is obtained from using the standard expression $I_{max} = ST_C/R$. The maximum cooling flux q_{max} is calculated by dividing the total cooling power Q_{max} by the top header area of the module, which

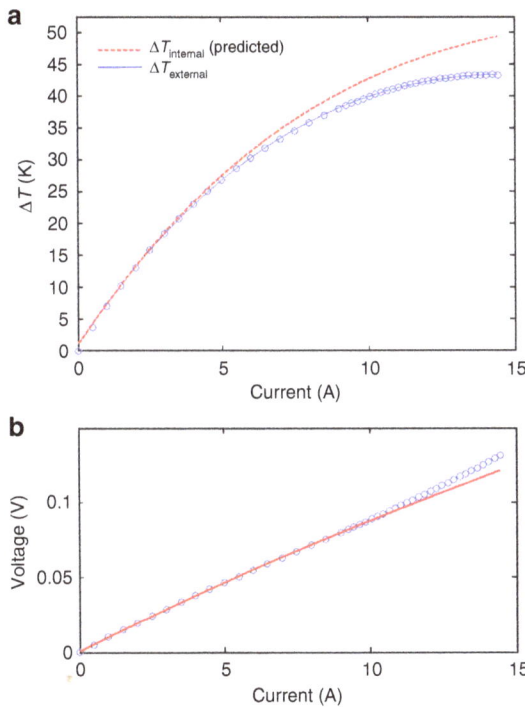

Figure 4 | Analysis of experimental data. (a) Experimental ΔT versus I data for the module with the largest effective contact diameter (230 μm) and the fit of equation 4 to the data. Blue dots represent experimentally measured data points, and the blue line represents the fit. The red dotted line indicates predicted internal temperature difference ($T_{hint} - T_C$) versus I. **(b)** Corresponding V versus I data for the same module. The blue dots represent experimentally measured data points, and the red line represents the theoretical fit.

Table 3 | Performance of thin-film thermoelectric modules of varying contact diameter (that is, packing fraction).

Diameter (μm)	ΔT_{max} (K)	I_{max} (A)	Q_{max}/A (W cm^{-2})			Coefficient of performance
			Predicted	Measured	% Difference	
130	45.58	9.91	183.3	158.3	− 13.6	0.86
180	40.49	11.99	213.9	184.1	− 13.9	0.40
200	36.06	14.32	247.2	213.7	− 13.6	0.72
230	43.54	14.78	294.4	257.6	− 12.5	0.51

is $600 \times 600\,\mu m^2$. The performance of four thin-film thermoelctric modules with different contact diameters (that is, the packing fraction) is summarized in Table 3. The table shows the measured maximum cooling flux ($q_{max} = Q_{max}/A_T$), as well as the predicted value and a discrepancy of 13% is observed between the predicted and measured values. The coefficient of performance (COP) values were calculated at q_{max}; that is for $\Delta T = 0$ K. With the

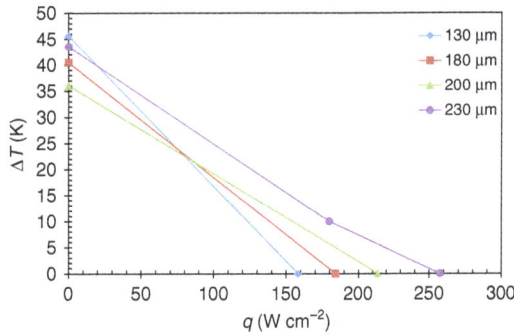

Figure 5 | Load lines for each of the four measured modules of this work. The load lines are determined by the measurement of q_{max} at $\Delta T = 0$ K, and the measurement of ΔT_{max} at $q = 0$ W m^{-2}. An additional data point for the case of $\Delta T = 10\,^\circ$C is included for the module with the largest contact diameter (230 μm).

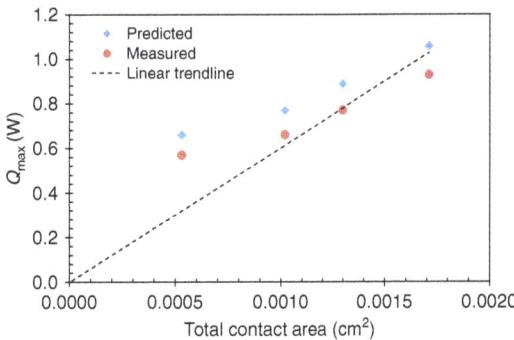

Figure 6 | Q_{max} versus total contact area data for the four thin-film thermoelectric modules of this work. Experimentally measured values are represented by red squares, and the corresponding theoretically predicted values are represented by blue diamonds. The dotted line represents a linear trendline fit to the experimental data. This trendline has been indexed to the origin and is expressed by the equation $y = 601.6x$.

establishment of the q_{max} value and ΔT values, the load lines can be determined, as shown in Fig. 5.

In the case of the largest module, an additional data point that was obtained at $\Delta T = 10\,^\circ$C is also shown. The maximum temperature difference ΔT_{max} should not be dependent on the module geometry, but the variation seen indicates a possible undetermined parasitic phenomenon may be present. Further studies are needed to identify this undetermined parasitic phenomenon.

It can be seen that a maximum cooling flux in excess of 250 W cm^{-2} is achieved in the thin-film Bi_2Te_3 superlattice thermoelectric module with a contact diameter of 230 μm (that is, 48% packing fraction). This value is 25 times higher than is typically observed in commercial-off-the-shelf bulk thermoelectric modules (http://www.marlow.com) and more than 2.5 times better than commercial-off-the-shelf thin-film modules (http://www.lairdtech.com). The maximum cooling flux of the module was measured using two different methods, the Q-meter method by RTI and the non-contact IR method by University of Maryland.

Contact diameter and packing fraction. Figure 6 shows the dependence of maximum cooling power Q_{max} as a function of the total element contact area A_c (including both n and p contacts). The top header area A_T is $600 \times 600\,\mu m^2$ for all of the modules tested. The packing fraction is the ratio of the total element contact area A_c to the top header area of the module A_T, and the module with a contact diameter of 230 μm has a packing fraction of 48%.

The calculated Q_{max} values shown in Fig. 6 are consistently 12–14% larger than what is experimentally observed, which is likely due to other internal parasitic thermal resistances not accounted for in the model shown in equations (4) and (6). Specifically, the internal thermal resistance values that are used in these calculations are obtained using ΔT_{max} measurements under the condition of no heat pumping ($Q_P = 0$). In this case the only internal thermal resistance that the measurement is sensitive to is located below the thermoelectric elements, as indicated by R_{th} in Fig. 3. However, when heat is being pumped by the module through the top header, as is the case in these heat flux measurements ($Q_P > 0$), any additional thermal resistances, such as thermal spreading resistance that occurs in the module regions above the thermoelectric elements will be important. This additional internal thermal resistance will further reduce the thermoelectric heat pumping in a manner similar to that seen for the lower internal thermal resistance. Capturing the additional

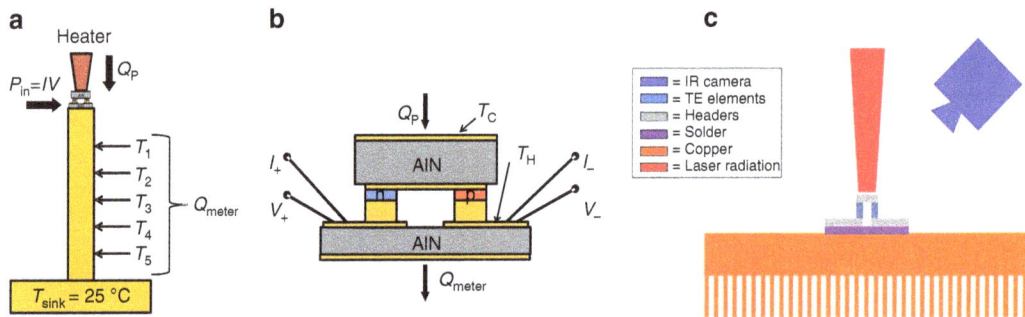

Figure 7 | Experimental arrangements for the measurement of heat pumped by the thermoelectric module. (a) Schematic of a thermoelectric module mounted on a Q-meter. T_1 through T_5 represent the thermocouple measurements along the length of the Q-meter. **(b)** Schematic of a thermoelectric module showing the heat pumped by the module (Q_P) and additional contributions of the electrical input power (P_{in}) and wire conduction/heat generation (P_{wire}) to the heat measured by the Q-meter (Q_{meter}). The placement of the thermocouples on the cold side and hot side of the thermoelectric module is indicated by T_C and T_H, respectively. **(c)** Testing apparatus used to perform non-contact measurements of cooling flux q and temperature difference ΔT of the thermoelectric modules.

internal thermal resistances in the model will reduce the calculated heat pumping and improve agreement between predicted and measured values.

Discussion

The basis for the larger cooling fluxes produced by thinner thermoelectric elements is the reduced electrical resistance of the thermoelectric structure, which in turn allows for the use of larger electrical currents. The higher currents produce more Seebeck heat pumping per unit area, as shown in equation (8).

$$Q_P = ST_C I - 0.5I^2 R - K(T_H - T_C)$$

$$= nsT_C I - 0.5I^2 \frac{nl\rho}{A} - \frac{knA}{l}(T_H - T_C) \qquad (8)$$

where Q_P is the amount of heat pumped; S, R and K are the Seebeck coefficient, electrical resistance and thermal conductance of the thermoelectric module, respectively; T_C and T_H are the heat source and heat sink temperatures, respectively; I is electric current; n is the number of thermoelectric couples in the module; l is the thickness of one semiconductor leg; A is the cross-sectional area of one semiconductor leg (n and p are assumed to have identical geometry); and s, $\rho l/A$ and kA/l are the per-couple Seebeck coefficient, electrical resistance and thermal conductance, respectively. Equation (8) does not consider the parasitic thermal and electric resistance in the module.

The input electrical current that corresponds to the case of max Q_P can be calculated from the first derivative of the $Q_P = f(I)$ function and is equal to ST_C/R. For bulk thermoelectric modules, the maximum value of Q_P is low, because bulk thermoelectric material cannot be thinned below a thickness of few hundred microns, limiting the maximum current due to the high module electric resistance R. The electrical current limitation results in bulk thermoelectric heat-pumping capabilities in the $1–10\,\mathrm{W\,cm}^{-2}$ range.

Epitaxial semiconductor films can be grown much thinner (for example, hundreds of nm), decreasing the electrical resistance and allowing larger electrical currents producing much higher cooling fluxes. For any given reduction in l by a factor of α, optimized I will increase by α. With all other things being equal, the cooling flux can be increased by a factor of 2, 5 or 10 times simply by decreasing the film thickness by a factor of 2, 5 or 10 times and increasing the electrical current accordingly. Theoretically, cooling fluxes, on the order of several hundred $\mathrm{W\,cm}^{-2}$, can be achieved simply by thinning the thermoelectric material. However, there are several parasitic effects that need to be overcome to achieve the full potential of thin-film thermoelectric modules. The most significant barrier to high cooling fluxes is electrical contact resistance between metal electrodes and the semiconductor layers. In thin thermoelectric modules, the magnitude of the contact electric resistance can be comparable to the values of the resistance of the thermoelectric element itself, which will increase the total electric resistance of the module and reduce its maximum cooling flux. In this work, a low electric contact resistivity ρ_c in the range of $1–2 \times 10^{-6}\,\Omega\,\mathrm{cm}^2$ has been achieved, using the evaporation of Cr/Ni/Au to fabricate the metal electrodes as well as δ-doped superlattice structures. Further reduction to the order of $10^{-8}\,\Omega\,\mathrm{cm}^2$ is desired for thin-film thermoelectric module with element thickness $<2\,\mu\mathrm{m}$.

Another barrier to wide adoption of thin-film thermoelectric modules is thermal resistances in the thermal management system. For heat flux values of several hundred $\mathrm{W\,cm}^{-2}$, low thermal resistances, on the order of $0.03–0.05\,\mathrm{cm}^2\,\mathrm{K\,W}^{-1}$, need to be present on both the hot and cold sides of the thermoelectric module for efficient integration of thermoelectric coolers into the system. In particular, a low thermal resistance present at the hot

side takes the higher priority, due to the greater heat flow through the hot side, with $Q_{hot} = Q_{cold}\,(1 + 1/\mathrm{COP})$. This requires more advanced heat exchangers and cold plates.

Methods

Q-meter method for characterization of thin-film thermoelectric modules. To evaluate the heat-pumping capacity, the thermoelectric modules are placed on top of a Q-meter (Fig. 7), a metal post of known dimensions and thermal conductivity, used to determine the total heat flow passing through it by measuring the temperature gradient along its length. The Q-meter base is mounted to a water-cooled heat sink and the thermoelectric module is mounted to the top. The thermoelectric module is energized using a four-wire current–voltage measurement system employing thin Cu wires, approximately 2.5 cm long (either 250 or 500 μm in diameter), to connect the thermoelectric module to external instrumentation. Heat is applied to the top of the thermoelectric module using a resistive heater. The top and bottom thermoelectric module temperatures T_C and T_H, respectively, are read using 25 μm diameter thermocouples precisely positioned on the thermoelectric module. The entire system is operated under vacuum ($P<1\,\mathrm{mTorr}$) to minimize convective losses. The water-cooled heat sink on the Q-meter is adjusted to maintain the thermoelectric hot-side temperature (T_H) at about 25 °C. Since the test chamber is evacuated, convective losses are minimized. Thus, the heat flowing down the Q-meter is the sum of heat being pumped by the thermoelectric module, Q_P, the electrical input power flowing into the thermoelectric module, P_{in} and the thermal contribution of the four wires connecting the thermoelectric module to the external circuit, P_{wire} ($Q_{meter} = Q_P + P_{in} + P_{wire}$).

Non-contact method for characterization of thin-film thermoelectric modules. The performance of the thermoelectric modules was also characterized using the non-contact IR method. Figure 7c shows a schematic of the apparatus used for testing, consisting of a heat sink, the thermoelectric module, a laser and an infrared camera. The testing procedure consisted of first providing the thermoelectric module with electrical power, creating a temperature difference across the module. The laser was then used to heat the cold side of the thermoelectric module, decreasing the module-level ΔT. The laser power was gradually increased until the ΔT across the module was 0 K, indicating that the maximum heat pumping for that current had been reached. The process was repeated at several electrical powers to determine the electrical current that produced the maximum module heat pumping. The output power of the laser was determined using a calibration curve, which was created in situ, taking into account all loses in lenses and optical equipment. The emissivity of the cold side of the thermoelectric module was determined using the infrared camera, and from Kirchhoff's law, the absorptivity of the cold side was assumed to be equal to the emissivity. Thus, the total amount of power being pumped through the module could be readily calculated.

Nomenclature. A Nomenclature table defining all of the variables used in this work is given in Supplementary Note 1.

References

1. Chowdhury, I. et al. On-chip cooling by superlattice-based thin-film thermoelectrics. *Nat. Nanotech.* **4**, 235–238 (2009).
2. Mahajan, R. et al. Cooling a microprocessor chip. *Proc. IEEE* **94**, 1476–1486 (2006).
3. Prasher, R. S. et al. Nano and micro technology-based next-generation package-level cooling solutions. *Intel Tech. J.* **9**, 285–296 (2005).
4. Meysenc, L. et al. Power electronics cooling effectiveness versus thermal inertia. *IEEE Trans. Power Electron.* **20**, 687–693 (2005).
5. Joo Goh, T. et al. Thermal investigations of microelectronic chip with non-uniform power distribution: temperature prediction and thermal placement design optimization. *Microelectron. Int.* **21**, 29–43 (2004).
6. Semenyuk, V. Miniature thermoelectric modules with increased cooling powerRotter, M. (Ed.)in *Proc. ICT'06 25th International Conference* 322–326 (2006).
7. Kim, S. et al. Dense dislocation arrays embedded in grain boundaries for high-performance bulk thermoelectrics. *Science* **348**, 109–114 (2015).
8. Poudel, B. et al. High-thermoelectric performance of nanostructured bismuth antimony telluride bulk alloys. *Science* **320**, 634–638 (2008).
9. Biswas, K. et al. High-performance bulk thermoelectrics with all-scale hierarchical architectures. *Nature* **489**, 414–418 (2012).
10. Bell, L. Cooling heating, generating power, and recovering waste heat with thermoelectric systems. *Science* **321**, 1457–1461 (2008).
11. Heremans, J. P. et al. Enhancement of thermoelectric efficiency in PbTe by distortion of the electronic density of states. *Science* **321**, 554–557 (2008).
12. Majumdar, A. Thermoelectricity in semiconductor nanostructures. *Science* **303**, 777–778 (2004).
13. Hsu, K. F. et al. Cubic AgPbmSbTe$_{2+m}$: bulk thermoelectric materials with high figure of merit. *Science* **303**, 818–821 (2004).

14. Harman, T. C. *et al.* Quantum dot superlattice thermoelectric materials and devices. *Science* **297,** 2229–2232 (2002).

15. Chung, D. Y. *et al.* CsBi₄Te₆: a high-performance thermoelectric material for low-temperature applications. *Science* **287,** 1024–1027 (2000).

16. Rowe, D. M. (ed.) *Thermoelectrics Handbook: Macro to Nano* (CRC Press, 2005).

17. Yang, B. & Wang, P. *Encyclopedia of Thermal Packaging, Volume 4: Thermoelectric Microcoolers* (World Scientific, 2013).

18. Venkatasubramanian, R. *et al.* Thin-film thermoelectric devices with high room-temperature figures of merit. *Nature* **413,** 597–602 (2001).

19. Jood, P. *et al.* Al-doped zinc oxide nanocomposites with enhanced thermoelectric properties. *Nano Lett.* **11,** 4337–4342 (2011).

20. Chen, X. *et al.* Effects of ball milling on microstructures and thermoelectric properties of higher manganese silicides. *J. Alloys Compd.* **641,** 30–36 (2015).

21. Kraemer, D. *et al.* High-performance flat-panel solar thermoelectric generators with high thermal concentration. *Nat. Mater.* **10,** 532–538 (2011).

22. Majumdar, A. Thermoelectric devices: helping chips to keep their cool. *Nat. Nanotechnol.* **4,** 214–215 (2009).

23. Su, X. L. *et al.* Self-propagating high-temperature synthesis for compound thermoelectrics and new criterion for combustion processing. *Nat. Commun.* **5,** 4908 (2014).

24. Wu, H. J. *et al.* Broad temperature plateau for thermoelectric figure of merit ZT > 2 in phase-separated PbTe₀.₇S₀.₃. *Nat. Commun.* **5,** 4515 (2014).

25. Shakouri, A. Nanoscale thermal transport and microrefrigerators on a chip. *Proc. IEEE* **94,** 1613–1638 (2006).

26. Zhang, Y. *et al.* in *ASME 2005 Pacific Rim Technical Conference and Exhibition on Integration and Packaging of MEMS, NEMS, and Electronic Systems collocated with the ASME 2005 Heat Transfer Summer Conference* 2189–2197 (2005).

27. Rowe, D. M. (ed.) *CRC Handbook of Thermoelectrics* (CRC Press, 1995).

28. Bulman, G. E. *et al.* Three-stage thin-film superlattice thermoelectric multistage microcoolers with a ΔT max of 102 K. *J. Electron. Mater.* **38,** 1510–1515 (2009).

29. Habbe, B. & Nurnus, J. Thin film thermoelectrics today and tomorrow. *Electron. Cooling* **17,** 24–31 (2011).

30. Wang, P., Yang, B. & Bar-Cohen, A. Mini-contact enhanced thermoelectric coolers for on-chip hot spot cooling. *Heat Transfer Eng.* **30,** 736–743 (2009).

31. Yang, B. *et al.* Mini-contact enhanced thermoelectric cooling of hot spots in high power devices. *IEEE Trans. Compon. Packag. Technol.* **30,** 432–438 (2007).

32. Lents, C. E. *et al.* DARPA ACM program final report. Contract No. N66001-11-C-4053 (2013).

33. Schroder, D. *Semiconductor Material and Device Characterization* (Wiley-IEEE Press, 2015).

34. Bulman, G. *et al.* Large external ΔT and cooling power densities in thin-film Bi₂Te₃-superlattice thermoelectric cooling devices. *Appl. Phys. Lett.* **89,** 122117 (2006).

35. Nolas, G. S., Sharp, J. & Goldsmid, H. J. *Thermoelectrics: Basic Principles and New Materials Developments* (Springer, 2001).

Acknowledgements

G.B. and P.B. acknowledge the financial support of RTI International. B.Y. and M.M. acknowledge the financial support by DARPA ICECool programme (award number 019266). P.B. acknowledge Chuck Lents and Joe Turney of United Technologies Research Center for their assistance in thermal conductivity modelling.

Author contributions

G.B., P.B. and B.Y. conceived this paper; G.B., P.B., J.L., M.M. and B.Y. contributed to the experiments; N.B. and A.B.-C. contributed intellectually; all authors discussed the results and contributed to the revision of the final manuscript.

Additional information

12

Giant photostriction in organic–inorganic lead halide perovskites

Yang Zhou[1,*], Lu You[1,*], Shiwei Wang[1], Zhiliang Ku[2], Hongjin Fan[2], Daniel Schmidt[3], Andrivo Rusydi[3], Lei Chang[1], Le Wang[1], Peng Ren[1], Liufang Chen[4], Guoliang Yuan[4], Lang Chen[5] & Junling Wang[1]

Among the many materials investigated for next-generation photovoltaic cells, organic–inorganic lead halide perovskites have demonstrated great potential thanks to their high power conversion efficiency and solution processability. Within a short period of about 5 years, the efficiency of solar cells based on these materials has increased dramatically from 3.8 to over 20%. Despite the tremendous progress in device performance, much less is known about the underlying photophysics involving charge–orbital–lattice interactions and the role of the organic molecules in this hybrid material remains poorly understood. Here, we report a giant photostrictive response, that is, light-induced lattice change, of >1,200 p.p.m. in methylammonium lead iodide, which could be the key to understand its superior optical properties. The strong photon-lattice coupling also opens up the possibility of employing these materials in wireless opto-mechanical devices.

[1] School of Materials Science and Engineering, Nanyang Technological University, Block N4.1-02-24, 50 Nanyang Avenue, Singapore 639798, Singapore. [2] School of Physical and Mathematical Sciences, Nanyang Technological University, Singapore 639798, Singapore. [3] Singapore Synchrotron Light Source, National University of Singapore, 5 Research Link, Singapore 117603, Singapore. [4] Department of Materials Science and Engineering, Nanjing University of Science and Technology, Nanjing 210094, China. [5] Department of Physics, South University of Science and Technology of China, Shenzhen 518055, China. * These authors contributed equally to this work. Correspondence and requests for materials should be addressed to L.Y. (email: mailyoulu@gmail.com) or to J.W. (email: jlwang@ntu.edu.sg).

The past few years witnessed the explosion of research on photovoltaic cells based on the hybrid organic–inorganic perovskites[1–5], in particular methylammonium lead iodide (MAPbI₃). Concomitantly, these materials have also been explored for lasers[6], light emitting diodes[7] and photodetectors[8]. Besides improving the photovoltaic cell efficiency, much work has been devoted to the mechanism behind their extraordinary performances. Anomalously long lifetime and diffusion length of photo-carriers have been observed and related to the high efficiency[9–12]. In solution processed thin films of mixed halide perovskites, carrier lifetime of longer than $1 \mu s$[13] and diffusion length exceeding $1 \mu m$[11] were reported. These values can be further increased in high-quality MAPbI₃ single crystals with greatly suppressed trap-state densities[14,15], suggesting an even higher attainable efficiency approaching the Shockley–Queisser limit[16]. However, to become commercially viable, the long-term stability of these materials has to be significantly improved. Despite the intensive research efforts, the origin of the long-carrier lifetime and diffusion length remains elusive. The bi-molecular recombination rate deviates from Langevin theory by orders of magnitude, which underlines the unique attribute of hybrid perovskites compared with conventional low-mobility semiconductors, but is reminiscent of disordered material systems such as amorphous Si and organic solar cells[17,18]. These clues lead us to ponder on the role of the organic molecules on the material's unusual photophysical properties. These molecules are dynamically tumbling inside the inorganic scaffold due to the small rotational energy barrier[19,20]. Even though the organic group is not directly involved in the electronic structure around the band edges, it may interact with the inorganic PbI₆ octahedron through its rotational degree of freedom, as revealed in recent density functional theory (DFT) calculations[21,22]. Experimentally, hydrogen bonding between the halides and the amine group at room temperature was confirmed by infrared spectroscopy[23], supporting the theoretical results.

Here we provide yet another evidence for the interaction between the organic and inorganic moieties in the hybrid perovskites. A giant photostrictive response, namely, light-induced lattice change, of $>1,200$ p.p.m. was observed in MAPbI₃. Careful analysis suggests that the strong photon-lattice coupling may arise from the weakening of the hydrogen bonding between N and I by the photo-generated carriers. Not only could it shed light on the anomalous photophysics, this discovery also opens up new possibilities for these fascinating materials, enabling novel device paradigms such as photo-driven microsensing and microactuation[24,25].

Results

Photostriction in MAPbI₃ single crystals. MAPbI₃ single crystals as large as $10 \times 8 \times 8$ mm are prepared using solution growth method. Photograph of a typical sample with well-defined facets used in this study is shown in Fig. 1a. Only those peaks corresponding to the tetragonal ($l00$) planes (Fig. 1b) are present in the X-ray diffraction scan (Fig. 1e), confirming the high quality of the single crystal. Based on the absorption coefficient deduced from the spectroscopic ellipsometry (Supplementary Fig. 1), an abrupt excitonic absorption at 1.575 eV has been determined (Fig. 1f), which is consistent with previous reports[14,15]. The topography and photo-induced dimension change are investigated using an atomic force microscope (AFM) over the atomically flat surface of the sample as shown in Fig. 1c,d.

Interestingly, when light is shining on the crystal, a sudden change in the dimension is observed. Figure 2a schematically depicts the experimental set-up for the measurements. Single crystals with various facets facing up are glued on the glass

substrates using silver paint. The crystals are then illuminated from top surface using a halogen lamp with a continuous spectrum ranging from 400 to 750 nm (Supplementary Fig. 2). The light goes through the AFM built-in optical system onto the surface of the sample. By placing the AFM tip at the surface, the sample height is recorded as a function of time and illumination conditions (see Methods). As shown in Fig. 2b, under 100 mW cm^{-2} white-light illumination, a reproducible change in the sample height is clearly observed. Both $(100)_T$- and $(010)_T$-oriented (subscript T denotes tetragonal index) single crystals produce a similar elongation of approximately 50 nm along the vertical direction. Considering the crystal thickness of approximately 1 mm, this translates into a photostriction (defined as the height change divided by the sample thickness $\Delta H/H$) of 5×10^{-5} (or 50 p.p.m.). To confirm that the observed height change is an intrinsic material property instead of a light-induced measurement artifact, we also tested Si and SrTiO₃ (STO) single crystals (Fig. 2b and Supplementary Fig. 3a), both of which should show negligible photostrictive effects. Instead of a sudden change of height, both samples exhibit a much slower response, whose magnitude scales with the illumination time. This suggests a possible thermal effect. However, the very different optical absorption and thermal expansion coefficients of Si and STO are at odds with the similar responses. Furthermore, this slow response can also be seen for MAPbI₃ as indicated in Fig. 2b. Thus, we infer that this material-independent response is a measurement artifact that results from the heating-induced bending of the AFM tip rather than the samples under test. To exclude the possible contribution from the thermal expansion of the sample, the temperature of the sample surface was monitored under the same illumination conditions (Supplementary Fig. 3b). Clearly, the temperature profile does not match the sudden height change. Besides, a brief estimation based on the material parameters of MAPbI₃ gives a thermal expansion on the order of 0.1 nm (Supplementary Note 1), much smaller than the response observed.

Further investigation shows that the photostrictive response is proportional to the light intensity (Fig. 2c), and no saturation is observed up to 100 mW cm^{-2}, which suggests a possible correlation between the photostriction and photo-generated carriers. To check the photoconductivity of the sample, we have measured the current under the same illumination condition by applying 1 V bias to the Pt electrodes coated on the two opposite facets of the crystal. As shown in Fig. 2d, the sample height change follows the same profile as the current change on illumination. Since the conductivity is proportional to the amount of free carriers (provided there is no significant change in the carrier mobilities), it implies that the photostriction is directly related to the photo-generated carriers, that is, of an electronic origin. The long tail of the height signal, after the light is turned off, is attributed to the slow thermal relaxation of the AFM tip. This is supported by the fact that it scales with the illumination time (Fig. 2b).

Photon energy dependence. If the photo-generated free carriers are indeed responsible for the dimension change, it must depend on the incident photon energy. As revealed by the energy-dependent absorption coefficient (Fig. 1f), the excitonic absorption of MAPbI₃ single crystal is 1.575 eV with the true band gap being a few tens of meV above that. When the lasers with photon energy higher than or close to the band gap of MAPbI₃ are used (460, 650 and 808 nm), efficient generation of electron–hole pairs can be expected and thus large photoconductivity is obtained (Fig. 3a, the bias applied is 1 V). When 980 nm laser is used, no photocurrent is observed since its photon

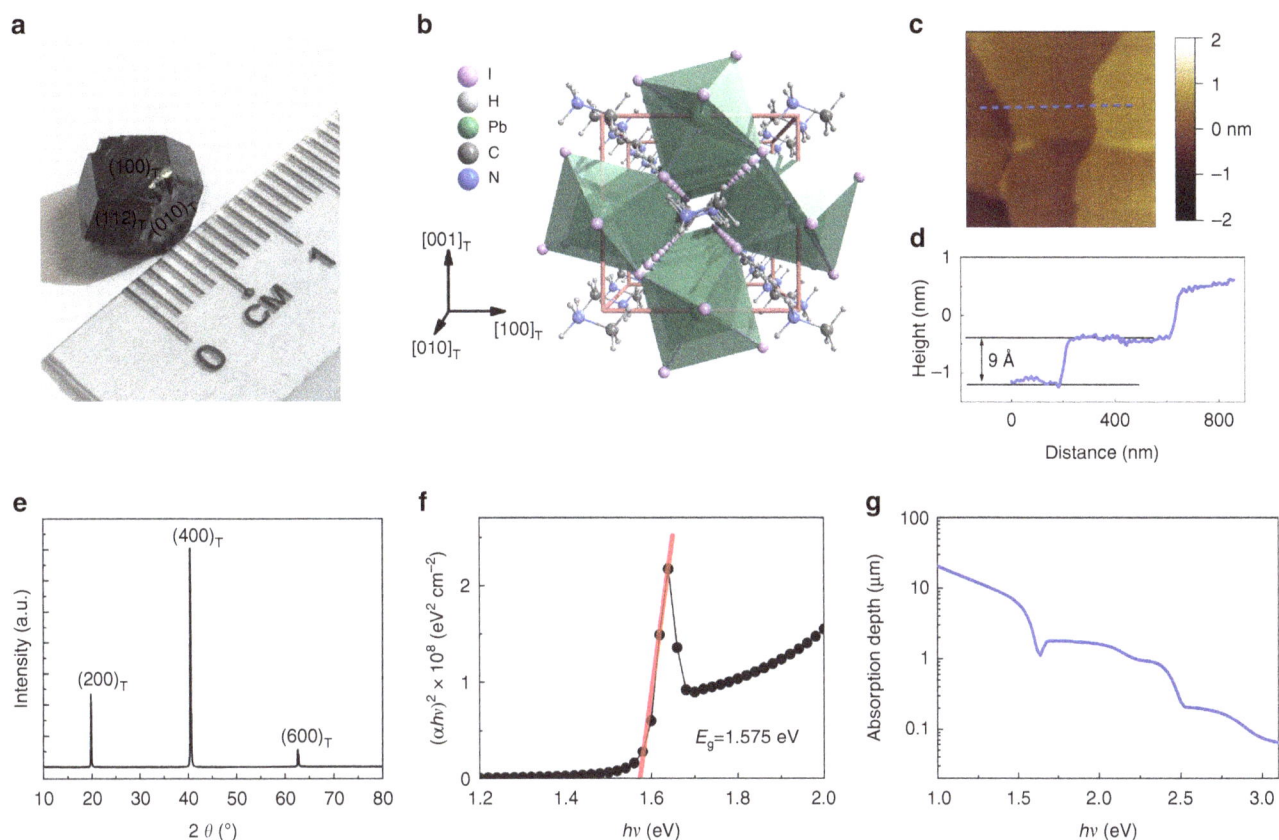

Figure 1 | Basic properties of the MAPbI₃ single crystals. (a) Photograph of a MAPbI$_3$ single crystal used in this study. **(b)** Perspective view of the unit cell of tetragonal MAPbI$_3$. **(c)** Typical topography of the single crystal obtained by AFM, showing smooth surface with unit-cell steps. The scan size is $1 \times 1\,\mu m$. **(d)** Height profile along the blue dash line denoted in **c**. **(e)** X-ray diffraction pattern of the crystal, which only shows tetragonal (*l*00) type peaks, indicating the high quality of the single crystal. **(f)** Direct-transition tauc plot according to the absorption coefficient of the crystal, from which a direct band gap of 1.575 eV is extracted. The red solid line is the linear fit. **(g)** The penetration depth deduced from the absorption coefficient.

energy is below the band gap of MAPbI$_3$. Similarly, the photo-strictive response shows clear photon energy dependence as well (Fig. 3b). When we plot the photocurrent and photostriction together as functions of the incident photon energy (Fig. 3c), the correlation is even clearer. This observation again confirms that the photostrictive response in MAPbI$_3$ single crystal is caused by photo-generated carriers.

Giant photostriction in MAPbI₃ film. According to the Beer–Lambert law, the intensity of incident light decays exponentially into the material's bulk. The strong absorption of MAPbI$_3$ in visible spectrum results in a penetration depth less than a few micrometres (Fig. 1g), leaving most of the crystal not illuminated and the effective photostriction ($\Delta H/H$) likely underestimated. To address this issue, we have carried out thickness-dependent photostriction measurement on small flakes cleaved from the same single crystal (Supplementary Fig. 4). It is found that for a 700-μm-thick flake, the height change of 50 nm is comparable to those of crystals with thicknesses greater than or equal to 1 mm, suggesting a saturated value. However, for a 150-μm-thick flake, the photostrictive response dramatically reduces to about 20 nm. Note that this thickness is already smaller than the carrier diffusion length estimated for single crystals[15]. Hence, we argue that although the light absorption depth is only a few micrometres, the photo-excited carriers can diffuse deep into the bulk, leading to a much thicker responsive layer. This conclusion helps us to further separate the photostriction signal into a fast near-surface response (absorption limited), a relatively slower contribution from the bulk limited by diffusion of the

photo-carriers, and finally the much slower plateau due to the thermal bending of the AFM tip (Supplementary Note 2). The calculated photostriction of the 150-μm-thick flake is greater than 100 p.p.m. Last, we prepare MAPbI$_3$ thin films of 4 μm thick on fluorine-doped tin oxide (FTO)-coated glass and measure its photostrictive response. The quality of the film is confirmed by the performance of a testing cell (FTO/TiO$_2$/MAPbI$_3$/Spiro-OMeTAD/Ag), in which an efficiency of 12.5% is obtained (Fig. 4a). Since the slow height response is likely due to the thermal effect of the AFM tip, we subtract it from the height change profile and obtain a thickness change of about 5 nm under 100 mW cm^{-2} white light, which translates into a photostrictive response of approximately 1,250 p.p.m. in the MAPbI$_3$ film (Fig. 4b). This is much larger than any of the reported intrinsic photostrictive effect in ferroelectric materials, polar and non-polar semiconductors[24].

Discussion

In the literature, photostriction has been reported for several non-polar semiconductors, polar materials (including ferroelectrics), chalcogenide glasses and organic materials. Following the work of Kundys[24], we analyse the experimentally observed photostrictive coefficients (normalized to the light intensity) of known materials as shown in Table 1 (refs 24,26–40). In non-polar semiconductors, it is related to the pressure susceptibility of the energy gap based on the deformation potential theory[41]. This is justified by the opposite signs of dE_g/dP in Ge and Si and the corresponding photostrictive responses[26,27]. Recently, it is reported that the band gap of CH$_3$NH$_3$PbBr$_3$ (MAPbBr$_3$) reduces with increasing

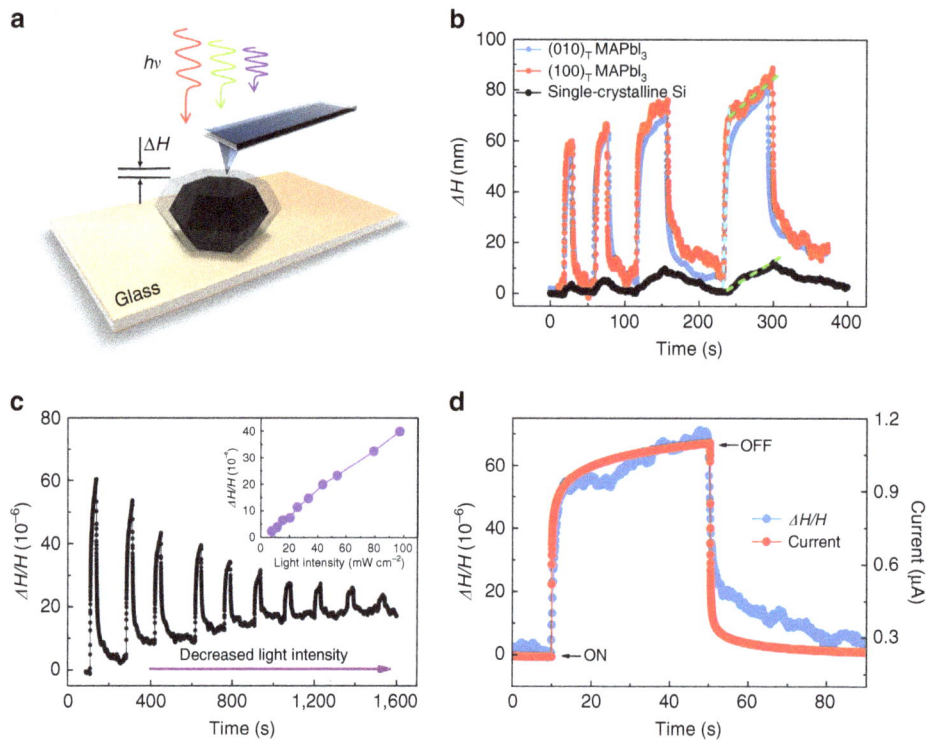

Figure 2 | Photostrictive effect of MAPbI₃ single crystals. (**a**) Schematic drawing of the experimental set-up for the photostrictive measurements. The AFM tip is fixed at one point on the sample to record the height as a function of time. (**b**) Photostriction in the MAPbI₃ single crystal. Both $(100)_T$ and $(010)_T$-oriented single crystals produce a similar elongation of ~ 50 p.p.m. under $100\,mW\,cm^{-2}$ white-light illumination. The height change of a Si single crystal is also measured as a comparison. Cyan and green dash lines delineate the fast and slow components of the height change, respectively. (**c**) Light intensity dependence of the photostrictive effect. The inset shows the proportional relationship between the photostriction and light intensity. (**d**) A comparison between the height and current changes on light irradiation.

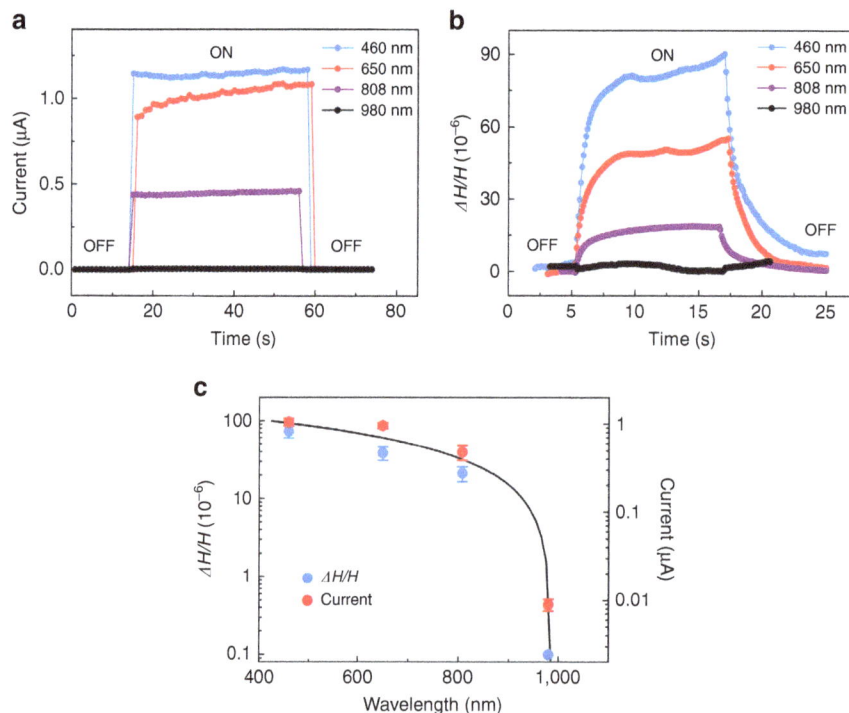

Figure 3 | Photon energy dependence of the photostriction. (**a**) When the photon energy is close to or above the band gap (that is, 460, 650 and 808 nm lasers), large current on illumination (under 1 V voltage) is observed, whereas negligible photocurrent is obtained when photon energy is below the band gap (that is, 980 nm laser). (**b**) The photostriction shows similar behaviour to that of photocurrent. (**c**) Photocurrents and height changes of the single crystal as functions of the incident photon energy, showing clear correlation between these two properties. The light intensity was kept at $\sim 100\,mW\,cm^{-2}$ for all lasers. The black solid line serves as a guide to the eye.

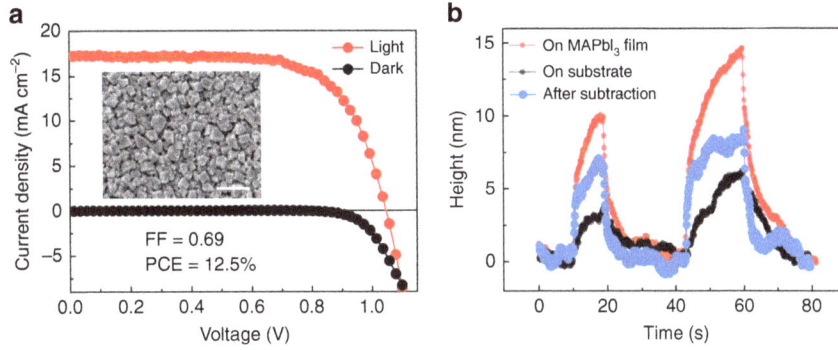

Figure 4 | Photovoltaic and photostrictive properties of MAPbI$_3$ thin films. (**a**) Typical current density–voltage characteristic of a MAPbI$_3$ thin-film photovoltaic cell (FTO/TiO$_2$/MAPbI$_3$/Spiro-OMeTAD/Ag) under simulated AM1.5 100 mW cm^{-2} illumination (red line) and under dark (black line). The power conversion efficiency can reach 12.5%. The inset shows the SEM image of the MAPbI$_3$ thin film. The scale bar, 1 μm. (**b**) Height change of the MAPbI$_3$ thin film (4 μm) on FTO-coated glass substrate under illumination. The net response from the film can be obtained by substracting the extrinsic contribution from the substrate. Under 100 mW cm^{-2} white light, about 5 nm height change can be observed, corresponding to a photostriction of 1,250 p.p.m.

Table 1 | Photostrictive coefficients of different materials.

		Photostriction, $\Delta L/L$	Light intensity, I (W m^{-2})†	$(\Delta L/L)/I$ (m^2 W^{-1})	Refs
Non-polar semiconductors	Si crystal	-6.4×10^{-6}	$8.47 \times 10^{10\dagger}$	-7.56×10^{-17}	26
	Ge crystal	7.84×10^{-10}	1,000*	7.84×10^{-13}	27
Polar semiconductors	CdS crystal	7.5×10^{-5}	1,000*	7.5×10^{-8}	24,28
	GaAs crystal	4×10^{-7}	1,000*	4×10^{-10}	29
Ferroelectric materials	SbSI crystal	4×10^{-5}	1,000*	4×10^{-8}	30
	La doped Pb(Zr$_x$Ti$_{1-x}$)O$_3$ ceramics	10^{-4}	150	6.67×10^{-7}	31
	BiFeO$_3$ crystal	3×10^{-5}	326	9.2×10^{-8}	32
	BiFeO$_3$ film (35 nm)	4.6×10^{-3}	$\sim 4 \times 10^{14\dagger}$	1.15×10^{-17}	33
	PbTiO$_3$ film (20 nm)	2.5×10^{-3}	$\sim 10^{15\dagger}$	2.5×10^{-18}	34
Chalcogenide glasses	As$_{40}$Se$_{25}$S$_{25}$Ge$_{10}$ film	4.5×10^{-4}	1,000*	4.5×10^{-7}	35
	As$_2$Se$_3$ film	6.4×10^{-2}	400	1.6×10^{-4}	36
	As$_2$S$_3$ film	5.4×10^{-2}	400	1.35×10^{-4}	36
	GeSe$_2$ film	-5.6×10^{-2}	400	-1.4×10^{-4}	36
	GeS$_2$ film	-1.1×10^{-1}	400	-2.75×10^{-4}	36
Organic materials	Poly-(4,4'-diaminoazoben-zenepyromelliti-mide) films	-1.2×10^{-2}	1,000*	-1.2×10^{-5}	37
	Nematic elastomers	2×10^{-1}	1,000*	2×10^{-4}	38
	Poly(ethylacrylate) networks with azo-aromatic crosslinks	2.5×10^{-3}	1,000*	2.5×10^{-6}	39
	Diarylethenes	-7×10^{-2}	5,200	-1.35×10^{-5}	40
Hybrid perovskites	MAPbI$_3$ crystal	5×10^{-5}	1,000	5×10^{-8}	Current work
	MAPbI$_3$ film (4 μm)	1.25×10^{-3}	1,000	1.25×10^{-6}	

*The light intensity was not reported in the references and we use 1,000 W m^{-2} (1 Sun) to calculate the photostrictive coefficients.
†The light sources were high-energy laser pulses.

hydrostatic pressure[42], which should lead to the contraction of the lattice under illumination, contrary to our results. Besides, this effect is usually small, inconsistent with the giant photostriction we observed. In polar materials, the photostriction is closely linked to the bulk photovoltaic effect and commonly interpreted as a consequence of the converse piezoelectric effect[24]. Owing to the non-centrosymmetric structure, photo-generated carriers are spontaneously separated to produce an effective electric field that deforms the lattice through the piezoelectric tensor. However, despite many theoretical and experimental studies advocating for the ferroelectricity in MAPbI$_3$ (refs 43–46), a strong proof remains lacking[47]. While the structure of the low-temperature orthorhombic phase was refined to be centrosymmetric[48,49], the high-temperature structure is difficult to be determined due to

the dynamic disorder of the CH$_3$NH$_3^+$ group. Considering the dynamic motion of the molecular dipole at room temperature, a long-range ferroelectric order can hardly exist, as confirmed by our polarization measurements, as well as piezoelectric force microscopy (Supplementary Figs 5 and 6). Even if polar nanoregions exist in MAPbI$_3$ as ferroelectric relaxors[50], a completely random distribution of the electric dipoles should cancel out the lattice change along all directions because of the volume conservation during piezoelectric deformation. For chalcogenide glasses and organic materials, both of them exhibit huge photostrictive response, which can be attributed to photo-induced bond modifications or photoisomerizations[51–53]. Strictly speaking, they should be classified as photochemical effects, most of which are irreversible after the irradiation. Besides, their responses are usually very slow, on the order of minutes or even hours, though some exceptions are recently reported[40,54]. It thus appears that none of these mechanisms can account for our observation.

Furthermore, it should be noted that the possible contribution from thermal expansion has been ruled out, and only the intrinsic effect is discussed here. By comparing the values listed in Table 1, it is clear that the photostrictive coefficient of MAPbI$_3$ happens to lie in between the inorganic and organic materials (chalcogenide glasses are an exception due to their amorphous nature), which is consistent with its hybrid character and mechanical properties[55]. The magnitude of the photostriction relies on how large a material's lattice can distort. In inorganic crystals, the atoms are closely packed with strong covalent or ionic bonds. Hence, it will be energetically costly for their lattices to deform. Organic materials, on the contrary, are at the other extreme. The small molecules or polymer chains are glued together by hydrogen bonds or van de Waals interactions, and their lattices can deform significantly due to the large intermolecular spacing. It suggests that the organic group may play a vital role in the giant photostriction effect observed in hybrid perovskites. In fact, hydrogen bonding between organic and inorganic moieties has long been studied in hybrid materials, which may lead to emergent properties[56–58]. Specifically in hybrid perovskites, hydrogen bonding between the amine group and the halide ions has been verified both theoretically and experimentally[23,59,60]. In this regard, we propose that it is the weakening of the hydrogen bonding by photo-generated carriers that results in the lattice dilation. As schematically shown in Fig. 5a, the hydrogen bonding between N and I is geometrically coupled to the buckling of the Pb–I–Pb bond and the tilting of the iodine octahedron. Under above-band-gap illumination, the first direct transition corresponds to the charge transfer from hybridized Pb $6s$–I $5p$ orbital to the Pb $6p$ orbital, forming weakly bound excitons that are easily dissociated by thermal energy[61–63]. The electronic transition directly leads to the reduction of electron density on the I site, and thereby reduces its Coulomb interaction with the amine group. This in turn straightens the Pb–I–Pb bond and results in a larger interatomic spacing (Fig. 5b).

If our analysis is correct, a couple of predictions can be made. First, the disordered orientations of the organic molecules, as well as the elastic multidomain of the crystal should lead to an isotropic photostrictive response, which is indeed observed in our orientation-dependent measurements shown in Fig. 5c. The result of the polycrystalline MAPbI$_3$ film provides another piece of evidence. Second, as the hydrogen bonding is intimately coupled to the octahedral tilt, divergence in the photostrictive response is expected around the tetragonal-cubic phase transition. Once again, our temperature-dependent measurements confirm this prediction (Fig. 5d,e). The photostriction measured at 60 °C is almost twice of that measured at room temperature, which can be interpreted by the enhanced lattice susceptibility at the structure transition boundary. The photostriction is reduced in the high-temperature cubic phase, but still with appreciable magnitudes. As pointed out by Quarti et al.[64], although the high-temperature phase appears cubic at a large scale, the local structure may strongly deviate from the nominal cubic one at any given time owing to the fluctuation of the organic molecule inside the inorganic framework. Thus, the interaction between them, though reduced, does not completely vanish.

The giant photostriction suggests strong photon-lattice coupling in MAPbI$_3$. It has important implications for understanding the exceptional photovoltaic performance of hybrid perovskites. For instance, the lattice expansion and reduced

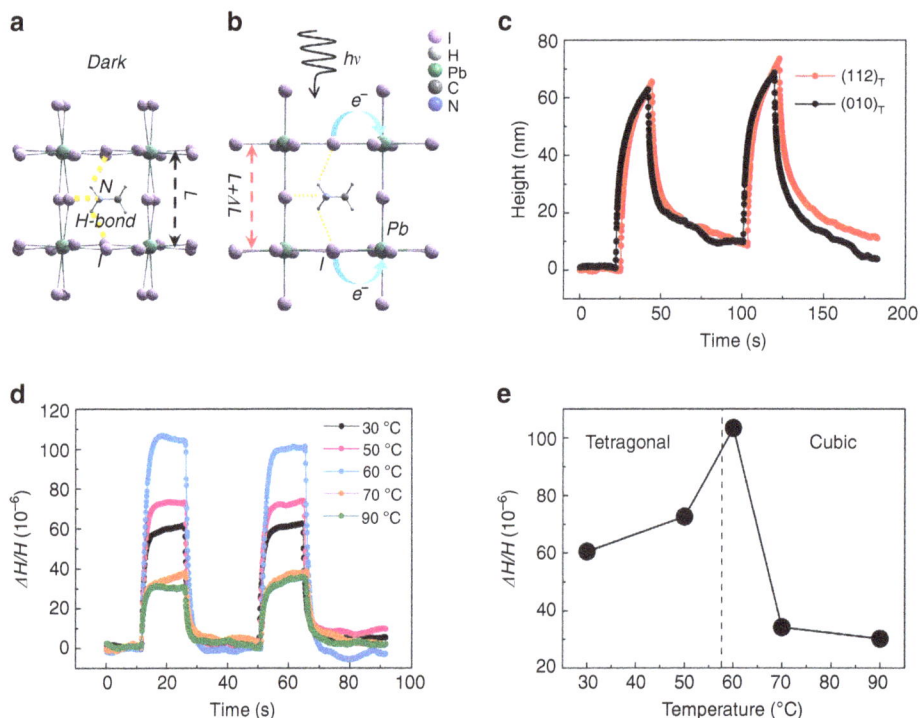

Figure 5 | Proposed mechansim for the giant photostriction in MAPbI$_3$. (**a,b**) Schematic illustrations (not to scale) showing that the weakening of the hydrogen bonding between the amine group and the iodine ion by photo-generated carriers leads to the lattice expansion. (**c**) Orientation dependence of photostriction of a MAPbI$_3$ single crystal shows similar magntidue due to the isotropic expansion of the lattice. (**d,e**) Temperature-dependent photostriction of a MAPbI$_3$ single crystal across the tetragonal-cubic phase transition under 100 mW cm^{-2} white light. The phase transition point is indicated by the black dash line. Enhanced lattice susceptibility around the phase transition boundary is likely responsible for the larger photostriction observed.

Coulomb interaction between the organic group and the inorganic framework will further reduce the rotational barrier for the molecule dipole. The enhanced tumbling of the electric dipole may lead to a dynamic change of the local band structure that suppresses the electron–hole recombination[22]. This scenario is in line with the decrease of the bi-molecular recombination rate at high temperatures[65]. Besides being important to explain the long-carrier lifetime and diffusion length in the hybrid lead halide perovskites, the giant photostriction also opens up new pathways for applications in opto-mechanical devices[24,25].

Methods

Growth of MAPbI₃ single crystal and structure characterization. CH_3NH_3I was first prepared according to previous report[66]. Typically, 27.8 ml methylamine (40% in methanol, Aldrich) was added into 30 ml hydroiodic acid (57 wt.% in water, Aldrich) and stirred for 2 h at 0 °C. After removing the solvent by rotary evaporating at 50 °C, the product was washed with diethyl ether and then recrystallized with ethanol. White crystals were obtained and dried in a vacuum box. PbI_2 (1.157 g, 99%, Aldrich) and the as prepared CH_3NH_3I (0.395 g) were mixed in 2 ml γ-butyrolactone at 60 °C with stirring. Then, 1 ml acetonitrile (Aldrich) was added into the solution and a clear pale yellow solution was obtained. The solution was then placed in a vial and kept in an oven at 70 °C for 20 min. Small MAPbI₃ single crystals then appeared at the bottom of the vial. One of these small crystals was picked out and put into another vial with the same solution for continuous growth. The crystals used for measurement were grown for 3 h. X-ray diffraction data were collected on a high-resolution diffractometer (Bruker D8 Discover) using Cu K_α radiation.

Thin-film solar cell testing. Photovoltaic cell with the structure of FTO/TiO₂/ MAPbI₃/Spiro-OMeTAD/Ag was prepared to measure the power conversion efficiency. Under illumination of solar-simulated AM1.5 sunlight at 100 mW cm⁻², the current–voltage curve was obtained using a pA meter/direct current (DC) voltage source (Hewlett Package 4140B).

Spectroscopic ellipsometry. The spectroscopic ellipsometry measurements have been performed using a commercially available rotating analyser instrument with compensator (V-VASE; J.A. Woollam Co., Inc.) within the spectral range from 0.6 to 6 eV. Data has been collected at two incidence angles (50° and 70°), while the sample was continuously purged with nitrogen gas to avoid degradation. The absorption coefficient was then calculated from the pseudodielectric function (Supplementary Fig. 1).

Atomic force microscopy and piezoelectric force microscopy. Two commercial AFMs were used to measure the photostrictive responses: Asylum Research (AR) MFP 3D and Park XE 150 under ambient condition (20–30% relative humidity, 25 °C). Although the surface of the crystal degrades slowly with time in ambient condition possibly due to surface hydration, it does not affect the photostrictive response significantly. We believe this is because the hydration product is an optically transparent layer, which does not affect the absorption and thus the photostriction of the bulk. During the photostriction test, the light (from either a halogen lamp or laser diodes) was guided through an optical fibre into the built-in optical microscope of the AFM. For the AR AFM system, the light has to go through its internal optics, which include a cold mirror that reflects only visible light. As such, only the white-light tests were carried out using AR AFM. The wavelength-dependent tests were performed on a Park AFM system, whose optics system includes only an optical microscope that allows all the studied wavelengths (460, 650, 808 and 980 nm) to pass through. In both systems, the light was shined from above the AFM tip. However, because the light had gone through an optical fibre, then been focused by the optical lens, the light was no more collimated. Therefore, the region right beneath the AFM tip was not shadowed (Supplementary Note 3). Furthermore, the beam size was about 0.1 cm² for all light sources. The height profile as a function of time, with the light periodically irradiated on the surface, was acquired. Both contact mode and tapping mode measurements gave similar results (Supplementary Fig. 7). The light intensity at the sample location was carefully calibrated using a commercial energy meter (Newport, 91,150 V). Piezoelectric force microscopy was carried out on AR MFP 3D mode under dual AC resonance tracking (DART) mode using a Pt/Ir coated tip with a spring constant of 2 N m⁻¹. The imaging voltage $V_{ac} = 1$ V.

Photocurrent and ferroelectric polarization measurements. We fabricated a simple device with the two opposite facets of the single crystal coated by semi-transparent Pt electrodes and measured the current under illumination through the top electrode. A low-noise probe station and a pA meter/DC voltage source were used and the voltage applied was 1 V. The ferroelectric polarization measurement was conducted using a commercial ferroelectric tester (Radiant Technologies, Precision LC) at different temperatures on a low-temperature probe station.

References

1. Gratzel, M. The light and shade of perovskite solar cells. *Nat. Mater.* **13**, 838–842 (2014).
2. Green, M. A., Ho-Baillie, A. & Snaith, H. J. The emergence of perovskite solar cells. *Nat. Photon.* **8**, 506–514 (2014).
3. Kojima, A., Teshima, K., Shirai, Y. & Miyasaka, T. Organometal halide perovskites as visible-light sensitizers for photovoltaic cells. *J. Am. Chem. Soc.* **131**, 6050–6051 (2009).
4. Stranks, S. D. & Snaith, H. J. Metal-halide perovskites for photovoltaic and light-emitting devices. *Nat. Nanotechnol.* **10**, 391–402 (2015).
5. National Renewable Energy Labs (NREL). Efficiency chart http://www.nrel.gov/ncpv/images/efficiency_chart.jpg (2015).
6. Xing, G. C. et al. Low-temperature solution-processed wavelength-tunable perovskites for lasing. *Nat. Mater.* **13**, 476–480 (2014).
7. Tan, Z. K. et al. Bright light-emitting diodes based on organometal halide perovskite. *Nat. Nanotechnol.* **9**, 687–692 (2014).
8. Dou, L. T. et al. Solution-processed hybrid perovskite photodetectors with high detectivity. *Nat. Commun.* **5**, 5404 (2014).
9. Xing, G. C. et al. Long-range balanced electron- and hole-transport lengths in organic–inorganic CH₃NH₃PbI₃. *Science* **342**, 344–347 (2013).
10. Wehrenfennig, C., Eperon, G. E., Johnston, M. B., Snaith, H. J. & Herz, L. M. High charge carrier mobilities and lifetimes in organolead trihalide perovskites. *Adv. Mater.* **26**, 1584–1589 (2014).
11. Stranks, S. D. et al. Electron–hole diffusion lengths exceeding 1 micrometer in an organometal trihalide perovskite absorber. *Science* **342**, 341–344 (2013).
12. Ponseca, C. S. et al. Organometal halide perovskite solar cell materials rationalized: ultrafast charge generation, high and microsecond-long balanced mobilities, and slow recombination. *J. Am. Chem. Soc.* **136**, 5189–5192 (2014).
13. deQuilettes, D. W. et al. Impact of microstructure on local carrier lifetime in perovskite solar cells. *Science* **348**, 683–686 (2015).
14. Shi, D. et al. Low trap-state density and long carrier diffusion in organolead trihalide perovskite single crystals. *Science* **347**, 519–522 (2015).
15. Dong, Q. F. et al. Electron–hole diffusion lengths > 175 μm in solution-grown CH₃NH₃PbI₃ single crystals. *Science* **347**, 967–970 (2015).
16. Shockley, W. & Queisser, H. J. Detailed balance limit of efficiency of p-n junction solar cells. *J. Appl. Phys.* **32**, 510–519 (1961).
17. Adriaenssens, G. J. & Arkhipov, V. I. Non-Langevin recombination in disordered materials with random potential distributions. *Solid State Commun.* **103**, 541–543 (1997).
18. Lakhwani, G., Rao, A. & Friend, R. H. Bimolecular recombination in organic photovoltaics. *Annu. Rev. Phys. Chem.* **65**, 557–581 (2014).
19. Bakulin, A. A. et al. Real-time observation of organic cation reorientation in methylammonium lead iodide perovskites. *J. Phys. Chem. Lett.* **6**, 3663–3669 (2015).
20. Leguy, A. M. A. et al. The dynamics of methylammonium ions in hybrid organic–inorganic perovskite solar cells. *Nat. Commun.* **6**, 7124 (2015).
21. Park, J.-S. et al. Electronic structure and optical properties of α-CH₃NH₃PbBr₃ perovskite single crystal. *J. Phys. Chem. Lett.* **6**, 4304–4308 (2015).
22. Motta, C. et al. Revealing the role of organic cations in hybrid halide perovskite CH₃NH₃PbI₃. *Nat. Commun.* **6**, 7026 (2015).
23. Glaser, T. et al. Infrared spectroscopic study of vibrational modes in methylammonium lead halide perovskites. *J. Phys. Chem. Lett.* **6**, 2913–2918 (2015).
24. Kundys, B. Photostrictive materials. *Appl. Phys. Rev.* **2**, 011301 (2015).
25. Kreisel, J., Alexe, M. & Thomas, P. A. A photoferroelectric material is more than the sum of its parts. *Nat. Mater.* **11**, 260–260 (2012).
26. Buschert, J. R. & Colella, R. Photostriction effect in silicon observed by time-resolved X-ray-diffraction. *Solid State Commun.* **80**, 419–422 (1991).
27. Figielski, T. Photostriction effect in germanium. *Phys. Status Solidi* **1**, 306–316 (1961).
28. Lagowski, J. & Gatos, H. C. Photomechanical effect in noncentrosymmetric semiconductors CdS. *Appl. Phys. Lett.* **20**, 14–16 (1972).
29. Lagowski, J. & Gatos, H. C. Photomechanical vibration of thin crystals of polar semiconductors. *Surf. Sci.* **45**, 353–370 (1974).
30. Tatsuzak, I., Itoh, K., Ueda, S. & Shindo, Y. Strain along c axis of SbSI caused by illumination in dc electric field. *Phys. Rev. Lett.* **17**, 198–200 (1966).
31. Takagi, K. et al. Ferroelectric and photostrictive properties of fine-grained PLZT ceramics derived from mechanical alloying. *J. Am. Ceram. Soc.* **87**, 1477–1482 (2004).
32. Kundys, B. et al. Wavelength dependence of photoinduced deformation in BiFeO₃. *Phys. Rev. B* **85**, 092301 (2012).
33. Schick, D. et al. Localized excited charge carriers generate ultrafast inhomogeneous strain in the multiferroic BiFeO₃. *Phys. Rev. Lett.* **112**, 097602 (2014).
34. Daranciang, D. et al. Ultrafast photovoltaic response in ferroelectric nanolayers. *Phys. Rev. Lett.* **108**, 087601 (2012).
35. Igo, T., Noguchi, Y. & Nagai, H. Photoexpansion and thermal contraction of amorphous-chalcogenide glasses. *Appl. Phys. Lett.* **25**, 193–194 (1974).

36. Kuzukawa, Y., Ganjoo, A. & Shimakawa, K. Photoinduced structural changes in obliquely deposited As- and Ge-based amorphous chalcogenides: correlation between changes in thickness and band gap. *J. Non-Cryst. Solids* **227**, 715–718 (1998).

37. Vanderve, G. & Prins, W. Photomechanical energy conversion in a polymer membrane. *Nature* **230**, 70–72 (1971).

38. Finkelmann, H., Nishikawa, E., Pereira, G. G. & Warner, M. A new opto-mechanical effect in solids. *Phys. Rev. Lett.* **87**, 015501 (2001).

39. Eisenbach, C. D. Isomerization of aromatic azo chromophores in poly(ethyl acrylate) networks and photomechanical effect. *Polymer* **21**, 1175–1179 (1980).

40. Kobatake, S., Takami, S., Muto, H., Ishikawa, T. & Irie, M. Rapid and reversible shape changes of molecular crystals on photoirradiation. *Nature* **446**, 778–781 (2007).

41. Thomsen, C., Grahn, H. T., Maris, H. J. & Tauc, J. Surface generation and detection of phonons by picosecond light pulses. *Phys. Rev. B* **34**, 4129–4138 (1986).

42. Wang, Y. *et al.* Pressure-induced phase transformation, reversible amorphization, and anomalous visible light response in organolead bromide perovskite. *J. Am. Chem. Soc.* **137**, 11144–11149 (2015).

43. Frost, J. M. *et al.* Atomistic origins of high-performance in hybrid halide perovskite solar cells. *Nano Lett.* **14**, 2584–2590 (2014).

44. Kutes, Y. *et al.* Direct observation of ferroelectric domains in solution-processed $CH_3NH_3PbI_3$ perovskite thin films. *J. Phys. Chem. Lett.* **5**, 3335–3339 (2014).

45. Stroppa, A. *et al.* Tunable ferroelectric polarization and its interplay with spin–orbit coupling in tin iodide perovskites. *Nat. Commun.* **5**, 5900 (2014).

46. Fan, Z. *et al.* Ferroelectricity of $CH_3NH_3PbI_3$ perovskite. *J. Phys. Chem. Lett.* **6**, 1155–1161 (2015).

47. Xiao, Z. *et al.* Giant switchable photovoltaic effect in organometal trihalide perovskite devices. *Nat. Mater.* **14**, 193–198 (2015).

48. Swainson, I. P., Hammond, R. P., Soullière, C., Knop, O. & Massa, W. Phase transitions in the perovskite methylammonium lead bromide, $CH_3ND_3PbBr_3$. *J. Solid State Chem.* **176**, 97–104 (2003).

49. Baikie, T. *et al.* Synthesis and crystal chemistry of the hybrid perovskite $(CH_3NH_3)PbI_3$ for solid-state sensitised solar cell applications. *J. Mater. Chem. A* **1**, 5628–5641 (2013).

50. Xu, G. Y., Zhong, Z., Bing, Y., Ye, Z. G. & Shirane, G. Electric-field-induced redistribution of polar nano-regions in a relaxor ferroelectric. *Nat. Mater.* **5**, 134–140 (2006).

51. Kugler, S., Hegedüs, J. & Kohary, K. in *Optical Properties of Condensed Matter and Applications* 143–158 (John Wiley & Sons, Ltd, 2006).

52. Iqbal, D. & Samiullah, M. Photo-responsive shape-memory and shape-changing liquid-crystal polymer networks. *Materials* **6**, 116–142 (2013).

53. Yu, H. Recent advances in photoresponsive liquid-crystalline polymers containing azobenzene chromophores. *J. Mater. Chem. C* **2**, 3047–3054 (2014).

54. Camacho-Lopez, M., Finkelmann, H., Palffy-Muhoray, P. & Shelley, M. Fast liquid-crystal elastomer swims into the dark. *Nat. Mater.* **3**, 307–310 (2004).

55. Rakita, Y., Cohen, S. R., Kedem, N. K., Hodes, G. & Cahen, D. Mechanical properties of $APbX_3$ (A = Cs or CH_3NH_3; X = I or Br) perovskite single crystals. *MRS Commun.* **5**, 623–629 (2015).

56. Mitzi, D. B. Templating and structural engineering in organic–inorganic perovskites. *J. Chem. Soc. Dalton Trans.* 1–12 (2001).

57. Jain, P. *et al.* Multiferroic behaviour associated with an order – disorder hydrogen bonding transition in metal – organic frameworks (MOFs) with the perovskite ABX_3 architecture. *J. Am. Chem. Soc.* **131**, 13625–13627 (2009).

58. Zhang, W. & Xiong, R.-G. Ferroelectric metal–organic frameworks. *Chem. Rev.* **112**, 1163–1195 (2012).

59. Swainson, I. *et al.* Orientational ordering, tilting and lone-pair activity in the perovskite methylammonium tin bromide, $CH_3NH_3SnBr_3$. *Acta Crystallogr. Sect. B* **66**, 422–429 (2010).

60. Lee, J.-H., Bristowe, N. C., Bristowe, P. D. & Cheetham, A. K. Role of hydrogen-bonding and its interplay with octahedral tilting in $CH_3NH_3PbI_3$. *Chem. Commun.* **51**, 6434–6437 (2015).

61. D'Innocenzo, V. *et al.* Excitons versus free charges in organo-lead tri-halide perovskites. *Nat. Commun.* **5**, 3586 (2014).

62. Saba, M. *et al.* Correlated electron–hole plasma in organometal perovskites. *Nat. Commun.* **5**, 5049 (2014).

63. Miyata, A. *et al.* Direct measurement of the exciton binding energy and effective masses for charge carriers in organic–inorganic tri-halide perovskites. *Nat. Phys.* **11**, 582–587 (2015).

64. Quarti, C. *et al.* Structural and optical properties of methylammonium lead iodide across the tetragonal to cubic phase transition: implications for perovskite solar cells. *Energy Environ. Sci.* **9**, 155–163 (2016).

65. Milot, R. L., Eperon, G. E., Snaith, H. J., Johnston, M. B. & Herz, L. M. Temperature-dependent charge-carrier dynamics in $CH_3NH_3PbI_3$ perovskite thin films. *Adv. Funct. Mater.* **25**, 6218–6227 (2015).

66. Kim, H. S. *et al.* Lead iodide perovskite sensitized all-solid-state submicron thin film mesoscopic solar cell with efficiency exceeding 9%. *Sci. Rep.* **2**, 591 (2012).

Acknowledgements

This work is supported by the Ministry of Education, Singapore under project No. MOE2013-T2-1-052 and AcRF Tier 1 RG126/14. We thank Dr Pio John S. Buenconsejo and Dr Fucai Liu for the help on XRD and photocurrent measurements, respectively.

Author contributions

Y.Z., L.Y. and J.W. conceived and designed the work. Z.K. and H.F. synthesized the single crystals. S.W. fabricated the films and corresponding photovoltaic cells. Y.Z. and L.Y. conducted the photostriction and photocurrent measurements with help from Lei Chang, L.W. and P.R. D.S. and A.R. performed the spectroscopic ellipsometry measurements. Liufan Chen and G.Y. independently confirmed the photostrictive response. Y.Z., L.Y. and J.W. co-wrote the manuscript. J.W. supervised the project. All authors discussed the results and commented on the manuscript.

Additional information

13

Engineering of frustration in colloidal artificial ices realized on microfeatured grooved lattices

Antonio Ortiz-Ambriz[1,2] & Pietro Tierno[1,2]

Artificial spin ice systems, namely lattices of interacting single domain ferromagnetic islands, have been used to date as microscopic models of frustration induced by lattice topology, allowing for the direct visualization of spin arrangements and textures. However, the engineering of frustrated ice states in which individual spins can be manipulated *in situ* and the real-time observation of their collective dynamics remain both challenging tasks. Inspired by recent theoretical advances, here we realize a colloidal version of an artificial spin ice system using interacting polarizable particles confined to lattices of bistable gravitational traps. We show quantitatively that ice-selection rules emerge in this frustrated soft matter system by tuning the strength of the pair interactions between the microscopic units. Via independent control of particle positioning and dipolar coupling, we introduce monopole-like defects and strings and use loops with defined chirality as an elementary unit to store binary information.

[1] Department of Structure and Constituents of Matter, University of Barcelona, Avinguda Diagonal 647, 08028 Barcelona, Spain. [2] Institute of Nanoscience and Nanotechnology, IN2UB, Universitat de Barcelona, 08028 Barcelona, Spain. Correspondence and requests for materials should be addressed to P.T. (email: ptierno@ub.edu).

Geometric frustration emerges in disparate physical and biological systems that span from disordered solids[1], to trapped ions[2], ferroelectrics[3], microgel particles[4], high-T_c superconductors[5], folding proteins[6] and neural networks[7]. When topological constraints between the individual elements impede the simultaneous satisfaction of all local interaction energies, the system is geometrically frustrated, featuring a low-temperature residual entropy and a large ground state (GS) degeneracy[8], as observed in water ice[9] and in rare-earth pyrochlore oxides, called spin ice[10–12]. On the theoretical side, topologically frustrated spin systems have been scrutinized for a long time, dating back to the work of Wannier[13] on the Ising model applied to a triangular lattice, in which the system cannot accommodate three spins on each plaquette in such a way that all antiferromagnetic couplings are minimized. In pyrochlore crystals, the situation is similar: the rare-earth ions carry a net magnetic moment, and they are located on the sites of a lattice of corner-sharing tetrahedra[10]. At each vertex, the moments can point either towards the centre of the tetrahedron or away from it, and pairs of spins align in the low-energy head-to-tail configuration. The degenerate GS of pyrochlore crystals follows the ice-rules[14] where two spins point towards the centre of the tetrahedron and two away from it. The ice rules were first introduced by Bernal and Fowler[14] to describe the proton ordering in water ice (ice I_h). In the hexagonal I_h, the lowest energy configuration is characterized by two protons near an oxygen ion in a tetrahedron and two away from it similar to the spin ice systems. To fulfil these rules, there are six equivalent atomic configuration at each tetrahedron and Pauling showed that this degeneracy was at the origin of the residual entropy of water at low temperature[15]. However, the ice rules can be locally violated[16] due to the presence of disorder or fluctuations in the system, giving rise to mobile excitations that mimic the behaviour of magnetic monopoles and Dirac strings[17–20].

Investigating the governing rule in frustrated systems is a key issue not only for understanding exotic phases in magnetism but also for providing guidelines to engineer new magnetic memory and data processing devices[21]. However, experimental investigations of bulk spin ice materials have often been restricted to averaged quantities, such as heat capacity curves[22], magnetic susceptibility[23] or neutron scattering data[24]. Artificial spin ice has recently been introduced as an alternative system that displays ice-like behaviour and allows for the direct visualization of the individual spins[25]. Such systems are composed of lithographically fabricated ferromagnetic nanostructures[25,26] or nanowires[27] arranged into periodic lattices that generate frustration by design. Given the experimentally accessible length-scale of a few nanometres, the spin orientation and the system GS can be visualized by using magnetic force microscopy, although monitoring the dynamics leading to the system degeneracy remains an elusive task because of the extremely fast spin flipping process.

Here we engineer a mesoscopic artificial spin ice that consists of an ensemble of interacting colloidal particles confined to lattices of gravitational traps. Our experimental system is inspired by a recent theoretical proposition[28], in which electrically charged colloids in a square lattice of bistable optical traps were observed to obey ice-rule ordering for strong electrostatic interactions. The high demand in laser power required to generate the necessary optical traps combined with the difficulty of tuning the surface charge in colloidal systems motivates the use of an alternative approach. We overcome these problems by using interacting microscopic magnetic particles confined to the lattices of bistable gravitational traps. The colloidal spin ice allows us to probe the equilibrium states and the dynamics of pre-designed frustrated lattices, and provides guidelines to engineer novel magnetic storage devices based on frustrated spin states.

Results

Realization of the colloidal spin ice.
We use paramagnetic particles with tunable interactions inside lithographically sculptured double-well traps, as shown in Fig. 1a. Each trap is fabricated by etching an elliptical indentation in a photocurable resin and leaving a small hill in the middle. We arrange these bistable traps into honeycomb or square lattices, although different lattice conformations can be easily implemented by lithographic design. The elliptical wells have an average length of 21 µm, width of 11 µm and we use a lattice constant of $a = 33$ µm for the honeycomb lattice and $a = 44$ µm for the square lattice. Paramagnetic microspheres of diameter $d = 10.4$ µm are dispersed in water and then allowed to sediment above the surface of the resin. Later, the particles are placed in the traps at a one-to-one filling ratio using optical tweezers, (see Methods section). Within the double wells, the colloidal particles are gravitationally trapped in one of the two low-energy states. A typical profile of the double wells obtained via an optical profilometer is shown in Fig. 1b. With this technique we measure an average barrier height within the well of $h = 0.43 \pm 0.04$ µm, where Fig. 1c shows a typical well of ~ 0.3 µm height. Given that the density mismatch is $\Delta\rho = 0.9$ g cm^{-3} between the particles and the suspending medium, we estimate a gravitational energy in the centre of the bistable trap of $U_g = 540 k_B T$, and an outer confining potential for each island of $\sim 3,000 k_B T$, where k_B is the Boltzmann constant and $T = 20$ °C is the experimental temperature. Thermal fluctuations are unable to induce spontaneous switching of the particle state unless either smaller particles or a smaller hill are used.

To tune the pair interaction between the magnetic colloids, the paramagnetic particles are doped with nanoscale iron oxide grains; as a result of doping, these particles are responsive to a magnetic field **B**. Under the applied field, the particles acquire a dipole moment $\mathbf{m} = V\chi\mathbf{B}/\mu_0$, where $V = (\pi d^3/6)$ is the particle volume, $\chi = 0.1$ the magnetic volume susceptibility and $\mu_0 = 4\pi 10^{-7}$ H m the susceptibility of vacuum. Pairs of particles interact via dipolar forces, and for a magnetic field applied perpendicular to the plane, the interaction potential is isotropic and is given by $U_d = \frac{\mu_0 m^2}{4\pi r_{ij}^3}$ where $r_{ij} = |\mathbf{r}_i - \mathbf{r}_j|$, r_i is the position of particle i. We apply a homogeneous field ranging from 0 to 25 mT with an accuracy of 0.1 mT. When the paramagnetic colloids cross the central hill, we find that the corresponding out-of-plane motion produces a negligible effect on the overall collective dynamics.

Spin configuration and vertex energy.
Two typical experimental realizations are shown in Fig. 1d,e for honeycomb and square lattices, respectively, both following exposure to a constant field of amplitude $B = 18$ mT for 60 s. Each experiment is initialized by loading one particle in each well using optical tweezers, and randomly flipping the position within the well according to a random number generator. The repulsive magnetic interactions force the particles to maximize their distance and are such that the particles can easily switch state but cannot escape from the gravitational trap. By assigning a vector (analogous to a spin) to each bistable trap pointing from the vacant site towards the side occupied by the particle, one can construct a set of ice rules for the colloidal spin ice system, equivalent to those for artificial spin ice[28]. At each vertex where three traps (in the case of honeycomb) or four traps (in the case of square) meet, the vector assigned to each trap can point either in when the colloidal particle is close to the vertex or out when it is far from the vertex, following the same classification scheme used in the three-dimensional pyrochlore tetrahedron. The vertices of the honeycomb lattice, sometimes called kagome ice since the spin midpoints are arranged in a kagome lattice, can have four different types of spin

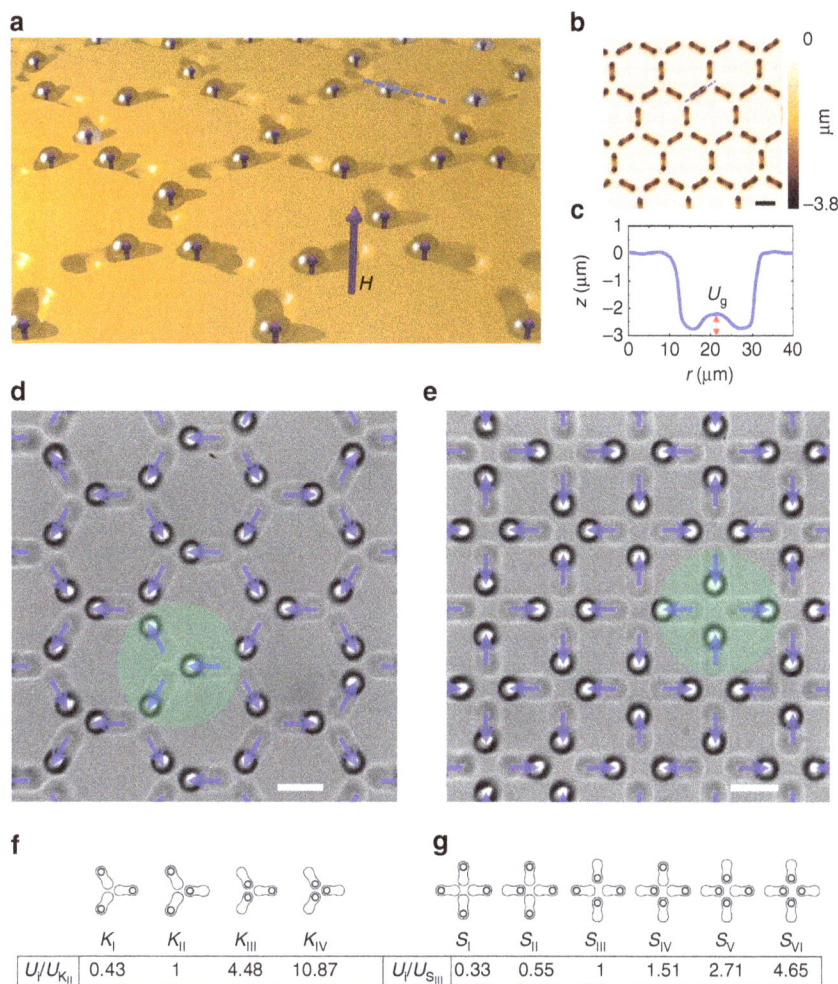

Figure 1 | Realization of the colloidal spin ice system. (**a**) Schematic view of the colloidal spin ice made by a honeycomb lattice of double-well islands filled with paramagnetic colloids. The applied field **B** perpendicular to the plane induces repulsive dipolar interactions between the particles. (**b**) Optical profilometer image of the honeycomb spin ice, and (**c**) the cross-section of a double well with a small central hill, giving a gravitational potential U_g. (**d,e**) Equilibrium state of a honeycomb ice (**d**) (lattice constant $a = 44\,\mu m$) and a square ice (**e**) (lattice constant $a = 33\,\mu m$). Blue arrows denote spin direction, while green circles highlight vertices of type K_{II} (in **d**) and S_{III} (in **e**). Scale bars, 20 μm for all images. (**f,g**) Vertex configurations for honeycomb (**f**) and square (**g**) ices. The lowest panel shows the normalized magnetostatic energy for each type of vertex.

arrangements. The highest energy configuration occurs when three particles are close to the vertex (K_{IV}), and the lowest energy configuration has three particles far from the vertex (K_I). In contrast, the square lattice has six types of spin configurations: the highest energy vertex is composed of four particles close to each other (S_{VI}), and the lowest energy vertex has all the particles far away (S_I). The corresponding energetic weight of all the vertices in both lattices is shown at the bottom of Fig. 1. In particular, for the elliptical wells used in Fig. 1, the magnetic interaction between nearest neighbours can vary from $U_d = 6425\,k_B T$ to $U_d = 450\,k_B T$ in the honeycomb lattice and from $U_d = 1675\,k_B T$ to $U_d = 228\,k_B T$ in the square lattice. These interactions increase as more particles are added at each vertex. Moreover, we notice that in contrast to the artificial spin ice, the colloidal system is characterized by mobile particles and their pair interaction depends on the relative distance between them. The energy at each vertex thus changes as the particles move, and the corresponding GS results from a collective effect between the interacting units.

Ice-selection rules in the colloidal spin ice. Systematic experiments performed by increasing the interaction strength via the

applied field reveal that the colloidal spin ice has a clear tendency to follow ice-selection rules for both the square and honeycomb configurations. Figure 2a,b shows that the fraction of low-energy vertices of type S_{III} and K_{II} increases up to ~ 0.8. We confirm both trends by performing Brownian dynamic simulations, shown as continuous lines in Fig. 2a,b. In these simulations, we neglect many-body effects due to the relatively large separation between the interacting particles at each vertex, more details can be found in the Methods. In the ferromagnetic artificial square ice, Wang et al.[25] found that as the interaction increases, the system is dominated by vertices of types S_{III} and S_{IV}. In contrast, we observe that when the applied field increases, the S_{III} vertices dominate over the S_{IV} vertices due to their slightly lower energetic weight. This is closer to the true GS of the square ice, which corresponds to a lattice that is fully covered by S_{III} vertices. To obtain the GS in the magnetic bar system, Zhang et al.[26] recently used a dedicated annealing procedure based on a rotating demagnetizing field, while for the colloidal spin ice system a long-range ordered GS arises by simply increasing the interactions strength between the magnetic particles. In the case of the honeycomb lattice, a different set of ice rules arises, in which the high-energy K_{IV} vertices and their topologically connected K_I disappear in favour of the K_{II} and K_{III} vertices. In

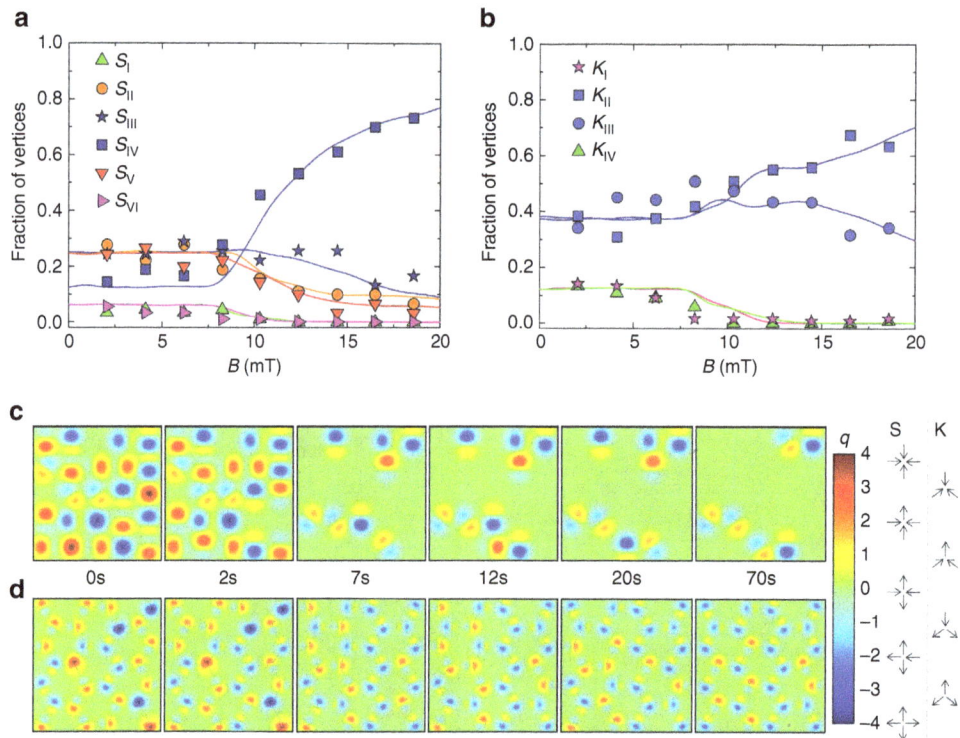

Figure 2 | Ice rules and dynamics in colloidal spin ice. (a,b) Average fraction of vertices at equilibrium for square (**a**) and honeycomb (**b**) ice versus the applied magnetic field. In both cases, ice-selection rules (blue) emerge for large interaction strengths. Scattered points are experimental data, continuous lines are results from Brownian dynamic simulation (Methods). The experimental points were averaged over 10 realizations, the numerical data over 1,000. (**c,d**) Evolution of the net vertex charge q in the square (**c**) and honeycomb (**d**) ice (experimental system) after the application of an external field with amplitude $B = 18.5$ mT. Point charges at each vertex were assigned by considering a positive value for an incoming spin and a negative value for an outgoing spin, relative to the centre of the vertex. The schematics on the side of the colour bar illustrate the total sum of the charges $q = 0$; ± 2; ± 4 for the square ice and $q = \pm 1$; ± 3 for the honeycomb ice.

the colloidal system, beyond a field of $B = 9$ mT, the K_{II} vertices begin to prevail, because they are energetically more favourable and maximize the average particle distance. In reality, at much higher strength, the number of observed K_{II} and K_{III} vertices should converge due to the isotropy of the honeycomb lattice vertices compared with the square one.

Particle dynamics above the square and honeycomb lattices.
One advantage of our mesoscopic system is that by using particle tracking routines, we can directly follow the colloid displacement and monitor the entire ordering process for a given interaction strength. The dynamics are better visualized by displaying the net topological charge q at each vertex, calculated according to the dumbbell model, in which each spin carries a pair of opposite charges at each of its ends[25]. Colour maps of the charge in the system at different times are shown in Fig. 2c,d for the square and honeycomb lattices, respectively. The GS for the square ice corresponds to vertices with a zero net charge. The highly frustrated honeycomb ice shows how low-energy vertices with net charge $q = \pm 1$ are preferred over high-energy vertices with $q = \pm 3$. Starting from an initial random distribution of particles within the traps, the square ice shifts towards a state without highly charged defects. However, some defects do remain frozen close to the sample edges, and the system converges to a low-energy metastable state. In contrast to the square ice, the honeycomb spin ice has two equivalent vertices and thus an inherent extensive degeneracy. It has been predicted to undergo a series of phases (Ising paramagnet, Ice I, Ice II and 'solid' ice) as the temperature decreases[27]. In Fig. 2d, we observe that the system organizes into a superlattice region of $+1$ and -1 charged vertices when a strong magnetic field is applied. This is

more similar to the Ice II phase with the presence of few defects located at the edges which break the long-range order present in the solid ice phase. We note that the order observed in both types of lattices can be further improved either by applying annealing protocols with dynamic fields or by applying a bias force obtained by a strong magnetic gradient[29].

Discussion
Our system allows us to manipulate individual particles within the wells using optical tweezers. This method can be used to introduce defects, which can be later erased by turning on the magnetic field and thus increasing the interaction between the particles. We demonstrate this feature with the square lattice, although we have the same degree of control over the honeycomb ice. For the colloidal square ice, the spin directions in the GS (Fig. 3c) define chiral cells, which alternate in chirality in a checkerboard pattern. Pairs of defects emerge in the form of achiral cells with a net excess of magnetic charge when flipping one spin from the GS configuration. Depending on the location of these cells within the array, defects can be made stable or unstable at a given applied field. In Fig. 3a–c, we observe the evolution of a pair of defects with opposite charges, separated by a string of achiral domains indicated by the green crosses in Fig. 3a. When the annihilation of these defects involves only a few spin flips, the energetic cost for the system to recover its GS is low, and the defects rapidly disappear as the field is turned on, as shown in the sequence in Fig. 3b. The defect annihilation process occurs via a stepwise flipping of the particle position rather than via a cooperative shift of all spins simultaneously, as seen in Fig. 3f. In contrast, in Fig. 3d,e, where we show a string of defects that acts

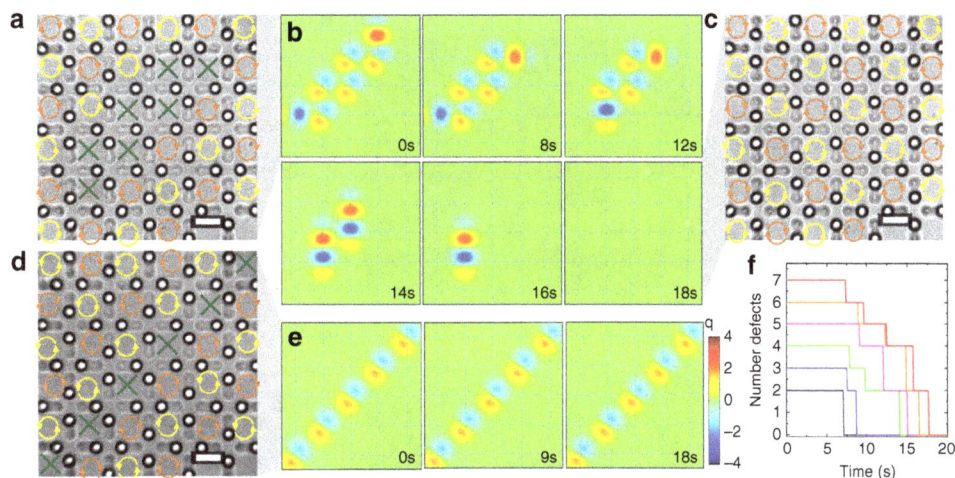

Figure 3 | Defect annihilation and dynamics. (**a**) Microscope image showing a square ice state prepared with a pair of $q = \pm 2$ charged defects separated by two parallel strings composed of seven achiral cells (green crosses). The string is introduced in a ground state region with cells that possess defined clockwise (orange arrows) or counterclockwise (yellow arrows) chirality. (**b**) Upon application of an external field ($B = 18.5$ mT), the defect pair rapidly annihilates via a sequence of particle flips, restoring the ground state after 18 s (**c**). (**d**) Square ice state in which a single grain boundary of achiral cells separates two incompatible ground state regions. Scale bars, 20 μm for all images. In **e**, the time evolution of the defect line when subjected to the same field condition as in **b** shows no change because ground state recovery requires the high-energy cost of flipping all the spins in one of the two regions. (**f**) Stepwise defect reduction as the field is applied, reflecting the consecutive spin flipping process that leads to the ground state in **b**.

as a domain wall separating two GS regions with unmatched chirality, defective vertices are more difficult to erase because they require an entire region to flip to escape this metastable state. As a result, the domain wall remains practically frozen in place. The presence of large GS regions separated by domain walls emerges as a natural low-energy metastable state in ferromagnetic spin ice[19], even after a subsequent thermally induced annealing process[30], given the elusive nature of the true long-range ordered GS.

The stability of domain walls between regions of different chirality suggests that one possible mechanism of information storage in the square system can be achieved by arranging the particles in the vertices in such a way as to maximize the number of spin flips required to reach the GS. For example, the triangular pattern of achiral cells shown in Fig. 4a is formed by flipping three of the four spins in a cell. The defects disappear by applying a field of 25 mT, which causes the three particles to shift consecutively, in a manner similar to that presented in Fig. 3b. In contrast, if we set a counterclockwise chiral cell in place of a clockwise chiral cell, surrounded by four achiral cells, fixing the defects requires a simultaneous four-spin reversal because each individual switch leads to a higher energy state. This simultaneous flipping is an energetically higher erasure process, as shown in Fig. 3c. Beyond 25 mT, the pair interactions can become stronger than the lateral confinement, and the colloids have been observed to escape from their gravitational trap in such a way that they rearrange into a triangular lattice. For this reason, we use numerical simulation to determine the threshold field B_c necessary to reset the GS, (details in the Methods). We confirm that the GS is reached by inverting the chirality of the central cell at a field of $B_c = 30$ mT. Cell writing with defined chirality can be used to store digital information in the form of 8 bit ASCII characters. Once written with optical tweezers, the chiral domains are stable below B_c. In addition, any low-energy defect, such as those shown in Fig. 4a, easily disappears. More examples of stable and unstable achiral cells are shown in Supplementary Figs 1 and 2, and commented in Supplementary Note 1.

In conclusion, we have engineered artificial colloidal spin ice states in which an external magnetic field fully controls the spin

interactions and the collective dynamics leading to the degenerate GS of the system. Unlike the original proposition of a completely optical system[28], our bistable gravitational traps are sculptured in soft-lithographic platforms, can be arranged into periodic lattices and can be designed with diverse geometries and lattice constants. Our geometrically frustrated soft matter system provides a robust approach for probing the effect of disorder by manually introducing defects into the lattice pattern. Disorder in the system can be also created by either leaving traps empty or, as recently proposed[31], by creating sites of double occupation which correspond to pairs of outward pointing spins, not possible with the nanoscale spin ice system. Finally, the strategy presented here can offer guidelines for designing similar experiments on nanoscopic systems or probing the stability of spin arrangements to record information in magnetic data storage devices[32].

Methods

Fabrication of the soft-lithographic platform. The pattern was written via direct write laser lithography (DWL 66, Heidelberg Instruments Mikrotechnik GmbH) on a 5-inch Cr mask with a $\lambda = 405$ nm diode laser at a 5.7 mm^2 min^{-1} writing speed. As shown in Supplementary Figs 3 and 4, the small hill in the centre of each trap was obtained by drawing a small constriction in the middle of the elliptical well. These structures were exposed on a ~ 100-μm-thick coverglass by a 2.8-μm-thick layer of a positive photoresist AZ-1512HS (Microchem, Newton, MA) deposited by spin coating (Spinner Ws-650Sz, Laurell) performed at 1,000 r.p.m. for 30 s. After the deposition, the photoresist was irradiated for 3.5 s with ultraviolet light at a power of 21 mW cm^{-2} (UV-NIL, SUSS Microtech). Later the exposed regions were eliminated by submerging the plate in AZ726MF developer solution for 7 s. Some representative images of the resulting structures are shown in the Supplementary Figs 3–5. The substrate was finally covered with a thin layer of polysodium 4-styrene sulfonate by using the layer-by-layer adsorption technique. More details are given in the Supplementary Note 2.

Magneto-optical set-up and experiments. The experimental set-up allows applying simultaneously and independently magnetic and optical forces. It is composed by an inverted homemade optical microscope equipped with a white light illumination LED (MCWHL5 from Thorlabs), a charge-coupled device (Basler A311f) and custom-made coil perpendicular to the sample cell such that the main axis points along the z-direction. The coil was connected to a programmable power supply (KEPCO BOP-20 10M), which is remotely controlled along with the image acquisition and recording with a custom-made LabVIEW programme. The photoresist is sensitive to ultraviolet light, so the white light of the LED is filtered with a long pass filter with a cutoff at 500 nm (FEL0500 Thorlabs). Optical tweezers are realized by tightly focusing a $\lambda = 975$ nm, $P = 330$ mW, Butterfly Laser Diode

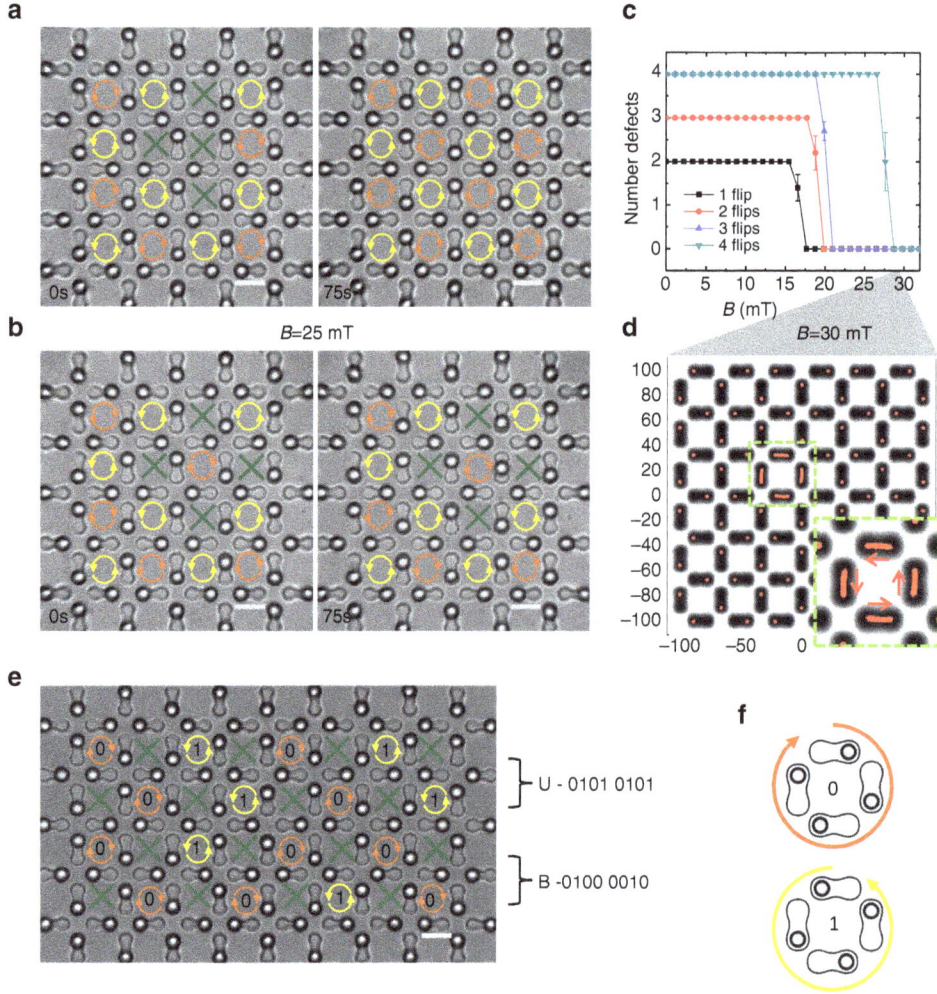

Figure 4 | Defect stability and binary writing procedure. (**a,b**) Microscope images showing the evolution with time of three (**a**) and four (**b**) achiral cells in a square lattice when subjected to a static field of $B = 25\,\text{mT}$. Flipping three spins is necessary in **a** to reach the ground state, while in **b** the defect arrangement requires the inversion of the chirality of the central cell, which is energetically more expensive. (**c**) Reduction of defects versus applied field obtained from numerical simulation. At a field of $B = 30\,\text{mT}$, (**d**) the simulation shows that it is possible to drive the system to its ground state by the simultaneous flipping of all four particles, inverting the cell's chirality. Inset shows an enlargement of the central cell. (**e**) Example showing the word 'UB' written in a square ice using the binary ASCII representation (8 bit). (**f**) Schematic showing how it is assigned a value of 0 (1) to a clockwise (counterclockwise) chiral cell. Bits in **e** are read from left to right and from top to bottom. Scale bars, $20\,\mu\text{m}$ for all images.

(Thorlabs) with a $100\times$ Achromatic microscope objective (Nikon, numerical aperture = 1.2) which is also used for observation purpose. During the experiments, a solution is prepared with $3.5\,\mu\text{l}$ of polystyrene paramagnetic particles (PS-MAG-S2874Microparticles GmbH) with $10\,\text{ml}$ of high-deionized water (MilliQ system, Millipore). A drop is placed on the soft lithography structures and after few minutes, the particles sediment due to density mismatch until they are suspended above the substrate due to the electrostatic repulsive interactions with the negative charged surface. We use the optical tweezers to fill all the bistable traps with exactly one particle, removing excess or aggregated colloids. After the initial setting, the particles are allowed equilibrate in their wells for $\sim 2\,\text{min}$ before applying the external field. We place a total of 84 particles in the square lattice and 64 in the honeycomb lattice in an experimental field of view of $325 \times 222\,\mu\text{m}^2$. The effect of the boundary on the experimental system is discussed in Supplementary Note 3 and shown in Supplementary Fig. 6.

Brownian dynamics simulation. We perform two-dimensional Brownian dynamics with periodic boundary conditions containing N particles arranged into an ensemble of double-well traps. A particle i at position $\vec{r}_i \equiv (x_i, y_i)$ obeys the set of overdamped equations:

$$\begin{cases} \eta \dot{x} = \vec{F}_{\text{tot}} \cdot \hat{e}_x + \xi_x(t) \\ \eta \dot{y} = \vec{F}_{\text{tot}} \cdot \hat{e}_y + \xi_y(t) \end{cases} \quad (1)$$

where η is the viscous friction and $\vec{F}_{\text{tot}}$ is the sum of external forces acting on the particle, composed by three terms $\vec{F}_{\text{ext}} = \vec{F}_g + \vec{F}_N + \vec{F}_M$. Here $\vec{F}_g$ is the gravitational force, $\vec{F}_N$ the normal force exerted by double-well confinement and $\vec{F}_M$ the magnetic interaction between particles. The gravitational force is given by

$\vec{F}_g = gV\Delta\rho$, where g is the gravitational acceleration, V is the volume of the particles and $\Delta\rho$ is the density mismatch. We assume the following shape for the double-well potential,

$$z(\delta\vec{r}) = h\left(\frac{2}{d}\right)^4 \left(\left(\delta\vec{r} \cdot \hat{e}_\parallel\right)^2 - \left(\frac{d}{2}\right)^2\right)^2 + k(\delta\vec{r}.\hat{e}_\perp)^2 \quad (2)$$

where $\delta\vec{r}$ the displacement vector from the centre of the trap, d the distance between the two stable positions, h the height of the central hill and k is the transverse width of the trap. The two unit vectors $\hat{e}_\parallel$ and $\hat{e}_\perp$ define the orientation of the trap; $\hat{e}_\parallel$ is the vector that joins the two stable positions and $\hat{e}_\perp$ is the transverse axis. The corresponding trap is shown in Supplementary Fig. 7a. Assuming the walls have a small inclination angle, the normal force can be calculated as $\vec{F}_N = |\vec{F}_g|\nabla_t z$, which gives:

$$\vec{F}_N = |\vec{F}_g|\left[h\left(\frac{d}{2}\right)^4 \left(4\delta r_\parallel^3 - \delta r_\parallel d^2\right)\hat{e}_\parallel + 2k\delta r_\perp \hat{e}_\perp\right] \quad (3)$$

where $\delta r_\parallel = \delta\vec{r} \cdot \hat{e}_\parallel$ and $\delta r_\perp = \delta\vec{r} \cdot \hat{e}_\perp$. The magnetic interaction is calculated assuming every particle has a magnetic moment $\vec{m} = \frac{|\vec{B}|\chi V}{\mu_0}\hat{z}$. Here $|\vec{B}|$ is the amplitude of the magnetic field, χ is the magnetic volume susceptibility, μ_0 is the permeability of the medium and V is the particle volume. The total magnetic force exerted on one particle is then given by: $\vec{F}_{M_i} = \sum_j \frac{3\mu_0}{2\pi|\vec{r}_{ij}|^4}|\vec{m}|^2\vec{r}_{ij}$ where $\vec{r}_{ij}$ is the vector that goes from particle i to particle j. Finally $\xi(t)$ in Equation (1) is a Gaussian white noise with zero mean, and a correlation function:

$\langle \xi(t)\xi(t') \rangle = 2\eta k_B T \delta(t - t')$. We numerically integrate Equation (1) using a finite time step of 0.01 s and substituting experimental parameters for most quantities. However, the height of the central hill h connecting the two circular traps in the double wells is modified to match the experimental data. Small discrepancies can arise in the hill elevations in photolithographic platforms realized during different fabrication process. First, to validate our theoretical model, we perform initial simulation to match the displacement observed between isolated pair of particles placed within two double well-oriented in a square and honeycomb lattice. The good comparison between experimental data (scattered points) and simulation results (continuous lines) is shown in Supplementary Fig. 7b. The step-like behaviour of the pair distance observed for the blue and green curves is due to the barrier overcoming of one particle. This process is energetically more expensive for isolated particles, reflecting that in the colloidal spin ice system the particle arrangement is a true collective effect rather than resulting from local energy minimization.

References

1. Ramirez, A. P. Geometric frustration: magic moments. *Nature* **421**, 483 (2003).
2. Kim, K. *et al.* Quantum simulation of frustrated Ising spins with trapped ions. *Nature* **465**, 590 (2010).
3. Choudhury, N., Walizer, L., Lisenkov, S. & Bellaiche, L. Geometric frustration in compositionally modulated ferroelectrics. *Nature* **470**, 513–517 (2011).
4. Yilong Han, Y., Shokef, Y., Alsayed, A. M., Yunker, P., Lubensky, T. C. & Yodh, A. G. Geometric frustration in buckled colloidal monolayers. *Nature* **456**, 898–903 (2008).
5. Anderson, P. W. The Resonating valence bond state in La2CuO4 and superconductivity. *Science* **235**, 1196–1198 (1987).
6. Bryngelson, D. & Wolynes, P. G. Spin glasses and the statistical mechanics of protein folding. *Proc. Natl Acad. Sci. USA* **84**, 7524–7528 (1987).
7. Dorogovtsev, S. N., Goltsev, A. V. & Mendes, J. F. Critical phenomena in complex networks. *Rev. Mod. Phys.* **80**, 1275–1335 (2008).
8. Moessner, R. & Ramirez, A. P. Geometrical frustration. *Phys. Today* **59**, 24–29 (2006).
9. Giauque, W. & Stout, J. The entropy of water and the third law of thermodynamics. the heat capacity of ice from 15 to 273°K. *J. Am. Chem. Soc.* **58**, 1144–1150 (1936).
10. Harris, M. J., Bramwell, S. T., Mc Morrow, D. F., Zeiske, T. & Godfrey, K. W. Geometric frustration in the ferromagnetic pyrochlore Ho$_2$Ti$_2$O$_7$. *Phys. Rev. Lett.* **79**, 2554 (1997).
11. Bramwell, S. T. & Gingras, M. J. P. Spin ice state in frustrated magnetic pyrochlore materials. *Science* **294**, 1495–1501 (2001).
12. Bramwell, S. T., Gingras, M. J. P. & Holdsworth, P. C. W. Spin Ice. *Frustrated Spin Systems.* , Vol. 7 (ed. Diep, H. T.) 367–451 (World Scientific, 2004).
13. Wannier, G. H. Antiferromagnetism. the triangular ising net. *Phys. Rev.* **79**, 357–364 (1950).
14. Bernal, J. D. & Fowler, R. H. A theory of water and ionic solution, with particular reference to hydrogen and hydroxyl ions. *J. Chem. Phys.* **1**, 515–548 (1933).
15. Pauling, L. The structure and entropy of ice and of other crystals with some randomness of atomic arrangement. *J. Chem. Phys.* **57**, 2680–2684 (1935).
16. Castelnovo, C., Moessner, R. & Sondhi, S. L. Magnetic monopoles in spin ice. *Nature* **451**, 42–45 (2008).
17. Morris, D. J. P. *et al.* Dirac strings and magnetic monopoles in spin ice Dy$_2$Ti$_2$O$_7$. *Science* **326**, 411–414 (2009).
18. Ladak, S., Read, D. E., Perkins, G. K., Cohen, L. F. & Branford, W. R. Direct observation of magnetic monopole defects in an artificial spin-ice system. *Nat. Phys.* **6**, 359–363 (2010).
19. Morgan, J. P., Stein, A., Langridge, S. & Marrows, C. H. Thermal ground-state ordering and elementary excitations in artificial magnetic square ice. *Nat. Phys.* **7**, 75–79 (2011).
20. Mengotti, E. *et al.* Real-space observation of emergent magnetic monopoles and associated Dirac strings in artificial kagome spin ice. *Nat. Phys.* **7**, 68–74 (2011).
21. Nisoli, C., Moessner, R. & Schiffer, P. Colloquium: artificial spin ice: designing and imaging magnetic frustration. *Rev. Mod. Phys.* **85**, 1473–1490 (2013).
22. Ramirez, A. P., Hayashi, A., Cava, R. J., Siddharthan, R. & Shastry, B. S. Zero-point entropy in 'spin ice'. *Nature* **399**, 333–335 (1999).
23. Blöte, H. W. J., Wielinga, R. F. & Huiskamp, W. J. Heat-capacity measurements on rare-earth double oxides R2M2O7. *Physica* **43**, 549–568 (1969).
24. Kadowaki, H., Ishii, Y., Matsuhira, K. & Hinatsu, Y. Neutron scattering study of dipolar spin ice Ho$_2$Sn$_2$O$_7$: frustrated pyrochlore magnet. *Phys. Rev* **B65**, 144421 (2002).
25. Wang, R. F. *et al.* Artificial 'spin ice' in a geometrically frustrated lattice of nanoscale ferromagnetic islands. *Nature* **439**, 303–306 (2006).
26. Zhang, S. *et al.* Crystallites of magnetic charges in artificial spin ice. *Nature* **500**, 553–557 (2013).
27. Branford, W. R., Ladak, S., Read, D. E., Zeissler, K. & Cohen, L. F. Emerging chirality in artificial spin ice. *Science* **335**, 1597–1600 (2012).
28. Libál, A., Reichhardt, C. & Reichhardt, C. J. O. Realizing colloidal artificial ice on arrays of optical traps. *Phys. Rev. Lett.* **97**, 228302 (2006).
29. Libál, A., Reichhardt, C. & Reichhardt, C. J. O. Hysteresis and return-point memory in colloidal artificial spin ice systems. *Phys. Rev. E* **86**, 021406 (2012).
30. Porro, J. M., Bedoya-Pinto, A., Berger, A. & Vavassori, P. Exploring thermally induced states in square artificial spin-ice arrays. *New J. Phys.* **15**, 055012 (2013).
31. Libál, A., Reichhardt, C. J. O & Reichhardt, C. Doped colloidal artificial spin ice. *New J. Phys.* **17**, 103010 (2015).
32. Heyderman, L. J. & Stamps, R. L. Artificial ferroic systems: novel functionality from structure, interactions and dynamics. *J. Phys.: Condens. Matter* **25**, 363201 (2013).

Acknowledgements

This work was supported by the European ResearchCouncil through Starting Grant No. 335040. P.T. acknowledges support from 'Ramon y Cajal' Programme No.RYC-2011-07605 and from Mineco (Grant Mo. FIS2013-41144-P). We thank T.M. Fischer and J. Ortín for fruitful discussions. R. Albalat is acknowledged for help with the AFM measurements and J. Löhr for the profilometer measurements.

Author contributions

P.T. conceived and supervised the research. A.O.-A. performed the experiments. P.T. and A.O.-A. analysed the data. Both authors wrote the manuscript and discussed all the implications.

Additional information

14

Cavity-excited Huygens' metasurface antennas for near-unity aperture illumination efficiency from arbitrarily large apertures

Ariel Epstein[1], Joseph P.S. Wong[1] & George V. Eleftheriades[1]

One of the long-standing problems in antenna engineering is the realization of highly directive beams using low-profile devices. In this paper, we provide a solution to this problem by means of Huygens' metasurfaces (HMSs), based on the equivalence principle. This principle states that a given excitation can be transformed to a desirable aperture field by inducing suitable electric and (equivalent) magnetic surface currents. Building on this concept, we propose and demonstrate cavity-excited HMS antennas, where the single-source-fed cavity is designed to optimize aperture illumination, while the HMS facilitates the current distribution that ensures phase purity of aperture fields. The HMS breaks the coupling between the excitation and radiation spectra typical to standard partially reflecting surfaces, allowing tailoring of the aperture properties to produce a desirable radiation pattern, without incurring edge-taper losses. The proposed low-profile design yields near-unity aperture illumination efficiencies from arbitrarily large apertures, offering new capabilities for microwave, terahertz and optical radiators.

[1] The Edward S. Rogers Department of Electrical and Computer Engineering, University of Toronto, Toronto, Ontario, Canada M5S 2E4. Correspondence and requests for materials should be addressed to A.E. (email: ariel.epstein@utoronto.ca) or to G.V.E. (email: gelefth@waves.utoronto.ca).

A chieving high directivity with compact radiators has been a major problem in antenna science since early days[1-3]. Still today, many applications, such as automotive radars and satellite communication, strive for simple and efficient low-profile antennas producing the narrowest beams[4-7]. Increasing the radiating aperture size enhances directivity, but only if the aperture is efficiently excited. To date, uniform illumination of large apertures is achievable with reflectors and lenses; however, these require substantial separation between the source and the aperture, resulting in a large overall antenna size[8,9]. High aperture illumination efficiencies can also be achieved using antenna arrays[10], but the elaborated feed networks increase complexity and cost, and can lead to high losses[11].

Contrarily, leaky-wave antennas (LWAs) can produce directive beams using a low-profile structure fed by a simple single source[12]. In Fabry–Pérot (FP) LWAs, a localized source is sandwiched between a perfect electric conductor (PEC) and a partially reflecting surface (PRS), forming a longitudinal FP cavity[2,13]. By tuning the cavity dimensions and source position, favourable coupling to a single waveguided mode is achieved, forming a leaky wave emanating from the source; typical device thicknesses lie around half of a wavelength. The leaky mode is characterized by a transverse wavenumber whose real-part k_t corresponds to the waveguide dispersion, and is accompanied by a small imaginary part α determined by the PRS. Assuming $|k_t| \gg \alpha$, this leads to a conical directive radiation through the PRS towards $\theta_{out} \approx \pm \arcsin(|k_t|/k)$, $k = \omega\sqrt{\mu\varepsilon}$ being the free-space wavenumber, with a beamwidth proportional to α. Broadside radiation is achieved when $|k_t|$ is small enough such that the splitting condition $|k_t| < \alpha$ is satisfied, and the peaks of the conical beam merge[14]. LWAs based on modulated metasurfaces (MoMetAs) are also compact and probe-fed, but utilize a surface wave $|k_t| > k$ guided on a PEC-backed dielectric sheet covered with metallic patches[15-18]. This mode is coupled to radiation via periodic modulation of the patch geometry; its leakage rate α is determined by the modulation depth[18].

Although FP-LWAs and MoMetAs have compact configurations, they suffer from a fundamental efficiency limitation for finite structures: designing a moderate leakage rate α yields uniform illumination but results in considerable losses from the edges; on the other hand, for large leakage rates only a portion of the aperture is effectively radiating[19-21].

To mitigate edge-taper losses, shielded FP-LWA structures have been recently proposed, using PEC side walls which form a lateral cavity[22-28]. Nevertheless, the tight coupling between the propagation of the leaky mode inside the FP cavity and the angular distribution of the radiated power manifested by $\theta_{out} \approx \arcsin(k_t/k)$ poses serious limitations on the achievable aperture illumination efficiency. This is most prominent for antennas radiating at broadside, in which only low-order lateral modes (satisfying the splitting condition) can be used. Consequently, such antennas are designed to excite exclusively the TE_{10} lateral mode, which inherently limits the aperture illumination efficiency, defined as the relative directivity with respect to the case of uniform illumination, to 81% (ref. 29). In addition, as the dominant spectral components of the cavity fields directly translate to prominent lobes in the radiation pattern, only a single mode should be excited to guarantee high directivity. However, suppression of parasitic cavity modes is a very difficult problem[30], especially for large apertures.

From the so far discussion it follows that it would be very beneficial if the fields inside the cavity and those formed on the aperture could be optimized independently. This would facilitate good aperture illumination without the necessity to meet excitation-related restricting conditions. But how to achieve such

a separation? The equivalence principle suggests that for a given field exciting a surface, desirable (arbitrary) aperture fields can be formed by inducing suitable electric and (equivalent) magnetic surface currents[29]. On the basis of this idea, the concept of Huygens' metasurfaces (HMSs) has been recently proposed, where subwavelength electric and magnetic polarizable particles (meta-atoms) are used to generate these surface currents in response to a known incident field[31-39]. In previous work, we have shown that if the reflected and transmitted fields are properly set, the aperture phase can be tailored by a passive and lossless HMS to produce prescribed directive radiation, for any given excitation source[40].

In this paper, we harness the equivalence principle to efficiently convert fields excited in a cavity by a localized source to highly directive radiation using a Huygens' metasurface: cavity-excited HMS antenna. The device structure resembles a typical shielded FP-LWA, with an electric line source surrounded by three PEC walls and a HMS replacing the standard PRS (Fig. 1). For a given aperture length L and a desirable transmission angle θ_{out}, we optimize the cavity thickness and source position to predominantly excite a high-order lateral mode, thus guaranteeing good aperture illumination. Once the source configuration is established, we stipulate the aperture fields to follow the power profile of the cavity mode, and impose a linear phase to promote radiation towards θ_{out}. With the cavity and aperture fields in hand, we invoke the equivalence principle and evaluate the (purely reactive[40]) electric surface impedance and magnetic surface admittance required to support the resultant field discontinuity[31,32,41,42]. As the power profile of the chosen high-order mode creates hot spots of radiating surface currents approximately half a wavelength apart, a uniform virtual phased array is formed on the HMS aperture; such excitation profile is expected to yield very high directivity with no grating lobes regardless of θ_{out} (ref. 10). Furthermore, in contrast to LWAs, the antenna directivity does not deteriorate significantly even if other modes are partially excited, as these would merely vary the amplitude of the virtual array elements, without affecting the phase purity. This semianalytical design procedure can be applied to arbitrarily large apertures, yielding near-unity aperture illumination efficiencies. With the PEC side walls, no power is lost via the edges, offering an effective way to overcome the efficiency tradeoff inherent to FP-LWAs and MoMetAs, while preserving the advantages of a single-feed low-profile antenna.

Results

Cavity-excited Huygens' metasurface antennas. To design the HMS-based antenna, we apply the general methodology developed in ref. 40 to the source configuration of Fig. 1; for

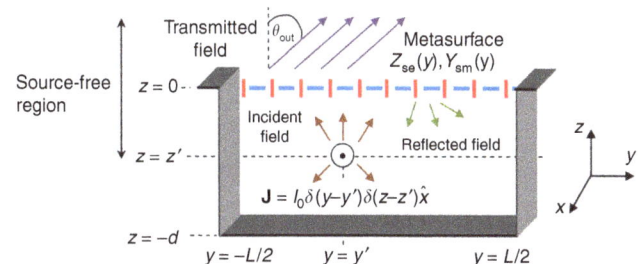

Figure 1 | Physical configuration of a cavity-excited Huygens' metasurface antenna. An electric line source is positioned at (y',z'), surrounded by three perfect-electric-conductor (PEC) walls at $z = -d, y = \pm L/2$, forming a lateral cavity. The cavity is covered by a Huygens' metasurface of aperture length L situated at $z = 0$, facilitating directive radiation towards θ_{out}.

completeness, we recall briefly its main steps. We consider a two-dimensional (2D) scenario ($\partial/\partial x = 0$) with the HMS at $z = 0$ and a given excitation geometry at $z \leq z' < 0$ embedded in a homogeneous medium ($k = \omega\sqrt{\epsilon\mu}$, $\eta = \sqrt{\mu/\epsilon}$). Under these circumstances, the incident, reflected and transmitted fields in the vicinity of the HMS can be expressed via their plane-wave spectrum[43]

$$
\begin{cases}
E_x^{\text{inc}}(y, z) = k\eta I_0 \mathcal{F}^{-1}\left\{\frac{1}{2\beta} f(k_t) e^{-j\beta z}\right\} \\
E_x^{\text{ref}}(y, z) = -k\eta I_0 \mathcal{F}^{-1}\left\{\frac{1}{2\beta} \Gamma(k_t) f(k_t) e^{j\beta z}\right\} \\
E_x^{\text{trans}}(y, z) = k\eta I_0 \mathcal{F}^{-1}\left\{\frac{1}{2\beta} \overline{T}(k_t) e^{-j\beta z}\right\},
\end{cases}
\tag{1}
$$

where $\mathcal{F}^{-1}\{g(k_t; z)\} \triangleq \frac{1}{2\pi}\int_{-\infty}^{\infty} dk_t g(k_t; z) e^{jk_t y}$ is the inverse spatial Fourier transform of $g(k_t; z)$ (ref. 44), $f(k_t)$ is the source spectrum, $\Gamma(k_t)$ is the HMS reflection coefficient, and $\overline{T}(k_t) \triangleq T(k_t)[1 + \Gamma(k_t)]$ is the transmission spectrum. As before, k_t denotes the transverse wavenumber and the longitudinal wavenumber is $\beta = \sqrt{k^2 - k_t^2}$. For simplicity, we only consider here transverse electric (TE) fields ($E_z = E_y = H_x = 0$); the nonvanishing magnetic field components H_y, H_z can be calculated from E_x via Maxwell's equations.

For a given source spectrum, it is required to determine the reflected and transmitted fields, through the respective degrees of freedom $\Gamma(k_t)$ and $T(k_t)$, that would implement the desirable functionality. Once the tangential fields on the two facets of the HMS are set, the equivalence principle is invoked to evaluate the required electric and magnetic surface currents to induce them[29]. The polarizable particles comprising the HMS are then designed such that the average fields acting on them effectively induce these surface currents[41,42]. Analogously, the HMS can be characterized by its electric surface impedance $Z_{\text{se}}(y)$ and magnetic surface admittance $Y_{\text{sm}}(y)$, relating the field discontinuity and the average excitation via the generalized sheet transition conditions (GSTCs)[31,32,40,41].

To promote directive radiation towards θ_{out} we require that the aperture (transmitted) fields approximately follow the suitable plane-wave-like relation (Supplementary Note 1)

$$
\begin{aligned}
E_x(\mathbf{r})|_{z \to 0^+} &\approx Z_{\text{out}} H_y(\mathbf{r})|_{z \to 0^+} \\
&\approx k\eta I_0 \mathcal{F}^{-1}\left\{\frac{1}{2\beta} T(k_t)\right\} \\
&\triangleq k\eta I_0 W_0(y) e^{-jky \sin\theta_{\text{out}}},
\end{aligned}
\tag{2}
$$

where $W_0(y)$ is the aperture window (envelope) function (yet to be determined) and $Z_{\text{out}} = 1/Y_{\text{out}} = \eta/\cos\theta_{\text{out}}$ is the TE wave impedance of a plane-wave directed towards θ_{out}.

In previous work[40], we have shown that if the wave impedance and the real power are continuous across the metasurface, then these aperture fields can be supported by a passive lossless HMS (purely reactive Z_{se} and Y_{sm}). The first condition, local impedance equalization, means that the total (incident + reflected) fields on the bottom facet of the metasurface should exhibit the same wave impedance as the aperture fields, that is, $E_x(\mathbf{r})|_{z \to 0^-} = Z_{\text{out}} H_y(\mathbf{r})|_{z \to 0^-}$; this is achieved by setting the reflection coefficient to a Fresnel-like form

$$
\Gamma(k_t) = \frac{k\cos\theta_{\text{out}} - \beta}{k\cos\theta_{\text{out}} + \beta},
\tag{3}
$$

determining the reflected fields everywhere, fixing our first degree of freedom.

To satisfy the second condition, local power conservation, we require that the aperture window function follows the magnitude

of the total (incident + reflected) fields at $z \to 0^-$, namely,

$$
\begin{aligned}
W_0(y) &= |E_x(\mathbf{r})|_{z \to 0^-} = \left|\mathcal{F}^{-1}\left\{\frac{1}{2\beta}[1 - \Gamma(k_t)]f(k_t)\right\}\right| \\
&= \left(\mathcal{F}^{-1}\left\{\left[\frac{1}{2\beta}(1 - \Gamma)f\right] \star \left[\frac{1}{2\beta}(1 - \Gamma)f\right]\right\}\right)^{1/2},
\end{aligned}
\tag{4}
$$

where $g^\star g$ is the autocorrelation of the spectral-domain function $g(k_t)$ (ref. 44); this determines the transmitted fields everywhere, fixing our second degree of freedom.

The absolute value operator in the last equality is of utmost significance: it indicates that the transmission spectrum of the aperture fields follows, up to a square root, the power spectral density of $E_x(\mathbf{r})|_{z \to 0^-}$, and not the spectral content of the incident and reflected fields. This is directly related to the balanced (plane-wave-like) contribution of the electric and magnetic fields to the power flow that we stipulated in equation (2), and results in a significantly favourable plane-wave spectrum, as will be discussed in detail in the next section.

Finally, we use these semianalytically predicted fields and the equivalence principle, manifested by the GSTCs, to calculate the required HMS surface impedance, yielding the desirable purely reactive modulation given by[40],

$$
\frac{Z_{\text{se}}(y)}{Z_{\text{out}}} = \frac{Y_{\text{sm}}(y)}{Y_{\text{out}}} = -\frac{j}{2}\cot\left[\frac{\phi_-(y) - \phi_+(y)}{2}\right]
\tag{5}
$$

where $\phi_\pm(y) \triangleq \angle E_x(y, z)|_{z \to 0^\pm}$ are the phases of the stipulated fields just above and below the metasurface.

Once the general design procedure is established, applying it to the configuration of Fig. 1, which includes an electric line source at (y', z') surrounded by PEC walls at $z = -d$, $y = \pm L/2$, is straightforward: it is reduced to finding the corresponding source spectrum. The latter is quantized due to the lateral cavity, and includes multiple reflections between the HMS at $z = 0$ and the PEC at $z = -d$; explicitly[43],

$$
f(k_t) = \frac{\pi}{2L}\sum_{n=-\infty}^{\infty}\left\{\frac{e^{-j\beta(d+z')} - e^{j\beta(d+z')}}{e^{j\beta d} - \Gamma(k_t)e^{-j\beta d}}\left[e^{-jk_t y'} + (-1)^{n+1}e^{jk_t y'}\right]\delta\left(k_t - \frac{n\pi}{L}\right)\right\}.
\tag{6}
$$

We refer to the sum of the fields corresponding to the n, $-n$ terms in the summation as the field of the nth mode of the lateral cavity, where $n \geq 0$.

Although this procedure is applicable for any transmission angle, we restrict ourselves from now on to the case of broadside radiation $\theta_{\text{out}} = 0$, where the performance of shielded and unshielded FP-LWAs is the most problematic due to the splitting condition[14] (design of oblique-angle radiators is addressed in the Supplementary Methods). For simplicity, we set the lateral position of the source to be $y' = 0$; with this choice, the even modes vanish, and the odd modes follow a cosine profile in the lateral dimension.

Optimizing the cavity excitation. One of the key differences between the cavity-excited HMS antenna and FP-LWAs is that by harnessing the equivalence principle we control the individual contributions of the electric and magnetic fields to the flow of power, expressed by the lateral distribution of the z-component of the Poynting vector on the aperture. More specifically, the resultant (transmitted) aperture fields corresponding to equation (4) actually follow the square root of the power profile dictated by the cavity mode, and not the profile of the cavity fields. This distinction is very important, as the power profile of a standing wave is always positive, whereas the field profile changes signs along the lateral dimension. Hence, the spectral content of

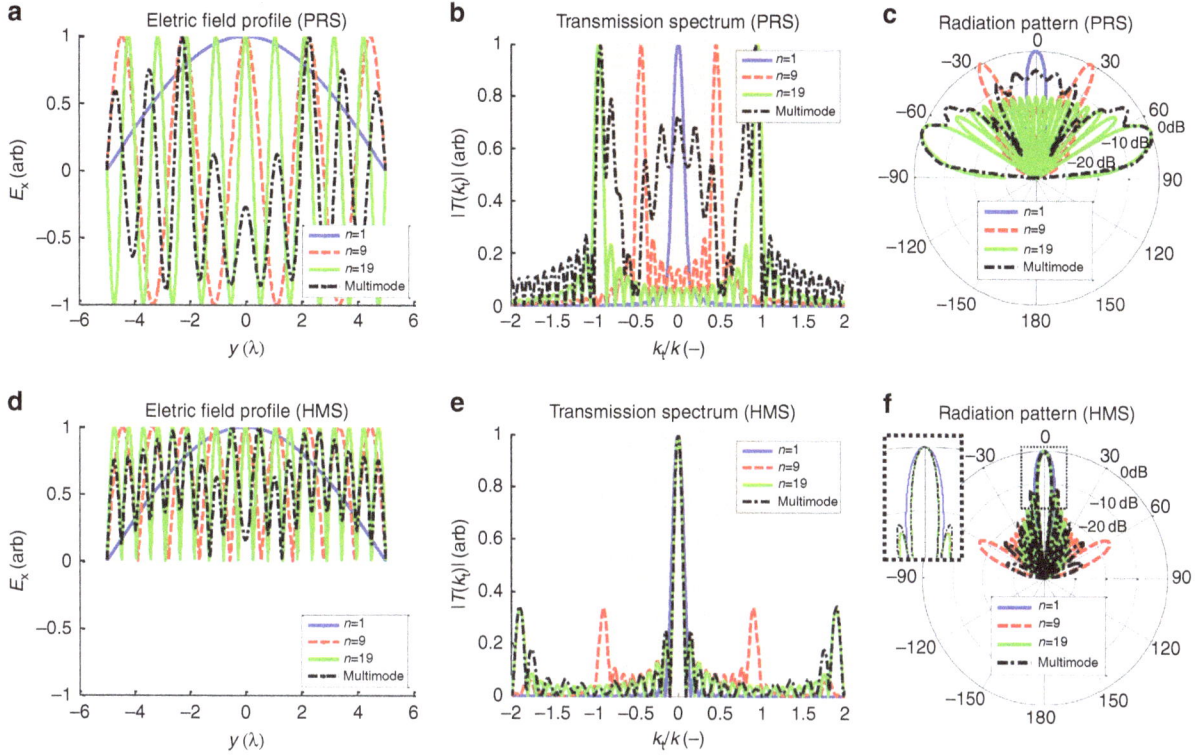

Figure 2 | Comparison between aperture profiles and radiation patterns of cavity-excited PRSs and cavity-excited HMSs. Single-mode excitations of the $n=1$ (blue solid line), $n=9$ (red dashed line), and $n=19$ (green solid line) modes of an aperture of length $L=10\lambda$ are compared to the multimode excitation corresponding to the HMS antenna presented in Fig. 1 with $L=10\lambda$, $z'=-\lambda$, and $d=1.61\lambda$ (black dash-dotted line). (**a,d**) Normalized spatial profile of the tangential electric field on the aperture. (**b,e**) Normalized spectral content of the aperture field; shaded region correspond to the visible part of the spectrum. (**c,f**) Normalized radiation patterns. Inset: close-up of the radiation pattern around $\theta=0$.

the aperture fields, which determines the far-field radiation pattern, is fundamentally different.

To illustrate this point, we compare the fields formed on the device aperture for a shielded FP-LWA, where a standard PRS is used, and for a cavity-excited HMS antenna with the same excitation. Figure 2 presents the spatial profile of the tangential electric field, its spatial Fourier transform, and corresponding radiation patterns (calculated following ref. 29), for single-mode excitation of the $n=1$ (blue solid lines), $n=9$ (red dashed lines) and $n=19$ (green solid lines) modes, for an aperture length of $L=10\lambda$. All plots are normalized to their maximum, as the radiation pattern is sensitive to the variation of the fields, and not to their magnitude.

As follows from equations (4) and (6), the spatial profile of the nth-mode aperture field is proportional to $\cos(n\pi y/L)$ for a standard PRS, but for an HMS it is proportional to $|\cos(n\pi y/L)|$ (Fig. 2a,d). Except for the lowest order mode $n=1$, for which the two functions coincide, the difference in the spatial profile translates into distinctively different features in the spectral content (Fig. 2b,e). For the nth mode, the transmission spectrum of the HMS aperture corresponds to the autocorrelation of the PRS aperture spectrum, leading to formation of peaks centred around the second harmonics ($k_t=\pm 2n\pi/L$) and d.c. ($k_t=0$). As both the right-propagating and left-propagating components of the standing wave coherently contribute to the d.c. peak, the latter dominates the transmission spectrum, and the radiation patterns corresponding to the HMS aperture exhibit highly directive radiation towards broadside (Fig. 2f). In contrast, the PRS-based devices exhibit conical radiation to angles determined by the dominant spectral components of the aperture fields, that is, towards $\theta=\pm\arcsin[n\lambda/(2L)]$ (Fig. 2c)[12,13].

The transverse wavenumber $k_t=\pi/L$ corresponding to the lowest order mode $n=1$ is small enough such that the two symmetric beams merge[14], which enables the PRS aperture to radiate a single beam at broadside. Indeed, small-aperture shielded FP-LWAs utilize this TE_{10} mode to generate broadside radiation. However, as demonstrated by ref. 29, the aperture illumination efficiency of this mode is inherently limited to 81%, due to the non-optimal cosine-shaped aperture illumination[22-28], leading to broadening of the main beam (inset of Fig. 2f). This highlights a key benefit of using an HMS-based antenna, as it is clear from Fig. 2f that we can use high-order mode excitations, which provide a more uniform illumination of the aperture, for generating narrow broadside beams with enhanced directivities.

In fact, as the mode index n increases, the autocorrelation of equation (4) drives the second harmonic peaks outside the visible region of the spectrum (shaded region in Fig. 2b,e), funnelling all the HMS-radiated power to the broadside beam, subsequently increasing the overall directivity. This improvement in radiation properties can be explained using ordinary array theory. As seen from Fig. 2d, the peaks of the field profile generated by the nth mode on the HMS aperture form hot spots of radiating currents separated by a distance of L/n. The radiation from such an aperture profile would resemble the one of a uniform array with the same element separation. As known from established array theory, to avoid grating lobes the element separation should be smaller than a wavelength[10]. For an aperture length of $L=N\lambda$, where N is an integer, the hot-spot separation satisfies this condition for mode indices $n>N$; specifically, for $N=10$ (Fig. 2), grating lobes would not be present in the radiation pattern for mode indices $n>10$. In agreement with this argument, Fig. 2f

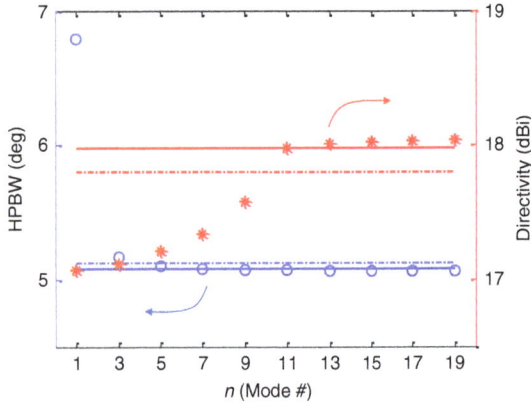

Figure 3 | Radiation characteristics of different lateral cavity modes.
Half-power beamwidth (HPBW, blue circles) and 2D directivity[58] (red asterisks) of an HMS aperture of length $L = 10\lambda$ excited by a single mode as a function of the mode index n. Solid lines denote the respective radiation characteristics of a uniformly excited aperture[29] and dash-dotted lines mark the HPBW (blue) and directivity (red) of *multimode* excitation corresponding to the HMS antenna of Fig. 1 with $L = 10\lambda$, $z' = -\lambda$, and $d = 1.61\lambda$.

shows that for $n = 9$ grating lobes still exist, while for the highest order fast mode $n = 19$ they indeed vanish.

These observations are summarized in Fig. 3, where the radiation characteristics of an HMS aperture of $L = 10\lambda$ excited by a single mode are plotted as a function of the mode index n (only fast modes $k_{t,n} = n\pi/L < k$ are considered). Indeed, it can be seen that the lowest order lateral mode exhibits the worst performance, and the performance improves as the mode index increases. While the half-power beamwidth (HPBW) saturates quickly, the directivity D continues to increase with n until the point in which grating lobes disappear $n = N = 10$ is crossed; for mode indices $n > 10$ the radiation characteristics of the HMS aperture are comparable to those of the optimal uniformly excited aperture (solid lines).

From an array theory point of view, excitation of the highest order fast mode is preferable, as the corresponding equivalent element separation approaches $\lambda/2$, implying that such aperture profile would be suitable for directing the radiation to large oblique angles $\theta_{out} \neq 0$ without generating grating lobes[10]. Furthermore, as the HMS reflection coefficient $\Gamma(k_t)$ grows larger with $k_t = n\pi/L$ (equation (3)), the power carried by the highest order fast-mode $n = 2N - 1$ is best-trapped in the cavity, guaranteeing uniform illumination even in the case of very large apertures.

Nevertheless, generating a single-mode excitation of a cavity via a localized source can be very problematic[30,45]. Fortunately, the cavity-excited HMS antenna can function very well also with multimode excitation, as long as high-order modes dominate the transmission spectrum. This is demonstrated by the dot-dashed lines in Figs 2 and 3, corresponding to a multimode excitation generated by the configuration depicted in Fig. 1 with $L = 10\lambda$, $z' = -\lambda$, and $d = 1.61\lambda$. As expected from the expression for the source spectrum (equation (6)), for a given aperture length L, the field just below the aperture due to a line source would be a superposition of lateral modes, the weights of which are determined by the particular source configuration, namely the cavity thickness d and source position z'. The multimode transmission spectrum in Fig. 2b indicates that for the chosen parameter values, high-order modes ($k_t \to \pm k$) predominantly populate the aperture spectrum, however, low-order modes ($k_t \to 0$) are present as well, to a non-negligible extent. Considering that the far-field angular power distribution $S(\theta)$ is

proportional to $\cos^2 \theta |T(k_t = k \sin \theta)|^2$, the multimode excitation of the PRS aperture results in a radiation pattern resembling the one corresponding to single-mode excitation of the highest order fast-mode ($n = 19$) but with significant lobes around broadside (Fig. 2c); consequently, the directivity is significantly deteriorated.

On the other hand, the same multimode excitation does not degrade substantially the performance of the HMS antenna. The autocorrelated spectrum results in merging of all spectral components into a sharp d.c. peak, with most grating lobes pushed to the evanescent region of the spectrum (Fig. 2e). This retains a beamwidth comparable to that resulting from a single-mode excitation of the highest order fast mode, with only slight deterioration of the directivity due to increased side-lobe level (Fig. 2f and inset). Continuing the analogy to array theory, such multimode excitation introduces slight variations to the magnitude of the array elements, forming an equivalent non-uniform array[10]. The corresponding multimode HPBW and directivity values marked by dash-dotted lines in Fig. 3 verify that, indeed, cavity-excited HMS antennas achieve near-unity aperture illumination efficiencies with a practical multimode excitation; this points out another key advantage of the cavity-excited HMS antenna with respect to shielded FP-LWAs.

We utilize these observations to formulate guidelines for optimizing the cavity excitation for maximal directivity. For a given aperture length $L = N\lambda$, with respect to equation (6), we maximize the coupling to the $n = 2N - 1$ mode (which exhibits the best directivity) by tuning the cavity thickness d as to minimize the denominator of the corresponding coupling coefficient; equally important, we minimize the coupling to the $n = 1$ mode (which exhibits the worst directivity) by tuning the source position z' as to minimize the numerator of the corresponding coupling coefficient. To achieve these with minimal device thickness we derive the following design rules

$$d = \frac{\lambda}{2}\frac{2N}{\sqrt{4N-1}} \xrightarrow{N \gg 1} \frac{\lambda}{2}\sqrt{\frac{L}{\lambda}}, \quad z' \approx -\left(d - \frac{\lambda}{2}\right). \quad (7)$$

Although this is somewhat analogous to the typical design rules for FP-LWAs[13], the key difference is that for HMS-based antennas we optimize the source configuration regardless of the desirable transmission angle θ_{out}. This difference is directly related to the utilization of the equivalence principle for the design of the proposed device, which provides certain decoupling between its excitation and radiation spectra (cf. Fig. 2b,e). This decoupling becomes very apparent when the HMS antenna is designed to radiate towards oblique angles $\theta_{out} \neq 0$, in which case the same cavity excitation yields optimal directivity as well (see Supplementary Methods, Supplementary Fig. 1, and Supplementary Table 1).

Two important comments are relevant when considering these design rules. First, even though following equation (7) maximizes the coupling coefficient of the highest order fast mode and minimizes the coupling coefficient of the lowest order mode, it does not prohibit coupling to other modes. The particular superposition of lateral modes exhibits a tradeoff between beamwidth and side-lobe level (as for non-uniform arrays[10]). Thus, final semianalytical optimization of the cavity illumination profile is achieved by fine tuning the source position z' for the cavity thickness d derived in equation (7). In fact, the source position z' is another degree of freedom that can be used to optimize the radiation pattern for other desirable performance features, such as minimal side-lobe level; this feature is further discussed in the Supplementary Methods and demonstrated in Supplementary Figs. 2 and 3, and Supplementary Table 2. Second, although the optimal device thickness increases with increasing aperture length, the increase is sublinear. Therefore, applying the proposed concept to very large apertures would still result in a relatively compact device, while efficiently utilizing the aperture for producing highly directive beams.

Physical implementation and radiation characteristics. We follow the design procedure and considerations discussed above to design cavity-excited HMS antennas for broadside radiation with different aperture lengths: $L = 10\lambda$, $L = 14\lambda$, and $L = 25\lambda$. The cavity thickness was determined via equation (7) to be $d = 1.61\lambda$, $d = 1.89\lambda$ and $d = 2.50\lambda$, respectively; the source position was set to $z' = -1.00\lambda$, $z' = -1.33\lambda$, and $z' = -1.94\lambda$, respectively, exhibiting maximal directivity.

The required electric surface impedance and magnetic surface admittance modulations are implemented using the 'spider' unit cells depicted in Fig. 4. At the design frequency $f = 20$ GHz ($\lambda \approx 15$ mm), the unit cell transverse dimensions are $\lambda/10 \times \lambda/10$ and the longitudinal thickness is 52 mil $\approx \lambda/12$. Each unit cell consists of three layers of metal traces defined on two bonded laminates of high-dielectric-constant substrate (see Methods). The two (identical) external layers provide the magnetic response of the unit cell, corresponding to the magnetic surface susceptance $B_{sm} = \Im\{Y_{sm}\}$, which is tuned by modifying the arm length L_m (affects equivalent magnetic currents induced by tangential magnetic fields H_y). Analogously, the middle layer is responsible for the electric response of the meta-atom, corresponding to the electric surface reactance $X_{se} = \Im\{Z_{se}\}$, which is tuned by modifying the capacitor width W_e (affects electric currents induced by tangential electric fields E_x). By controlling L_m and W_e, these unit cells can be designed to exhibit Huygens source behaviour, with balanced electric and magnetic responses ranging from $B_{sm}\eta = X_{se}/\eta = -3.1$ to $B_{sm}\eta = X_{se}/\eta = 0.9$ (see Methods and Supplementary Fig. 4).

To experimentally verify our theory, we have fabricated and characterized the $L = 14\lambda$ cavity-excited HMS antenna, based on the simulated spider cell design at $f = 20$ GHz (see Methods). The antenna is composed of a one unit-cell-wide metastrip excited by a coaxial-cable-fed short dipole positioned inside an Aluminium cavity, forming the suitable 2D excitation configuration (Fig. 5). The aperture fields were allowed to radiate into (3D) free-space; the far-field radiation measured in the $\hat{y}z$ plane then corresponds to the theoretically predicted 2D radiation patterns.

Figure 6 presents the design specifications, field distributions, and radiation patterns for the three cavity-excited HMS antennas; Table 1 summarizes the antenna performance parameters (for reference, parameters for uniformly excited apertures[29] are also included). The semianalytical predictions[40] are compared with full-wave simulations conducted with commercially available finite-element solver (ANSYS HFSS), as well as to experimental measurements where applicable (see Methods). As demonstrated by Fig. 6a–c, the realized unit cells are capable of reproducing the required surface impedance modulation, except maybe around large values of $B_{sm}\eta = X_{se}/\eta$; however, such discrepancies usually have little effect on the performance of HMSs[46].

The results in Fig. 6 and Table 1 indicate that the fields and radiation properties predicted by the semianalyical formalism are in excellent agreement with the full-wave simulations for a wide range of aperture lengths. The utilization of realistic (lossy) models for the conductors and dielectrics in the simulated device, as well as other deviations from the assumptions of the design procedure (Supplementary Note 1), result in some discrepancies between the full-wave simulations and predicted performance; however, these mostly affect radiation to large angles (Fig. 6d–f). While this contributes to a minor quantitative difference in the directivity, the properties of the main beam and the side lobes follow accurately the semianalytical results (Table 1), indicating that the theory can reliably predict the dominant contributions to the radiation pattern, as discussed in reference to Fig. 2.

This conclusion is further supported by the experimental results presented for the $L = 14\lambda$ antenna at $f = 20.04$ GHz, where good agreement between theoretical and measured radiation patterns is observed (Fig. 6e). The experimental values of the HPBW, directivity and side-lobe level and position documented in Table 1 also agree quite well with the simulated ones. The slightly higher side-lobe levels and the broadening of the side lobe at $\theta = -7.2°$ contribute to a smaller measured directivity value, and can be attributed to fabrication errors. Nevertheless, the fact that the main features of the radiation pattern are reproduced well with only negligible deviation of 0.2% from the design frequency, and the fact that the predicted, simulated and measured main beams practically coincide, forms a solid validation of our theory.

The measured frequency response of the antenna presented in Supplementary Fig. 5a also compares very well with the simulated

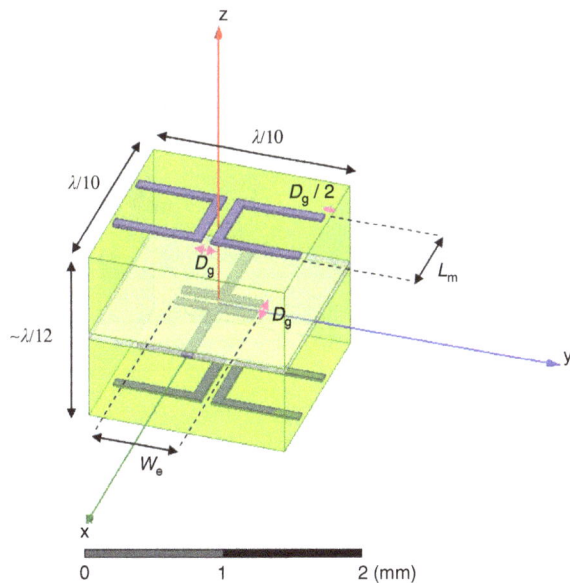

Figure 4 | Spider unit cells. Physical configuration of the meta-atoms used for implementing the HMS at a frequency of $f = 20$ GHz ($\lambda \approx 15$ mm). The electric response is controlled by the capacitor width W_e of the electric dipole, while the magnetic response is determined by the magnetic dipole arm length L_m. The gap size (magenta) and copper trace width are fixed to $D_g = 3$ mil ≈ 76 μm to comply with standard printed-circuit board fabrication techniques.

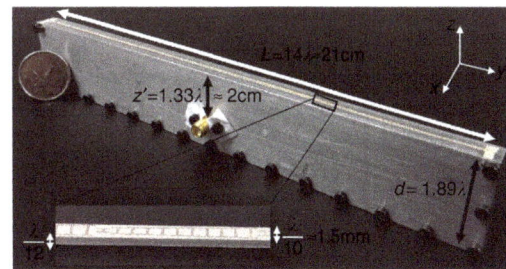

Figure 5 | Fabricated cavity-excited HMS antenna. Probe-fed cavity excites a $\lambda/10$-wide $L = 14\lambda$-long metastrip implementing the simulated design corresponding to Fig. 6b, based on the spider unit cells. The two metallic walls parallel to the $\hat{y}z$ plane form a 2D excitation environment, while the two (shorter) metallic walls parallel to the $\hat{x}z$ plane form the lateral cavity. Inset: close-up of a section of the metastrip before integration with the cavity.

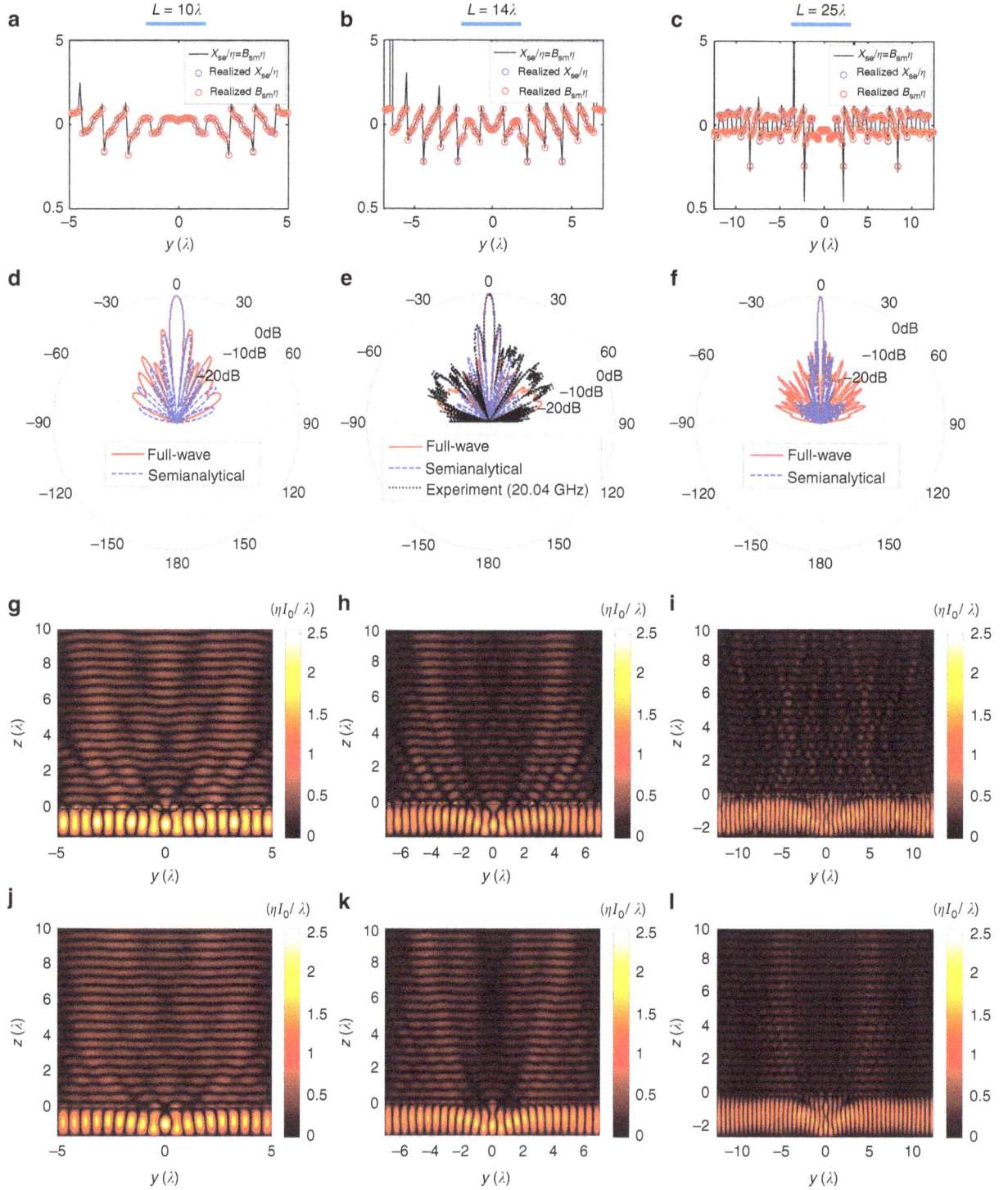

Figure 6 | Performance of cavity-excited HMS antennas. Results are presented for devices with aperture lengths of $L = 10\lambda$, $L = 14\lambda$, and $L = 25\lambda$. (**a–c**) HMS design specifications $X_{se}(y)/\eta = B_{sm}(y)\eta$ (black solid line) derived from equation (5), and the realized electric surface reactance (blue circles) and magnetic surface susceptance (red circles) using the spider unit cells. (**d–f**) Radiation patterns produced by semianalytical formalism (blue dashed line) and full-wave simulations (red solid line). For $L = 14\lambda$, the experimentally obtained radiation pattern is presented in black dotted line. (**g–i**) Field distribution $|\Re\{E_x(y,z)\}|$ produced by full-wave simulations. (**j–l**) Semianalytical prediction of $|\Re\{E_x(y,z)\}|$ (ref. 40).

results, indicating a measured 3- and 1-dB directivity bandwidths of 5.5% and 1.3%, respectively. These values are not very high, but this is expected due to the resonant nature of the cavity and metasurface[24,38]. Bandwidth enhancement can be likely achieved by using stacked HMSs or increasing the number of layers comprising the unit cells[24,47]. The measured 10-dB return-loss bandwidth is 0.2% (Supplementary Fig. 5b), comparable to the

values reported for cavity-based antennas[48]; nonetheless, it could be improved by using suitable matching circuitry[49]. The measured (3D) realized gain at $f = 20.04$ GHz is 16.12 ± 1 dBi, corresponding to a gain of 17.21 ± 1 dBi (with 2.7% 3-dB bandwidth). From the 3D directivity estimated from the measurements (Supplementary Methods and Supplementary Fig. 6) we evaluate a radiation efficiency of $75 \pm 17\%$, in a

Table 1 | Radiation characteristics of cavity-excited HMS antennas.

| | $L = 10\lambda$ ($d = 1.61\lambda$, $|z'| = 1.00\lambda$) | | | $L = 14\lambda$ ($d = 1.89\lambda$, $|z'| = 1.33\lambda$) | | | | $L = 25\lambda$ ($d = 2.50\lambda$, $|z'| = 1.94\lambda$) | | |
	Full-wave	Semianalytic	Uniform	Experiment	Full-wave	Semianalytic	Uniform	Full-wave	Semianalytic	Uniform
HPBW	5.11°	5.13°	5.08°	3.63°	3.83°	3.64°	3.63°	2.09°	2.13°	2.03°
Directivity (2D) (dBi)	17.42	17.84	17.98	18.20	18.79	19.15	19.44	21.33	21.75	21.96
First side-lobe	8.6°	8.6°	8.2°	−7.2°|6.0°	6.4°	6.3°	5.9°	3.4°	3.4°	3.3°
Side-lobe level (dB)	−10.4	−12.0	−13.5	−9.1	−11.1	−10.4	−13.5	−14.6	−14.0	−13.5

HMS, Huygens' metasurface; HPBW, half-power beamwidth; 2D, two dimensional.

reasonable agreement with the ≈15% conductor and dielectric losses predicted by full-wave simulations (Supplementary Fig. 7). These results indicate that if impedance matching can be achieved via suitable circuitry, then cavity-excited HMS antennas could exhibit reasonably high antenna efficiencies. Lastly, the measured cross-polarization gain was below the noise floor, indicating at least 30-dB cross-polar discrimination at broadside.

For all cases considered, the excitation of the highest order lateral fast mode is clearly visible (Fig. 6g–l), leading to beamwidth and (2D) directivity values comparable to the ones achieved by uniform excitation of the aperture (Table 1). In particular, the simulated (measured) radiation patterns yield aperture illumination efficiencies $\eta_{\text{apt}} \triangleq D/(2\pi L/\lambda)$ of 88%, 86% (75%) and 87% for the $L = 10\lambda$, $L = 14\lambda$ and $L = 25\lambda$ devices, respectively, retained even when the aperture length is very large. In terms of HPBW, often taken as a measure for effective exploitation of the aperture[19], the device performance is even closer to that of a uniformly excited aperture, with pencil-beam HPBWs reaching 99%, 95% (100%) and 97% of the optimal beamwidth, for $L = 10\lambda$, $L = 14\lambda$ and $L = 25\lambda$ devices, respectively. Similar to FP-LWAs[13], the simulated 3-dB bandwidths of the antennas reveal a tradeoff between directivity and bandwidth, reflected by an approximately constant value of 3.1 for the directivity-bandwidth product (Supplementary Fig. 8).

It should be stressed that even though optimized TE_{10} shielded FP-LWAs can reach aperture illumination efficiencies of 81%, their HPBWs are limited to about 75% of the optimal beamwidth[29]; more importantly, their PRS-based design requires single-mode excitation to achieve this performance, thus preventing practical realization of large-aperture devices. Just recently, a low-profile metasurface-based lens antenna with a smaller aperture length (diameter of 6.6λ) operating at a lower frequency (10 GHz), has been reported[50], also exhibiting a high aperture illumination efficiency. Nevertheless, this antenna utilizes a metasurface designed to convert Bessel beams to Gaussian beams; hence, to excite it, a Bessel beam launcher has to be separately designed and fabricated for each application, in addition to the metasurface lens. In contrast, our design approach and rationale can be straightforwardly applied to apertures of any size, and yields an integrated air-filled device requiring only a single simple feed.

While the main purpose of the proposed solution is to maximize antenna directivity, this design goal is suitable in cases where the environment clatter is relatively controlled (for example, point-to-point communications[4]), as uniformly illuminated apertures incur relatively high side-lobe levels; for many applications lower side-lobe levels are required[7]. Nonetheless, as discussed in the Supplementary Methods and demonstrated in Supplementary Fig. 3, a desirable compromise between directivity and side-lobe level can be achieved in the framework of our theory, by simple tuning of the source position z'. Setting the cavity thickness d following equation (7) still forms a virtual antenna array with linear phase, while the different values of z' effectively vary the magnitude

of the array elements, facilitating side-lobe control[10]. The example presented in Supplementary Fig. 2 and Supplementary Table 2 harnesses this scheme to design ($L = 10\lambda$)-long cavity-excited HMS antennas with side-lobe level of −20 dB, still retaining a rather high aperture illumination efficiency of 81%.

Discussion

We have introduced a novel design for low-profile single-fed antennas exhibiting beamwidth and directivity values comparable with uniformly excited apertures. To that end, we harness the equivalence principle to devise a cavity-excited Huygens' metasurface, setting the source configuration, HMS reflection coefficient, and aperture fields such that (1) the highest order (lateral) fast mode is predominantly excited, which guarantees that the aperture is well-illuminated; (2) the aperture fields follow the incident power profile and not the incident field profile, which forms an array-like aperture profile with favourable transmission spectrum; and (3) the power flow and wave impedance are continuous across the metasurface, which ensures the design can be implemented by a passive and lossless HMS[40]. The possibility to control the field discontinuities using the electric and (equivalent) magnetic currents induced on the HMS allows us to optimize separately the cavity excitation and the radiated fields, thus overcoming the fundamental tradeoff existing in FP-LWAs between aperture illumination efficiency and edge-taper losses.

It should be emphasized that the general design procedure formulated and demonstrated herein facilitates further optimization of such devices for various applications. The extensive freedom one has in choosing the source configuration, combined with the efficient semianalytical approach, allows explorations of other excitation sources, for example, with different orientations and current distributions, to further tailor the aperture fields (extension to 3D configurations and polarization control are discussed in Supplementary Note 2); once the source spectrum is characterized, the rest of the procedure is straightforward, and the fields and radiation patterns are readily predicted. In addition, considering the recent demonstrations of metasurfaces in general and Huygens' metasurfaces in particular, operating at terahertz and optical frequencies[36,39,51–56], the proposed methodology could be applied to realize compact and efficient pencil beam radiators across the electromagnetic spectrum, extending the range of applications even further.

Methods

Spider unit-cell modelling. The spider unit cells depicted in Fig. 4 were defined in ANSYS Electromagnetic Suite 15.0 (HFSS 2014) with two 25-mil-thick (≈0.64 mm) Rogers RT/duroid 6010LM laminates (green boxes in Fig. 4) bonded by 2-mil-thick (≈51 μm) Rogers 2929 bondply (white box in Fig. 4). The electromagnetic properties of these products at 20 GHz, namely, permittivity tensor and dielectric loss tangent, as were provided to us by Rogers Corporation, have been inserted to the model. Specifically, a uniaxial permittivity tensor with $\varepsilon_{xx} = \varepsilon_{yy} = 13.3\varepsilon_0$, $\varepsilon_{zz} = 10.81\varepsilon_0$ and loss tangent of tan $\delta = 0.0023$ were considered for Rogers RT/duroid 6010LM laminates, while an isotropic permittivity of $\varepsilon = 2.94\varepsilon_0$ and loss tangent tan$\delta = 0.003$ were considered for Rogers 2929 bondply. The copper traces corresponded to 1/2 oz. cladding, featuring a thickness of 18 μm; the standard value of $\sigma = 58 \times 10^6$ S m^{-1} bulk conductivity was used in the model. To comply with standard printed-circuit board manufacturing processes, all copper traces were 3 mil ($\approx 76\,\mu m$) wide, and a minimal distance of 3 mil was kept between adjacent traces (within the cell or between adjacent cells). This implies that the fixed gaps between the capacitor traces (along the x axis) of the electric dipole in the middle layer, as well as between the two arms (along the y axis) of the magnetic dipole in the top and bottom layer (Fig. 4), were fixed to a value of $D_g = 3$ mil ($\approx 76\,\mu m$); the distance from the arm edge to the edge of the unit cell was fixed to $D_g/2 = 1.5$ mil ($\approx 38\,\mu m$).

Unit cells with different values of magnetic dipole arm length L_m and electric dipole capacitor width W_e were simulated using periodic boundary conditions; HFSS Floquet ports were placed at $z = \pm\lambda$ and used to characterize the scattering of a normally incident plane wave off the periodic structure (the interface between the bondply and the bottom laminate was defined as the $z = 0$ plane). For each combination of L_m and W_e, the corresponding magnetic surface susceptance B_{sm} and electric surface reactance X_{se} were extracted from the simulated impedance matrix of this two-port configuration, following the derivation in ref. 57.

The magnetic response B_{sm} was found to be proportional to the magnetic dipole arm length L_m, with almost no dependency in W_e (ref. 37). Thus, to create an adequate lookup table for implementing broadside-radiating HMSs, we varied L_m by constant increments, and for a given L_m, plotted $B_{sm}\eta$ and X_{se}/η as a function of W_e. The value of W_e for which the two curves intersected corresponded to a balanced-impedance point ($Z_{se}/Z_{out} = Y_{sm}/Y_{out}$), where the unit cell acts as a Huygens source, and thus suitable for implementing our metasurface. A lookup table composed of (B_{sm}, X_{se}) pairs and the corresponding unit cell geometries (L_m, W_e) was constructed, and refined through interpolation. The interpolated unit cell geometries were eventually simulated again, to verify the interpolation accuracy and finalize the lookup table entries, as presented in Supplementary Fig. 4. Finally, for a given HMS with prescribed surface impedance modulation ($B_{sm}(y)$, $X_{se}(y)$), a corresponding structure could be defined in HFSS using the unit cells ($L_m(y)$, $W_e(y)$) found via the lookup table in terms of least squares error.

Antenna simulations. To verify our semianalytical design via full-wave simulations, each of the cavity-excited HMS antennas designed in this paper was defined in HFSS using a single strip of unit cells implementing the metasurface, occupying the region $|x| \leq \lambda/20$, $|y| \leq L/2$ (L being the aperture length of the antenna), and $-0.64\,mm \leq z \leq 0.69\,mm$ (in correspondence to the laminate and bondply thicknesses). The simulation domain included $|x| \leq \lambda/20$, $|y| \leq L/2 + 2.5\lambda$, and $-d \leq z \leq 10\lambda$ (d being the cavity thickness), where PEC boundary conditions were applied to the $x = \pm\lambda/20$ planes to form the equivalence of a 2D scenario. PEC boundary conditions were also applied to the $z = -d$ plane, and to two 18-μm-thick side-walls at $y = \pm L/2$, forming the cavity. The line-source excitation was modelled by a $\lambda/20$-wide 1A current sheet at $z = z'$, with the current aligned with the x axis. Radiation boundary conditions were applied to the rest of the simulation space boundaries, namely $z = 10\lambda$, and $y = \pm(L/2 + 2.5\lambda)$, allowing proper numerical evaluation of the fields surrounding the antenna.

To reduce the computational effort required to solve this configuration, we utilized the symmetries of our TE scenario. Specifically, we placed a perfect-magnetic-conductor (PMC) symmetry boundary conditions at the $\widehat{xz}$ plane, and a PEC symmetry boundary conditions at the $\widehat{yz}$ plane. We also noticed that adding a thin layer (1 mil $\approx 25\,\mu m$) of copper between the electric dipole edges and the PEC parallel plates at $x = \pm\lambda/20$ enhanced the convergence of the simulation results. With that minor modification, all of the simulated antennas converged within <40 iterations (maximum refinement of 10% per pass), where the stop conditions was three consecutive iterations in which ΔEnergy<0.03.

Antenna realization and measurement. The simulated design corresponding to the cavity-excited HMS antenna with $L = 14\lambda$ aperture (Fig. 6b) has been realized and measured to obtain experimental validation of our theory. As in the full-wave simulations, the fabricated metasurface consisted of two 25-mil-thick (≈ 0.64 mm) Rogers RT/duroid 6010LM laminates. The copper traces' geometry used in the simulations to implement the spider unit cells was exported to standard grbr files, later used by Candor Industries Inc. to accordingly etch the electrodeposited 1/2 oz. copper foils covering the laminates. The etched laminates were then bonded using 2-mil-thick ($\approx 51\,\mu m$) Rogers 2929 bondply, forming the desirable three-layer metasurface. Lastly, unit-cell-wide ($\lambda/10 \approx 1.5$ mm) metastrips were achieved via routing (see inset of Fig. 5).

The cavity was realized by replacing the PEC walls utilized in simulations by 4-mm-thick Aluminium. The resultant five-faceted box was split along the $\widehat{yz}$ plane into two parts to enable accurate fabrication using computerized numerical control (CNC) machines in the University of Toronto; the two parts of the box were attached using multiple metallic screws along the box perimeter (Fig. 5). A subminiature version A (SMA) female connector flange with an exposed pin was mounted using screws on the front facet to feed the current source exciting the structure. The current source was created by extending the exposed pin using a soldered copper wire, forming an electric dipole with an overall length of $\lambda/10$, in accordance to the simulated configuration.

Finally, the metastrip was vertically positioned using two dowel pins inserted at the edges of the cavity aperture, where the electric field is predicted to be negligible. When assembling the antenna, the horizontal alignment of the metastrip was insured by the use of masking tape strips while the box peripheral screws were

tightened; the tape was removed before measurements were conducted. To guarantee good coupling between the middle layer copper traces (Fig. 4) and the metallic walls, a thin adhesive copper film was attached to the metallic walls where they made contact with the the metastrip facets.

Antenna measurements were conducted in the far-field measurement chamber at the University of Toronto, calibrated using Quinstar Technology Inc. QWH-KPRS00 standard-gain horn antennas following the gain comparison method. The HMS antenna was then mounted on a stage situated in the far field of a transmitting horn antenna, and radiation patterns were obtained by rotating the stage in steps of 0.2°. The received power was spectrally resolved with a frequency resolution of 0.02 GHz within the frequency range $f \in [18.1, 21.9\,GHz]$. The procedure for evaluating the antenna gain, directivity, radiation efficiency, and aperture illumination efficiency out of the measured radiation patterns is addressed in detail in the Supplementary Methods.

References

1. Kraus, J. The corner-reflector antenna. *Proc. IRE* **28,** 513–519 (1940).
2. von Trentini, G. Partially reflecting sheet arrays. *IRE Trans. Antennas Propag.* **4,** 666–671 (1956).
3. Hansen, R. Communications satellites using arrays. *Proc. IRE* **49,** 1066–1074 (1961).
4. Franson, S. & Ziolkowski, R. Gigabit per second data transfer in high-gain metamaterial structures at 60 GHz. *IEEE Trans. Antennas Propag.* **57,** 2913–2925 (2009).
5. Tichit, P.-H., Burokur, S. N. & de Lustrac, A. Design and experimental demonstration of a high-directive emission with transformation optics. *Phys. Rev. B* **83,** 155108 (2011).
6. Jiang, Z. H., Gregory, M. D. & Werner, D. H. Experimental demonstration of a broadband transformation optics lens for highly directive multibeam emission. *Phys. Rev. B* **84,** 165111 (2011).
7. Lier, E., Werner, D. H., Scarborough, C. P., Wu, Q. & Bossard, J. A. An octave-bandwidth negligible-loss radiofrequency metamaterial. *Nat. Mater.* **10,** 216–222 (2011).
8. Imbriale, W. A. *Modern Antenna Handbook*. Ch. 5 (Wiley, 2008).
9. Hum, S. V. & Perruisseau-Carrier, J. Reconfigurable reflectarrays and array lenses for dynamic antenna beam control: A review. *IEEE Trans. Antennas Propag.* **62,** 183–198 (2014).
10. Tsoilos, G. V. & Christodoulou, C. G. *Modern Antenna Handbook*. Ch. 11 (Wiley, 2008).
11. Haupt, R. & Rahmat-Samii, Y. Antenna array developments: A perspective on the past, present and future. *IEEE Antennas Propag. Mag.* **57,** 86–96 (2015).
12. Jackson, D. R. & Oliner, A. A. *Modern antenna handbook*. Ch. 7 (Wiley, 2008).
13. Jackson, D. R. *et al.* The fundamental physics of directive beaming at microwave and optical frequencies and the role of leaky waves. *Proc. IEEE* **99,** 1780–1805 (2011).
14. Lovat, G., Burghignoli, P. & Jackson, D. R. Fundamental properties and optimization of broadside radiation from uniform leaky-wave antennas. *IEEE Trans. Antennas Propag.* **54,** 1442–1452 (2006).
15. Fong, B., Colburn, J., Ottusch, J., Visher, J. & Sievenpiper, D. Scalar and tensor holographic artificial impedance surfaces. *IEEE Trans. Antennas Propag.* **58,** 3212–3221 (2010).
16. Minatti, G., Caminita, F., Casaletti, M. & Maci, S. Spiral leaky-wave antennas based on modulated surface impedance. *IEEE Trans. Antennas Propag.* **59,** 4436–4444 (2011).
17. Patel, A. M. & Grbic, A. A printed leaky-wave antenna based on a sinusoidally-modulated reactance surface. *IEEE Trans. Antennas Propag.* **59,** 2087–2096 (2011).
18. Minatti, G. *et al.* Modulated metasurface antennas for space: Synthesis, analysis and realizations. *IEEE Trans. Antennas Propag.* **63,** 1288–1300 (2015).
19. Sievenpiper, D. Forward and backward leaky wave radiation with large effective aperture from an electronically tunable textured surface. *IEEE Trans. Antennas Propag.* **53,** 236–247 (2005).
20. Komanduri, V. R., Jackson, D. R. & Long, S. A. in *Proceedings of IEEE International Symposium on Antennas and Propagation (APSURSI)*, 1–4 (Toronto, ON, Canada, 2010).
21. Garca-Vigueras, M., Delara-Guarch, P., Gómez-Tornero, J. L., Guzmán-Quirós, R. & Goussetis, G. in *Proceedings of the 6th European Conference on Antennas and Propagation (EuCAP)*, 247-251 (Prague, Czech Republic, 2012).
22. Feresidis, A. P. & Vardaxoglou, J. C. in *Proceedings of the 1st European Conference on Antennas and Propagation (EUCAP)*, 3–7 (Nice, France, 2006).
23. Ju, J., Kim, D. & Choi, J. Fabry-Pérot cavity antenna with lateral metallic walls for WiBro base station applications. *Electron. Lett.* **45,** 141–142 (2009).
24. Muhammad, S. A., Sauleau, R. & Legay, H. Small-size shielded metallic stacked Fabry-Pérot cavity antennas with large bandwidth for space applications. *IEEE Trans. Antennas Propag.* **60,** 792–802 (2012).
25. Muhammad, S. A. & Sauleau, R. in *Proceedings of the 11th International Bhurban Conference on Applied Sciences and Technology (IBCAST)*, 14–17 (Islamabad, Pakistan, 2014).

26. Kim, D., Ju, J. & Choi, J. A mobile communication base station antenna using a genetic algorithm based Fabry-Pérot resonance optimization. *IEEE Trans. Antennas Propag.* **60**, 1053–1058 (2012).

27. Hosseini, S. A., Capolino, F. & De Flaviis, F. in *Proceedings of the IEEE International Symposium on Antennas and Propagation (APSURSI)*, 746–747 (Orlando, FL, USA, 2013).

28. Haralambiev, L. A. & Hristov, H. D. Radiation characteristics of 3D resonant cavity antenna with grid-oscillator integrated inside. *Int. J. Antennas Propag.* **2014**, 479189 (2014).

29. Balanis, C. *Antenna Theory: Analysis and Design.* Ch. 12 (Wiley, 1997).

30. Muhammad, S. A., Sauleau, R. & Legay, H. in *Proceedings of the 5th European Conference on Antennas and Propagation (EUCAP)*, 1526–1530 (Rome, Italy, 2011).

31. Pfeiffer, C. & Grbic, A. Metamaterial Huygens' surfaces: tailoring wave fronts with reflectionless sheets. *Phys. Rev. Lett.* **110**, 197401 (2013).

32. Selvanayagam, M. & Eleftheriades, G. V. Discontinuous electromagnetic fields using orthogonal electric and magnetic currents for wavefront manipulation. *Opt. Express* **21**, 14409–14429 (2013).

33. Selvanayagam, M. & Eleftheriades, G. V. Polarization control using tensor Huygens surfaces. *IEEE Trans. Antennas Propag.* **62**, 6155–6168 (2014).

34. Ra'di, Y., Asadchy, V. S. & Tretyakov, S. A. One-way transparent sheets. *Phys. Rev. B* **89**, 075109 (2014).

35. Zhu, B. O. *et al.* Dynamic control of electromagnetic wave propagation with the equivalent principle inspired tunable metasurface. *Sci. Rep.* **4**, 4971 (2014).

36. Pfeiffer, C. *et al.* Efficient light bending with isotropic metamaterial Huygens' surfaces. *Nano Lett.* **14**, 2491–2497 (2014).

37. Wong, J. P. S., Selvanayagam, M. & Eleftheriades, G. V. Design of unit cells and demonstration of methods for synthesizing Huygens metasurfaces. *Photon. Nanostruct.* **12**, 360–375 (2014).

38. Wong, J. P. S., Selvanayagam, M. & Eleftheriades, G. V. Polarization considerations for scalar Huygens metasurfaces and characterization for 2-D refraction. *IEEE Trans. Microwave Theor. Technol.* **63**, 913–924 (2015).

39. Decker, M. *et al.* High-efficiency dielectric Huygens' surfaces. *Adv. Opt. Mater.* **3**, 813–820 (2015).

40. Epstein, A. & Eleftheriades, G. V. Passive lossless Huygens metasurfaces for conversion of arbitrary source field to directive radiation. *IEEE Trans. Antennas Propag.* **62**, 5680–5695 (2014).

41. Kuester, E., Mohamed, M., Piket-May, M. & Holloway, C. Averaged transition conditions for electromagnetic fields at a metafilm. *IEEE Trans. Antennas Propag.* **51**, 2641–2651 (2003).

42. Tretyakov, S. *Analytical Modeling in Applied Electromagnetics* (Artech House, 2003).

43. Felsen, L. B. & Marcuvitz, N. *Radiation and Scattering of Waves,* 1st edn (Prentice-Hall, 1973).

44. Howell, K. B. *The Transforms and Applications Handbook.* Ch. 2, 2nd edn (CRC Press, 2000).

45. Chung-Shu, K. S. Cavity-backed spiral antenna with mode suppression. (U.S. Patent 3,555,554, 1971).

46. Epstein, A. & Eleftheriades, G. V. Floquet-Bloch analysis of refracting Huygens metasurfaces. *Phys. Rev. B* **90**, 235127 (2014).

47. Abadi, S. M. A. M. H. & Behdad, N. Design of wideband, FSS-based multibeam antennas using the effective medium approach. *IEEE Trans. Antennas Propag.* **62**, 5557–5564 (2014).

48. Martinis, M., Mahdjoubi, K., Sauleau, R., Collardey, S. & Bernard, L. Bandwidth behavior and improvement of miniature cavity antennas with broadside radiation pattern using a metasurface. *IEEE Trans. Antennas Propag.* **63**, 1899–1908 (2015).

49. Pozar, D. M. *Microwave Engineering.* Ch. 5, 4th edn (John Wiley and Sons, Inc., 2011).

50. Pfeiffer, C. & Grbic, A. Planar lens antennas of subwavelength thickness: collimating leaky-waves with metasurfaces. *IEEE Trans. Antennas Propag.* **63**, 3248–3253 (2015).

51. Monticone, F., Estakhri, N. M. & Alù, A. Full control of nanoscale optical transmission with a composite metascreen. *Phys. Rev. Lett.* **110**, 203903 (2013).

52. Cheng, J. & Mosallaei, H. Optical metasurfaces for beam scanning in space. *Opt. Lett.* **39**, 2719–2722 (2014).

53. Yu, N. & Capasso, F. Flat optics with designer metasurfaces. *Nat. Mater.* **13**, 139–150 (2014).

54. Lin, D., Fan, P., Hasman, E. & Brongersma, M. L. Dielectric gradient metasurface optical elements. *Science* **345**, 298–302 (2014).

55. Campione, S., Basilio, L. I., Warne, L. K. & Sinclair, M. B. Tailoring dielectric resonator geometries for directional scattering and Huygens' metasurfaces. *Opt. Express* **23**, 2293 (2015).

56. Aieta, F., Kats, M. A., Genevet, P. & Capasso, F. Multiwavelength achromatic metasurfaces by dispersive phase compensation. *Science* **347**, 1342–1345 (2015).

57. Selvanayagam, M. & Eleftheriades, G. V. Circuit modelling of Huygens surfaces. *IEEE Antennas Wirel. Propag. Lett* **12**, 1642–1645 (2013).

58. Lovat, G., Burghignoli, P., Capolino, F., Jackson, D. & Wilton, D. Analysis of directive radiation from a line source in a metamaterial slab with low permittivity. *IEEE Trans. Antennas Propag.* **54**, 1017–1030 (2006).

Acknowledgements

Financial support from the Natural Sciences and Engineering Research Council of Canada (NSERC) is gratefully acknowledged. A.E. gratefully acknowledges the support of the Lyon Sachs Postdoctoral Fellowship Foundation as well as the Andrew and Erna Finci Viterbi Postdoctoral Fellowship Foundation of the Technion - Israel Institute of Technology, Haifa, Israel. He thanks M. Selvanayagam, T. Cameron and L. Liang for valuable advice and assistance with the experimental set-up.

Author contributions

A.E. performed formulation, analysis, simulations, physical design, experiments and generation of the results, J.P.S.W. conceived the unit-cell structure and performed simulations, and G.V.E. supervised all stages. A.E. and G.V.E. contributed to conceiving the idea, and writing and editing the manuscript.

Additional information

15

Acoustic trapping of active matter

Sho C. Takatori[1,*], Raf De Dier[2,3,*], Jan Vermant[2] & John F. Brady[1]

Confinement of living microorganisms and self-propelled particles by an external trap provides a means of analysing the motion and behaviour of active systems. Developing a tweezer with a trapping radius large compared with the swimmers' size and run length has been an experimental challenge, as standard optical traps are too weak. Here we report the novel use of an acoustic tweezer to confine self-propelled particles in two dimensions over distances large compared with the swimmers' run length. We develop a near-harmonic trap to demonstrate the crossover from weak confinement, where the probability density is Boltzmann-like, to strong confinement, where the density is peaked along the perimeter. At high concentrations the swimmers crystallize into a close-packed structure, which subsequently 'explodes' as a travelling wave when the tweezer is turned off. The swimmers' confined motion provides a measurement of the swim pressure, a unique mechanical pressure exerted by self-propelled bodies.

[1] Division of Chemistry and Chemical Engineering, California Institute of Technology, Pasadena, California 91125, USA. [2] Department of Materials, ETH Zürich, Vladimir-Prelog-Weg 5, Zürich 8093, Switzerland. [3] Department of Chemical Engineering, KU Leuven, Leuven B-3001, Belgium. * These authors contributed equally to this work. Correspondence and requests for materials should be addressed to S.C.T. (email: Takatori@caltech.edu).

The study of active matter systems such as swimming bacteria and molecular motors in geometrically confined environments plays an essential role in understanding many cellular and biophysical processes. Many studies have demonstrated that self-propelled bodies exhibit intriguing phenomena in confined spaces, such as accumulation at boundaries[1-5]. In addition to confinement by a physical boundary, confinement of active particles via a harmonic external field can engender many useful properties of active systems[6-8]. However, this has remained an experimental challenge because most biological trapping instruments are too weak to confine active systems over a large spatial extent, as swimmers' run lengths can easily exceed $100\,\mu m$.

In this work, we overcome this challenge by developing a custom-built acoustic tweezer to confine self-propelled Janus particles in a two-dimensional (2D) near-harmonic trap. Analogous to optical or magnetic tweezers, acoustic traps employ sound waves to move objects to special regions of the acoustic radiation field. One-dimensional or 2D standing wave devices[9,10] utilize axial radiation forces of multiple transducers and/or reflectors to manipulate objects into pressure nodes or antinodes. Unlike studies that impose a strong trap to manipulate structures and place objects at specified locations, our goal is to interrogate the motion and behaviour of the trap constituents themselves (that is, active microswimmers) under confinement.

Invoking the transverse acoustic forces of a single-beam transducer[11], in the present work we use a 2D device capable of confining active particles over a wide range of trapping forces and spatial extents of the trap, from larger than the run length of the swimmers to much smaller than it. In addition to having a significantly stronger trapping force compared with optical tweezers, acoustic traps preclude potential laser damage to biological specimens ('optication')[12] especially for the high power intensities required for our study. Our judicious choice of the trap strength enables us to study the non-equilibrium behaviours of active particles in varying degrees of confinement: for weak traps, the probability density is Boltzmann-like and can be mapped to that of classical equilibrium Brownian systems, and for strong traps, the density is peaked along the trap perimeter. We discover that the external trap behaves as an 'osmotic barrier' confining swimmers inside the trapping region, analogous to semipermeable membranes that confine passive Brownian solutes inside a boundary. From the swimmers restricted motion inside the trap, we calculate the unique swim pressure generated by active systems originating from the mechanical force required to confine them by boundaries. Finally, we investigate the crossover from ballistic to diffusive behaviour of active microswimmers by observing the 'explosion' of an active crystal.

Results

Active Janus particles in acoustic confinement. We fabricate Janus particles (platinum/polystyrene) that swim near the interface in hydrogen peroxide solution via self-diffusiophoresis[13,14]. These active Brownian spheres translate with an intrinsic swim velocity U_0, tumble with a reorientation time τ_R, and experience a hydrodynamic drag ζ from the surrounding continuous Newtonian fluid. The tumbling of the swimmer from rotational Brownian motion results in a random walk process for $t > \tau_R$ with translational diffusivity $D^{swim} = U_0^2 \tau_R / 2$ in 2D (Swimmers' translational Brownian diffusivity $D_0 = k_B T / \zeta \sim \mathcal{O}(0.1)\,\mu m^2\,s^{-1}$ is small compared with $D^{swim} = U_0^2 \tau_R / 2 \sim \mathcal{O}(100)\mu m^2\,s^{-1}$, and D_0 is not considered here.). We choose active synthetic Janus particles as a model living system with narrow distributions in swim velocity

and reorientation time, but our tweezer setup can accommodate bacteria and other biological microswimmers.

Our acoustic trap exerts a Gaussian trap force[11] with spring constant k and width w, $F^{trap}(r) = -kr\exp(-2(r/w)^2)$, which is well-approximated by a harmonic trap $F^{trap}(r) \approx -kr$ for small departures $r \ll w$. The Janus particles explore the interior of the trap with their intrinsic swimming motion; however, they are confined to remain within a circular boundary because they cannot travel beyond a certain distance from the trap centre where the magnitude of their self-propulsive force equals that of the trapping force, $F^{trap}(r) = F^{swim} \equiv \zeta U_0$; F^{swim} is the swimmers' propulsive force that can be interpreted as the force necessary to hold the swimmer fixed in space. For a harmonic trap, the swimmers are confined within a radius $R_c \equiv \zeta U_0 / k$ of the trap centre. We measure the positions and mean-square displacement (MSD) of the swimmers in both weak and strong confinement as they explore the trapping region (see Methods section for further experimental detail). We also conduct Brownian dynamics computer simulations to corroborate the experimental measurements (see Methods section for simulation detail).

Probability distribution. A passive Brownian particle confined in a harmonic trap has the familiar Boltzmann probability distribution: $P(r) \sim \exp(-V(r)/(\zeta D))$, where D is the translational diffusivity. Since the active Brownian motion of a swimmer can be interpreted as a random walk, the distribution of swimmers in a trap is also Boltzmann with the swim diffusivity $D = D^{swim} = U_0^2 \tau_R / 2$:

$$P(r) = \frac{k}{\pi \zeta U_0^2 \tau_R} \exp\left(-\frac{kr^2}{\zeta U_0^2 \tau_R}\right), \qquad (1)$$

or by non-dimensionalizing, $P(\bar{r})(U_0\tau_R)^2 = (\alpha/\pi)\exp(-\alpha\bar{r}^2)$, where $\bar{r} = r/(U_0\tau_R)$ and $\alpha \equiv k\tau_R/\zeta$ is the nondimensional trap stiffness that dictates the swimmer behaviour inside the trap. For a weak trap, $\alpha < 1$, the swimmers are allowed to explore and reorient freely before reaching the 'ends' of the well (Fig. 1); the maximum density occurs at the trap centre $r = 0$. Equation 1 is valid for $\alpha \lesssim 1$ since the swimmers must be allowed to undergo a random walk process within the confines of the well[6,8]. As shown in Fig. 2a, equation 1 agrees with both experiments and Brownian dynamics simulations. The uniform density far away from the trap, $P(\infty)$, has been subtracted in the experiments. The active Janus particles have a range of activity levels due to variations in the platinum coating during fabrication. With a weak trap, strong swimmers are able to swim straight past the trap without getting confined, whereas the weaker swimmers struggle to escape the vicinity of the trap centre. The swimmer properties (speed U_0 and reorientation time τ_R) in equation 1 are the average of those particles that are confined within the trapping region.

In the other limit of a strong trap, $\alpha > 1$, the swimmer sees the 'ends' of the well before it is able to reorient (that is, $R_c < U_0\tau_R$), so the swimmer will be stuck at R_c until it reorients and then run quickly to the other side and again wait there[7]. As shown in Fig. 2b we observe a peak in the probability distribution near $R_c = \zeta U_0/k$, and the Boltzmann distribution no longer applies. To observe Brownian-like motion the spring must be weak, that is, $\alpha = k\tau_R/\zeta < 1$, so that the particle can undergo a random walk before it discovers the ends of the well. Along the trap perimeter for large α, the particles are on average oriented radially outward relative to the trap centre. This directional alignment is most pronounced near $r = R_c$—the particles want to swim away but the trapping force confines them to remain inside. The correlation between the distance from the trap centre and the particles' orientation is directly related to the mechanical swim pressure exerted by these particles (see later subsection). Videos

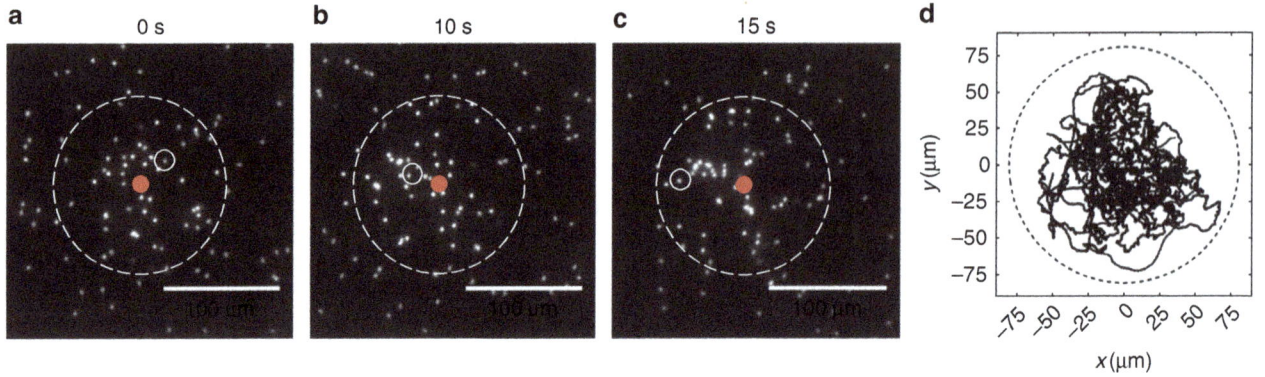

Figure 1 | Active Janus particles in a weak acoustic trap. (a–c) Snapshots of 2 μm swimmers in an acoustic trap. The solid red spot indicates the trap centre and the dashed white circle delineates the outer edge of the well. The swimmer shown inside the solid white circle undergoes active Brownian motion while exploring the confines of the trap. (d) Two-dimensional trajectories of several particles inside the trap.

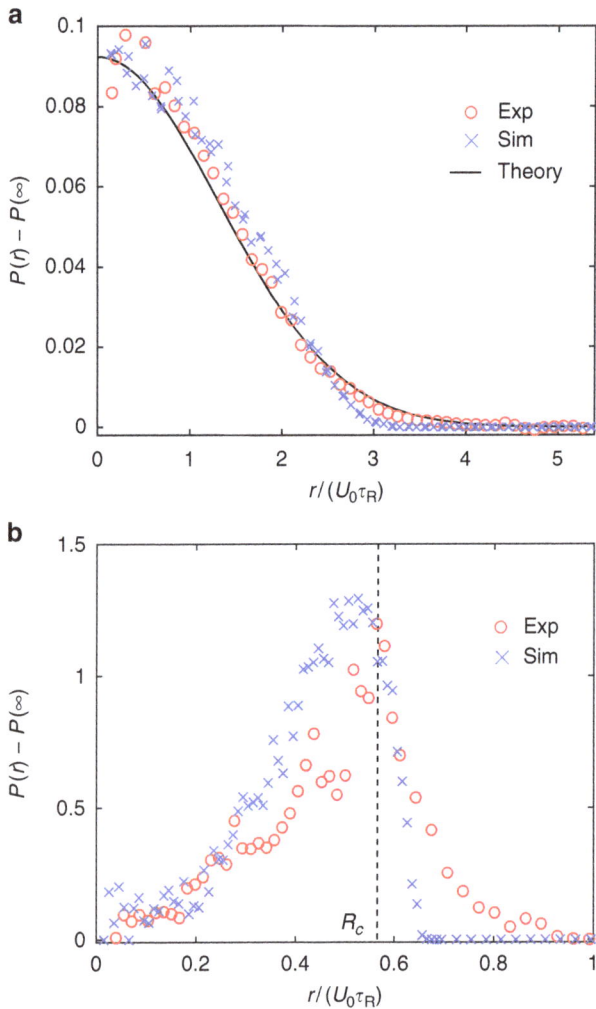

Figure 2 | Probability distribution of confined active Janus particles.
(a) 2 μm swimmers with $\alpha \equiv k\tau_R/\zeta = 0.29$ follow a Boltzmann distribution (solid black curve is the analytical theory, equation 1). (b) Distribution of 3 μm swimmers with $\alpha = 1.76$ has a peak near $R_c = \zeta U_0/k$ (vertical dashed black line) and decreases to zero for $r > R_c$. In both **a,b**, the red and blue symbols are data from experiment and Brownian dynamics simulations, respectively. Data are averages of measurements of over 500 snapshots for a duration of 50 s, each frame consisting ≈100 and 20 particles for the $\alpha = 0.29$ and $\alpha = 1.76$ cases, respectively.

of active particles in confinement are available in Supplementary Movies 1–3.

Explosion of a 'swimmer-crystal'. We have focused thus far on a dilute concentration of swimmers subjected to a relatively weak trap. Using a stronger trap, all swimmers that wander within the trapping region (∼ 150 μm radius from trap centre) are pulled towards the trap centre and form a dense close-packed 2D crystal (see Fig. 3a,e).

When the trap is subsequently turned off, the crystal quickly 'melts' or 'explodes' and the constituent particles swim away (see progression in Fig. 3). Videos of the accumulation, crystal formation, and melting process are available in Supplementary Movies 4–6. On first glance, this process resembles the melting of an active crystal due to the constituents' sudden loss of motility[15]. Palacci et al.[15] use polymer/hematite particles that self-propel and interact with each other via long-ranged phoretic attraction in the presence of blue–violet light. Due to concentration–field interaction, the particles cluster and form crystals in the presence of light. When light is shut off, the crystal melts because the particles' motility and concentration–field interactions are turned off, and the now-passive particles spread with their translational Brownian diffusivity—the entire melting process is diffusive. In contrast, our active crystal explodes due to a sudden loss of an external trap forcing the particles together, not the slow diffusion process caused by a loss of swimmer motility. Thus, the motion of the spreading swimmers is still that of active particles—translating with speed U_0 in randomly oriented directions that relax with the reorientation timescale τ_R.

We observe three time regimes in the explosion process. For times very short after release, only the swimmers positioned along the periphery of the crystal, escape the crystal. Particles in the centre obstruct each others's paths and are unable to escape the crystal, so the density is peaked at the origin (Fig. 3b,f). During the second regime the escaped particles move ballistically outwards in the direction given by their random initial orientation. The swimmers move ballistically because they have not yet reoriented sufficiently to be diffusive (that is, times $t \lesssim \tau_R$). The result is a depletion of particles from the origin (given the initial crystal is small; see below) and a peak in the density that propagates outward like a travelling wave (Fig. 3c,g). Finally, for times $t \gg \tau_R$ the swimmers have reoriented sufficiently to behave diffusively (Fig. 3d,h) characterized by the translational diffusivity $D = D_0 + D^{swim}$ where D_0 is the Stokes–Einstein–Sutherland diffusivity and $D^{swim} = U_0^2 \tau_R/2$. In this regime, the

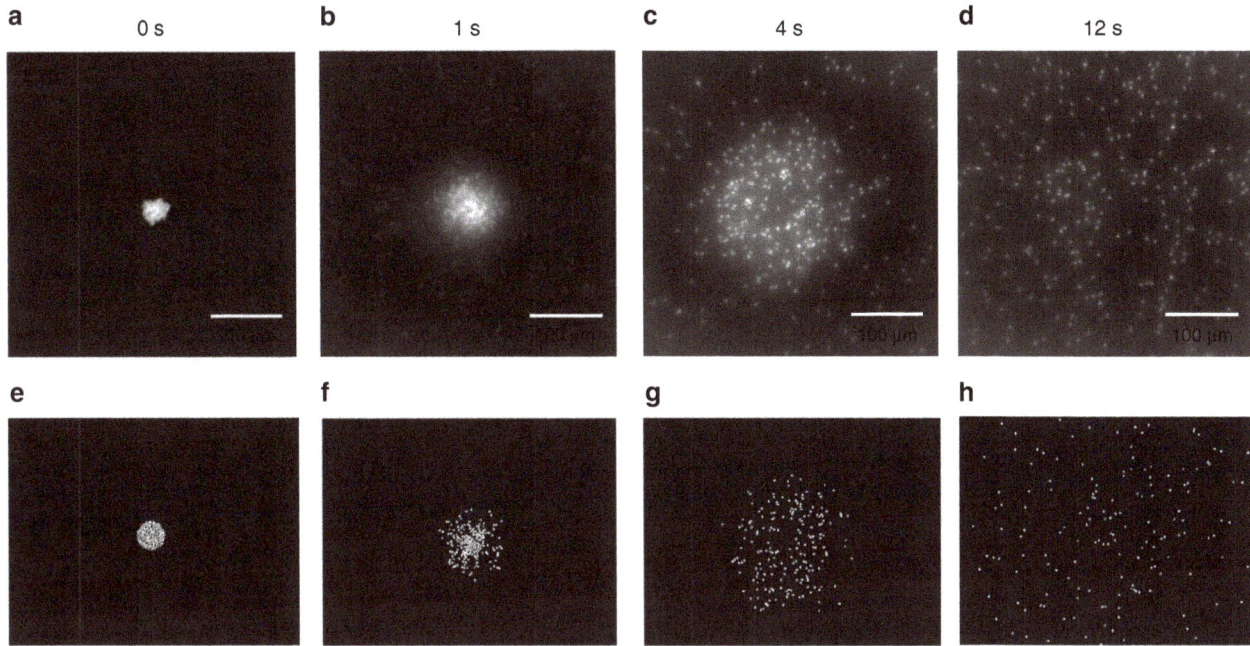

Figure 3 | 'Explosion' of active crystal. Explosion of swimmer crystal in (**a–d**) experiments and (**e–h**) Brownian dynamics simulations. (**a,e**) A strong trapping force draws the swimmers into a dense close-packed 2D crystal. (**b,f**) A subsequent release of the trap frees the swimmers, causing the crystal to explode. (**c,g**) At later times, a ballistic shock propagates outward like a travelling wave. (**d,h**) At long times, the swimmers spread diffusively.

spreading process is analogous to the classic diffusion of an instantaneous point source, where the transient probability distribution is

$$P(r,t) = \frac{1}{4\pi t D} \exp\left(-\frac{r^2}{4Dt}\right), \text{ for } t \gg \tau_R, \quad (2)$$

where t is the time after the point source is introduced and D is the translational diffusivity of the constituent 'solute' particles. For the swimmer crystal, the diffusivity of the constituent particles $D \approx D^{\text{swim}} = U_0^2 \tau_R / 2$.

Figure 4 corroborates our observations that show the three distinct time regimes in experiments and Brownian dynamics simulations. At short times, particles inside the crystal spread slightly faster in the experiment than the simulation, perhaps due to the sudden release of a strong acoustic force causing the particles to relax and loosen the initial packing of the crystal (see 1 s in Fig. 4). As predicted, equation 2 is valid only for times $t \gg \tau_R$ when the explosion process is diffusive (shown as dashed curves in Fig. 4 for comparison). This is a distinguishing feature of our explosion experiments compared with the melting of a passive Brownian crystal[15]. Because the swimmers' reorientation time can be large compared with the measurement times (as opposed to Brownian momentum relaxation times that are small), we are able to observe the crossover between ballistic and diffusive explosion of the active crystal. There is no ballistic regime for the melting of passive Brownian particles.

We conduct this experiment with different particle activities and initial crystal sizes. When the initial crystal is large and/or the swimmers reorient rapidly (that is, τ_R is small), the ballistic shock and depletion of swimmers from the crystal origin was less noticeable. For a large initial crystal, the particles in the centre must wait long times to break free of the cluster; by this time, swimmers initially positioned along the crystal periphery have had enough time to reorient and explore back towards the origin to fill the void.

Inside the crystal there is an effective interaction of the concentration fields due to the swimmers' competition for 'fuel' (hydrogen peroxide). However, for a system that can be assumed

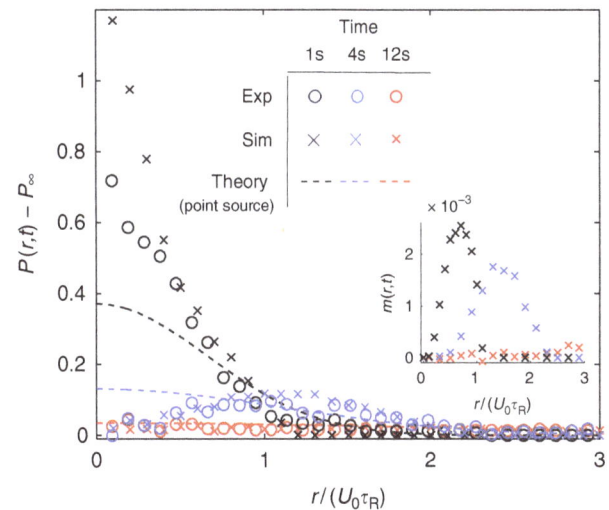

Figure 4 | Evolution of active crystal 'explosion'. Transient probability density of 2 μm swimmers as they explode from the crystal, drawn at three representative times as described in the text. Dashed curves are the analytical theory of diffusion of a point source, equation 2 (for 1 s and 4 s, drawn as a reference for comparison), and the circles and crosses are the experiment and Brownian dynamics simulation, respectively. Inset shows the polar order of the swimmers $m(r, t)$ as the peaks spread outward. We average over four independent explosion measurements for a duration of 30 s after release; each run consists of ≈150 spreading particles.

to be a 2D monolayer, fuel is being supplied from the third dimension (that is, the bulk) and thus the competition is reduced. Data near the trap centre are difficult to analyse in the experiments (especially at short time), since we cannot accurately differentiate between individual particles in the large crystal. Due to the finite size of the initial crystal, theoretical prediction for diffusion of a step function (as opposed to a delta function) would appear to be more accurate at very short times after release, but equation 2 is valid for $t \gg \tau_R$, so there is little difference between the two solutions in this regime.

This experiment provides a macroscopic method to measure the diffusivity of an active system using tweezers. The size or width of the spreading swimmer crystal is related to the diffusivity by $L^2 = 2Dt$, so one can measure the size of the spreading system (ignoring details about the motion of individual swimmers) for times $t \gg \tau_R$ to infer the diffusivity from $L \sim \sqrt{Dt}$.

These 'explosion' experiments show the development of local polar order of the swimmers as they spread outward. Orientational polar order is established when the swimmers' motion is directionally aligned. Inset of Fig. 4 shows the average swimming orientation $m(r, t) = \|\langle q \rangle\|(r, t)$ at three representative times, where $q = (\cos\theta, \sin\theta)$ is a unit orientation vector defined by the swimmer's direction of self-propulsion (see Methods section for further detail), and the brackets $\langle \cdot \rangle$ indicate an average over all particles at a distance r from the crystal centre. One interpretation of $m(r, t)$ is the local average orientation distribution, $\langle \cos(\theta') \rangle$, where θ' is the angle of the swimmer orientation relative to the outward normal from the crystal centre. Local polar order is peaked along the perimeter of the crystal, and spreads radially outward with time like a wave front. For short times when the system is ballistic, there is coherent motion of particles in the outward radial direction. This directed behaviour cannot be seen for a purely passive Brownian system, which exhibits 'biased-diffusion' for all times. At longer times when the crystal melts completely, there is no polar order.

Finally, a possible interpretation of the explosion process is that the particles are spreading from regions of large mechanical pressure (centre of crystal) to small mechanical pressure (far away from crystal). Spatial gradients in the active mechanical pressure results in an outward flux of constituent particles[16]—in the next section we compute the unique mechanical swim pressure exerted by active systems.

Swim pressure. In addition to the probability density and diffusivity of confined swimmers, the acoustic trap can be used as a measurement of the swim pressure[16], a unique mechanical pressure that all self-propelled bodies exert as a result of their self-motion. The origin of the swim pressure is that all active bodies exert a mechanical, self-propulsive force on the surrounding boundaries that confine them per unit area, $\Pi^{swim} = F^{wall}/A$. A swimmer's self-propulsive force is given by $F^{swim} \equiv \zeta U_0$ where ζ is the drag factor and U_0 is the swimming velocity. This is the force required to hold the swimmer fixed, at every instance in time. The swim pressure differs from the swim force, and also differs from the random thermal Brownian osmotic pressure due to the different intrinsic timescale. An active particle that collides into a wall will not reorient as a result of the collision—it continues to 'push' against the wall until it finally reorients from its intrinsic reorientation mechanism. These continuous collisions of magnitude F^{swim} against the wall over the time duration $t \sim \mathcal{O}(\tau_R)$ is what gives rise to the swim pressure.

Although many theoretical studies[16–21] have analysed the swim pressure of active matter, there is a dearth of experimental corroboration. A recent study on sedimentation[22] gives an indirect measurement based on density profiles of active particles under gravity. Because the acoustic trap behaves as an invisible 'semipermeable membrane' that confines the swimmers (but allows the solvent to pass through), our experiment allows us to determine the swim pressure using principles analogous to the osmotic pressure of colloidal solutes in solution.

We use the virial theorem[23] to express the swim pressure as a force moment[16]: $\Pi^{swim} = 1/(2A) \sum_i^N \langle x_i \cdot F_i^{swim} \rangle$ in 2D where N is the number of particles, A is the system area, and the swim forces

of each particle $F_i^{swim} \equiv \zeta_i U_{i,0}$. As shown in the Methods section, the first moment of the equation of motion for an active particle gives $\Pi^{swim} \equiv n/2\langle x \cdot F^{swim} \rangle = (n\zeta/4) d\langle x \cdot x \rangle/dt - n/2\langle x \cdot F^{trap} \rangle$, where n is the number density of swimmers. In the absence of the trap, $F^{trap} = 0$, we have $\Pi^{swim} = n\zeta D^{swim}$, where we have expressed the first term on the right using the diffusivity, $D^{swim} = (1/4)d\langle x \cdot x \rangle/dt$. Here the swimmer undergoes an entropic, random walk in unbounded space with $D^{swim} = U_0^2 \tau_R/2$, giving the 'ideal-gas' swim pressure $\Pi^{swim} = n\zeta U_0^2 \tau_R/2$ (ref. 16). In the presence of a trap force, F^{trap}, the long-time diffusivity $D^{swim} = 0$ because the swimmers are confined and cannot translate in unbounded space, and the MSD achieves a constant value (that is, does not grow with time). Therefore, at steady-state $d\langle x \cdot x \rangle/dt = 0$ and the acoustic trap acts as a confining force that is equal and opposite to the swim force, enabling us to determine the swim pressure via the known trap force: $\Pi^{swim} = -n/2\langle x \cdot F^{trap} \rangle$.

For a harmonic trap where $F^{trap} = -kx$, the swim pressure can be obtained directly by measuring the MSD of the swimmer inside the trap:

$$\frac{\Pi^{swim}}{n\zeta U_0^2 \tau_R/2} = \alpha \langle \bar{x} \cdot \bar{x} \rangle. \tag{3}$$

where $\alpha \equiv k\tau_R/\zeta$ and $\bar{x} = x/(U_0 \tau_R)$ is the nondimensional position vector of the swimmer relative to the trap centre. This elegant result reveals that the MSD contains information about the mechanical pressure exerted by self-propelled particles. It comes directly from the virial theorem, where the force moment becomes a MSD for a harmonic trap.

Solving the Langevin equation analytically for a swimmer confined in a trap (see Methods section), we obtain

$$\frac{\Pi^{swim}}{n\zeta U_0^2 \tau_R/2} = \frac{1}{1 + \alpha}. \tag{4}$$

The swim pressure depends only on the parameter $\alpha \equiv U_0 \tau_R/R_c$, a ratio of the swimmers' run length $U_0 \tau_R$ to the size of the trap $R_c = \zeta U_0/k$. Therefore, this expression gives the container-size dependent swim pressure—for a weak trap, $\alpha \rightarrow 0$, and we obtain the 'ideal-gas' swim pressure $\Pi^{swim} = n\zeta U_0^2 \tau_R/2$, whereas a strong trap causes the fictitious 'container' to shrink and decrease Π^{swim} because the distance the swimmers travel between reorientations decreases. Because the swim pressure is a force moment, the distance the swimmers travel between reorientations (that is, within a time τ_R) is the 'moment arm' that determines the magnitude of the swim pressure. For a weak trap the particles are allowed to travel the full run length $U_0 \tau_R$, set by the particles' intrinsic swimming speed U_0, whereas a strong trap establishes a small trap size $R_c < U_0 \tau_R$ that obstructs and causes the particles to travel a distance smaller than $U_0 \tau_R$.

Figure 5 shows the swim pressure computed in the experiments and Brownian dynamics simulations using equation 3. The swim pressure has transient behaviour for small times and we require data for times $t > \tau_R$ to observe a steady state (Methods section). At steady state, all curves approach the theoretical prediction given by equation 4.

As a swimmer wanders away from the trap centre, it may explore regions of the trap that are not strictly in the linear Hookean regime. Therefore one may expect the MSD to be slightly higher than the linear theory. However, although the swimmer concentration away from the trap is dilute, near the trap centre swimmers accumulate and cluster, which obstructs the motion of free swimmers trying to swim across to the other end of the trap, decreasing the MSD. Hydrodynamic interactions may also play a role near the trap centre where the density of swimmers is higher. The analytical theory is valid for a dilute

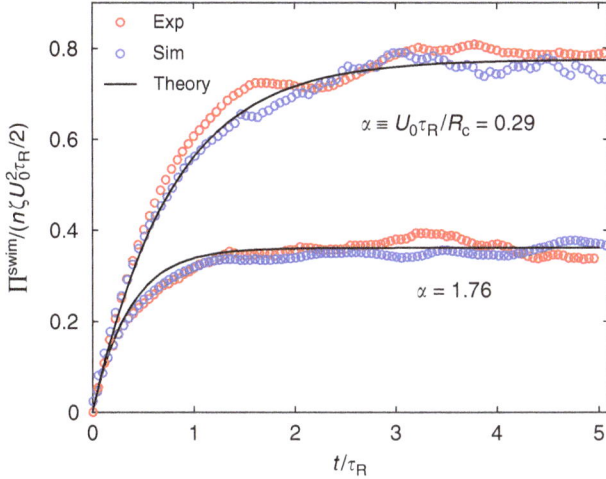

Figure 5 | Swim pressure of Janus particles in different degrees of confinement. The parameter $\alpha \equiv U_0\tau_R/R_c$ is a ratio of the particles' run length to the trap size $R_c = \zeta U_0/k$. The solid black curves are the theoretical prediction with a harmonic trap approximation, and the red and blue symbols are results from experiments and Brownian dynamics simulations, respectively. A smaller trap size diminishes the distance the particles travel between reorientations and decreases the swim pressure. The experimental and simulation data are averages of 150 and 90 independent particle trajectories for a duration of 40s for the $\alpha = 0.29$ and $\alpha = 1.76$ cases, respectively.

system of swimmers in a linear harmonic trap without hydrodynamic interactions, but we find that the linear approximations are sufficient. In addition to the MSD and equation 3 (which come from a linear approximation of F^{trap}), we also compute the full correlation using a Gaussian F^{trap} (Methods section) and the results have minor quantitative differences.

We scale the swim pressure in Fig. 5 using the average activity of the swimmers confined within the trapping region. The number density n is given by the number of trapped particles N divided by the area of the trapping region.

When a weak trap is present for a long time ($\gtrsim 20$–30 min) there is a gradual accumulation of swimmers inside the trap because those located initially far away wander near the trap and become confined. This induces a slow variation in the number density inside the trap $n(t)$ over time. We use a dilute system of swimmers (total area fraction $\phi_A \leq 0.001$) and the timescale for the change in number density (≥ 5 min) is large compared with the swimmers' reorientation time ($\tau_R \sim 2$–10s). Since the important timescale in our problem is the swimmers' reorientation time τ_R, we only require data over a timespan of several τ_R and the effect of swimmer accumulation is negligible in our results.

Discussion

We do not observe a self-pumping state[7] induced by hydrodynamic interactions, perhaps due to the dilute concentrations of this study. Although we do not observe large-scale coherent motion, the precise manipulation of swimmers towards special regions may provide a method to study the collective motion of living systems in a controlled manner. Further experiments using acoustic traps may give insight into the origin of polar order, how and why living organisms align, and the advantages of such collective behaviour.

Our measurement of the swim pressure may support forthcoming applications in biophysics and molecular cell biology, as researchers are becoming increasingly concerned with the mechanical forces, pressures, and stresses generated by active constituents inside a living cell. In addition, experimental determination of the swim pressure may engender real-life

engineering applications, such as fabrication of novel soft materials using active swimmers.

Methods

Equations of motion. The Langevin equation for a dilute system of swimmers in a trap is given by

$$0 = -\zeta U(t) + F^{swim} + F^{trap}, \qquad (5)$$

where $U(t)$ is the velocity, ζ is the hydrodynamic drag factor, $F^{swim} \equiv \zeta U_0$ is the self-propulsive swim force of a swimmer, $U_0 = U_0 q$ is the intrinsic swim velocity of an isolated swimmer, $q(t)$ is the swimmer's unit orientation vector defined by their swimming direction, and $F^{trap} = -\nabla V(x)$ is the restoring force caused by the trap with potential $V(x)$. The left-hand side of equation 5 is zero because inertia is negligible for a colloidal dispersion.

Transverse trapping with an acoustic tweezer results in a Gaussian trap[11] with stiffness k and width w, $F^{trap} = -kr \exp(-2(r/w)^2)\hat{r}$, which we independently verify. We use Janus particles in the absence of hydrogen peroxide (that is, inactive particles) to calibrate k and w of the acoustic trap by measuring the position and velocity of the particles in the trap. For a trap with large spatial extent (large w), a linear force $F^{trap} \approx -kr\hat{r}$ approximates the trapping force. As a swimmer wanders far away from the focus of the trap, there is a critical radius $\sim R_c = \zeta U_0/k$ at which the swimmer cannot move any farther. At this position the swimmer's self-propulsive force F^{swim} exactly cancels the trapping force F^{trap} and the swimmer does not move. The swimmer is 'stuck' in this position for a time of order $\sim \tau_R$ until the swimmer changes its orientation. The trapping force does not affect particles located far away from the trap origin because $F^{trap}(r \to \infty) \to 0$.

Experimental methods. We fabricate active Janus particles from 2 and 3 μm diameter sulfate latex particles (Life Technologies, Carlsbad, CA, USA). We coat half of the particle surface with a ~ 7-nm thick layer of platinum using a BAL-TEC SCD 050 sputter coater (Leica Microsystems GmbH, Wetzlar, Germany). When deposited in a hydrogen peroxide solution, the particles self-propel via diffusiophoresis near the air-water interface. The particles initially deposited into the bulk rise towards the air/solution interface because the platinum half is heavy and orients with gravity. The particles swim towards the non-platinum face, and begin to move in 2D once they reach the interface. The particles at the interface do not diffuse back into the third dimension (the bulk). For the 2 and 3 μm particles, they have a swim speed of $U_0 \sim 15$–25 μm s^{-1} and ~ 8–15 μm s^{-1} with a reorientation time of $\tau_R \sim 2$–5 s and ~ 5–10 s, respectively. We compute the reorientation time by analysing the swimmers' orientation autocorrelation: $\langle(q(t) - q(0))^2\rangle = 2(1 - \exp(-t/\tau_R))$. We verify that the swimmers undergo active Brownian motion characterized by the swim diffusivity $D^{swim} = U_0^2\tau_R/2$.

To confine the swimmers in the transverse direction we develop a custom-built acoustic tweezer setup. We excite a 0.25-inch diameter immersion type transducer (UTX Inc., Holmes, NY, USA) in a continuous sinusoidal signal at 25 MHz with variable voltages from 0 to $10V_{pp}$ using an AM300 Dual Arbitrary Generator (Rohde & Schwarz, Munich, Germany). We immerse the transducer in the solution in an inverted position (face up to the air/solution interface) and deposit Janus particles on this interface. We adjust the focal point of the transducer at a distance of 12 mm from the air/solution interface and hold it fixed in place throughout the experiment using an XY positioner and a tilt stage (Thorlabs Inc, Newton, NJ, USA). Although the transducer also has a radiation force in the axial direction, the effect of particles in the bulk being pushed to the interface is negligible because the Janus particles remain on the interface and do not diffuse in three dimension into the bulk. We control the strength of the trap by changing the input voltage from the function generator. We connect a $\times 50$ objective (Leica Microsystems GmbH, Wetzlar, Germany) to a sCMOS digital camera (ORCA-Flash 4.0, Hamamatsu, Japan) to obtain images and a glass fibre ringlight (Volpi AG, Schlieren, Switzerland) to provide lighting. Most commonly researchers use acoustic tweezers to generate a standing field to confine objects primarily in one dimension in acoustic pressure nodes or antinodes, depending on the properties of the objects (density, compressibility)[9]. Here we develop a 2D device which generates a near-harmonic potential using the transverse radiation forces of single-beam transducer.

We deposit Janus particles on the air-water interface of a 0.5 wt% hydrogen peroxide solution, and their activity remains constant for at least 1 h (each experimental run last $\sim$ few minutes). We turn on the acoustic transducer and observe the motion of the swimmers in confinement within the trapping region. For the small voltages we apply to the transducer (0–3V_{pp}) and the particle sizes (2 and 3 μm) used in our swim pressure measurements, we do not detect any acoustic streaming. The acoustic tweezer exerts a Gaussian trapping force on the particles; a linear Hookean spring force approximates the trapping force since the width w is large compared with the swimmers' run lengths. We identify the centre of the trap at the end of each experiment by applying a strong trapping force to collect all of the swimmers to the trap centre. We use a modified particle tracking script[24] in the analysis.

Brownian dynamics simulations. In our Brownian dynamics simulations we evolve the particles following equation 5. Although the concentration of swimmers far away from the trap centre is dilute, we may have an accumulation of swimmers near the trap centre that may obstruct the motion of free swimmers trying to swim across to the other end of the trap. To more accurately model the experimental system, our Brownian dynamics simulations include the interparticle force F^P in equation 5. Nondimensionalizing the force by ζU_0 and position by $U_0\tau_R$, equation 5 (with the interparticle force) becomes $0 = -u(t) + q + \bar{F}^{\mathrm{trap}} + F^P$, where $u \equiv U/U_0$ is the particle velocity, $\bar{F}^{\mathrm{trap}}(\bar{r}) = -\alpha\bar{r}\exp(-2\gamma^2\bar{r}^2)\hat{r}$ is the trapping force, $\bar{r} \equiv r/(U_0\tau_R)$ is the radial position of the swimmer, $\gamma \equiv U_0\tau_R/w$ is a ratio of the swimmers' run length to the trap width, and $\alpha \equiv \tau_R/(\zeta/k)$ is a ratio of the swimmers' reorientation time to the timescale of the trap. We can also interpret $\alpha \equiv (U_0\tau_R)/(\zeta U_0/k)$, the ratio of the swimmers' run length to the 'size' of the container (set by the trap). For Figs 2 and 5 we use $\alpha = 0.29$, $\gamma = 0.08$ and $\alpha = 1.76$, $\gamma = 0.25$ for a weak and strong trap, respectively. We vary the number of particles from 20 to 500 to match the experimental measurements. We use a hard-disk interparticle force $F^P = F^{\mathrm{HS}}$ that prevents particle overlap in our simulations[25,26]. We evolve the swimming orientation of the swimmers $q = (\cos\theta, \sin\theta)$ following $d\theta/dt = \sqrt{2/\tau_R}\Lambda(t)$ where $\Lambda(t)$ is a unit random deviate.

Derivation of the swim pressure in a harmonic trap. For a harmonic trapping force $F^{\mathrm{trap}} = -kx$, we can solve equation 5 for the position $x(t)$ and compute the MSD:

$$\frac{\langle x(t)x(t)\rangle}{(U_0\tau_R)^2} = \frac{(-1+\alpha) - 2\alpha e^{-(1+\alpha)t/\tau_R} + (1+\alpha)e^{-2t/\tau_{\mathrm{trap}}}}{\alpha(-1+\alpha^2)}\left(\frac{I}{2}\right), \quad (6)$$

where I is the isotropic tensor and $\tau_{\mathrm{trap}} = \zeta/k$ is the characteristic timescale of the trap. For small times the MSD grows quadratically in time, and for $\alpha = 0$ we obtain the long-time self diffusivity of an active swimmer: $D^{\mathrm{swim}} = 1/2\lim_{t\to\infty}d(\langle x(t)x(t)\rangle_{\alpha=0})/dt = U_0^2\tau_R I/2$. Most importantly, for $\alpha \neq 0$ and times long compared with both τ_R and τ_{trap} the MSD becomes a constant

$$\lim_{t\to\infty}\frac{\langle x(t)x(t)\rangle}{(U_0\tau_R)^2} = \frac{1}{\alpha(1+\alpha)}\frac{I}{2}. \quad (7)$$

This is a main result that we will use later.

Multiplying equation 5 by nx and taking the average we obtain

$$\sigma^{\mathrm{swim}} \equiv -n\langle xF^{\mathrm{swim}}\rangle = -\frac{n\zeta}{2}\frac{d\langle xx\rangle}{dt} + n\langle xF^{\mathrm{trap}}\rangle, \quad (8)$$

where we use the definition of the swim stress $\sigma^{\mathrm{swim}} \equiv -n\langle xF^{\mathrm{swim}}\rangle$ and n is the number density of swimmers[16]. As shown in equation 7, for times long compared to both τ_R and τ_{trap} the MSD becomes a constant and its time derivative is zero: $d/dt(\lim_{t\to\infty}\langle x(t)x(t)\rangle) = 0$. Therefore, the swim pressure $\Pi^{\mathrm{swim}} = -\mathrm{tr}\sigma^{\mathrm{swim}}/2$ (in 2D) is

$$\Pi^{\mathrm{swim}} = -\frac{n}{2}\langle x\cdot F^{\mathrm{trap}}\rangle, \quad (9)$$

which is a general result valid in principle for any trapping force F^{trap}. For a harmonic trap $F^{\mathrm{trap}} = -kx$, the swim pressure can be determined from a simple MSD measurement as given in equation 3 of the main text. Substituting equation 7 into equation 3, we obtain the theoretical result $\Pi^{\mathrm{swim}}/(n\zeta U_0^2\tau_R/2) = (1+\alpha)^{-1}$ as given in equation 4 of the main text.

For times not large compared to τ_R and τ_{trap}, the slope of MSD is not zero and the swim pressure has a transient start up period:

$$\frac{\Pi^{\mathrm{swim}}}{n\zeta U_0^2\tau_R/2} = \frac{1}{1+\alpha}\left(1 - e^{-\left(\tau_R^{-1} + \tau_{\mathrm{trap}}^{-1}\right)t}\right). \quad (10)$$

This expression is exact and valid for all times t. On taking times $t > \tau_R$ and $t > \tau_{\mathrm{trap}} = \zeta/k$, this result agrees with equation 4. Therefore, measuring the MSD $\langle xx\rangle$ is an easy and simple method to quantify the swim pressure in an experimental system.

For nonlinear traps with a general form of F^{trap}, we must evaluate equation 9 directly. For a Gaussian trap with stiffness k and width w, $F^{\mathrm{trap}}(r) = -kr\exp(-2(r/w)^2)\hat{r}$, we have

$$\Pi^{\mathrm{swim}} = \frac{n}{2}\langle kr^2\exp(-2(r/w)^2)\rangle. \quad (11)$$

For a large well (large w), the trapping force becomes harmonic and we get back the previous result in equation 3, where the MSD $\langle r^2\rangle$ gives the swim pressure.

References

1. Rothschild, L. Non-random distribution of bull spermatozoa in a drop of sperm suspension. *Nature* **198**, 1221–1222 (1963).
2. Berke, A. P., Turner, L., Berg, H. C. & Lauga, E. Hydrodynamic attraction of swimming microorganisms by surfaces. *Phys. Rev. Lett.* **101**, 038102 (2008).
3. Sokolov, A., Apodaca, M. M., Grzybowski, B. A. & Aranson, I. S. Swimming bacteria power microscopic gears. *Proc. Natl Acad. Sci. USA* **107**, 969–974 (2010).
4. Fily, Y., Baskaran, A. & Hagan, M. F. Dynamics of self-propelled particles under strong confinement. *Soft Matter* **10**, 5609–5617 (2014).
5. Yang, X., Manning, M. L. & Marchetti, M. C. Aggregation and segregation of confined active particles. *Soft Matter* **10**, 6477–6484 (2014).

6. Wang, Z., Chen, H.-Y., Sheng, Y.-J. & Tsao, H.-K. Diffusion, sedimentation equilibrium, and harmonic trapping of run-and-tumble nanoswimmers. *Soft Matter* **10**, 3209–3217 (2014).
7. Nash, R. W., Adhikari, R., Tailleur, J. & Cates, M. E. Run-and-tumble particles with hydrodynamics: Sedimentation, trapping, and upstream swimming. *Phys. Rev. Lett.* **104**, 258101 (2010).
8. Tailleur, J. & Cates, M. E. Sedimentation, trapping, and rectification of dilute bacteria. *Europhys. Lett.* **86**, 60002 (2009).
9. Evander, M. & Nilsson, J. Acoustofluidics 20: applications in acoustic trapping. *Lab Chip* **12**, 4667–4676 (2012).
10. Ding, X. *et al.* On-chip manipulation of single microparticles, cells, and organisms using surface acoustic waves. *Proc. Natl Acad. Sci. USA* **109**, 11105–11109 (2012).
11. Lee, J. *et al.* Transverse acoustic trapping using a gaussian focused ultrasound. *Ultrasound. Med. Biol.* **36**, 350–355 (2010).
12. Rasmussen, M. B., Oddershede, L. B. & Siegumfeldt, H. Optical tweezers cause physiological damage to *Escherichia coli* and listeria bacteria. *Appl. Environ. Microbiol.* **74**, 2441–2446 (2008).
13. Howse, J. R. *et al.* Self-motile colloidal particles: from directed propulsion to random walk. *Phys. Rev. Lett.* **99**, 048102 (2007).
14. Cordova-Figueroa, U. M. & Brady, J. F. Osmotic propulsion: the osmotic motor. *Phys. Rev. Lett.* **100**, 158303 (2008).
15. Palacci, J., Sacanna, S., Steinberg, A. P., Pine, D. J. & Chaikin, P. M. Living crystals of light-activated colloidal surfers. *Science* **339**, 936–940 (2013).
16. Takatori, S. C., Yan, W. & Brady, J. F. Swim pressure: stress generation in active matter. *Phys. Rev. Lett.* **113**, 028103 (2014).
17. Fily, Y., Henkes, S. & Marchetti, M. C. Freezing and phase separation of self-propelled disks. *Soft Matter* **10**, 2132–2140 (2014).
18. Solon, A. P. *et al.* Pressure and phase equilibria in interacting active brownian spheres. *Phys. Rev. Lett.* **114**, 198301 (2015).
19. Mallory, S. A., Saric, A., Valeriani, C. & Cacciuto, A. Anomalous thermomechanical properties of a self-propelled colloidal fluid. *Phys. Rev. E* **89**, 052303 (2014).
20. Takatori, S. C. & Brady, J. F. Swim stress, motion, and deformation of active matter: effect of an external field. *Soft Matter* **10**, 9433–9445 (2014).
21. Takatori, S. C. & Brady, J. F. Towards a thermodynamics of active matter. *Phys. Rev. E* **91**, 032117 (2015).
22. Ginot, F. *et al.* Nonequilibrium equation of state in suspensions of active colloids. *Phys. Rev. X* **5**, 011004 (2015).
23. Goldstein, H. *Classical Mechanics* 2nd edn (Addison-Wesley, Reading, 1990).
24. Crocker, J. C. & Grier, D. G. Methods of digital video microscopy for colloidal studies. *J. Colloid Interface Sci.* **179**, 298–310 (1996).
25. Foss, D. R. & Brady, J. F. Brownian dynamics simulation of hard-sphere colloidal dispersions. *J. Rheol.* **44**, 629–651 (2000).
26. Heyes, D. M. & Melrose, J. R. Brownian dynamics simulations of model hard-sphere suspensions. *J. Non-Newton. Fluid Mech.* **46**, 1–28 (1993).

Acknowledgements

S.C.T. is supported by a Gates Millennium Scholars fellowship and a National Science Foundation (NSF) Graduate Research Fellowship (No. DGE-1144469). R.D.D. is supported by a doctoral fellowship of the fund for scientific research (FWO-Vlaanderen). This work is also supported by NSF Grant CBET 1437570.

Author contributions

All authors participated in designing the project and performing the research. S.C.T. and R.D.D. performed the experiments and numerical simulations. All authors participated in writing the paper.

Additional information

16

Substantial bulk photovoltaic effect enhancement via nanolayering

Fenggong Wang[1], Steve M. Young[2], Fan Zheng[1], Ilya Grinberg[1] & Andrew M. Rappe[1]

Spontaneous polarization and inversion symmetry breaking in ferroelectric materials lead to their use as photovoltaic devices. However, further advancement of their applications are hindered by the paucity of ways of reducing bandgaps and enhancing photocurrent. By unravelling the correlation between ferroelectric materials' responses to solar irradiation and their local structure and electric polarization landscapes, here we show from first principles that substantial bulk photovoltaic effect enhancement can be achieved by nanolayering $PbTiO_3$ with nickel ions and oxygen vacancies $((PbNiO_2)_x(PbTiO_3)_{1-x})$. The enhancement of the total photocurrent for different spacings between the Ni-containing layers can be as high as 43 times due to a smaller bandgap and photocurrent direction alignment for all absorption energies. This is due to the electrostatic effect that arises from nanolayering. This opens up the possibility for control of the bulk photovoltaic effect in ferroelectric materials by nanoscale engineering of their structure and composition.

[1] The Makineni Theoretical Laboratories, Department of Chemistry, University of Pennsylvania, Philadelphia, Pennsylvania 19104-6323, USA. [2] Center for Computational Materials Science, United States Naval Research Laboratory, Washington, DC 20375, USA. Correspondence and requests for materials should be addressed to F.W. (email: fenggong@sas.upenn.edu) or to A.M.R. (email: rappe@sas.upenn.edu).

Studies of photo-induced effects in ferroelectrics have experienced a revival due to the demonstration of a variety of fascinating physical effects, and in particular an increase in the power conversion efficiency of ferroelectric-based solar cells[1-5]. In particular, recent experiments on $BiFeO_3$ thin films with absorption at the visible-light edge have revealed phenomena that were not observed in classic studies of bulk ferroelectric photoresponses[6-9]. Advances in the power conversion efficiency of ferroelectric-based solar cells and the discovery of visible-light-absorbing ferroelectric materials have provided further encouragement for this field[3,4,10,11].

The recent experimental discoveries and progress in thin film fabrication have been accompanied by significant advances in the understanding of the unique light–matter interactions enabled by the presence of a spontaneous, switchable polarization in ferroelectric materials. In particular, it has been demonstrated that the very large open-circuit photovoltage observed in $BiFeO_3$ films is due to the bulk photovoltaic effect (BPVE)[12-15,7]; in turn, BPVE has been shown to arise from the shift current mechanism, in which electrons are continuously excited under sustained illumination to quasiparticle coherent states that have intrinsic momenta, generating a spontaneous direct short-circuit photocurrent[16,17]. While the fundamental mechanism underlying BPVE has been determined, it is not yet completely understood how the BPVE depends on the structure of ferroelectric materials. In particular, a thorough understanding of how the BPVE can be controlled through adjustment of the atomic and electronic compositions and change of the chemical bonding properties is highly desired from the fundamental standpoint as well as for future development of BPVE-based devices.

Nano heterostructured materials have been shown to exhibit intriguing physics that are inaccessible in their bulk material counterparts due to the quantum confinement effect, the sensitivity of functional properties to structure, as well as the unique properties at interfaces, such as in the case of the two-dimensional electron gas at the $LaAlO_3/SrTiO_3$ interface[18]. Furthermore, it has been established that there is a connection between materials' local chemistry and structural properties, their fundamental and functional properties, and their responses to external stimuli[19,20]. Therefore, superior properties, specifically an enhancement of the photovoltaic performance, may be obtained by tuning the local structure and the landscape of local electric polarization in oxides with an engineered bandgap[21]. Motivated by the strong effect of electrostatics on materials' bandgaps and electronic transitions[22,23], here we explore the possibility of nano-ordering photoresponsive ferroelectric materials to control and enhance the BPVE, highlighting the importance of length scale (layer thickness), electronics and electrostatics for photovoltaic performance.

We use first-principles calculations to investigate the relationship between local displacements, electronic structure and the BPVE response in the nanolayered Ni + vacancy-substituted $PbTiO_3$ material $Pb(Ni_xTi_{1-x})O_{3-x}$ (called $(PbNiO_2)_x$ $(PbTiO_3)_{1-x}$ or PNT for brevity)[24,25]. We choose Ni because it is a late-transition metal, and so it is relatively electronegative, with filled and empty d orbitals near in energy; accordingly, Ni substitution induces a low-lying conduction band (CB). The combination of Ni and vacancy substitution guarantees that Ni ions are found in the $+2$ charge states. The Ni^{2+} state is also preferred to the Ni^{3+} state, because Ni^{3+} ion has a small ionic radius (0.60 Å) and is only stable in perovskites with small La^{3+} ions on the A site, whereas the larger Ni^{2+} ion is known to be stable in ferroelectric compounds[26]. Because it is more computationally convenient, we choose PNT to serve as a prototype for the recently developed KBNNO material that for the first time showed excellent light absorption in the visible

range by an oxide ferroelectric, breaking the decades-long restriction of ferroelectric oxides to only near-ultraviolet and ultraviolet absorption[10]. PNT is also closely related to the Ni-doped $Pb(Zr_{1-x}Ti_x)O_{3-x}$ (PZT) material (PNZT) synthesized in thin film form and shown to exhibit improved light-absorption properties relative to the parent PZT material[27]. We show that the BPVE response over a wide range of frequencies is sensitive to the changes in the electronic structure that are driven by changes in the local cation displacements. The PNT nanostructures exhibit several unexpected and counterintuitive effects that enable the tuning of the absorption properties and a marked enhancement (by a factor of 43) of the BPVE response by nanoscale (1–2 nm) layering. Our results suggest that nanoscale control of composition and structure is an attractive route for manipulating the BPVE in ferroelectric oxides.

Results

Oxygen vacancy and cation arrangement. We start from exploring the relative stabilities of different O vacancy sites by choosing a $1 \times 1 \times 2$ supercell with one Ti atom substituted by Ni (Ni_{Ti}) and one O atom removed (V_O), corresponding to the $(PbNiO_2)(PbTiO_3)$ composition ($x = 1/2$). Total energy analysis indicates that O vacancies prefer to form adjacent to Ni (Supplementary Fig. 1), because this provides more effective charge compensation. In addition, under normal conditions the equatorial O vacancies are more stable than the apical O vacancies. However, the apical O vacancies become more stable by 29 meV per atom with 1% in-plane compressive strains. Furthermore, the apical O vacancies are preferred in pure $PbTiO_3$ (ref. 28), and it is likely that a large proportion of the apical O vacancies are preserved after Ni incorporation.

The location of O vacancy can significantly affect the polarity of the structure and the composition of the contributing electronic states, and in turn may also affect the photocurrent. The shift current response of the $1 \times 1 \times 2$ supercell with equatorial O vacancies is almost negligible. This is because when equatorial O vacancies are introduced, the relaxed structure with NiO_4 complexes in the vertical yz (or xz) plane is nearly centrosymmetric, whereas the presence of apical O vacancies enlarges the lattice asymmetry. While emphasizing that a larger lattice asymmetry does not always induce a larger shift current, the obtained structural change is nevertheless consistent with the shift current results.

The $x = 1/2$ composition may also take the rock salt B-cation arrangement (Supplementary Fig. 1), in which the O vacancy has no preferred site. Even though the rock salt cation arrangement is less energetically favourable than its layered counterpart, it exhibits a more substantial photocurrent response (Table 1). Comparison of their total and projected densities of states shows that the Pb $6p$ orbitals contribute the most to the bottom of the CB in the rock salt structure, but in the layered structure the Ni $3d$ orbitals contribute the most (Supplementary Fig. 1). This occurs because in the rock salt structure the Ni–V_O–Ni network is interrupted by the Ti and Pb atoms, making the Ni $3d$ orbitals mainly localized at an energy position of 1 eV higher than the bottom of the CB. Electronic transitions involving the more delocalized Pb $6p$ orbitals facilitate the motion of the photocurrent carriers and lead to stronger photocurrent response.

Photocurrent cancellation in PNT. Inspection of the shift current results (Fig. 1a; we focus only on the most substantial, xxZ component) for the $PbTiO_3$ with 33% Ni + vac composition ($1 \times 1 \times 3$ superlattice) shows that the overall current produced by illumination in the visible range is small, due to a strong cancellation of the contributions for light absorption of photons

Table 1 | Comparison of the photocurrent and electronic properties of different compositions of PNT.

Structures	E_{On}^{σ} (eV)	σ_{Max} $(10^{-4} V^{-1})$	G_{Max} $(10^{-9} cm V^{-1})$	P $(C\,m^{-2})$	E_g^{HSE} (eV)	ΔE meV
$PbTiO_3$	3.6	3.7	12.0	0.81	2.71	—
$1 \times 1 \times 2$	2.1	1.5	3.6	0.73	1.58	0
Rock salt	2.1	6.1	11.0	0.96	2.05	+30
$1 \times 1 \times 3$	1.9	3.9	14.0	0.91	1.58	—
$1 \times 1 \times 4$	1.6	4.7	29.7	1.10	1.33	—
$1 \times 1 \times 5$	1.5	4.5	36.9	1.07	1.28	—
$1 \times 1 \times 6$	1.5	6.7	46.8	1.05	1.08	—

Calculated shift current onset energy E_{On}^{σ}, maximum shift current susceptibility σ_{Max}, maximum Glass coefficient G_{Max}, polarization P and Heyd-Scuseria-Ernzerhof (HSE) hybrid functional bandgap E_g^{HSE} of the layered and rock salt PNT. ΔE (meV per atom) is the relative stability of different cation arrangements that have the same composition. $PbTiO_3$ values are also shown for comparison. Maximum values are for energies smaller than 3.0 eV ($E < 4.0$ eV for $PbTiO_3$).

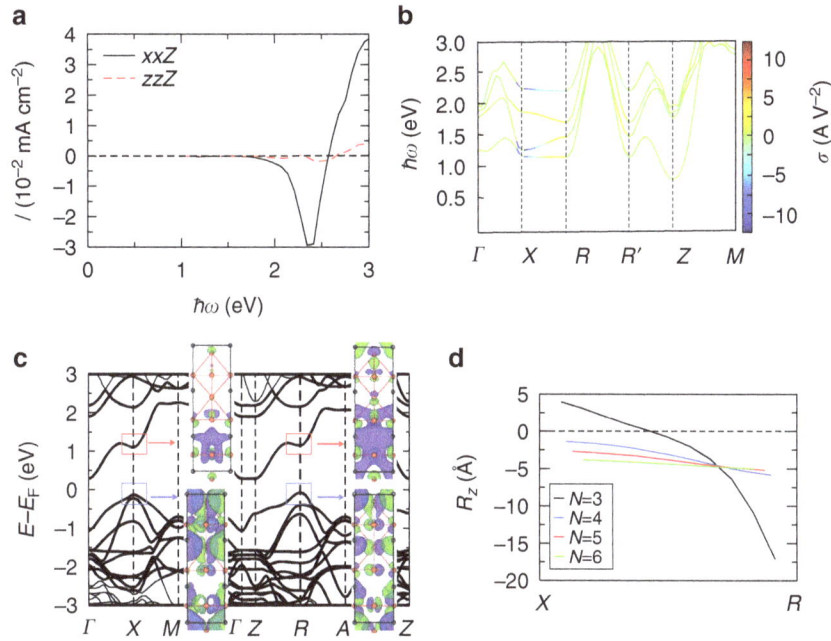

Figure 1 | The photocurrent cancellation in layered PNT and its electronic and wavefunction origins. The (**a**) photocurrent (assuming $0.1 \, W \, cm^{-2}$ light absorption), (**b**) k-resolved photocurrent and (**c**) band structure and real-space wavefunction isosurfaces corresponding to the states indicated by the rectangular regions, of the $1 \times 1 \times 3$-layered PNT. In **b**, the colour gives the value of the photocurrent response σ ($A V^{-2}$). The near-gap response is dominated by the region around the line from X to R, and changes direction along this path. R' indicates R plus a reciprocal lattice vector, so that R' traverses the Brillouin zone. (**d**) The shift vector for the k points along the X–R line in the Brillouin zone of the $1 \times 1 \times N$-layered PNT.

in the 2–2.5 eV range and in the 2.5–3.0 eV range. Such a cancellation was also found by our theoretical calculations for the KBNNO material[29]. This is unlike the shift currents calculated for $PbTiO_3$ and $BiFeO_3$ that exhibit a uniform sign of the photoresponse throughout the photon spectrum. It is likely that the shift current cancellation observed in KBNNO and PNT is related to the greater diversity of electronic orbital character caused by the presence of dissimilar Ni and Ti/Nb local environments. Examination of the k-resolved shift current (Fig. 1b) shows that the electronic transitions occurring near the $X(0, 0.5, 0)$ and $R(0, 0.5, 0.5)$ k points and along the X–R line induce the most substantial photocurrent responses and are thus the most important for the band-edge electronic transitions. However, due to the changes in the direction of the shift vector (Fig. 1d), the contributions from the transitions at X and at R largely counteract each other. This suggests that the BPVE response can be significantly increased by aligning the sign of the photocurrent throughout the Brillouin zone.

To elucidate the origin of the photocurrent cancellation, we consider the details of the electronic states at the X and R points.

The calculated PNT band structure (Fig. 1c) shows that the bandgap is reduced due to the introduction by nickel of a low-lying CB into the bulk $PbTiO_3$ gap; as a result, the band-edge electronic transitions, which will be responsible for the shift current response generated by the visible light, occur only between the valence bands and this particular CB. Real-space wavefunction distribution analysis for the highest valence band shows that it is essentially delocalized at both X and R, with the wavefunction extending to the whole supercell because of the overlap among the Pb 6s, Ti 3d, Ni 3d and O 2p orbitals. However, the wavefunction distribution for the lowest CB shows substantial differences between X and R. At X, it arises mainly from the Ni $3d_{x^2-y^2}$ and the nearby Pb 6p orbitals, and is more localized, whereas at R it becomes much more delocalized and has larger orbital contributions from the Pb 6p, Ti $3d_{x^2-y^2}$ and Ti $3d_{z^2}$ orbitals. The greater delocalization at the R point is directly related to the wavefunction character; compared with the Ni $3d_{x^2-y^2}$ orbital, the Pb 6p and Ti $3d_{x^2-y^2}$ orbitals are more delocalized, while the Ti $3d_{z^2}$ orbitals extend the wavefunction along the photocurrent direction. The increasing delocalization of the CB minimum

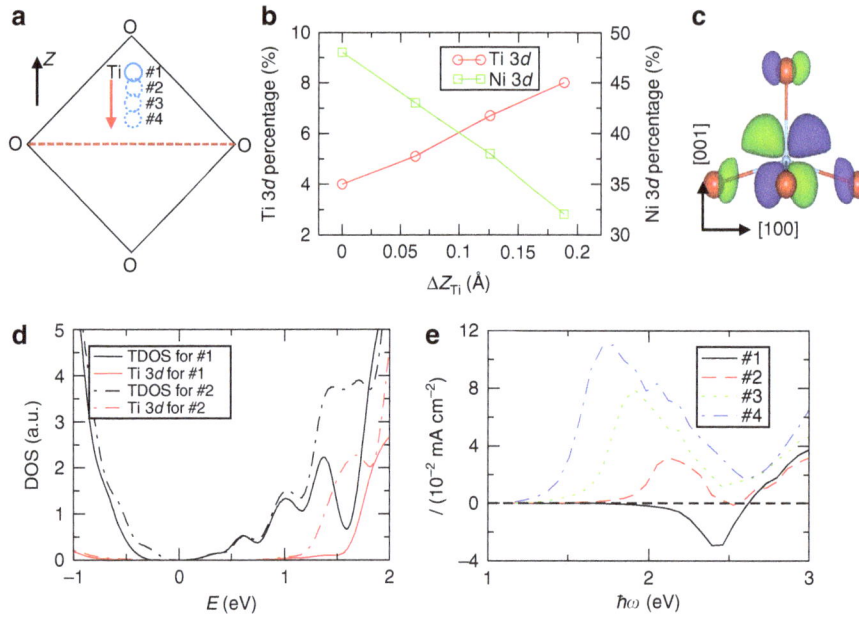

Figure 2 | The photocurrent enhancement by artificial local structure displacement and the corresponding change of the electronic composition. (**a**) The schematic representation of how the Ti atoms are moved with respect to the fully relaxed structure (structure #1), (**b**) the percentages of the Ti and Ni 3d orbitals at the CB ($\sum_m |\langle \psi_{CB,X} | \phi_{32m} \rangle|^2$) at X k point as a function of the antiparallel displacement (ΔZ_{Ti}) of Ti atoms, (**c**) the orbital character of the Ti–O orbital overlap at the CB, (**d**) total and projected density of states and (**e**) photocurrent (assuming for the 1,000 W m^{-2} light absorption), of the $1 \times 1 \times 3$-layered PNT. The Ti atoms are moved antiparallel to the polarization, reducing but not entirely eliminating their off-centre displacements. The inset label #n in **a,d** and **e** denotes the structures with different Ti antiparallel displacements. For example, #1 denotes the fully relaxed structure and #2, #3 and #4 denote the structures with Ti atoms displaced antiparallel by 0.06, 0.12 and 0.18 Å, respectively, with respect to their positions in #1. Note that the Ti off-centre displacements in #1 are much larger than those in plain PbTiO$_3$. In **b** and **c**, when Ti moves towards the O$_4$ plane, the Ti–O interaction becomes more of nonbonding and less of antibonding character, moving the Ti 3d orbitals downward and enhancing their contribution to the band-edge electronic states. Also see Fig. 3a for the Glass coefficient change under artificial Ti atom antiparallel displacement, and Supplementary Fig. 2 for the photocurrent change under artificial Pb atom displacement in the rock salt cation arrangement.

(CBM) wavefunction from X to R is accompanied by the change in the sign of the shift current and leads to a bigger shift vector magnitude (shift vector can be viewed as an indicator of the nonthermal carrier mobility) (Fig. 1d).

Photocurrent enhancement by artificial manipulation. The relationship between the CBM wavefunction character, delocalization and shift current sign suggests that eliminating or weakening the differences between the X and R point wavefunctions can largely eliminate the cancellation of counter-propagating currents and substantially increase the total BVPE response. To verify this, we engineer the CBM wavefunction by changing the chemical bonding in PNT. The bottom of the CB is composed of the Ni 3d, Ti 3d, Pb 6p and O 2p orbitals, with the Ni 3d orbitals making a major contribution (Supplementary Fig. 1). Since the peak of the Ti 3d orbitals is located at a higher energy than that of the Ni 3d peak, the contribution of the Ti 3d states to the band-edge electronic transitions can be enhanced by a downshift of the Ti 3d orbital energy at the CB (Fig. 2d). A reduced overlap between the Ti 3d and O 2p orbitals would give rise to a weaker Ti–O bond and a smaller energy level difference between the bonding (valence band) and antibonding (CB) Ti–O orbitals, resulting in a downshift of the CBM Ti 3d orbitals relative to the those of the relaxed structure[30,31]. Therefore, to shift the Ti 3d states to lower energies, we increase the Ti–O distances in $1 \times 1 \times 3$ PNT by moving Ti atoms antiparallel to the overall polarization, reducing, but not entirely eliminating the local Ti off-centre displacements found in the relaxed structure (Fig. 2).

The resulting changes in the Ti–O bonding have three effects. First, analysis of the CBM density of states shows that the contribution of the Ti 3d (and also Pb 6p, not shown) orbitals increases while the proportion of the Ni 3d orbitals decreases steadily with Ti atomic movements (Fig. 2b), confirming that CBM electronic structure can be effectively controlled by changes in the local structure and Ti–O bonding. Second, the calculated bandgap decreases, leading to smaller photocurrent onset energy with greater artificial Ti sublattice displacement antiparallel to the polarization. Third, as the Ti sublattice is moved, this first leads to a change in the sign of the shift current direction, followed by a substantial enhancement of the shift current magnitude (Figs 2e and 3a). Following Planck's law of black-body radiation for solar power distribution over light frequency at 6,000 K, the total photocurrent is enhanced by a factor of 97, compared with the ground-state structure. This demonstrates that fairly gentle changes in the structure can significantly affect the CBM character and lead to a markedly enhanced BPVE response.

Photocurrent enhancement via nanolayering. The design principle of greater CBM delocalization elucidated by the study of artificial lattice adjustment can be realized in nanoscale heterostructures. They are essentially PbNiO$_2$-doped PbTiO$_3$, but can be viewed as a nanolayer heterostructure of PbNiO$_2$ and PbTiO$_3$. While a more rigorous temperature-dependent thermodynamic analysis is required for investigating its experimental viability, our 0 K density functional theory (DFT) stability analysis shows that these structures are very likely experimentally accessible in the Pb(Zr$_{0.2}$Ti$_{0.8}$)$_{0.7}$Ni$_{0.3}$O$_{3-\delta}$ and

KBNNO materials for which PNT serves as a prototype. We find that nanolayered structures are preferred to the rock salt-ordered PNT and that under 1% in-plane compressive strains, the $1 \times 1 \times 2$ superlattice with apical O vacancies is more stable than its equatorial counterpart. Thus, the enhancement of the photo-

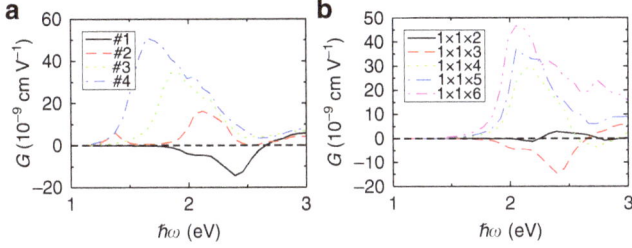

Figure 3 | The substantial enhancement of the Glass coefficient. The Glass coefficients of (**a**) the $1 \times 1 \times 3$ superlattice with different Ti antiparallel displacements and (**b**) the $1 \times 1 \times N$ superlattices. The inset label #n in **a** denotes the structures with different Ti antiparallel displacements. For example, #1 denotes the fully relaxed structure (no artificial antiparallel displacement) and #2, #3 and #4 denote the structures with Ti atoms displaced antiparallel by 0.06, 0.12 and 0.18 Å, respectively, with respect to their positions in #1. Note that the Ti off-centre displacements in #1 are much larger than those in plain PbTiO$_3$.

voltaic response found by our calculations can be realized through the growth of an epitaxially strained thin film or by layer deposition methods.

In PNT, the Ni$''_{Ti}$O$_2$ layer has -2 charge, and is adjacent to the PbV$_O^{··}$ layer with a $+2$ charge. Thus, Ni$''_{Ti}$ + V$_O^{··}$ substitution gives rise to charge separation. The Ni-vacancy layers also provide a polarizing field on the PbTiO$_3$ layers, with this effect reduced as the Ni-vacancy layers are spaced further apart. Approximating the layers as simple charged sheets, we obtain a periodic series of regions with electric field, analogous to the polar catastrophe in LaAlO$_3$/SrTiO$_3$ (ref. 18). As the separation between the Ni increases with increasing layer numbers N, the electric field in the bulk PbTiO$_3$ layers becomes smaller under short-circuit boundary conditions, and bound charges screen the field over a greater length, delocalizing the electron densities (Fig. 4a). This mechanism for the control of wavefunction delocalization and photocurrent is confirmed by the analysis of the electronic structure and the calculated response for $1 \times 1 \times N$ ($N = 2$–6) PNT supercells with one NiO$_2$ plane per supercell. To show the changes due to the delocalization, we plot the layer-averaged probability density as a function of layer normal coordinate for the relevant CBM and valence band maximum (VBM) states (Fig. 4b,c). Going from $N = 3$ to 4, these states, especially the valence band state, become more delocalized, corresponding to the enhancement of the shift current magnitude for a certain

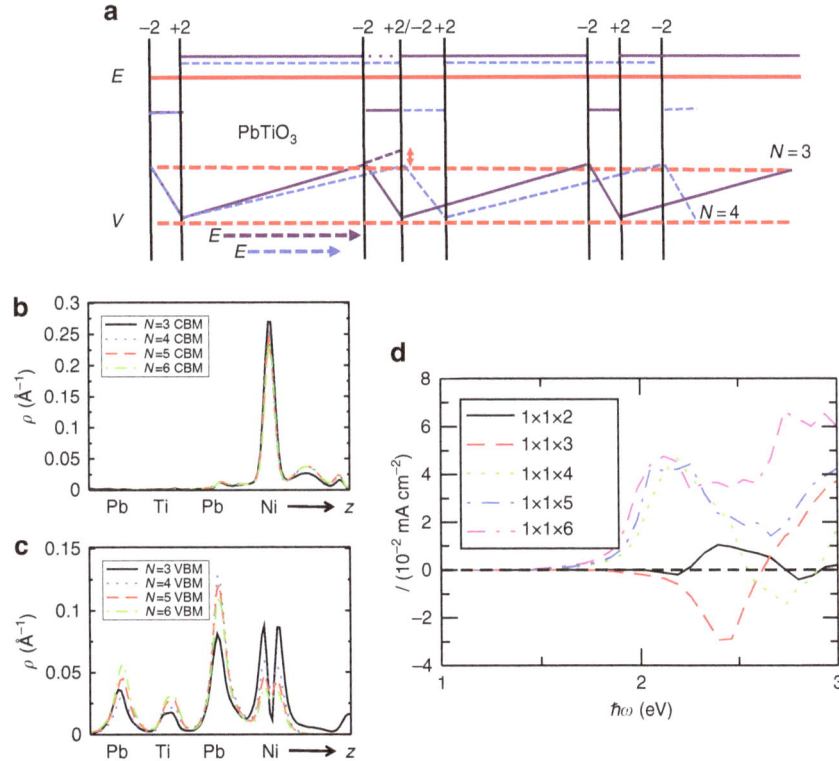

Figure 4 | The photocurrent enhancement via nanolayering and the corresponding change of the electric field landscape and state delocalization. (**a**) The electric field and potential inside PNT for $N = 3$ and 4. The Ni$''_{Ti}$ + V$_O^{··}$ substitution results in adjacent planes of charge. As N increases, less screening over a greater distance is generated in response to the potential changes introduced by this charge separation. The electric field magnitude over the PbTiO$_3$ layers is approximately $\frac{2e}{(2N-1)\varepsilon_0}$, decreasing as N increases. (**b,c**) The real-space probability density distribution summed over the Cartesian x and y coordinates $\left(\rho(z) = \int dxdy \, | \langle \mathbf{r} | \, \psi_{n_0,R} \rangle |^2 \right)$ for the CBM and valence band maximum (VBM) states at R k point. N is the number of layers. For $N > 3$, the valence band wavefunction is localized within the supercell, yielding qualitatively distinct behaviour (as seen in Fig. 1c). The decreased electric field allows the wavefunction to decay more slowly, shifting the probability density of the valence band away from nickel, increasing the intracell delocalization. (**d**) The photocurrent (assuming 0.1 W cm^{-2} light absorption) of the layered PNT with varying superlattice periodicity. As N increases, the near-gap response stabilizes and the response at 3 eV increases; this represents an across-the-board increase in response per light-absorbing Ni layer due to greater delocalization. Increasing the superlattice N reduces the polarizing field on PbTiO$_3$, aligning the shift current response across the visible range and increasing the current per absorbed photon.

photon energy. In addition, this delocalization is only moderately enhanced from $N = 4$ to 6, resulting in the relatively flat progression of shift current at a particular photon energy observed for $N > 3$.

Comparison of the calculated shift current spectra shows that contrary to naive expectations, reduction of the Ni fraction actually increases the effectiveness of substitution, increasing the response and reducing the bandgap as the number of layers is increased (Figs 3b and 4d). This enhancement of the total photocurrent is due both to the enhancement of photocurrent magnitude at a certain photon energy and to the alignment of the photocurrent sign at different photon energies throughout the spectrum. The effect for visible-light illumination is quite marked, with an increase by a factor of 43 due to a simple insertion of additional bulk PbTiO$_3$ layer(s). Such extraordinary sensitivity underscores the importance and usefulness of nanoscale ordering for the BPVE engineering. We find that the total shift current under visible-light illumination is also over 10 times greater than that for the BiFeO$_3$ material that has been the standard for experiments on the photoresponse of ferroelectrics. The proximate reason for the enhancement in response becomes clear on looking at the response along the X–R line (Fig. 1d and Supplementary Fig. 3). The difference between the $1 \times 1 \times 3$ and $1 \times 1 \times 4$ is due to the disappearance of the sign change versus k point in the shift vector and the resulting shift current response. The shift current magnitude then further increases with the number of PbTiO$_3$ layers. As N increases further, the effect of eliminating cancellation of counter-propagating photocurrents at different photon energies starts to dominate, and this further enhances the total photo-current for the whole visible-light spectrum.

Discussion

The nanoscale layered heterostructures have another potential benefit. Recent experimental work has highlighted the importance of conventional transport characteristics for BPVE[7.] Essentially, the performance of the bulk photovoltaic materials depends not only on their current generation capability but also on the photovoltage that they can sustain. This latter quantity is determined by the resistance of the current leaking back through the material, and the design of useful devices will therefore require control over conventional conductivity. In BiFeO$_3$, domain walls can serve as barricades against such leakage, substantially increasing the resistance and, consequently, the maximum photovoltage. We note that the present system similarly features a nanoscale heterogeneous layered structure and may be viewed as a nanoscale composite with alternating photocurrent-generating layers and insulating layers, ideal for a BPVE device.

The foregoing holds important lessons for engineering bulk photovoltaic materials. First, it indicates that the problems of bandgap reduction and state delocalization can be partially separated. Nickel substitution provides lower-energy CB states that tend to be localized along the layer normal direction and are not favourable for a large shift current. Nevertheless, these states may participate in transitions with what are essentially PbTiO$_3$ states that are delocalized and thus favour larger shift currents. We have chosen a material with robust shift current potential but a large gap (PbTiO$_3$), and introduced a modification that solves the latter without losing the former. Second, it emphasizes the importance of not only the magnitude of photocurrent as a factor in response, but consistency in direction, which is sensitive to relatively small structural changes, and therefore amenable to manipulation. Third, the electrostatic effects of the Ni$_{Ti}'' + V_O^{\cdot\cdot}$ substitution on the electronic structure of bulk PbTiO$_3$ are pronounced and most apparent for lower nickel fractions. We

hope that our work will motivate further studies of photoresponse in nanoscale ferroelectrics.

The fact that a substantial enhancement of the BPVE is found in an ordered nanostructure highlights the importance of order/disorder effects in photoferroelectrics. It has been long accepted that most ferroelectrics and related materials exhibit local deviations from their average structures, whether in the form of disordered B-cation displacements in the conventional BaTiO$_3$ and KNbO$_3$ ferroelectrics, in relaxor ferroelectrics or in solid solutions[20,32,33.] The disorder in BaTiO$_3$ and KNbO$_3$ is due to the presence of ordered chains of displacements and that in the relaxor-like material Na$_{1/2}$Bi$_{1/2}$TiO$_3$ is related to planar defects, and both of these structural features are found in our system[19,32.] This order/disorder scenario leads to intriguing physics of electronic structure, phase transition, ferroelectricity, piezoelectricity, ferroelasticity, magnetism, and, in our case, the ferroelectric photovoltaic effect. Therefore, to obtain a good ferroelectric photovoltaic, attention should be paid to the effects of order/disorder and to the manipulation of the electric field landscape and bandgap via ordering/disordering; the present work serves as a good starting point for this approach.

Finally, we point out that our finding is not specific to the particular system studied here, but is rather general to a variety of ferroelectric oxide systems incorporating Ni (for example, PbNiO$_2$-doped PZT and KBNNO) and other $3d$ transition metals. This is because the contributing electronic states in such a complex system are typically mixtures of different atomic orbitals that can be tailored by changes in structure and electrostatics, and by ordering/disordering. However, due both to the complexity of the relationship between photocurrent, electronic structure and atomic structure, and to the current paucity of visible-light-absorbing ferroelectric photovoltaic materials, a careful and detailed analysis will be required to estimate the magnitude of the nanolayering effects for each particular system.

Methods

First-principles calculation. A previously developed approach implemented in our in-house code that yields good agreement with experiment for shift current magnitude and spectral profile was used to calculate the shift current[17,34.] The QUANTUM-ESPRESSO code was used to perform DFT calculations with the local density approximation functional[35–37.] All elements are represented by norm-conserving, optimized nonlocal pseudopotentials[38.] For structural optimizations, Monkhorst–Pack grids[39] with at least $6 \times 6 \times 6$ k-point meshes were used, while for the self-consistent and non-self-consistent calculations, finer k-point grids up to $40 \times 40 \times 40$ were used to get a well-converged shift current response. The DFT + U method was used to improve the description of d-orbital electrons, with the Hubbard U parameterized by the linear-response approach[40.] The calculated U values are ≈ 4.7 and ≈ 8.9 eV for Ti and Ni, respectively. All photon energies are shifted by 1.05 eV so that PbTiO$_3$ is at its experimental bandgap (3.6 eV). Heyd-Scuseria-Ernzerhof hybrid functional calculations[41] were also used to corroborate the bandgap trends obtained by DFT + U calculations.

The shift current susceptibility tensor. Tetragonal PbTiO$_3$ belongs to the $P4mm$ space group, corresponding to the C_{4v} point group; in this case, the third rank shift current response tensor must have the form

$$\sigma = \begin{bmatrix} 0 & 0 & 0 & 0 & \sigma_{zyY} & 0 \\ 0 & 0 & 0 & \sigma_{zyY} & 0 & 0 \\ \sigma_{xxZ} & \sigma_{xxZ} & \sigma_{zzZ} & 0 & 0 & 0 \end{bmatrix} \qquad (1)$$

where the lower- and upper-case letters represent directions of the light polarization and photocurrent, respectively. The Glass coefficient

$$G_{rrq} = \frac{\sigma_{rrq}(\omega)}{\alpha_{rr}(\omega)} \qquad (2)$$

describes the current response in a thick sample and includes the light attenuation effect due to the absorption coefficient $\alpha_{rr}(\omega)$. We choose our polarization direction to be normal to the Ni layers, so that all analysed systems have the same symmetry properties, and (for unpolarized light) only the xxZ and zzZ components are relevant. In PNT, the shift current induced by the perpendicularly polarized light (xxZ) is much stronger than that induced by the parallel polarized light (zzZ)

because of their different absorption efficiencies. In the present work, we focus only on the more substantial xxZ component.

References

1. Choi, T., Lee, S., Choi, Y. J., Kiryukhin, V. & Cheong, S.-W. Switchable ferroelectric diode and photovoltaic effect in BiFeO₃. *Science* **324**, 63–66 (2009).
2. Kundys, B., Viret, M., Colson, D. & Kundys, D. O. Light-induced size changes in BiFeO₃ crystals. *Nat. Mater.* **9**, 803–805 (2010).
3. Qin, M., Yao, K. & Liang, Y. C. High efficient photovoltaics in nanoscaled ferroelectric thin films. *Appl. Phys. Lett.* **93**, 122904 (2008).
4. Cao, D. *et al.* High-efficiency ferroelectric-film solar cells with an n-type Cu₂O cathode buffer layer. *Nano Lett.* **12**, 2803–2809 (2012).
5. Alexe, M. & Hesse, D. Tip-enhanced photovoltaic effects in bismuth ferrite. *Nat. Commun.* **2**, 256 (2011).
6. Fridkin, V. M. Bulk photovoltaic effect in noncentrosymmetric crystals. *Crystallogr. Rep.* **46**, 654–658 (2001).
7. Bhatnagar, A., Chaudhuri, A. R., Kim, Y. H., Hesse, D. & Alexe, M. Role of domain walls in the abnormal photovoltaic effect in BiFeO₃. *Nat. Commun.* **4**, 2835 (2013).
8. Yang, S. Y. *et al.* Above-bandgap voltages from ferroelectric photovoltaic devices. *Nat. Nanotechnol.* **5**, 143 (2010).
9. Yang, S. Y. *et al.* Photovoltaic effects in BiFeO₃. *Appl. Phys. Lett.* **95**, 062909 (2009).
10. Grinberg, I. *et al.* Perovskite oxides for visible-light-absorbing ferroelectric and photovoltaic materials. *Nature* **503**, 509–512 (2013).
11. Nechache, R. *et al.* Bandgap tuning of multiferroic oxide solar cells. *Nat. Photon.* **9**, 61–67 (2015).
12. Glass, A. M., von der Linde, D. & Negran, T. J. High-voltage bulk photovoltaic effect and the photorefractive process in LiNbO₃. *Appl. Phys. Lett.* **1925**, 233–235 (1974).
13. Kraut, W. & von Baltz, R. Anomalous bulk photovoltaic effect in ferroelectrics: a quadratic response theory. *Phys. Rev. B* **19**, 1548–1554 (1979).
14. Chynoweth, A. G. Surface space-charge layers in barium titanate. *Phys. Rev.* **102**, 705–714 (1956).
15. Ji, W., Yao, K. & Liang, Y. C. Bulk photovoltaic effect at visible wavelength in epitaxial ferroelectric BiFeO₃ thin films. *Adv. Mater.* **22**, 1763–1766 (2010).
16. von Baltz, R. & Kraut, W. Theory of the bulk photovoltaic effect in pure crystals. *Phys. Rev. B* **23**, 5590–5596 (1981).
17. Young, S. M. & Rappe, A. M. First principles calculation of the shift current photovoltaic effect in ferroelectrics. *Phys. Rev. Lett.* **109**, 116601 (2012).
18. Chtomo, A. & Hwang, H. Y. A high-mobility electron gas at the LaAlO₃/SrTiO₃ heterointerface. *Nature* **427**, 423–426 (2004).
19. Kreisel, J. *et al.* High-pressure x-ray scattering of oxides with a nanoscale local structure: Applications to Na₁/₂Bi₁/₂TiO₃. *Phys. Rev. B* **68**, 014113 (2003).
20. Grinberg, I., Cooper, V. R. & Rappe, A. M. Relationship between local structure and phase transitions of a disordered solid solution. *Nature* **419**, 909 (2002).
21. Kreisel, J., Alexe, M. & Thomas, P. A. A photoferroelectric material is more than the sum of its parts. *Nat. Mater.* **11**, 260 (2012).
22. Balachandran, P. V. & Rondinelli, J. M. Massive band gap variation in layered oxides through cation ordering. *Nat. Commun.* **6**, 6191 (2015).
23. Liu, S. *et al.* Ferroelectric domain wall induced band gap reduction and charge separation in organometal halide perovskites. *J. Phys. Chem. Lett.* **6**, 693–699 (2015).
24. Bennett, J. W., Grinberg, I. & Rappe, A. M. New highly polar semiconductor fer-roelectrics through d^8 cation-O vacancy substitution into PbTiO₃: a theoretical study. *J. Am. Chem. Soc.* **130**, 17409–17412 (2008).
25. Gou, G. Y., Bennett, J. W., Takenaka, H. & Rappe, A. M. Post density functional theoretical studies of highly polar semiconductor $Pb(Ti_{1-x}Ni_x)O_{3-x}$ solid solutions: effects of cation arrangement on band gap. *Phys. Rev. B* **83**, 205115 (2011).
26. Kondo, M. & Kurihara, K. Sintering behavior and surface microstructure of PbO-rich PbNi₁/₃Nb₂/₃O₃-PbTiO₃-PbZrO₃ ceramics. *J. Am. Ceram. Soc.* **84**, 2469–2474 (2001).
27. Kumari, S., Crtega, N., Kumar, A., Scott, J. F. & Katiyar, R. S. Ferroelectric and photovoltaic properties of transitional metal doped $Pb(Zr_{0.14}Ti_{0.56}Ni_{0.30})O_{3-\delta}$ thin films. *AIP Adv.* **4**, 037101 (2014).
28. Zhang, Z., Wu, P., Lu, L. & Shu, C. Study on vacancy formation in ferroelectric PbTiO₃ from ab initio. *Appl. Phys. Lett.* **88**, 142902 (2006).
29. Wang, F. & Rappe, A. M. First-principles calculation of the bulk photovoltaic effect in KNbO₃ and (K,Ba)(Ni,Nb)O₃₋δ. *Phys. Rev. B* **91**, 165124 (2015).
30. Wang, F., Grinberg, I. & Rappe, A. M. Band gap engineering strategy via polarization rotation in perovskite ferroelectrics. *Appl. Phys. Lett.* **104**, 152903 (2014).
31. Eng, H. W., Barnes, P. W., Auer, B. M. & Woodward, P. M. Investigation of the electronic structure of d^0 transition metal oxides belonging to the perovskite family. *J. Solid State Chem.* **175**, 94–109 (2003).
32. Comes, R., Lambert & Guinier, A. The chain structure of BaTiO₃ and KNbO₃. *Solid State Commun.* **6**, 715 (1968).
33. Burns, G. & Dacol, F. H. Crystalline ferroelectrics with glassy polarization behavior. *Phys. Rev. B* **28**, 2527 (1983).
34. Young, S. M., Zheng, F. & Rappe, A. M. First-principles calculation of the bulk photovoltaic effect in bismuth ferrite. *Phys. Rev. Lett.* **109**, 236601 (2012).
35. Giannozzi, P. *et al.* Quantum Espresso: a modular and open-source software project for quantum simulations of materials. *J. Phys. Condens. Matter* **21**, 395502 (2009).
36. Kohn, W. & Sham, L. J. Self-consistent equations including exchange and correlation effects. *Phys. Rev.* **140**, A1133–A1138 (1965).
37. Perdew, J. P. & Zunger, A. Self-interaction correction to density-functional approximations for many-electron systems. *Phys. Rev. B* **23**, 5048–5079 (1981).
38. Rappe, A. M., Rabe, K. M., Kaxiras, E. & Joannopoulos, J. D. Optimized pseudopotentials. *Phys. Rev. B* **41**, 1227–1230 (1990).
39. Monkhorst, H. J. & Pack, J. D. Special points for Brillouin-zone integrations. *Phys. Rev. B* **13**, 5188–5192 (1976).
40. Cococcioni, M. & de Gironcoli, S. Linear response approach to the calculation of the effective interaction parameters in the LDA + U approach. *Phys. Rev. B* **71**, 035105 (2005).
41. Heyd, J., Scuseria, G. E. & Ernzerhof, M. Hybrid functionals based on a screened Coulomb potential. *J. Chem. Phys.* **118**, 8207–8215 (2003).

Acknowledgements

F.W. was supported by the National Science Foundation, under Grant DMR-1124696. S.M.Y. was supported by the Department of Energy Office of Basic Energy Sciences under Grant DE-FG02-07ER46431, and a National Research Council Research Asso-ciateship Award at the US Naval Research Laboratory. F.Z. was supported by the National Science Foundation under Grant CMMI-1334241. I.G. was supported by the Office of Naval Research under Grant N00014-12-1-1033. A.M.R. was supported by the Office of Naval Research under Grant N00014-11-1-0664. Computational support was provided by the High-Performance Computing Modernization Office of the Department of Defense and the National Energy Research Scientific Computing Center of the Department of Energy.

Author contributions

F.W. performed the calculations; F.W. and S.M.Y. performed the data processing and analysis; F.W., S.M.Y. and I.G. wrote the manuscript; A.M.R. supervised the work; all authors contributed to the conception of ideas, to discussions, and to manuscript modification.

Additional information

17

Size-dependent phase transition in methylammonium lead iodide perovskite microplate crystals

Dehui Li[1], Gongming Wang[1,2], Hung-Chieh Cheng[3], Chih-Yen Chen[1], Hao Wu[3], Yuan Liu[3], Yu Huang[2,3] & Xiangfeng Duan[1,2]

Methylammonium lead iodide perovskite has attracted considerable recent interest for solution processable solar cells and other optoelectronic applications. The orthorhombic-to-tetragonal phase transition in perovskite can significantly alter its optical, electrical properties and impact the corresponding applications. Here, we report a systematic investigation of the size-dependent orthorhombic-to-tetragonal phase transition using a combined temperature-dependent optical, electrical transport and transmission electron microscopy study. Our studies of individual perovskite microplates with variable thicknesses demonstrate that the phase transition temperature decreases with reducing microplate thickness. The sudden decrease of mobility around phase transition temperature and the presence of hysteresis loops in the temperature-dependent mobility confirm that the orthorhombic-to-tetragonal phase transition is a first-order phase transition. Our findings offer significant fundamental insight on the temperature- and size-dependent structural, optical and charge transport properties of perovskite materials, and can greatly impact future exploration of novel electronic and optoelectronic devices from these materials.

[1]Department of Chemistry and Biochemistry, University of California, 607 Charles E. Young Drive East, Los Angeles, California 90095, USA. [2]California Nanosystems Institute, University of California, Los Angeles, California 90095, USA. [3]Department of Materials Science and Engineering, University of California, Los Angeles, California 90095, USA. Correspondence and requests for materials should be addressed to X.D. (email: xduan@chem.ucla.edu).

The hybrid organic–inorganic methylammonium lead iodide perovskite (CH$_3$NH$_3$PbI$_3$, denoted as MAPbI$_3$) is emerging as one of most promising solution-processable light absorber for solar cells and thus has attracted intensive recent interest[1-9]. With long carrier diffusion length[10-12] and low non-radiative recombination rate, solution processed perovskite materials have been demonstrated to deliver a certified power-conversion efficiency as high as 20.1% in the past a few years[13]. The excellent optical properties of MAPbI$_3$ perovskite enables it to be applied in a wide range of optoelectronic devices such as photodetectors[14], lasers[15-17] and light-emitting diodes[18]. Despite the tremendous interest in MAPbI$_3$ perovskite, its charge transport properties remain elusive because of the ion motion, which leads to a very large hysteresis and prevents the observation of the intrinsic field-effect mobility at the room temperature[19,20]. In addition, it has been proven that the solar cell efficiency strongly depends on the size of cuboids of perovskites[21]. Therefore, it is expected that the size could significantly influence the optical and charge transport properties of MAPbI$_3$, yet there is no systematic investigation of size-dependent optical and charge transport properties in perovskite materials.

The structural phase transitions can significantly alter the optical and electronic properties of materials[22], both of which are essential to understand the underlying photophysics[23]. The temperature-dependent studies such as photoluminescence (PL) spectroscopy[23], neutron powder diffraction[24], calorimetric and infrared spectroscopy[25] have been utilized to investigate the structural phase transitions in bulk MAPbI$_3$. The MAPbI$_3$ adopts the simple cubic perovskite structure above 330 K, transits to the tetragonal phase at 330 K (refs 26,27), and further evolves into an orthorhombic phase as the temperature is reduced to 160 K (ref. 27). All those phase transitions have been proven to be of first order[25]. Previous studies have shown that the physical size of a material can be an important variable in determining the phase transition points in addition to pressure, temperature and compositions[28-30]. Therefore, it is important to investigate how the size alters the phase transition points in MAPbI$_3$, which in turn affects its optical and electronic properties[31,32]. Nevertheless, the size-dependent phase transition in MAPbI$_3$ remains elusive. Here we report a systematic investigation of size-dependent structural phase transitions in individual MAPbI$_3$ microplate crystals by using temperature-dependent charge transport measurements, PL spectroscopy and transmission electron microscopy and electron diffraction.

Results

Temperature-dependent electrical measurement. To investigate the fundamental charge transport properties of individual perovskite microplates, we have constructed field-effect transistors (FETs) using the perovskite crystals and measured their transistor characteristics from the room temperature to liquid nitrogen temperature in dark. Figure 1a shows a schematic of a typical FET device configuration we used. The individual perovskite microplates served as the semiconducting channel of FET devices bridging two pre-fabricated Cr/Au electrodes as the source-drain electrodes on 300 nm SiO$_2$/Si substrate (as both the gate dielectrics and gate electrode). The inset of Fig. 1b displays an optical image of a typical FET device, where the thickness of the perovskite microplate is around 400 nm. Figure 1b shows a set of typical output characteristics (source-drain current I_{sd} versus source-drain voltage V_{sd}) of a single perovskite microplate FET device under various gate voltages (V_g) at 77 K. The large positive gate voltage induces a higher source-drain current, which indicates an n-type conduction behaviour of perovskite microplate

(Fig. 1b). The slight nonlinearity of I_{sd}–V_{sd} curves near zero bias suggests that the contact is not fully optimized. The transfer characteristics exhibit dominant n-type behaviour with a slight p-type conductance at negative gate voltage (Fig. 1c). The maximum on/off ratio is nearly six orders of magnitude, which is better than recently reported perovskite thin-film transistors[19].

Strong hysteresis is commonly observed in perovskite thin-film transistors, which prevents fully understanding the charge transport properties and exact determination of carrier mobility in such perovskite materials. The origin of the hysteresis has been attributed to ferroelectricity, ion motion within the perovskite material and trapping/de-trapping of charge carriers at the interfaces[19]. However, no conclusive explanation is available to date. In our microplate devices, considerable hysteresis has been observed for all temperatures from 296 to 77 K, which reduces with decreasing temperature. It has been proven that the ion motion in halide perovskite is a thermally activated process[33] and the ion migration rate exponentially reduces as the temperature decreases. The contribution from ion motion to hysteresis is expected to negligibly small at lower temperatures (for example, 77 K). The presence of hysteresis at 77 K suggests that the ion motion only partly contributes to the hysteresis and other factors such as trap states and surface dipoles may play important roles as well[19]. It is important to note that the hysteresis in our microplate device only increases slightly when the temperature is increased from 77 to 296 K (Supplementary Fig. 1), and the hysteresis at 296 K is considerably smaller than that observed in thin-film perovskite FET devices, where the presence of huge hysteresis prevents the observation of field-effect behaviour above 258 K (ref. 19).

Based on the transfer characteristics, the field-effect carrier mobility can be extracted. The existence of hysteresis in transfer characteristics may lead to systematic errors in mobility

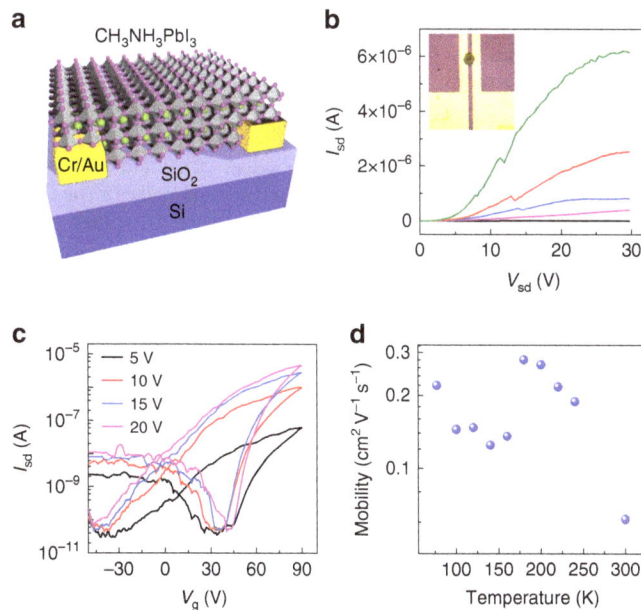

Figure 1 | Field-effect transistors based on individual perovskite microplate. (**a**) Schematic of the bottom-gate, bottom-contact halide perovskite microplate field-effect transistor fabricated on a 300-nm SiO$_2$/Si substrate with 5 nm Cr/50 nm Au as contact. (**b,c**) The output (V_g = 0, 20, 40, 60, 80 V; from bottom to top; **b**) and transfer (V_{sd} = 5, 10, 15, 20 V; **c**) characteristics of a field-effect transistor based on a perovskite crystal microplate at 77 K. The inset of **b** shows an optical image of a typical device. The channel length is around 8 μm. (**d**) The temperature-dependent field-effect electron mobility measured with a source-drain voltage of 20 V.

determination, with possible underestimation in positive sweeping direction, overestimation in negative sweeping direction and scan rate dependence (Supplementary Fig. 2). To this end, we have determined carrier mobility based on both the positive and negative sweepings. Nevertheless, both the positive and negative sweepings give the exactly same trend of the mobility versus temperature. For the simplicity of discussion, we focus on the mobility values derived from negative sweeping here. The field-effect electron mobility continuously increases with the decreasing temperature from 300 to 180 K, and then shows a sudden decrease when the temperature is reduced from 180 to 160 K (Fig. 1d). Afterwards with further decreasing temperature, the field-effect electron mobility starts to increase again. Qualitatively similar temperature-dependent characteristics have been observed in all devices except that the transition temperature varies with the thickness of microplates, which we will discuss below in detail. As there is a structural phase transition from the tetragonal phase to the orthorhombic phase at 160 K (ref. 23), we attribute this sudden decrease of field-effect mobility to the structural phase transition. The structural phase transition would induce the change of effective mass and dielectric constant[19,34,35], both of which could contribute to the change of field-effect mobility[19,36].

The theoretical calculation based on semi-classical Boltzman transport theory predicates that the mobility of orthorhombic phase should be larger than that of tetragonal phase[19]. In contrast, we observed a sudden decrease of the mobility when the $MAPbI_3$ transits from the tetragonal phase to the orthorhombic phase. Previous optical studies indicate that there are small inclusions of the tetragonal phase domains within the orthorhombic phase even when the temperature is much lower than the tetragonal phase to orthorhombic phase transition temperature (T_{t-o}), likely due to the strain imposed by the thermal expansion and change of the in-plane lattice constant during the phase transition[23]. Our PL studies also demonstrate the presence of such small inclusions (see below). It is very likely that such small inclusions introduce more boundaries and thus increases the carrier scattering, which also contribute to the sudden decrease of the mobility upon the phase transition. Although the improvement of field-effect mobility with the decreasing temperature can be attributed to the electron–phonon interaction and ion drift under applied electric field within the tetragonal phase[19,37], the origins of the rapid increase of the field-effect mobility within the orthorhombic phase are much more complicated. Previous studies have shown that the minimum phonon energy related to the methylammonium (MA) cation is estimated to be 15 meV (refs 38,39). Therefore, the interaction of carriers with phonons associated with MA libration should be quenched below 170 K. It has been shown that the quench of the carrier–phonon interaction related to the MA libration modes led to a weaker temperature dependence of carrier mobility below 198 K in halide perovskite thin-film transistors[19], indicating the strong interaction between the carriers and MA libration modes. Without the contribution from interaction with MA libration modes < 170 K, the increase rate of the mobility with decreasing temperature should be slowed if only carrier–phonon interaction contributes to the decrease of the mobility after the phase transition. On the contrary, we observed a different picture: a more rapid increase with decreasing temperature (Fig. 1d). Therefore, we suggest that in addition to the carrier–phonon scattering, the decrease of the small inclusions of tetragonal phase domains and the reduction of the grain boundaries might partly contribute to the rapid increase of mobility with the decreasing temperature. The inclusions of tetragonal phase near the transition point within the orthorhombic phase are also confirmed by our temperature-dependent selected area electron diffraction (SAED) studies (see below).

It should be noted that the electron field-effect mobility we extracted here is smaller than those measured by THz spectroscopy in perovskite films ($\sim 8\ cm^2\ V^{-1}\ s^{-1}$ at room temperature)[40], Hall measurement in perovskite single crystals ($\sim 66\ cm^2\ V^{-1}\ s^{-1}$)[27] and electrical measurement in the space-charge-limited current regime or time-of-flight measurement in perovskite single crystals (~ 2.5–$25\ cm^2\ V^{-1}\ s^{-1}$)[10,11], but compares favourably with the field-effect mobility reported in perovskite thin-film transistors ($\sim 0.1\ cm^2\ V^{-1}\ s^{-1}$)[19]. Time-resolved THz spectroscopy probes short-time dynamics up to a few nanoseconds and thus measures local carrier transport phenomena, whereas electrical measurements focus long-time (μs) or longer conduction processes occurring over several micrometres length-scale of a device[41,42]. Electrical measurements are therefore more sensitive to grain size, boundaries and interfacial effects because of the carrier transport over the device dimension. Thus, it is not surprising that the mobility measured by THz spectroscopy is larger than that measured by electrical measurements. Although Hall measurements, time-of-flight technique and space-charge-limited current method directly measure intrinsic charge transport, the field-effect mobility is extremely sensitive to the dielectric/semiconductor interfaces as well as the source, drain contact resistance[43]. Such extrinsic factors in field-effect measurement can often lead to an underestimation of the carrier mobility in FETs.

Thickness-dependent phase transition. To investigate how the thickness of the microplates influences the field-effect mobility, we systematically carried out the temperature-dependent transport measurement with different microplate thickness, with the transfer curves (120–200 K) of three representative devices shown in Fig. 2a–c. All three devices exhibit dominant n-type behaviour regardless of the thickness. The field-effect electron mobility extracted from the transfer curves shows a common trend with the temperature for devices with different thickness: as the temperature decreases, the electron mobility first increases, suddenly decreases at the structural phase transition point and increases again with further reducing temperature (Fig. 2d). It is noted that the structural phase transition temperature T_{t-o} strongly depends on the thickness of the perovskite microplates: the thicker the microplates are, the higher the structural phase transition temperature T_{t-o} is. For the 30-nm-thick microplate, the structural phase transition temperature T_{t-o} falls around 130 K, which increases to ~ 150 K for the 90 nm microplate, and to ~ 170 K for the 400-nm-thick microplate. Furthermore, we have carried out temperature-dependent transport measurement using different metal contact including Pt and graphene (Fig. 2e), on a hexagonal boron nitride (hBN) substrate (Supplementary Fig. 3 and Supplementary Note 1) and in a device with very long channel length (40 μm; Supplementary Fig. 4). It is found that the structural phase transition occurs with phase transition temperature T_{t-o} relying only on the thickness of perovskite microplates regardless of the contact materials, substrate and channel length, indicating that the structural phase transition is an intrinsic property of the perovskite microplates.

To precisely locate the tetragonal-to-orthorhombic phase transition point, we scan the temperature range with a higher resolution around the phase transition point for a 200-nm-thick microplate device (Fig. 2f). Similar to the hysteretic behaviour observed in the optical density of perovskite thin films[26] and dielectric and resistance measurement of $MAPbI_3$ crystals[27,35], an apparent hysteresis is observed with a temperature span of 15 K between the cooling cycle and heating cycle. The sharp decrease of the mobility near the phase transition temperature and the presence of the broad hysteresis confirm that the tetragonal phase to the orthorhombic phase transition is a first-order solid–solid phase transition[29,44,45].

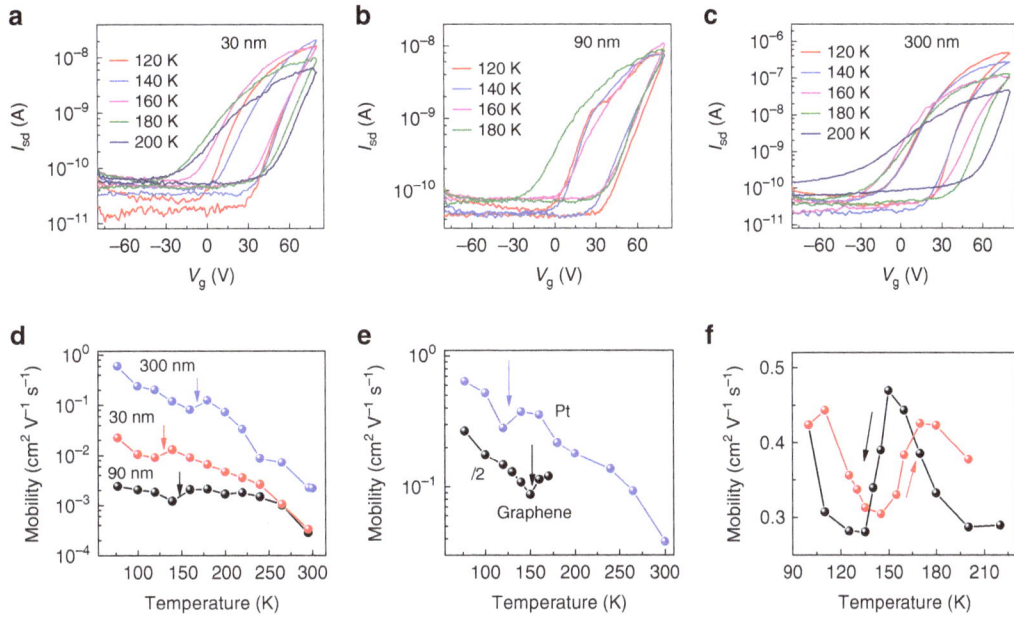

Figure 2 | Size-dependent transition of the orthorhombic phase to tetragonal phase. (**a–c**) The temperature-dependent transfer characteristics of field-effect transistors made of 30 nm (**a**), 90 nm (**b**) and 300 nm (**c**) thick individual perovskite crystal microplates. The applied source-drain voltage is 20 V and the channel length is 8 μm. (**d**) The temperature-dependent field-effect electron mobility for three different thickness devices extracted from **a–c**. The arrows indicate the temperature where the phase transition occurs. The mobility is measured for the heating cycle. (**e**) The temperature-dependent field-effect electron mobility for devices with Pt contact and graphene contact. The thickness of the microplates is around 35 nm and the channel length is 40 μm for Pt contact. For graphene contact, the thickness of the microplates is around 120 nm and the channel length is around 15 μm. The mobility is measured during the heating cycle. (**f**) The temperature-dependent field-effect mobility of a perovskite microplate device (5 nm Cr/50 nm Au as contact) with a thickness of around 200 nm measured for both heating and cooling cycle under a source-drain voltage of 20 V. The channel length of the device is 40 μm.

Temperature-dependent SAED studies. The tetragonal-to-orthorhombic phase transition can be directly confirmed by the temperature-dependent SAED studies. Transmission electron microscopy (TEM) image shows that the converted perovskite microplates we used to acquire the SAED patterns largely retain hexagonal shape similar with the PbI_2 microplate before conversion (Fig. 3a). The SAED pattern at room temperature (Fig. 3b) shows a single set of fourfold symmetric diffraction spots that can be indexed to the tetragonal structure of the perovskite crystals along [001] zone axis. Both first-order and second-order diffraction spots can be clearly distinguished, indicating excellent crystalline quality of the perovskite microplate. Decreasing the temperature to 90 K, the SAED pattern shows a set of fourfold symmetric diffraction spots and a few dispersedly distributed spots indicated by red circles (Fig. 3c). Although the fourfold symmetric diffraction spots can be indexed to the first-order diffraction of the orthorhombic structure along [001] zone axis, the dispersedly distributed spots likely belong to a different set of diffraction patterns. Increasing the temperature to 200 K again, the microplate completely transits from orthorhombic phase back to tetragonal phase. The SAED pattern shows similar features as those obtained at room temperature initially, with the dispersedly distributed spots disappeared. The fourfold symmetric spots can be indexed to the tetragonal structure of the perovskite crystals along [001] zone axis (Fig. 3d), implying that those dispersedly distributed spots at 90 K appear only after the tetragonal-to-orthorhombic phase transition completes. Therefore, those dispersedly distributed spots might be originated from the tetragonal phase along different zone axis because of the inclusions of tetragonal domains within orthorhombic phase or from the orthorhombic domains with different crystalline orientation. Nevertheless, based on lattice spacings analysis it is more likely that the inclusions of tetragonal domains contribute to those dispersedly distributed spots. As mentioned

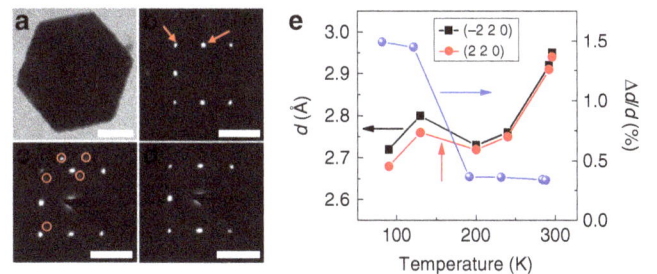

Figure 3 | Temperature-dependent selected area electron diffraction (SAED) patterns. (**a**) Low-resolution TEM image of a perovskite microplate. The scale bar is 1 μm. (**b–d**) The SAED patterns of the microplate at 296 K (**b**), 90 K (**c**) and 200 K (**d**) along [001] zone axis. The scale bar is 5 nm^{-1}. The red circles in **c** indicate the dispersive distributed spots. (**e**) Lattice spacings of (−2 2 0) plane (black squares) and (2 2 0) plane (red squares) of a perovskite microplate. The percentage of lattice spacing difference between those two planes (defined as $(d(-2\,2\,0) - d(2\,2\,0))/d(2\,2\,0)$) is displayed as well (right axis). The increasing difference between these two lattice spacings with the reducing temperature indicates the transitions from a tetragonal phase to orthorhombic phase.

above, even when the temperature is much lower than the phase transition temperature, there still are inclusions of tetragonal domains within orthorhombic phase, suggesting that single-crystal tetragonal phase may break into smaller grains of tetragonal phase and orthorhombic phase when the phase transition occurs. Although most of those small grains maintain almost same orientation, some of them significantly deviate, leading to the dispersedly distributed spots. The presence of dispersedly distributed spots also indicates the degradation of the crystalline quality after the tetragonal-to-orthorhombic phase transition.

The tetragonal-to-orthorhombic phase transition can also be identified by carefully analysing the temperature-dependent lattice spacing change. We extracted the lattice spacings for (− 2 2 0) plane and (2 2 0) plane of a perovskite microplate from SAED patterns and found that a sudden change of the lattice spacings between 130 and 200 K, indicating the phase change occurs in this temperature regime (Fig. 3e). As the (− 2 2 0) plane and (2 2 0) plane are perpendicular to each other, the changing ratio of the lattice spacings between these two planes can be used to identify the crystalline structure as well. Above 200 K, the lattice spacings for those two planes are almost same, indicating that the perovskite microplate has a tetragonal phase. In contrary, there is clear difference between the lattice spacings for those two planes below 200 K, implying the presence of orthorhombic structure (right axis of Fig. 3e). Those results agree with those obtained from transport measurement above.

Temperature- and thickness-dependent PL studies. To further probe the size-dependent tetragonal-to-orthorhombic phase transition, we have also studied temperature-dependent PL. Figure 4a–c displays the PL spectra for four perovskite microplates with various thicknesses at 270, 140 and 77 K. Only one broad emission peak was observed for all four microplates at 270 K and the emission peak shows a blueshift with the reducing thickness (Fig. 4a), which will be discussed in detail below. The PL spectra at 140 K show an extra emission peak at the higher energy for thicker microplates while still exhibits a single peak for the thinner ones (Fig. 4b). Further decreasing the temperature to 77 K, two emission peaks are observed for all microplates but the intensity ratio of the higher energy emission peak (P2) to the lower energy emission peak (P1) decreases with decreasing microplate thickness (Fig. 4c). The higher energy emission peak can be attributed to the orthorhombic phase, whereas the lower energy emission peak is due to the tetragonal phase domains within the orthorhombic phase[23], consistent with theoretical

calculations that the tetragonal phase has a smaller bandgap than that of the orthorhombic phase[46]. Therefore, the emergence of the two emission peaks signifies the occurrence of phase transition, and our PL studies also suggest that the phase transition temperature T_{t-o} is higher for the thicker microplates (Fig. 4b).

Excitation power-dependent PL spectra. Our excitation power-dependent PL studies also confirms the existence of small tetragonal inclusions below the phase transition temperature T_{t-o}, which is supported by the fact that both the emission peak position and intensity are extremely sensitive to the excitation power. We have collected excitation power-dependent PL spectra for a 20-nm perovskite microplate in orthorhombic phase (77 K), near orthorhombic-to-tetragonal phase transition point (140 K) and tetragonal phase (180 K), respectively, extracted the emission peak energy and plotted against the excitation power for each peak at 77, 140 and 180 K (Fig. 4d–f and Supplementary Figs 5 and 6). At 77 K, the lower energy emission peak P1 originating from tetragonal phase domains shows an obvious blueshift with the increasing of excitation power, whereas the higher energy peak P2 from orthorhombic phase shows little change. It is also noted that the P1 emission saturates at high excitation power. As the tetragonal phase has a smaller band gap, the photogenerated carriers prefer to occupy the small tetragonal phase inclusions within the orthorhombic phase. As a result, a large number of carriers are trapped and recombine within those small tetragonal inclusions. As the excitation power increases, the quasi-Fermi levels of the photogenerated carriers move into the conduction band and valence band, resulting in a band filling effect. As the size of the tetragonal inclusions is extremely small, the large blueshift of P1 emission peak and saturation of P1 intensity can be observed. This sort of blueshift of emission peaks has been commonly observed in quantum wells and other confined heterostructures[47,48]. Increasing the temperature to 140 K, similar

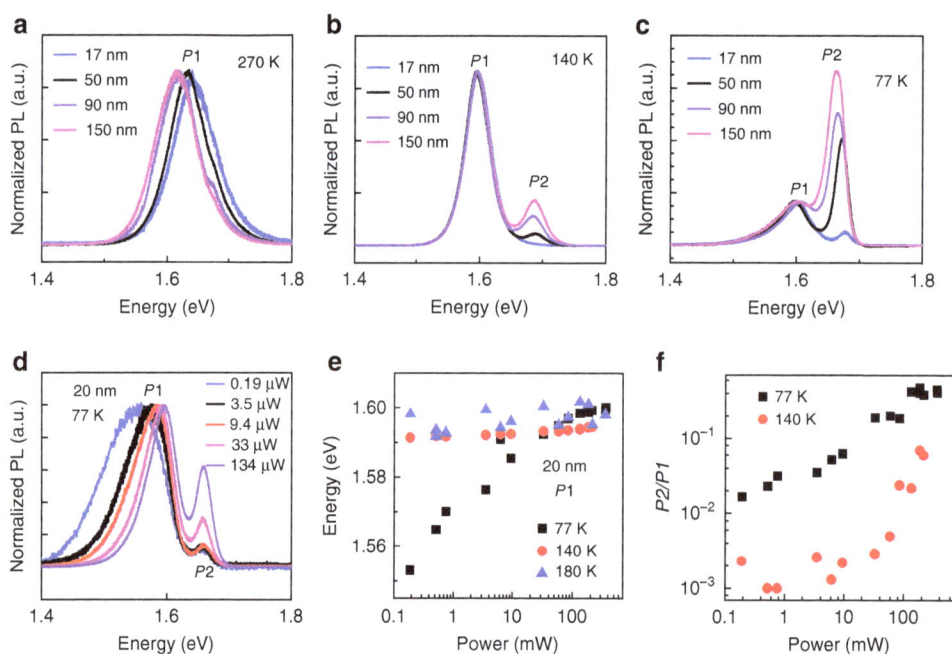

Figure 4 | Thickness and excitation power-dependent photoluminescence studies. (a–c) The photoluminescence spectra for four different thickness halide perovskite microplates at 270 K (a), 140 K (b) and 77 K (c). A 488-nm laser with a power of 3.5 μW was used as the excitation source. All spectra are normalized by the low-energy peak P1 in order to easily compare among each other. (d) The excitation power-dependent PL spectra for a 20-nm-thick microplate at 77 K. The spectra have been normalized by the low-energy emission peak P1. (e) The excitation power-dependent emission peak position of the tetragonal phase for the 20-nm perovskite microplate. (f) The P2/P1 ratios extracted from their corresponding PL spectra under different excitation power at 77 K (black squares) and 140 K (red dots).

trend was observed except that the blueshift of $P1$ becomes negligibly small, which is probably due to the increasing size of tetragonal inclusions. With the increasing size of the tetragonal inclusions, the density of states of the tetragonal inclusions increases accordingly, making it more difficult to observe the band filling effect. At 180 K when the orthorhombic-to-tetragonal phase transition has already completed, no noticeable blueshift was observed. It is expected as the excitation power we used is not big enough such that the band filling effect cannot occur in unconfined systems. The excitation power-dependent $P2/P1$ ratios also clearly demonstrate the band filling effect (Fig. 4f). At 77 K, the $P2/P1$ ratios are very sensitive to the excitation power and show monotonously increases with the increasing excitation power, which indicates the small grains of tetragonal inclusions. At 140 K, the $P2/P1$ ratios are always smaller than that at 77 K and only slightly increase with the excitation power, indicating the increasing size of tetragonal inclusions, which renders the band filling effect hard to be observed. Based on the above discussions, we concluded that the presence of the two emission peaks at low temperatures is due to the small inclusions of tetragonal phase domains within the orthorhombic phase.

The emission peak energy shows a blueshift with the decreasing microplate thickness both for tetragonal phase above 155 K and the orthorhombic phase below 140 K (Figs 4 and 5, and Supplementary Figs 7 and 8), which has been observed in solution-processed perovskite nanocrystals[49,50]. As the thickness of our microplates is much larger than the bulk exciton Bohr radius (2.2 nm) (ref. 51), the quantum confinement effect is unlikely to be the primary factor responsible for this blueshift. Surface effect has been previously proposed to explain such blueshift beyond the quantum confinement regime[52]. In brief, the surface charge-induced depletion electric field near the surfaces or interfaces modifies the confinement potential, leading to a potential well smaller than the actual geometric thickness of the microplates. Our observed blueshift in halide perovskite microplates might be due to the surface effect as well. Nevertheless, the exact underlying mechanism is still unclear and demands further investigation.

Thickness-dependent phase transition in PL spectra. The $P2/P1$ ratios can be used to identify the degree of the phase transition. From $P2/P1$ ratios (Fig. 5a,b and Supplementary Fig. 8), we can conclude that the orthorhombic-to-tetragonal phase transition occurs at a lower transition temperature $T_{t\text{-}o}$ in the thinner microplates, which is consistent with the conclusion obtained from the charge transport measurement. The temperature-dependent PL spectra indicates that the portion of tetragonal inclusions within the orthorhombic phase decreases with the decreasing temperature for all four different thickness microplates, which is supported by the increases of the $P2/P1$ intensity ratio with decreasing temperature (Fig. 5a and Supplementary Fig. 8). The transition temperature $T_{t\text{-}o}$ decreases with the decreasing thickness: $T_{t\text{-}o} < 140$ K for the thickness smaller than 40 nm, and between 140 and 150 K for the thickness around 40–200 nm. Within the respective tetragonal phase (155–290 K) and orthorhombic phase (77–140 K), the emission peak shows a redshift and the full-width at half-maximum (FWHM) narrows as the temperature decreases (Fig. 5c–f). The counter-intuitive redshift of the emission peak with decreasing temperature is strikingly different from the traditional semiconductors, where the emission peak blushifts with the decreasing temperature. This anomalous temperature-dependent band gap remains elusive and demands further investigations. The reducing FWHM with the decreasing temperature can be attributed to the weaker electron–phonon interaction at the lower temperature. For the tetragonal inclusions within the orthorhombic phase, the emission peak energy shows blueshift with the decreasing temperature (Fig. 5c), which is probably due to the increasing quantum confinement effect in the small tetragonal domains. As the temperature decreases, the size of the tetragonal inclusions decreases, leading to a stronger quantum confinement effect and thus a blueshift of emission peak. Furthermore, the FWHM of tetragonal inclusions increases with the decreasing temperature (Fig. 5e), which might be due to the size variation of the tetragonal inclusions. Therefore, the trend of the peak position and FWHM can be used to identify the phase transition points as well (Fig. 5c,e).

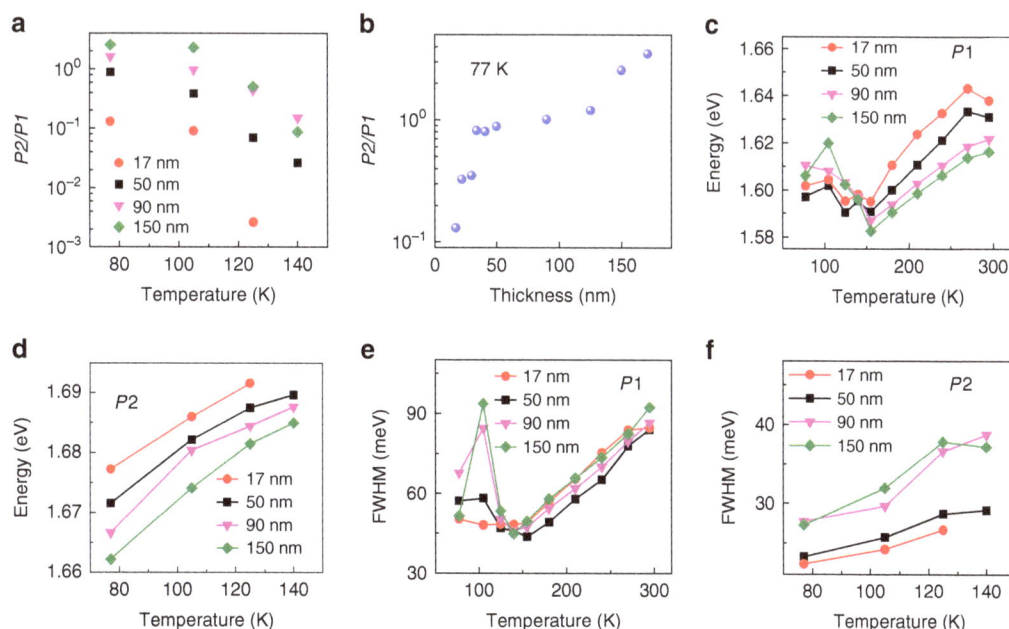

Figure 5 | The temperature- and thickness-dependent photoluminescence studies. (a) The temperature-dependent $P2/P1$ ratios for a 17-, 50-, 90- and 150-nm-thick halide perovskite microplates excited by a 488-nm laser with a power of 3.5 μW. (b) The $P2/P1$ ratios for halide perovskite microplates with various thicknesses at 77 K. (c,d) The temperature-dependent emission energy for tetragonal phase (c) and orthorhombic phase (d). (e,f) The temperature-dependent full-width at half-maximum (FWHM) for tetragonal phase (e) and orthorhombic phase (f).

Discussion

The size-dependent shift in the phase transition temperature has been extensively investigated and observed in confined systems such as nanocrystal[30,53] and two-dimensional layered materials (NbSe$_2$ (ref. 54) and TaS$_2$ (ref. 55)). A reduction in the nanomaterial size can lead to the decrease of the structural phase transition temperature. Many competing theories have been proposed to explain this behaviour, which includes the lack of nucleation sites, internal pressure and surface energy difference between polymorphs[53]. The rough surface in our microplates can exclude the possibility of the lack of nucleation sites. The internal strain would lead to the broadening of the emission peak and thus a larger FWHM in the thinner microplates. Nevertheless, the similar or smaller FWHM in the thinner microplates (Fig. 5e,f) implies that the strain effect should not play a dominant role in our case. The thickness-dependent phase change temperature $T_{t\text{-}o}$ observed here is more likely due to the surface energy difference between the polymorphs. As the thickness decreases, the surface-to-volume ratio increases, resulting in the lower transition temperature $T_{t\text{-}o}$ in the thinner microplates. This explanation is consistent with the blueshift of the PL emission peak with the decreasing of the thickness due to the surface effect (Supplementary Fig. 7).

In summary, we have systematically investigated the size-dependent phase transition in individual MA lead iodide microplate using temperature-dependent PL spectroscopy, charge transport measurement and TEM studies. Our studies demonstrate that the orthorhombic-to-tetragonal phase transition temperature $T_{t\text{-}o}$ decreases with the decreasing thickness of the perovskite microplates, and confirm the phase transition is a first-order solid–solid phase transition. In addition to the fundamental importance, the thickness-dependent structural phase transition has important practical implications in the fields ranging from electronics, optoelectronics to materials sciences. Our findings on the thickness- and temperature-dependent optical and electric properties can shed light on the development of electronic and optoelectronic devices not just at room temperature but also at low temperature, which would have important applications in airplanes and satellites[56].

Methods

Sample preparations. A dilute PbI$_2$ aqueous solution (0.1 g per 100 ml) prepared at 80 °C was cooled to room temperature, which leads to the formation of suspended PbI$_2$ microplates. For the PL measurement samples, we dipped the substrates (Si substrates with 300 nm SiO$_2$) with pre-fabricated markers by photolithography into the aqueous solution for a few seconds. After taking the substrate from the solution, we can find various thickness microplates by chance. For the FET samples, the 5 nm Cr/50 nm Au (Pt) electrodes with channel lengths of 8 and 40 μm were defined by photolithography and followed by thermal evaporation and lift-off. Then PbI$_2$ microplates were grown onto the pre-fabricated electrodes by randomly dispersion. The prepared PbI$_2$ microplates were converted into CH$_3$NH$_3$PbI$_3$ by vapour phase intercalation. The intercalation source (MA iodide powder) was synthesized by a solution method[57]. The MA iodide source was placed at the centre of a quartz tube and the substrate with PbI$_2$ microplates was placed 5–6 cm away downstream. Before conversion, the tube furnace was vacuumed and refilled with argon for at least three times to completely remove the air in the quartz tube. The conversion was conducted at a pressure of 100 mbar with 100 s.c.c.m. argon flow as carrier gas for several hours. The actual temperature is 140 °C at the MA iodide source region and 120 °C at the PbI$_2$ micro-plate substrate region measured by a thermocouple probe. Finally, the tube was naturally cooled down to room temperature.

Fabrication of graphene-contact FETs. To fabricate the graphene contact devices, graphene strips (as electrodes) was peeled on clean silicon/silicon oxide (300 nm) substrate, whereas the PbI$_2$ plates and the top hexagonal boron nitride (hBN) was peeled on polymer stack PMMA/PPC (polypropylene carbonate) spun on a silicon wafer. First, the peeled PbI$_2$ plate was aligned and transferred onto the graphene strips and the PMMA/PPC was dissolved by using chloroform solution. Then, the PbI$_2$ plate was converted to CH$_3$NH$_3$PbI$_3$ by using the vapour phase intercalation. Afterwards, the top BN was aligned and transferred to protect the perovskite for the following electrode fabrication processes. To make edge contact to graphene,

the windows on hBN were first defined by electron-beam lithography exactly upon the graphene stripes and followed by the plasma etching to remove hBN. Afterwards, electron beam lithography was used to pattern the edge contact to graphene and followed by thermal evaporation and lift-off.

Microscopic and optical characterizations. The thickness of the perovskite microplates was determined by tapping-mode atomic force microscopy (Vecco 5,000 system). TEM images and SAED patterns were acquired in an FEI Titan high-resolution transmission microscopy. The PL measurement was conducted under a confocal micro-Raman system (Horiba LABHR) equipped with a 600 g mm^{-1} grating in a backscattering configuration excited by an Ar ion laser (488 nm). For the low-temperature measurement, a liquid nitrogen continuous flow cryostat (Cryo Industry of America) was used to control the temperature from 77 to 300 K.

Electrical measurements. Temperature-dependent FET device measurements were carried out in a probe station ((Lakeshore, TTP4) coupled with a precision source/measurement unit (Agilent B2902A). The scanning rate for the transport measurement is 20 V s^{-1} and the devices were pre-biased at the opposite voltage for 30 s before each measurement.

References

1. Green, M. A., Ho-Baillie, A. & Snaith, H. J. The emergence of perovskite solar cells. *Nat. Photon* **8**, 506–514 (2014).
2. Lee, M. M. *et al.* Efficient hybrid solar cells based on meso-superstructured organometal halide perovskites. *Science* **338**, 643–647 (2012).
3. Liu, M., Johnston, M. B. & Snaith, H. J. Efficient planar heterojunction perovskite solar cells by vapour deposition. *Nature* **501**, 395–398 (2013).
4. Burschka, J. *et al.* Sequential deposition as a route to high-performance perovskite-sensitized solar cells. *Nature* **499**, 316–319 (2013).
5. Zhou, H. *et al.* Interface engineering of highly efficient perovskite solar cells. *Science* **345**, 542–546 (2014).
6. Liu, D. & Kelly, T. L. Perovskite solar cells with a planar heterojunction structure prepared using room-temperature solution processing techniques. *Nat. Photon* **8**, 133–138 (2014).
7. Jeon, N. J. *et al.* Solvent engineering for high-performance inorganic–organic hybrid perovskite solar cells. *Nat. Mater* **13**, 897–903 (2014).
8. Grätzel, M. The light and shade of perovskite solar cells. *Nat. Mater* **13**, 838–842 (2014).
9. Lin, Q. *et al.* Electro-optics of perovskite solar cells. *Nat. Photon* **9**, 106–112 (2015).
10. Dong, Q. *et al.* Electron-hole diffusion lengths > 175 μm in solution-grown CH$_3$NH$_3$PbI$_3$ single crystals. *Science* **347**, 967–970 (2015).
11. Shi, D. *et al.* Low trap-state density and long carrier diffusion in organolead trihalide perovskite single crystals. *Science* **347**, 519–522 (2015).
12. Nie, W. *et al.* High-efficiency solution-processed perovskite solar cells with millimeter-scale grains. *Science* **347**, 522–525 (2015).
13. Salim, T. *et al.* Perovskite-based solar cells: impact of morphology and device architecture on device performance. *J. Phys. Chem. A* **3**, 8943–8969 (2015).
14. Dou, L. *et al.* Solution-processed hybrid perovskite photodetectors with high detectivity. *Nat. Commun* **5**, 5404 (2014).
15. Xing, G. *et al.* Low-temperature solution-processed wavelength-tunable perovskites for lasing. *Nat. Mater* **13**, 476–480 (2014).
16. Zhang, Q. *et al.* Room-temperature near-infrared high-Q perovskite whispering-gallery planar nanolasers. *Nano Lett.* **14**, 5995–6001 (2014).
17. Zhu, H. *et al.* Lead halide perovskite nanowire lasers with low lasing thresholds and high quality factors. *Nat. Mater* **14**, 636–642 (2015).
18. Tan, Z.-K. *et al.* Bright light-emitting diodes based on organometal halide perovskite. *Nat. Nanotechnol* **9**, 687–692 (2014).
19. Chin, X. Y. *et al.* Lead iodide perovskite light-emitting field-effect transistor. *Nat. Commun* **6**, 7383 (2015).
20. Mei, Y., Zhang, C., Vardeny, Z. & Jurchescu, O. Electrostatic gating of hybrid halide perovskite field-effect transistors: balanced ambipolar transport at room-temperature. *MRS Commun* **5**, 1–5 (2015).
21. Im, J.-H. *et al.* Growth of CH$_3$NH$_3$PbI$_3$ cuboids with controlled size for high-efficiency perovskite solar cells. *Nat. Nanotechnol* **9**, 927–932 (2014).
22. Loi, M. A. & Hummelen, J. C. Hybrid solar cells: perovskites under the Sun. *Nat. Mater* **12**, 1087–1089 (2013).
23. Wehrenfennig, C. *et al.* Charge carrier recombination channels in the low-temperature phase of organic-inorganic lead halide perovskite thin films. *APL Mater* **2**, 081513 (2014).
24. Weller, M. T. *et al.* Complete structure and cation orientation in the perovskite photovoltaic methylammonium lead iodide between 100 and 352K. *Chem. Commun* **51**, 4180–4183 (2015).
25. Onoda-Yamamuro, N., Matsuo, T. & Suga, H. Calorimetric and IR spectroscopic studies of phase transitions in methylammonium trihalogenoplumbates (II)†. *J. Phys. Chem. Solids* **51**, 1383–1395 (1990).

26. Baikie, T. *et al.* Synthesis and crystal chemistry of the hybrid perovskite ($CH_3 NH_3$) PbI_3 for solid-state sensitised solar cell applications. *J. Phys. Chem. A* **1**, 5628–5641 (2013).

27. Stoumpos, C. C., Malliakas, C. D. & Kanatzidis, M. G. Semiconducting tin and lead iodide perovskites with organic cations: phase transitions, high mobilities, and near-infrared photoluminescent properties. *Inorg. Chem.* **52**, 9019–9038 (2013).

28. Chen, C.-C., Herhold, A., Johnson, C. & Alivisatos, A. Size dependence of structural metastability in semiconductor nanocrystals. *Science* **276**, 398–401 (1997).

29. Tolbert, S. & Alivisatos, A. Size dependence of a first order solid-solid phase transition: the wurtzite to rock salt transformation in CdSe nanocrystals. *Science* **265**, 373–373 (1994).

30. Rivest, J. B. *et al.* Size dependence of a temperature-induced solid–solid phase transition in copper (I) sulfide. *J. Chem. Phys. Lett.* **2**, 2402–2406 (2011).

31. Zhang, Y. Gate-tunable phase transitions in thin flakes of 1T-TaS$_2$. *Nat. Nanotechnol* **10**, 270–276 (2015).

32. Wu, K. *et al.* Temperature-dependent excitonic photoluminescence of hybrid organometal halide perovskite films. *Phys. Chem. Chem. Phys.* **16**, 22476–22481 (2014).

33. Eames, C. *et al.* Ionic transport in hybrid lead iodide perovskite solar cells. *Nat. Commun* **6**, 7497 (2015).

34. Frost, J. M., Butler, K. T. & Walsh, A. Molecular ferroelectric contributions to anomalous hysteresis in hybrid perovskite solar cells. *APL Mater* **2**, 081506 (2014).

35. Onoda-Yamamuro, N., Matsuo, T. & Suga, H. Dielectric study of $CH_3NH_3PbX_3$ (X = Cl, Br, I). *J. Phys. Chem. Solids.* **53**, 935–939 (1992).

36. Siemons, W. *et al.* Dielectric-constant-enhanced Hall mobility in complex oxides. *Adv. Mater.* **24**, 3965–3969 (2012).

37. Xiao, Z. *et al.* Giant switchable photovoltaic effect in organometal trihalide perovskite devices. *Nat. Mater* **14**, 193–198 (2015).

38. Quarti, C. *et al.* The Raman spectrum of the $CH_3NH_3PbI_3$ hybrid perovskite: interplay of theory and experiment. *J. Chem. Phys. Lett.* **5**, 279–284 (2013).

39. Brivio, F. *et al.* Lattice dynamics and vibrational spectra of the orthorhombic, tetragonal, and cubic phases of methylammonium lead iodide. *Phys. Rev. B* **92**, 144308 (2015).

40. Wehrenfennig, C. *et al.* High charge carrier mobilities and lifetimes in organolead trihalide perovskites. *Adv. Mater.* **26**, 1584–1589 (2014).

41. Esenturk, O., Melinger, J. S. & Heilweil, E. J. Terahertz mobility measurements on poly-3-hexylthiophene films: device comparison, molecular weight, and film processing effects. *J. Appl. Phys.* **103**, 023102 (2008).

42. Vukmirović, N. *et al.* Insights into the charge carrier terahertz mobility in polyfluorenes from large-scale atomistic simulations and time-resolved terahertz spectroscopy. *J. Chem. Phys. C* **116**, 19665–19672 (2012).

43. Jang, J., Liu, W., Son, J. S. & Talapin, D. V. Temperature-dependent Hall and field-effect mobility in strongly coupled all-inorganic nanocrystal arrays. *Nano Lett.* **14**, 653–662 (2014).

44. Ni, N. *et al.* First-order structural phase transition in CaFe$_2$As$_2$. *Phys. Rev. B* **78**, 014523 (2008).

45. Sethna, J. P. *et al.* Hysteresis and hierarchies: dynamics of disorder-driven first-order phase transformations. *Phys. Rev. Lett.* **70**, 3347–3350 (1993).

46. Even, J., Pedesseau, L. & Katan, C. Analysis of multivalley and multibandgap absorption and enhancement of free carriers related to exciton screening in hybrid perovskites. *J. Chem. Phys. C* **118**, 11566–11572 (2014).

47. Li, D. *et al.* Strain-induced spatially indirect exciton recombination in zinc-blende/wurtzite CdS heterostructures. *Nano Res* **8**, 3035–3044 (2015).

48. Liu, Q. *et al.* Evidence of type-II band alignment at the ordered GaInP to GaAs heterointerface. *J. Appl. Phys.* **77**, 1154–1158 (1995).

49. Di, D. *et al.* Size-dependent photon emission from organometal halide perovskite nanocrystals embedded in an organic matrix. *J. Chem. Phys. Lett.* **6**, 446–450 (2015).

50. D'Innocenzo, V. *et al.* Tuning the light emission properties by band gap engineering in hybrid lead-halide perovskite. *J. Am. Chem. Soc.* **136**, 17730–17733 (2014).

51. Tanaka, K. *et al.* Comparative study on the excitons in lead-halide-based perovskite-type crystals $CH_3NH_3PbBr_3$ $CH_3NH_3PbI_3$. *Solid State Commun.* **127**, 619–623 (2003).

52. Li, D., Zhang, J. & Xiong, Q. Surface depletion induced quantum confinement in CdS nanobelts. *ACS Nano* **6**, 5283–5290 (2012).

53. Mayo, M., Suresh, A. & Porter, W. Thermodynamics for nanosystems: grain and particle-size dependent phase diagrams. *Rev. Adv. Mater. Sci.* **5**, 100–109 (2003).

54. Xi, X. *et al.* Strongly enhanced charge-density-wave order in monolayer NbSe$_2$. *Nat. Nanotechnol* **10**, 765–769 (2015).

55. Yu, Y. *et al.* Gate-tunable phase transitions in thin flakes of 1T-TaS$_2$. *Nat. Nanotechnol* **10**, 270–276 (2015).

56. La-o-vorakiat, C. *et al.* Elucidating the role of disorder and free-carrier recombination kinetics in $CH_3NH_3PbI_3$ perovskite films. *Nat. Commun* **6**, 7903 (2015).

57. Heo, J. H. *et al.* Efficient inorganic-organic hybrid heterojunction solar cells containing perovskite compound and polymeric hole conductors. *Nat. Photonics* **7**, 487–492 (2013).

Acknowledgements

We acknowledge the support from the US Department of Energy, Office of Basic Energy Sciences, Division of Materials Science and Engineering through Award DE-SC0008055.

Author contributions

X.D. and Y.H. designed the experiments. D.L. performed most of the experiments including device fabrication, electric, PL measurement and data analysis. G.W. synthesized the materials. H.-C.C., H.W. and Y.L. contributed to device fabrication. C.-Y.C. conducted the TEM studies. X.D. and D.L. co-wrote the paper. All authors discussed the results and commented on the manuscript.

Additional information

18

Efficient linear phase contrast in scanning transmission electron microscopy with matched illumination and detector interferometry

Colin Ophus[1], Jim Ciston[1], Jordan Pierce[2], Tyler R. Harvey[2], Jordan Chess[2], Benjamin J. McMorran[2], Cory Czarnik[3], Harald H. Rose[4] & Peter Ercius[1]

The ability to image light elements in soft matter at atomic resolution enables unprecedented insight into the structure and properties of molecular heterostructures and beam-sensitive nanomaterials. In this study, we introduce a scanning transmission electron microscopy technique combining a pre-specimen phase plate designed to produce a probe with structured phase with a high-speed direct electron detector to generate nearly linear contrast images with high efficiency. We demonstrate this method by using both experiment and simulation to simultaneously image the atomic-scale structure of weakly scattering amorphous carbon and strongly scattering gold nanoparticles. Our method demonstrates strong contrast for both materials, making it a promising candidate for structural determination of heterogeneous soft/hard matter samples even at low electron doses comparable to traditional phase-contrast transmission electron microscopy. Simulated images demonstrate the extension of this technique to the challenging problem of structural determination of biological material at the surface of inorganic crystals.

[1] National Center for Electron Microscopy, Molecular Foundry, Lawrence Berkeley National Laboratory, 1 Cyclotron Road, Berkeley, California 94720, USA. [2] Department of Physics, University of Oregon, 1585 E 13th Avenue, Eugene, Oregon 97403, USA. [3] Gatan Inc., 5794 W Las Positas Boulevard, Pleasanton, California 94588, USA. [4] Department of Physics, Center for Electron Microscopy, Ulm University, Albert-Einstein-Allee 11, 89069 Ulm, Germany. Correspondence and requests for materials should be addressed to C.O. (email: cophus@gmail.com) or to P.E. (email: percius@lbl.gov).

Structural analysis at atomic resolution is commonly used to provide deep insight into the functionality of structures in both biological and physical sciences. Recent examples include the enormous progress in dynamic structural biology[1], direct imaging of screw dislocations via optical sectioning[2], measurements of TiO_6 octahedra in perovskite superlattices[3] and many others. Transmission electron microscopy (TEM) and scanning TEM (STEM) are ubiquitous techniques for high-resolution analysis of both hard and soft matter structures due to the focusing capabilities of electron optics. The theoretical resolution limit for traditional TEM and STEM is 1–2 Å, which has further been extended to 0.5 Å with aberration correction[4–6]. This has significantly improved quantitative structural analysis in both TEM and STEM to sub-Å resolution and single-picometre precision in materials science where many materials can tolerate very high electron doses. Compared with STEM, TEM phase-contrast imaging is overwhelmingly preferred by the biological community because it provides an efficient means of imaging weak-phase objects at doses at or below $10\,e^-\,Å^{-2}$. Resolution for biological materials is limited by the achievable signal-to-noise ratio (SNR) before the target structure is damaged or destroyed, rather than TEM information transfer[7–11]. The primary method currently used to solve the structure of dose-sensitive samples is single-particle reconstruction, from cryo-electron microscopy (cryo-EM). This method is very effective, but typically requires many thousands of identical particles isolated from each other so that the defocused signals from adjacent particles do not interfere. Further, information transfer depends on the defocus used, which has a large nonlinear effect on the contrast in the final image[12–15]. This effect is especially prevalent for phase-contrast high-resolution TEM (HRTEM) and requires careful inspection, numerical aberration correction and/or computer simulation for direct structural interpretation.

Image interpretation in STEM is typically simpler. The contrast and efficiency of STEM is controlled by the geometry of the post specimen, monolithic detectors which simply integrate over specific scattering angles in reciprocal space. The two most common STEM imaging techniques are annular dark field (ADF) that can produce incoherent image contrast roughly proportional to the projected mass thickness of the sample and bright field (BF), which can produce coherent image contrast similar to traditional TEM[15,16]. An alternative method called annular bright field is a new technique designed to directly image weakly scattering materials such as lithium and oxygen, but this imaging technique is difficult to optimize since the detector inner/outer angle ratio is set by the physical detector size[17,18]. ADF-STEM with incoherent image contrast is commonly used to produce interpretable images at atomic resolution based on high-angle scattering, but this process is relatively inefficient per incident electron, commonly requiring significantly more dose per unit area compared with TEM. This is especially limiting for light elements such as carbon, which scatter electrons very weakly. Thus, STEM is more commonly used for materials science and infrequently in biological sciences[19,20].

Coherent, phase-contrast imaging in STEM is also possible using a differential measurement of the BF centre disk as initially proposed by Rose[21]. Dekkers and De Lang, Rose and Haider et al.[22–24] also proposed using a STEM probe aberration corrector to form a probe containing a reference wave, which can be directly interpreted using a segmented detector. Ptychography is another dose-efficient phase-contrast method that utilizes full images of the transmitted electron diffraction pattern to reconstruct both the complex (real and imaginary) probe image and complex sample potential. Ptychography was demonstrated at atomic resolution for the first time by Nellist et al.[25],

with recent improvements in both computer algorithms and detectors[26–28]. One recent example of how phase-contrast imaging can be achieved in STEM is given by the work of Pennycook et al.[29,30]. Their method implements an elegant ptychographic reconstruction algorithm that uses subregions recorded on a pixel array detector to form efficient phase-contrast images.

STEM experiments can also be expanded using methods other than advanced detector geometries and computational algorithms. One example is the recent use of structured phase in electron microscopy, typically performed by placing a phase or amplitude plate in the probe-forming aperture to produce an electron probe with the desired properties. Diffraction gratings have been used in STEM to create vortex beams with orbital angular momentum[31–33] and Bessel beams for very long depth-of-field imaging[34].

Although all of these methods are important steps to improving STEM imaging beyond the simple mass-thickness contrast of ADF-STEM, it is highly desirable to develop a high-resolution imaging technique with directly interpretable contrast that can also operate with high efficiency to reduce beam damage. This could expand the use of STEM to solve important questions in the biological field, as well as hybrid hard/soft materials[35,36].

In this manuscript, we present simulations and a proof-of-principle experiment for a new kind of phase-contrast electron microscopy called matched illumination and detector interferometry (MIDI)–STEM. MIDI–STEM combines the concepts of phase gratings, aberration correction, high-speed pixelated direct electron detectors and phase reconstruction using an interference pattern (such as in electron holography). The MIDI–STEM method produces almost ideal linear phase-contrast images over a wide range of spatial frequencies with very high efficiency and could potentially be used to image soft matter and beam-sensitive samples at atomic resolution.

Results

Description of MIDI–STEM experimental set-up. A simplified diagram comparing a conventional STEM (Fig. 1a) and MIDI–STEM (Fig. 1b) experimental set-up is shown in Fig. 1. In conventional STEM experiments, the probe is formed by a plane wave incident on a circular condenser aperture. Lens elements are used to create a circular electron beam with (approximately) constant phase, which converges to an atomic-scale probe at the sample plane. Electromagnetic deflectors scan the probe over the sample surface in a 2D grid pattern. As shown in Fig. 1a, post specimen, monolithic detectors integrate over regions of the scattered (dark field) or unscattered (bright field) electron diffraction pattern. Two common detector configurations are shown, an ADF detector and a BF detector.

In a MIDI–STEM experiment, diagrammed in Fig. 1b, a patterned phase plate is placed at the probe-forming aperture position. The phase plate consists of alternating concentric trenches with equal area where a thin SiN film has been patterned by a focused ion beam. Each alternating ring applies either a 0 or $\pi/2$-phase shift due to the local SiN thickness. This phase plate generates a probe with a built-in reference wave, which is then scanned across the sample as in traditional STEM imaging. In this case, the high-angle scattering signal is recorded by a traditional ADF detector, and a pixelated direct electron detector is used to record an image of the transmitted centre beam at each scanned position. The ADF detector produces a signal that is very similar to a conventional ADF-STEM experiment. The images of the transmitted centre beam are processed by fitting a virtual detector to match the

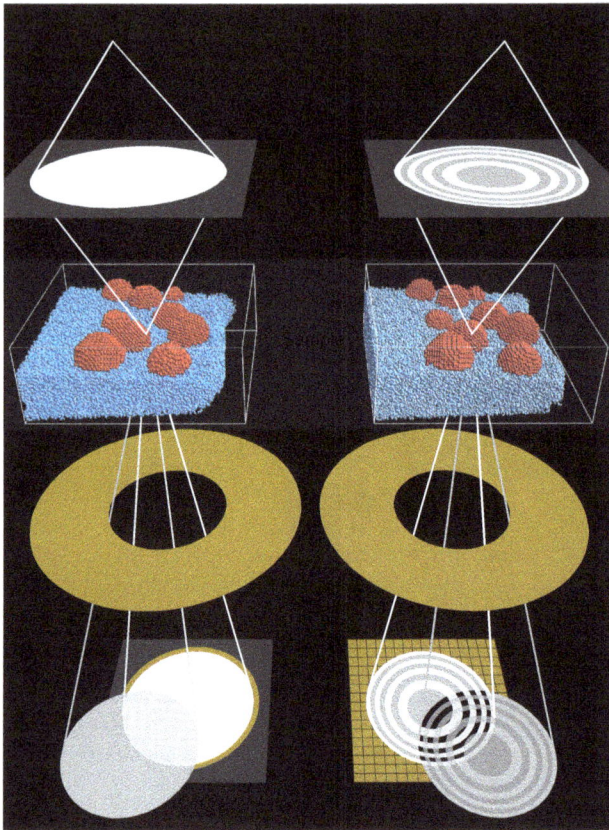

Figure 1 | Experimental set-up for STEM experiments. (a) Conventional set-up, with a round probe-forming aperture and monolithic, single-pixel ADF and BF detectors below the sample. **(b)** MIDI–STEM set-up, with a patterned phase plate placed in the probe-forming aperture and a pixelated detector below the sample.

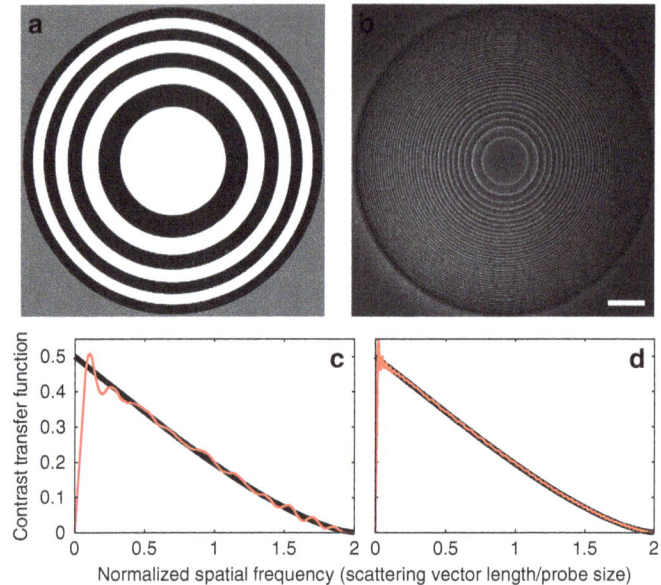

Figure 2 | MIDI–STEM phase plates and resulting CTFs. (a) Schematic of a phase plate with 4 ring pairs, and **(b)** scanning electron microscopy image of the patterned phase plate with 20 ring pairs used in this study. Scale bar, 5 µm. The calculated CTFs for (**a,b**) are plotted in (**c,d**), respectively. Black diagonal lines show the CTF for an ideal phase-contrast STEM experiment.

geometry of the phase plate producing an approximately linear phase signal. The virtual detector consists of the even and odd-numbered annular rings formed by the phase plate, where the phase signal is given by the difference between the sum of all odd ring intensities minus the sum of all even ring intensities. Precise alignment of the virtual detector rings can be achieved using an image of the centre beam in vacuum or by averaging all diffraction pattern images and fitting ellipses to the ring edges[37]. The ability to match the virtual detector to the phase-plate geometry using post processing makes MIDI–STEM highly flexible to compensate for any errors in the phase plate itself or in the scanning electronics. It is also capable of utilizing almost any pre-specimen phase-plate design. Further technical details of the MIDI–STEM model are given in Supplementary Note 1 and Supplementary Fig. 1.

The contrast transfer function (CTF) of a microscopy technique describes the measured contrast as a function of the scattering angle or spatial frequency[15]. A monotonically decreasing CTF that passes both low and high spatial frequencies is desirable for easy image interpretation. The CTF of MIDI–STEM can be calculated from the overlap region of the 0 and $\pi/2$ regions of the probe for the unscattered centre disk and a scattered disk. An example of a MIDI–STEM CTF is plotted in Fig. 2c, where the phase-plate geometry is shown in Fig. 2a. The scanning electron microscopy image in Fig. 2b shows the exact phase plate used in this study, which produces the CTF plotted in Fig. 2d. The geometric construction used to calculate MIDI–STEM CTFs is shown in more detail in Supplementary Figs 2, 3 and 4 and Supplementary Note 2.

A MIDI–STEM experiment. We performed an MIDI–STEM experiment to image a highly heterogeneous sample consisting of randomly oriented cuboctahedral gold nanoparticles (NPs) supported on a thin amorphous carbon film to demonstrate the linear imaging capabilities of this technique. The average image of the centre beam from all probe positions is plotted in Fig. 3a, with the fitted edges of the virtual detector overlaid on the right half of the image as red lines. Note that the contrast has been scaled up to make the phase-plate rings visible. The ADF detector is not visible at this contrast level, but was positioned such that the inner detector angle was just beyond the edge of the outer-most phase-plate ring at 20 mrad.

In Fig. 3b,c, we show the MIDI–STEM image reconstructed using the matched virtual detector and the simultaneously recorded ADF detector image, respectively. The ADF image shows strong contrast for the gold NPs, and the atomic planes are visible in several NPs typically with the (111) plane spacing. The carbon support is very faintly visible in the ADF image, and though it can be distinguished from vacuum, no structural information can be obtained.

The MIDI–STEM image shown in Fig. 3b also shows good contrast for the NPs, with a similar SNR for the atomic planes as the ADF image in Fig. 3c. Furthermore, the MIDI–STEM image also shows very strong contrast for the carbon support, especially at the vacuum edge. The ADF image was used to color the area occupied by the gold NPs in Fig. 3d to emphasize the surrounding carbon structure. Inside the carbon film, we observe regions of correlated intensity between adjacent pixels in both the fast (horizontal) and slow (vertical) directions. Because each image pixel is a separate probe position representing a completely independent measurement, we ascribe these features to the atomic clustering characteristics of filament-like structures known to exist in amorphous carbon[38]. The gold NPs also have significant additional contrast that we interpret as amorphous carbon clustering around the particles. The source of this carbon could be from the sample fabrication process, from the surrounding substrate, or contamination from previous STEM

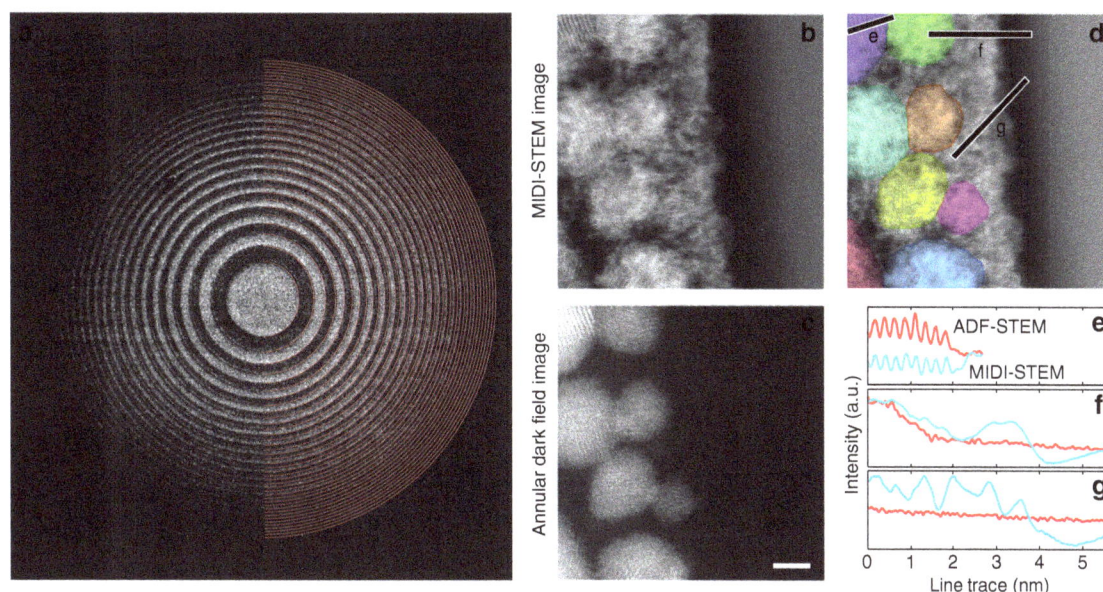

Figure 3 | MIDI–STEM experiment of a heterogeneous sample. (a) The image of the centre beam averaged from all diffraction patterns. Simultaneously, recorded images of gold NPs on a thin carbon support (**b**) using a virtual detector (edges outlined by red lines) shown in (**a,c**) using a conventional ADF-STEM detector. Scale bar, 2 nm. (**d**) Same as (**b**), with gold NPs shaded with random colours to emphasize the surrounding carbon. (**e–g**) Line traces with positions shown in (**d**) for images (**b,c**).

scans used for focusing. This additional contrast is not visible in the ADF images (except perhaps as some weak 'fuzziness') demonstrating that ADF imaging suppresses weakly scattering atoms such as carbon, which makes samples appear to be cleaner than they really are. ADF-STEM is essentially biased towards highly scattering materials, and this experiment demonstrates the capabilities of MIDI–STEM to simultaneously image low- and high-scattering materials.

Three line traces taken from the MIDI–STEM and ADF-STEM images of Fig. 3b,c are plotted in Fig. 3e–g. Figure 3e shows that both MIDI- and ADF-STEM are sensitive to the atomic lattice planes of the NPs, with approximately the same SNR. The left side of Fig. 3f shows that both methods show thickness contrast for an off-zone-axis NP, but the carbon substrate edge at the vacuum is essentially invisible in the ADF-STEM image, while strongly visible in the MIDI–STEM image. Finally, the trace in Fig. 3g along only amorphous carbon shows strong structural fluctuations in MIDI–STEM, and again no contrast in ADF-STEM. The MIDI–STEM image shows slowly varying intensity in the vacuum because it is a differential phase technique (similar to a high-pass filter) that cannot retrieve the d.c. component or very low spatial frequencies. This is evident in the CTF curves plotted in Fig. 2.

Contrast transfer of MIDI-STEM images from simulation. To validate our experimental results, we have simulated a MIDI–STEM experiment of a similar sample using the multislice method. The projected potential of the sample and atomistic model are shown in Fig. 4a,b. The simulated sample consists of randomly oriented cuboctahedral gold NPs attached to a wedge-shaped substrate of amorphous carbon. The realistic amorphous carbon atomic coordinates from[38] were tiled into a wedge with a maximum thickness of 5 nm on the left side and a minimum thickness at the substrate/vacuum edge of 3 nm. No additional carbon was added on top of the gold NPs as seen in the experimental results. The simulated STEM scan of 300×300 probe positions with a probe position spacing of 0.5 Å was confined to the green box overlaid on the projected potential.

Three detector configurations are considered in Fig. 4c–k; a BF-STEM detector constructed by summing over the central probe disk from 0 to 17 mrad, a (low angle) ADF-STEM detector from 20 to 95 mrad and a MIDI–STEM virtual detector consisting of the electrons recorded on odd rings minus those recorded on even rings. Detector geometries were chosen to match the experimental set-up presented earlier although only experimental ADF- and MIDI–STEM signals were experimentally available. All simulated images are plotted using infinite dose (no noise) and with an intermediate dose of $500\,e^-\,Å^{-2}$. In addition, the images are quantitatively evaluated by plotting the measured signal intensity at infinite dose as a fraction of the total incident electrons versus the projected potential of the sample at each probe position. The pixels are separated into two groups corresponding to probe positions at only amorphous carbon (blue) or probe positions including both carbon and gold (red) in projection. A polynomial trend line (black) was fitted to all points as a guide for the eye. Note that at an accelerating voltage of 300 kV, a projected potential of 1,500 V Å roughly corresponds to a π-phase shift of the incident electron wave, and therefore this specimen does not obey the weak-phase approximation.

The BF- and ADF-STEM simulations are essentially complementary, as expected when using such a low inner angle for the ADF detector. Both show strong contrast for the gold NPs and weak contrast for the carbon substrate at infinite electron dose. However, when using a dose of only $500\,e^-\,Å^{-2}$, the BF-STEM signal is overwhelmed by noise; only faint outlines of the NPs are visible and the substrate is almost invisible. The ADF-STEM image produces relatively better contrast at lower dose, as the NPs show high contrast both for atomic columns and atomic planes. In this ADF-STEM image, the substrate can be differentiated from the vacuum, but no structural information can be obtained. A more typical high-angle ADF-STEM image with a large detector inner angle (>50 mrad) of this sample produces a slightly better image of the gold, but significantly less contrast for the carbon substrate.

The simulated MIDI–STEM image by comparison shows very strong contrast for both the NPs and the amorphous substrate. Even at an electron dose much lower than typical STEM

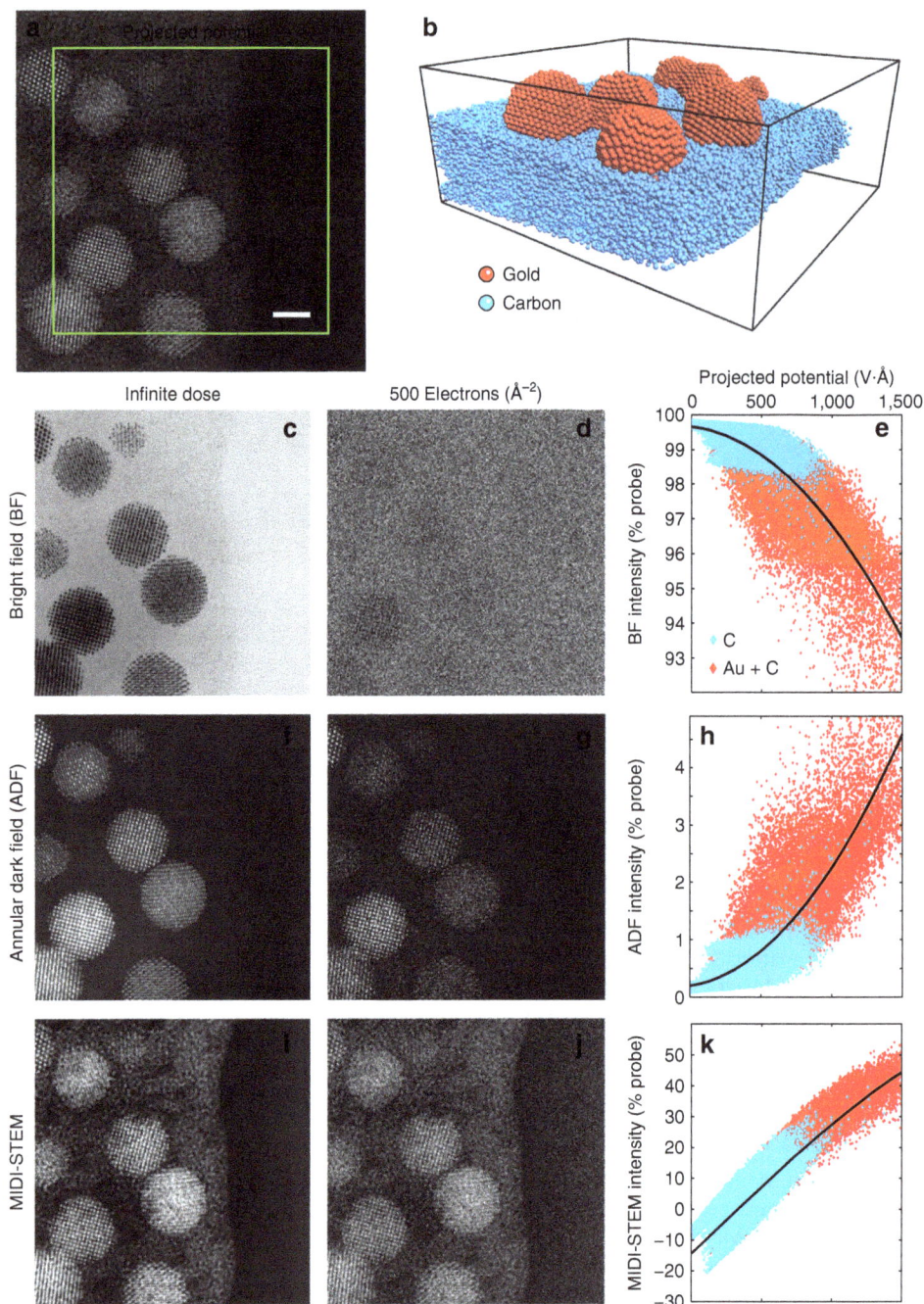

Figure 4 | MIDI–STEM simulations of a heterogeneous sample. (**a**) The projected potential and (**b**) corresponding atomistic three-dimensional model. Scale bar, 2 nm. Images were generated from scanning over the area enclosed in the green box using two common STEM imaging modes, (**c–e**) BF-STEM and (**f–h**) ADF-STEM, and (**i–k**) MIDI–STEM. Simulations are shown for (**c,f,i**) infinite and (**d,g,j**) moderate dose. (**e,h,k**) Quantitative comparison between the projected potential and the measured infinite dose signal shows that MIDI–STEM is a significantly more linear measurement than traditional methods.

experiments, atomic positions and planes are visible in the NPs, with contrast roughly equivalent to the ADF-STEM images. However, the contrast of the amorphous substrate has been significantly increased, and the atomic-scale details of the projected potential are visible in the infinite dose images. The finite dose MIDI–STEM image also shows many of the same structural details when comparing with the fine structure of the projected potential. Qualitatively, the efficiency of MIDI–STEM is explained by its ability to measure small scattering events due to the alternating rings of the phase plate. The plot of projected potential versus measured MIDI–STEM signal shows it is far more linear than BF- or ADF-STEM and has

a much tighter distribution of measurements. Importantly, both the carbon and gold plus carbon signals fall on the same roughly linear curve. The primary sources of non-linearity in the MIDI–STEM measurement are the high-pass filtering effect of MIDI–STEM (a minor effect on the scale of a 20 nm field of view) and the decreased number of electrons available to scatter for thicker regions of the sample.

Discussion

It is important to emphasize that the key strengths of this method are its applicability for beam-sensitive soft materials and the advantages of linear contrast transfer towards the study of hard/soft interfaces in materials science. We have reported both the experimental and theoretical validity of MIDI–STEM for a sample of gold NPs on an amorphous carbon support as a very general case of a highly heterogeneous sample as a proof of principle. To explore the limits of the MIDI–STEM method, we performed a multislice simulation of a DNA snippet connecting two gold NPs on a single layer of graphene, plotted in Fig. 5. DNA was chosen for its well-known structure with weak scattering and moderate dose sensitivity. The same microscope parameters and detectors as Fig. 4 were used, and electron doses of infinity, 500 and $100\,e^{-}\text{Å}^{-2}$ were simulated. However, unlike in Fig. 4c–h, in Fig. 5c–h the BF- and ADF-STEM simulations were performed using a conventional STEM probe without a phase plate. As above, neither the BF- nor ADF-STEM images show any appreciable contrast in the DNA section at a non-infinite electron dose. Conversely, the MIDI–STEM images (Fig. 5i–k) show linear contrast even at fairly low electron doses. Even at such low doses, MIDI–STEM produces enough contrast to identify not only the presence or absence of a bio-molecule, but also the shape envelope and orientation, while being in focus. MIDI–STEM is therefore a promising technique for imaging relatively radiation hard bio-molecules and heterostructures such as hard/soft interfaces. For the most dose-sensitive bio-molecules, highly defocused cryo-EM is a more efficient imaging method, but many samples cannot meet the cryo-EM requirements of many well-separated identical structures, with no strongly scattering components. The same atomic coordinates are used for a comparison with phase contrast, defocused HRTEM imaging in Supplementary Note 3 and Supplementary Fig. 5, using the deconvolution methods described in ref. 14. These simulations show that more information can be recovered from HRTEM imaging, but this requires large defocus values that can produce delocalization artifacts.

In summary, we have experimentally demonstrated the MIDI–STEM imaging method with great promise for improving the contrast in STEM images for weakly scattering materials. We also performed multislice simulations of a sample realistically modelled after our experiment to confirm our interpretation of the experimental results. In this experiment, we imaged gold NPs on an amorphous carbon support, using a pixelated direct electron detector to construct the virtual detectors necessary for MIDI–STEM while simultaneously recording an ADF-STEM image. The MIDI–STEM image simultaneously showed atomic-plane contrast for highly scattering gold NPs and the amorphous structure of carbon regions. Structural features on the near-atomic scale were clearly visible in the amorphous carbon film, showing that MIDI–STEM is a promising candidate to directly image samples consisting of both hard and soft matter at atomic or near-atomic resolution using relatively low electron doses. The primary advantages of MIDI–STEM are high signal efficiency, good transfer of low spatial frequency information and the ability to image while in focus to minimize signal delocalization. MIDI–STEM should also allow the possibility of post-acquisition software aberration correction using pychographic methods.

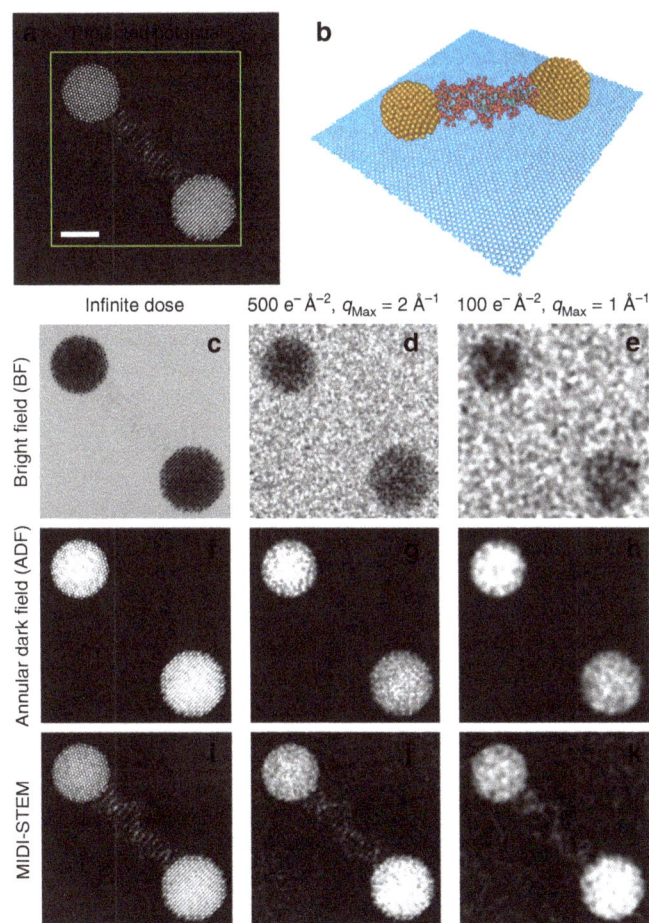

Figure 5 | MIDI–STEM simulations of a biological–inorganic interface.
(**a**) Projected potential of a DNA snippet connecting two gold NPs on a single layer of graphene substrate, and (**b**) corresponding atomistic three-dimensional model. Scale bar, 2 nm. Images were generated from scanning over the area enclosed in the green box using (**c–e**) BF-STEM, (**f–h**) ADF-STEM and (**i–k**) MIDI–STEM at infinite and two low electron doses. Conventional STEM probes were used for BF- and ADF-STEM, while 20 ring pairs were used for MIDI–STEM.

Methods

Experimental. All experimental results presented in this paper were recorded on TEAM I, an aberration-corrected FEI Titan 80–300 operated in STEM mode at 300 kV with a convergence semi-angle of 17.2 mrad. The phase-plate geometry used to form the MIDI–STEM probes were equal-area Fresnel zone plates with 20 ring pairs, fabricated using focused ion beam milling of a SiN membrane[33]. The transmitted electron diffraction pattern at each probe position was recorded using a Gatan K2 IS direct electron detector with $3{,}840 \times 3{,}712$ pixels, operated at 400 frames per second and binned by 2. The camera acquisition and probe scanning were synchronized using a Gatan Digiscan. The probe was scanned over the 14.5 nm field of view with 256×256 probe positions to create a $256 \times 256 \times 1{,}920 \times 1{,}792$ four-dimensional STEM data set consisting of 420 GB of raw images.

Analysis and Simulation. Post processing to fit the virtual detector was done using custom scripts in MATLAB. All multislice simulations were performed using custom MATLAB codes that follows the methods of Kirkland[15], using the same microscope parameters as in the experiment and eight frozen phonon configurations. STEM probes were spaced by 0.5 Å, and the simulation pixel size was 0.2 Å. Experimental and simulated microscope parameters were optimized to give the highest contrast. A model detailing the geometric CTF calculations is given in the Supplementary Notes 1 and 2.

References

1. van den Bedem, H. & Eraser, J. S. Integrative, dynamic structural biology at atomic resolution – it's about time. *Nat. Methods* **12**, 307–318 (2015).
2. Yang, H. *et al.* Imaging screw dislocations at atomic resolution by aberration-corrected electron optical sectioning. *Nat. Commun.* **6**, 7266 (2015).
3. Zhu, Y., Withers, R. L., Bourgeois, L., Dwyer, C. & Etheridge, J. Direct mapping of Li-enabled octahedral tilt ordering and associated strain in nanostructured per-ovskites. *Nat. Mat.* **14**, 1142–1149 (2015).
4. Batson, P. E., Dellby, N. & Krivanek, O. L. Sub-angstrom resolution using aberration corrected electron optics. *Nature* **418**, 617–620 (2002).
5. Rose, H. Prospects for aberration-free electron microscopy. *Ultramicroscopy* **103**, 1–6 (2005).
6. Dahmen, U. *et al.* Background, status and future of the transmission electron aberration-corrected microscope project. *Philos. Trans. R. Soc. Lond. A* **367**, 3795–3808 (2009).

Infinite dose　　　$500\,e^{-}\text{Å}^{-2}$, $q_{Max} = 2\,\text{Å}^{-1}$　　　$100\,e^{-}\text{Å}^{-2}$, $q_{Max} = 1\,\text{Å}^{-1}$

Bright field (BF)

Annular dark field (ADF)

MIDI–STEM

7. Glaeser, R. M. Limitations to significant information in biological electron microscopy as a result of radiation damage. *J. Ultrastruct. Res.* **36**, 466–482 (1971).

8. Isaacson, M., Johnson, D. & Crewe, A. V. Electron beam excitation and damage of biological molecules; its implications for specimen damage in electron microscopy. *Rad. Res.* **55**, 205–224 (1973).

9. Henderson, R. The potential and limitations of neutrons, electrons and x-rays for atomic resolution microscopy of unstained biological molecules. *Quart. Rev. Biophys.* **28**, 171–193 (1995).

10. Egerton, R. E., Li, P. & Malac, M. Radiation damage in the tem and sem. *Micron* **35**, 399–409 (2004).

11. Egerton, R. E. Mechanisms of radiation damage in beam-sensitive specimens, for tem accelerating voltages between 10 and 300 kv. *Micro. Res. Tech.* **75**, 1550–1556 (2012).

12. Erank, J. Single-particle imaging of macromolecules by cryo-electron microscopy. *Ann. Rev. Biophys. Biomol. Struct.* **31**, 303–319 (2002).

13. Frank, J. *Three-dimensional electron microscopy of macromolecular assemblies: visualization of biological molecules in their native state* (Oxford Univ. Press, 2006).

14. Downing, K. H. & Glaeser, R. M. Restoration of weak phase-contrast images recorded with a high degree of defocus: the twin image problem associated with CTE correction. *Ultramicroscopy* **108**, 921–928 (2008).

15. Kirkland, E. J. *Advanced computing in electron microscopy* (Springer Science & Business Media, 2010).

16. Pennycook, S. J. & Nellist, P. D. *Scanning transmission electron microscopy: imaging and analysis* (Springer Science & Business Media, 2011).

17. Eindlay, S. D. *et al.* Dynamics of annular bright field imaging in scanning transmission electron microscopy. *Ultramicroscopy* **110**, 903–923 (2010).

18. Hovden, R. & Muller, D. A. Efficient elastic imaging of single atoms on ultrathin supports in a scanning transmission electron microscope. *Ultramicroscopy* **123**, 59–65 (2012).

19. Muller, D. A. Structure and bonding at the atomic scale by scanning transmission electron microscopy. *Nat. Mat.* **8**, 263–270 (2009).

20. Williams, D. B. & Carter, C. B. *Transmission electron microscopy: a textbook for materials science* Vol. 3 (Springer Science & Business Media, 2009).

21. Rose, H. Phase contrast in scanning transmission electron microscopy. *Optik* **39**, 416–436 (1974).

22. Dekkers, N. H. & De Lang, H. Differential phase contrast in a STEM. *Optik* **41**, 452–456 (1974).

23. Rose, H. Nonstandard imaging methods in electron microscopy. *Ultramicroscopy* **2**, 251–267 (1977).

24. Haider, M., Epstein, A., Jarron, P. & Boulin, C. A versatile, software configurable multichannel STEM detector for angle-resolved imaging. *Ultramicroscopy* **54**, 41–59 (1994).

25. Nellist, P. D., McCallum, B. C. & Rodenburg, J. M. Resolution beyond the'information limit'in transmission electron microscopy. *Nature* **374**, 630–632 (1995).

26. Hue, F., Rodenburg, J. M., Maiden, A. M., Sweeney, F. & Midgley, P. A. Wave-front phase retrieval in transmission electron microscopy via ptychography. *Phys. Rev. B* **82**, 121415 (2010).

27. Humphry, M. J., Kraus, B., Hurst, A. C., Maiden, A. M. & Rodenburg, J. M. Ptychographic electron microscopy using high-angle dark-field scattering for sub-nanometre resolution imaging. *Nat. Commun.* **3**, 730 (2012).

28. Ou, X., Horstmeyer, R., Yang, C. & Zheng, G. Quantitative phase imaging via fourier ptychographic microscopy. *Opt. Lett.* **38**, 4845–4848 (2013).

29. Pennycook, T. J. *et al.* Efficient phase contrast imaging in STEM using a pixelated detector. part 1: experimental demonstration at atomic resolution. *Ultramicroscopy* **151**, 160–167 (2015).

30. Yang, H., Pennycook, T. J. & Nellist, P. D. Efficient phase contrast imaging in STEM using a pixelated detector. part ii: optimisation of imaging conditions. *Ultramicroscopy* **151**, 232–239 (2015).

31. Verbeeck, J., Tian, H. & Schattschneider, P. Production and application of electron vortex beams. *Nature* **467**, 301–304 (2010).

32. McMorran, B. J. *et al.* Electron vortex beams with high quanta of orbital angular momentum. *Science* **331**, 192–195 (2011).

33. Harvey, T. R. *et al.* Efficient diffractive phase optics for electrons. *New J. Phys.* **16**, 093039 (2014).

34. Grillo, V. *et al.* Generation of nondiffracting electron bessel beams. *Phys. Rev. X* **4**, 011013 (2014).

35. Peplow, M. Materials science: the hole story. *Nature* **520**, 148–150 (2015).

36. Hallinan, Jr D. T. & Balsara, N. P. Polymer electrolytes. *Ann. Rev. Mat. Res.* **43**, 503–525 (2013).

37. Fitzgibbon, A., Pilu, M. & Fisher, R. B. Direct least square fitting of ellipses. *IEEE Trans. Pat. Anal. Mach. Int.* **21**, 476–480 (1999).

38. Ricolleau, C., Le Bouar, Y., Amara, H., Landon-Cardinal, O. & Alloyeau, D. Random vs realistic amorphous carbon models for high resolution microscopy and electron diffraction. *J. Appl. Phys.* **114**, 213504 (2013).

Acknowledgements

Work at the Molecular Foundry was supported by the Office of Science, Office of Basic Energy Sciences, of the US Department of Energy under Contract No. DE-AC02-05CH11231. Work at University of Oregon was supported by the US Department of Energy, Office of Science, Basic Energy Sciences under Award No. DE-SC0010466.

Author contributions

C.O. developed the MIDI–STEM method to combine a structured phase electron probe with a matching virtual detector. H.H.R. initially proposed the alternating-zone interference imaging mode used in MIDI–STEM. Ercius performed the experiments, using STEM coil synchronization, recording scripts and four-dimensional STEM experimental protocols developed by J.Ci. and C.C. J.P., T.R.H., J.Ch. and B.J.M. fabricated and tested the phase plates. C.O. ran all multislice simulations and performed the analysis of experiments and simulations. C.O., P.E. and J.Ci. wrote the manuscript with input from all other authors. C.O. wrote the Supplementary Notes with input from T.R.H., J.Ch. and B.J.M.

Additional information

19

Dynamic heterogeneity and non-Gaussian statistics for acetylcholine receptors on live cell membrane

W. He[1], H. Song[2], Y. Su[2], L. Geng[3], B.J. Ackerson[4], H.B. Peng[3] & P. Tong[2]

The Brownian motion of molecules at thermal equilibrium usually has a finite correlation time and will eventually be randomized after a long delay time, so that their displacement follows the Gaussian statistics. This is true even when the molecules have experienced a complex environment with a finite correlation time. Here, we report that the lateral motion of the acetylcholine receptors on live muscle cell membranes does not follow the Gaussian statistics for normal Brownian diffusion. From a careful analysis of a large volume of the protein trajectories obtained over a wide range of sampling rates and long durations, we find that the normalized histogram of the protein displacements shows an exponential tail, which is robust and universal for cells under different conditions. The experiment indicates that the observed non-Gaussian statistics and dynamic heterogeneity are inherently linked to the slow-active remodelling of the underlying cortical actin network.

[1] Nano Science and Technology Program, Hong Kong University of Science and Technology, Clear Water Bay, Kowloon, Hong Kong. [2] Department of Physics, Hong Kong University of Science and Technology, Clear Water Bay, Kowloon, Hong Kong. [3] Division of Life Science, Hong Kong University of Science and Technology, Clear Water Bay, Kowloon, Hong Kong. [4] Department of Physics, Oklahoma State University, Stillwater, Oklahoma 74078, USA. Correspondence and requests for materials should be addressed to P.T. (email: penger@ust.hk).

Cell membranes, which define cell boundaries and maintain communication with the outside world, display an intriguing array of structural complexes of lipids/cholesterols and various proteins essential to the existence and functioning of the cell. In the original fluid mosaic model[1], the cell membrane was thought of as a quasi-two-dimensional fluid layer in which proteins are dispersed randomly at a low concentration and can float unencumbered. From the wealth of new data obtained in recent years, our general view of membrane architecture has evolved into a new paradigm in which the membrane has variable patchiness and thickness and a higher protein occupancy than previously thought[2]. The lipids and proteins on the membrane are not ideally mixed, and form molecular complexes ranging from nano-scale 'lipid rafts'[3,4] and protein clusters to micron-sized stable domains such as caveolae, microvilli and focal adhesions.

Moving in a structured membrane, the proteins do not enjoy continuous and unrestricted lateral diffusion as was originally envisioned[5]. Instead, proteins diffuse in a very complex landscape with considerable lateral heterogeneity in the membrane[6,7]. Transmembrane proteins also interact strongly with the underlying cytoskeletal cortex[4]. Using single-particle tracking (SPT) techniques[7,8], one can directly observe and follow the motion of individual proteins. The measured protein trajectories have been found to be quite heterogeneous[6,7,9], with some moving fast and appearing to diffuse freely while others are transiently confined to small membrane domains. A main issue in the continuing discussion is whether the dynamic heterogeneity of the transmembrane proteins is caused by the effect of clustering imposed by membrane clusters[3,10], such as lipid rafts, or by membrane partitions generated by interactions with the underlying cortical actin network[4], such as 'membrane-skeleton fences'[7,11,12]. Most of the theoretical discussions assumed that membrane organization is governed by equilibrium processes, such as critical thermal fluctuations and ligand-binding equilibrium.

The available SPT data are not conclusive because the protein trajectories were sampled over a relatively short time (due to the finite lifetime of the florescent probes used), and thus heavily influenced by the surrounding molecules without revealing their long-time behaviour and their interactions with distant molecules on the membrane[13]. In addition, the current analysis of protein motion often focuses on identifying only a few targeted single molecular events, while ignoring other molecular events of possibly equal importance owing to the lack of systematic statistical analysis. Such analysis is extremely important, because stochastic fluctuations at the single molecular level are significant[14]. The lack of a systematic analysis of the protein motion is partially due to the fact that direct measurement of the statistical properties, such as the probability density function (normalized histogram or PDF) $P(\Delta x)$ of the protein displacement Δx, often requires a large volume of individual protein trajectories, which are difficult to obtain from living cells. As a result, most previous studies in this area only measured the mean-squared displacement (the lowest moment of $P(\Delta x)$), which requires less statistics but is not adequate to describe the complex motion of proteins in a living cell[15].

In this paper, we report a systematic study of the lateral motion of a transmembrane protein on live muscle cell membranes cultured from *Xenopus* embryos. The protein chosen for the study is acetylcholine receptor (AChR), which is a well characterized neurotransmitter receptor for the study of neuromuscular junctions[16,17]. The lateral mobility of AChRs plays an essential role in determining the response of the postsynaptic membrane to neurotransmitter stimuli. The individual AChRs are labelled by bright and photostable fluorescent quantum dots (QDs). With the

help of an advanced single-molecule tracking algorithm, we are able to obtain a significantly large volume of individual AChR trajectories from more than 360 live cells over a wide range of sampling rates (up to 80 Hz) and long durations (up to 200 s). A central finding of this investigation is that the moving trajectories of the individual AChRs do not follow the Gaussian statistics for normal Brownian diffusion. Instead, we show for the first time that the measured PDF $P(\Delta x)$ has an exponential tail, which is robust and universal for cells under different conditions. A theoretical model is developed to explain why the structurally identical AChRs have very different dynamic behaviours with an exponential-like distribution in their diffusion coefficient.

Results

Characterization of the AChR trajectories. In the experiment, we obtain the AChR trajectories from consecutive images of the QDs, and find their position $\mathbf{r}(t)$ (and hence the position of AChRs) at time t using a homemade SPT program with a spatial resolution of ~ 20 nm. Because the viscosity of the plasma membrane is $\sim 1,000$ times higher than that of the extracellular medium, the motion of the QD-labelled AChRs is determined primarily by their transmembrane domains[18]. From the AChR trajectories, we compute the statistics of the two-dimensional displacement vector, $\Delta\mathbf{r}(\tau) = \mathbf{r}(t+\tau) - \mathbf{r}(t)$, over delay time τ, such as the mean-squared displacement (MSD) $\langle \Delta r^2(\tau) \rangle$ and the PDF $P(\Delta x)$ of the x-component of $\Delta\mathbf{r}$. We also compute the radius of gyration R_g of the AChR trajectories $R_g^2(\tau) = (1/N)\sum_i^N [(x_i - \langle x \rangle)^2 + (y_i - \langle y \rangle)^2]$, where N is the total number of time steps in each trajectory, x_i and y_i are the projection of the position of each trajectory step on the x- and y-axis, respectively, and $\langle x \rangle$ and $\langle y \rangle$ are their mean values. Physically, R_g quantifies the size of an AChR trajectory generated during the time lapse τ.

Figure 1a shows a representative collection of 130 AChR trajectories over a time interval of 60 s. These identical AChRs exhibit a huge amount of dynamic heterogeneity as evidenced by the large variation in trajectory sizes; some being mobile (red trajectories) and others nearly immobile (black trajectories). Among the mobile AChRs, some move fast (with a large trajectory size) and others slower (with a smaller trajectory size). The situation shown in Fig. 1a is in great contrast with the Brownian motion of colloidal particles in a simple fluid, as show in Fig. 1b. The distribution of the Brownian trajectories is much more uniform than that of the AChR trajectories.

To have a quantitative description of the AChR trajectories, we calculate their normalized radius of gyration $R_g' = R_g/\langle R_g \rangle$, where $\langle R_g \rangle$ is the mean value of R_g. For Brownian diffusion, one has $R_g(\tau) = [(2/3)D_0\tau]^{1/2}$ with D_0 being the diffusion coefficient (see Supplementary Note 1 for more details). For live cells, we define $\langle R_g \rangle = [(2/3)\langle D_L \rangle \tau]^{1/2}$, where $\langle D_L \rangle$ is the long-time diffusion coefficient averaged over 365 cells (see more discussions on Fig. 4 below). The use of the normalized R_g' allows us to compare the AChR trajectories taken over different τ and/or under different sample conditions. Figure 2 shows the measured PDF (normalized histogram) $h(R_g')$ of R_g' for the AChR trajectories taken under different sample conditions. All of the measured $h(R_g')$'s collapse onto a single master curve, once the normalized R_g' is used. The PDFs from different frogs and embryos and from cells cultured for different days and sampled at different τ exhibit a universal form (for clarity, some of the curves are not shown here). The measured $h(R_g')$ has a peak at $R_g' \simeq 0.15$ followed by an exponential tail (black solid line). For silica spheres undergoing the Brownian motion, their $h(R_g')$ has a narrow distribution peaked at $R_g' \simeq 1$ (blue dashed line, see Supplementary Note 1 for more discussions). Figure 2 reveals that there are many fast moving AChRs, whose R_g' is larger than that

a

b

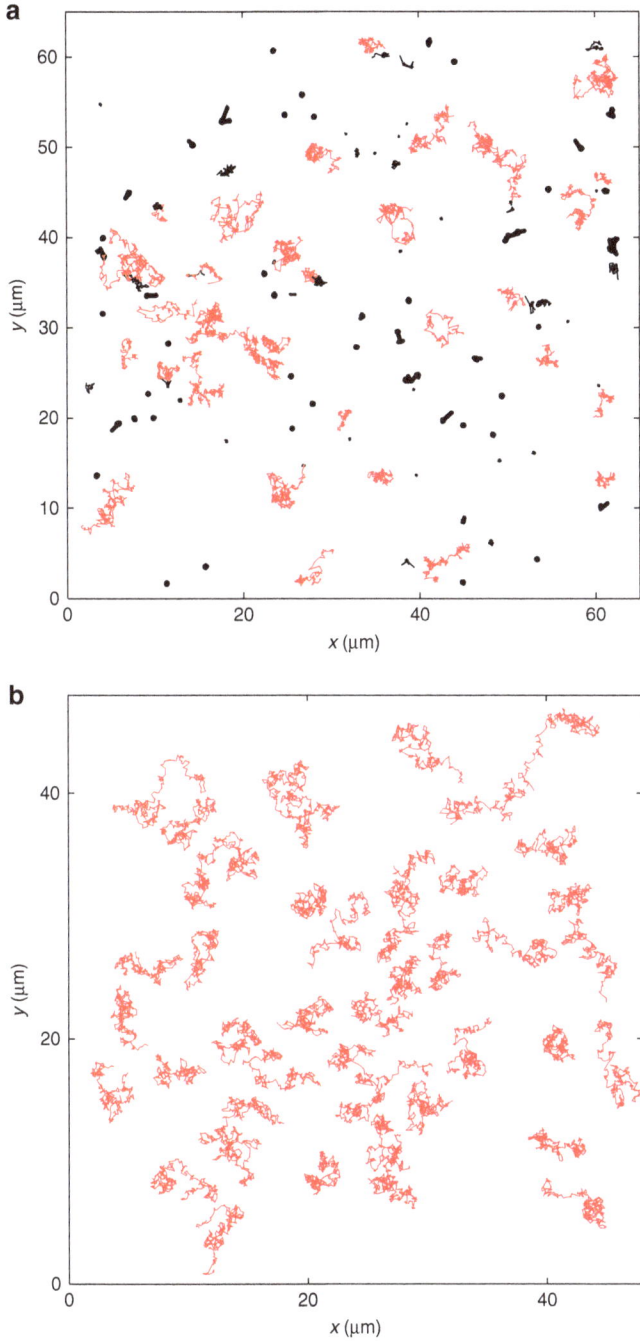

Figure 1 | Observed dynamic heterogeneity in the AChR trajectories.
(**a**) Overall 130 representative AChR trajectories with 300 time steps
(60 s). These trajectories are taken from the bottom membrane of a
Xenopus muscle cell. Red trajectories indicate fast moving AChRs and black
ones indicate 'nearly immobile' AChRs. (**b**) A total of 52 representative
trajectories of silica spheres 2.14 μm in diameter undergoing Brownian
diffusion in water over a flat substrate with 1,000 time steps (47 s).

**Figure 2 | Normalized histogram of the radius of gyration of AChR
trajectories.** Measured PDF $h(R'_g)$ of the normalized R'_g for the AChR
trajectories taken under different sample conditions: cultured for 1 day after
dissection (red circles), cultured for 4 days (magenta triangles), and
cultured for 8 days (green diamonds). Each $h(R'_g)$ is obtained by averaging
the data from 10 cells cultured under the same condition. The black circles
are obtained by averaging the data from 70 cells. Their statistics is
considerably improved with small error bars indicating the standard
deviations. The solid black line shows the exponential function,
$h(R'_g) \simeq 1.1 \exp(-1.35 R'_g)$. The dashed blue line shows the measured $h(R'_g)$
for silica spheres undergoing Brownian diffusion. The vertical red line
indicates the cutoff value $(R'_g)_c = 0.3$ used to define the immobile
trajectories.

for normal Brownian diffusion. The vertical red line indicates the
cutoff value $(R'_g)_c = 0.3$ used in the experiment, below which the
AChR trajectories are treated as immobile ones (black trajectories
in Fig. 1).

Mean-squared displacement. Figure 3 shows the measured MSD
$\langle \Delta \mathbf{r}^2(\tau) \rangle$ as a function of τ for the AChR trajectories taken at two
sampling rates of 80 and 5 frames per second (fps). The red and

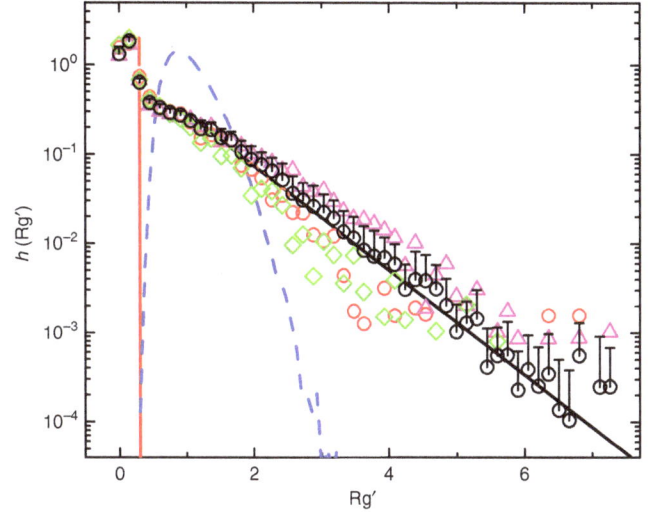

black dashed lines obtained at the two different sampling rates do
not superimpose with each other in the common region of τ
between 0.2 and 3 s. To achieve the higher sampling rate, the
viewing area of the camera is cropped. Because of the spatial
inhomogeneity of the immobile AChR distribution, we find the
number ratio γ of the mobile AChR trajectories to the total
number of trajectories obtained at the two sampling rates is
different. Once the immobile trajectories are removed from the
ensemble average, the measured $\langle \Delta \mathbf{r}^2(\tau) \rangle$ becomes reproducible
and the two curves (red and black circles) superimpose well with
each other. The final MSD curve (circles) in the log–log plot is not
a linear function and goes as $\langle \Delta \mathbf{r}^2(\tau) \rangle \sim \tau^\alpha$ with $0.4 < \alpha < 0.9$ in
the small-τ range 0.0125–1 s. Only at the long-time limit ($\tau > 4$ s),
does the measured MSD become diffusive with $\alpha \simeq 1$ (blue solid
line). In this case, $\langle \Delta \mathbf{r}^2(\tau) \rangle \simeq 4 D_L \tau$, where D_L is the long-time
diffusion coefficient of the AChRs.

The long-time behaviour of $\langle \Delta \mathbf{r}^2(\tau) \rangle$ is best presented in the
linear plot as shown in the inset of Fig. 3. Because of the high
efficiency of our tracking algorithm (see Supplementary Methods
for more details), we are able to obtain long-time trajectories of
the AChRs with adequate statistics. About 1,160 samples are used
to obtain $\langle \Delta \mathbf{r}^2(\tau) \rangle$ at the largest delay time $\tau \simeq 160$ s. The statistics
for $\langle \Delta \mathbf{r}^2(\tau) \rangle$ at smaller values of τ are even better with the error
bars smaller than the size of the symbols used in Fig. 3. In the
wide range of τ between 4 and 160 s in which the AChRs have
diffused more than 750 times of their own diameter, the
measured $\langle \Delta \mathbf{r}^2(\tau) \rangle$ can be well described by a linear function of
τ (red solid line). From the slope of the fitted solid line, we obtain
$D_L = 0.05 \ \mu m^2 \ s^{-1}$.

Figure 4 and its inset show, respectively, the final statistics of
the measured D_L and the mobile ratio γ for the AChRs
from different frogs and from cells cultured with different days.

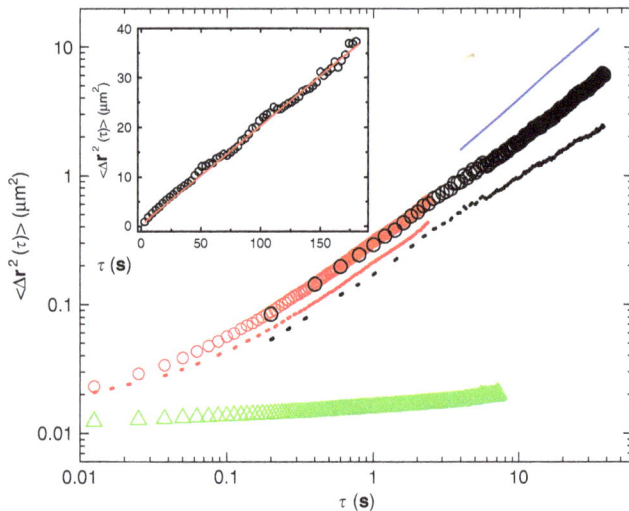

Figure 3 | Crossover from sub-diffusion to normal diffusion observed from the MSD curve. Measured $\langle\Delta\mathbf{r}^2(\tau)\rangle$ as a function of delay time τ for the AChR trajectories taken at two sampling rates of 80 fps (red dashed line and circles) and 5 fps (black dashed line and circles). Data from a single cell are used in the ensemble average. The red and black dashed lines are obtained when both the mobile and immobile trajectories are included in the calculation. The red and black circles are obtained when only the mobile trajectories are included in the ensemble average. The green triangles are obtained when only the immobile trajectories are included in the ensemble average. The blue solid line indicates the relationship $\langle\Delta\mathbf{r}^2(\tau)\rangle \sim \tau$ with a slope of unity in the log–log plot. Inset shows a linear plot of the measured $\langle\Delta\mathbf{r}^2(\tau)\rangle$ as a function of τ and the red solid line is a linear fit to the data points.

The value of D_L has a fairly narrow distribution with $\langle D_L\rangle = 0.041 \pm 0.015\,\mu m^2\,s^{-1}$. The distribution of γ is broader with $\langle\gamma\rangle = 0.64 \pm 0.17$. These mean values are obtained from 365 cells. The value of γ tends to be smaller for unhealthy cells and for cells cultured over a long period of time. We also examined the MSD curves obtained in different regions of the membrane in the same cell, both on the upper (away from the substrate) and lower (facing the substrate) portions of the membrane. The measured MSD in different regions remains approximately the same, suggesting that the AChRs on the same membrane have approximately the same value of D_L, which can be used as a parameter to characterize the mobility of membrane proteins in living cells. Because the bottom portion of the membrane has a large planer view with more than 100 QDs in each frame, we used this imaging setup to collect more data for a better statistical analysis.

Probability density function. Another important quantity to characterize the motion of AChRs is the PDFs (normalized histograms), $P(\Delta x')$ and $P(\Delta y')$, of the x- and y- components of $\Delta\mathbf{r}(\tau)$ at a fixed value of τ. Here, $\Delta x' = \Delta x/(2D_L\tau)^{1/2}$ and $\Delta y' = \Delta y/(2D_L\tau)^{1/2}$ are the displacements normalized by the diffusion length $(2D_L\tau)^{1/2}$. Figure 5 shows the measured $P(\Delta x')$ and $P(\Delta y')$ for mobile AChRs obtained under different sample conditions. Although the measured D_L varies considerably under different cell conditions, the PDFs collapse onto a single master curve, once the normalized $\Delta x'$ (and $\Delta y'$) is used. Except for a sharp peak near the origin, all of the PDFs have an exponential tail (red solid line). The error bars show the standard deviation of the black circles averaged over 10 cells. Because of the reduced number of data points, the diamonds have relatively larger experimental uncertainties. Figure 5 thus reveals that AChRs have a heavy-tailed distribution in their mobility, and this distribution

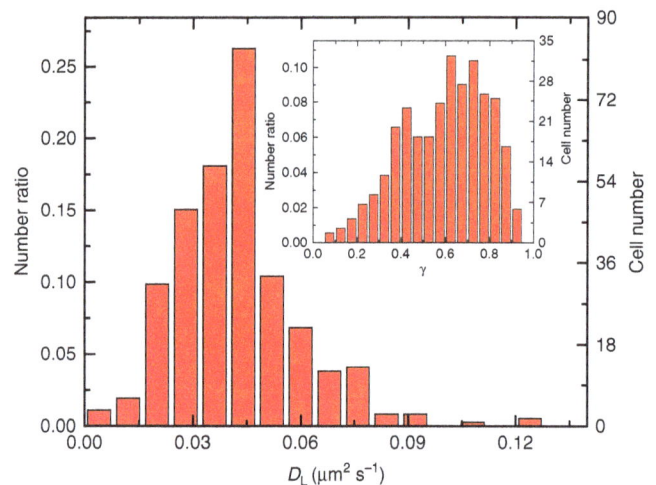

Figure 4 | Cell-to-cell variations in the measured long-time diffusion coefficient D_L and mobile ratio γ of the AChRs. Distribution of the measured D_L of AChRs. Inset shows the distribution of the measured γ of the AChRs. The total number of cells used in the statistics is 365.

is universal among the cells under different sample conditions. Similar $P(\Delta x')$ (and $P(\Delta y')$) are also found for the AChRs on the upper portion of the membrane away from the substrate (see Supplementary Fig. 7 for more details). Evidently, the exponential PDF is a leptokurtic distribution, which has a higher peak and a heavier tail compared with the Gaussian PDF[19].

Immobile trajectories and statistical sampling conditions. Figure 3 reveals that the measured $\langle\Delta\mathbf{r}^2(\tau)\rangle$ under two different sampling conditions (red and black dashed lines) has different values in the common region of τ. We find that the immobile trajectories have a dominant role in determining the value of $\langle\Delta\mathbf{r}^2(\tau)\rangle$. The measured $\langle\Delta\mathbf{r}^2(\tau)\rangle$ for the immobile AChRs (green triangles) is about two orders of magnitude smaller than the value of $\langle\Delta\mathbf{r}^2(\tau)\rangle$ in the long-time regime ($\tau \gtrsim 4\,s$) for the mobile AChRs (black circles) and thus contributes many near-zero values to the ensemble average. At the higher sampling rate (80 fps), the viewing area of the camera is cropped and the number of immobile trajectories recorded in the movie becomes different from that obtain at the lower sampling rate (5 fps). This is caused by the spatial inhomogeneity of the immobile AChR distribution. As shown in Fig. 3, once the immobile trajectories are removed from the ensemble average, the measured $\langle\Delta\mathbf{r}^2(\tau)\rangle$ under two different sample conditions becomes identical in the common region of τ (red and black circles).

By carefully examining the AChR trajectories, we also find that even the mobile trajectories still have some immobile segments of varying lengths (durations). As shown in the Supplementary Movie, the AChRs often move for a while and are transiently confined to a small region on the membrane for a different amount of time (up to seconds) and then move again. The transient confinement of AChRs is also reflected in the measured $h(R_g')$ in Fig. 2, $P(\Delta x')$ in Fig. 5 and PDF $f(\delta)$ of the 'instantaneous' diffusion coefficient δ in Fig. 6 below. For a fixed delay time τ, the immobile segments in a mobile trajectory tend to have smaller values of R_g', $\Delta x'$, and δ and hence give rise to a peak in the corresponding histograms at small values of R_g', $\Delta x'$ and δ. We believe that the transient confinement of AChRs is caused by the transient binding of the AChRs to the underlying cortical actin network. A similar effect was also observed for the Kv21 channel proteins in HEK 293 cells[20].

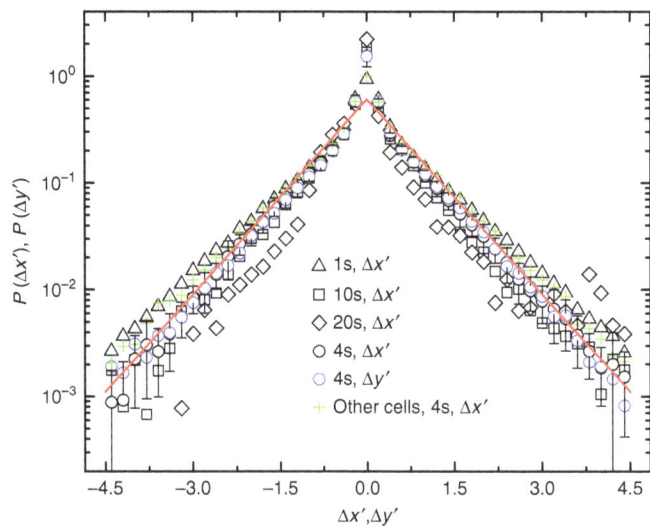

Figure 5 | Non-Gaussian behaviour of the normalized histogram of AChR's displacements. Measured PDFs $P(\Delta x')$ and $P(\Delta y')$ of the normalized displacements $\Delta x'$ and $\Delta y'$ for the trajectories of mobile AChRs. Data are obtained from 10 cells under each sample condition: (i) $\Delta x'(\tau)$ with $\tau = 1$ s (black triangles), 4 s (black circles), 10 s (black squares) and 20 s (black diamonds); (ii) $\Delta y'(\tau)$ with $\tau = 4$ s (blue circles) and (iii) $\Delta x'(\tau)$ with $\tau = 4$ s for a group of 10 cells from a different frog (green crosses). The red solid line is an exponential fit to the data points, $P(\Delta x') \simeq a \exp(-\beta|(\Delta x')|)$, with $a = 0.6$ and $\beta = 1.4$.

Figure 6 | Normalized histogram of the 'instantaneous' diffusion coefficient of mobile AChRs. Measured PDF $f(\delta')$ of the normalized diffusion coefficient $\delta' = \delta/D_L$ for the mobile AChR trajectories. Data are obtained from three groups of 10 cells each cultured for (i) 1 day (black circles), (ii) 3 days (blue triangles) and (ii) 6 days (green diamonds). The error bars show the standard deviation of the blue triangles averaged over 10 cells. The red solid line is an exponential fit to all data points, $f(\delta') \simeq 0.22 \exp(-0.75\delta')$. The blue dashed line shows the measured $f(\delta')$ for the silica spheres undergoing normal Brownian diffusion.

The above observation of non-reproducible $\langle \Delta \mathbf{r}^2(\tau) \rangle$ resulting from different sampling of the immobile AChR trajectories may shed light on the problem of nonergodicity between the time- and space-averaged MSDs, which has been observed for a number of molecules in live cells, such as Kv21 channel proteins[20], messenger RNAs in *Escherichia coli*[21] and lipid granules in yeast cells[22]. The origin of the nonergodicity in live cells has remained elusive[20–24]. The immobile trajectories may have an important role in determining the difference between the time- and space-averaged $\langle \Delta \mathbf{r}^2(\tau) \rangle$, because the immobile trajectories are typically included in the space-averaged $\langle \Delta \mathbf{r}^2(\tau) \rangle$, whereas in the time-averaged $\langle \Delta \mathbf{r}^2(\tau) \rangle$, one usually only samples the mobile trajectories[20–22].

Crowding effects and anomalous diffusion. In the original model of membrane diffusion[5], the cell membrane was considered as a continuum fluid layer, which is true only if the membrane is homogenous and the diffusing particle is much larger than the surrounding membrane molecules. For AChRs, however, their size is comparable to that of the surrounding membrane proteins and lipids, and their motion is hindered by the direct interactions with the surrounding macromolecules. In a crowded molecular solution, a tracer molecule faces a heterogeneous environment and its MSD is no longer a simple linear function of τ. Instead, the MSD often exhibits a sub-diffusive behaviour with $\langle \Delta \mathbf{r}^2(\tau) \rangle \sim \tau^\alpha$, where $\alpha < 1$ (refs 15,24,25). Such anomalous diffusion has been observed in a variety of dense fluid systems, such as colloidal diffusion near its glass transition[26,27] and over an external random potential[28].

Membrane proteins in live cells were also found to exhibit anomalous diffusion[7,11,20,29]. Because of the limited number and time span of the protein trajectories obtained, the measured MSD in some previous studies, however, only revealed a sub-diffusive regime without showing a crossover to the long-time diffusion. Some of the measurements also suffered relatively large statistical uncertainties at large delay times τ. The MSD shown in Fig. 3

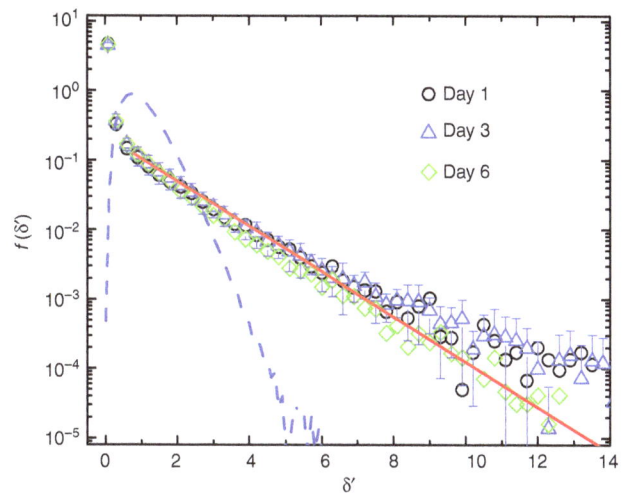

clearly reveals a crossover behaviour from sub-diffusion to long-time diffusion with the crossover time $\tau_L \simeq 4$ s. In the long-time diffusion regime as shown in the inset of Fig. 3, the measured MSD remains as a linear function of τ up to the longest tracking time 160 s, indicating that the membrane is very fluidic for AChRs. During this time, the AChRs diffuse more than 6 μm (or about 860 times of their own diameter) and no permanent fence is found at this length scale to confine the motion of AChRs.

In the short-time sub-diffusion regime ($\tau < 4s$), the measured $\langle \Delta \mathbf{r}^2(\tau) \rangle$ decreases with decreasing τ and reaches an asymptotic value $\langle \Delta \mathbf{r}^2 \rangle_0 \simeq 24.4 \times 10^{-3} \mu m^2$ for the mobile AChRs. The value of $\langle \Delta \mathbf{r}^2 \rangle_A$ for the immobile AChRs is about half the value for the mobile AChRs with $\langle \Delta \mathbf{r}^2 \rangle_A \simeq 12.4 \times 10^{-3} \mu m^2$. At the $\tau \to 0$ limit, the measured MSD becomes the mean square fluctuation, $\langle \Delta \mathbf{r}^2 \rangle_0 = \sum_i (\langle \Delta \mathbf{r}^2 \rangle_0)_i$, which is a sum of $(\langle \Delta \mathbf{r}^2 \rangle_0)_i$ from all independent fluctuation sources i. By subtracting out the background noise, $\langle \Delta \mathbf{r}^2 \rangle_B \simeq 3.11 \times 10^{-3} \mu m^2$, from the immobile QDs, which are physically attached to the glass substrate, we find the immobile AChRs jiggle in a typical range $R_0 = [\langle \Delta \mathbf{r}^2 \rangle_A - \langle \Delta \mathbf{r}^2 \rangle_B]/4]^{1/2} \simeq 48.2$ nm. This 48.2-nm-ranged jiggling may result from the agitation of the underlying cortical actin network, which the immobile AChRs are bound to (see more discussions below). With this understanding, one may define the net MSD of the mobile AChRs without the influence of independent agitations of the actin network as (ref. 30), $\langle \Delta \mathbf{r}^2(\tau) \rangle_{AChR} = \langle \Delta \mathbf{r}^2(\tau) \rangle - \langle \Delta \mathbf{r}^2 \rangle_A$, which differs from the measured $\langle \Delta \mathbf{r}^2(\tau) \rangle$ in Fig. 3 only in the small-τ range ($\tau \lesssim 0.5$ s). The resulting $\langle \Delta \mathbf{r}^2(\tau) \rangle_{AChR}$ still goes as τ^α, but the value of α varies in a narrower range 0.7–0.9 with a typical value $\alpha \approx 0.8$. We note that the above analysis can only remove the effect of independent agitations from the actin network and the correlated effect with the cortical network still remains in the measured MSD of the mobile AChRs.

Theoretical models of anomalous diffusion of membrane proteins have considered the effects of diffusion obstruction by permanent or transient obstacles and confinement by transient binding of diffusing proteins to a hierarchy of traps[7,15,25,31]. In the latter case, the time that the protein molecules are confined in the traps was assumed to have a power-law distribution[15,24,32], which gives rise to a non-converging mean trapping time. While these models can predict certain aspects of anomalous diffusion, such as the sub-diffusion exponent α, the present experiment reveals some new features of membrane diffusion, which have not been considered in the previous models. The new features of membrane diffusion include the persistent exponential tail in the measured $P(\Delta x')$ (and $P(\Delta y')$), which is invariant with delay time τ, and a crossover to apparently normal diffusion (in terms of MSD) at long-delay times ($\tau > 4s$) but with non-Gaussian statistics. One could introduce a crossover to normal diffusion by assuming that the trapping time of the protein molecules has an upper bound at equilibrium and thus their correlation time is finite. In this case, the protein trajectories would eventually be randomized at the long-time limit, and their displacement $\Delta x'$ (and $\Delta y'$) would follow the Gaussian statistics. Therefore, a new crossover mechanism is needed in order to explain the non-Gaussian diffusion dynamics of AChRs on live cell membrane in both the short- and long-time regimes.

Dynamic heterogeneity and non-Gaussian statistics. Figure 5 clearly demonstrates that the lateral motion of AChRs on the live cell membrane does not follow the Gaussian statistics. For the first time, we have obtained a universal PDF with its amplitude varied by more than three decades. With such a large number of statistics, we are able to pin down the functional form of $P(\Delta x')$ (and $P(\Delta y')$). The fit shown in Fig. 5 (red solid line) reveals that the measured $P(\Delta x')$ has a simple exponential form, $P(\Delta x') \simeq \exp(-\beta|\Delta x'|)$, with $\beta = 1.4$ (which is a straight line in the semi-log plot). The obtained exponential PDFs are found to be universal independent of delay time τ, the measured value of D_L, and the origin and cultured days of the cells.

In fact, the observed exponential form of $P(\Delta x')$ is directly linked to the dynamic heterogeneity in the diffusion coefficient. Assuming that the entire ensemble of mobile AChRs can be divided into independent subgroups, and that each subgroup obeys the Gaussian statistics with a diffusion coefficient δ:

$$g(\Delta x; \delta) = \frac{1}{(4\pi\delta\tau)^{1/2}} e^{-\Delta x^2/(4\delta\tau)}, \qquad (1)$$

and let δ have an exponential-like distribution, $f_0(\delta) = (1/D_L)\exp(-\delta/D_L)$, where D_L is the mean value of δ measured in the inset of Fig. 3. The ensemble averaged $P(\Delta x')$ then takes the form,

$$P(\Delta x') = \int_0^\infty g(\Delta x; \delta) f_0(\delta) d\delta = \frac{1}{(4D_L\tau)^{1/2}} e^{-\sqrt{2}|\Delta x'|}, \qquad (2)$$

where $\Delta x' = \Delta x/(2D_L\tau)^{1/2}$ is the normalized displacement. Equation (2) thus explains the exponential PDF shown in Fig. 5. The predicted decay rate $\beta = \sqrt{2}$ is in excellent agreement with the measured $\beta = 1.4$.

To further test equation (2), we directly measure the 'instantaneous' diffusion coefficient $\delta = \langle \Delta r^2(\tau) \rangle_t/(4\tau)$ with the delay time set at $\tau = 1 s$ and the averaging time $t = 4.2 s$, above which the measured MSD becomes diffusive (see Fig. 3). Figure 6 shows the measured PDF $f(\delta')$ of the normalized diffusion coefficient $\delta' = \delta/D_L$ for three groups of cells under different culture conditions. The measured PDFs for the cells from different embryos or cultured on different days all collapse onto a single master curve, once the normalized δ' is used in the plot. They have a universal shape with a sharp peak for small values of

δ' followed by an exponential-like tail (red solid line). Figures 5 and 6 thus confirm the theoretical prediction in equation (2).

Because of sampling fluctuations, the measured δ' (or δ) will have its own distribution $f(\delta')$ even for Brownian diffusion without any dynamic heterogeneity[33,34]. The blue dashed line in Fig. 6 shows the measured $f(\delta')$ for the silica spheres undergoing normal Brownian diffusion, which is a narrowly peaked function with its most probable value at $\delta' \simeq 1$. It is found that the measured $f(\delta')$ for Brownian diffusion obeys the χ^2-distribution, which depends sensitively on the number $2N$ of degrees of freedom for the statistical variable δ' (see Supplementary Fig. 4 for more details). In our case, we have $2N = 8$, where $N = (t - \tau_0)/\tau = 4$ and $\tau_0 = 0.2 s$ is the sampling time used in the experiment. Compared with Brownian diffusion, the measured $f(\delta')$ for the AChRs reveals a heavier tail with many AChR trajectories having large values of δ'. In addition, the measured $f(\delta')$ for the AChRs is found to be insensitive to the change of $2N$ (see Supplementary Fig. 5 for more details). These findings further confirm that the exponential-like distribution of AChR's diffusion coefficient, as shown in Fig. 6, has its own dynamic origin and does not result from the sampling statistics (see Supplementary Discussion for more discussions).

Discussion

Figure 6 reveals that the AChRs on live cell membrane have an exponential-like distribution in their diffusion coefficient δ, even though they are structurally identical. There are two possible causes for the observed dynamic heterogeneity in δ. One is that the AChRs form equilibrium clusters (or domains) with the surrounding proteins/lipids; and the other is that the motion of AChRs involves some active (non-equilibrium) process with a long correlation time to which the central limit theorem does not apply. There are several hypotheses in the literature on equilibrium membrane organization. One is the lipid raft model, which conceives the membrane to be compartmentalized by cholesterol organized glycolipoprotein nano-domains[3]. The typical size of the lipid rafts was estimated as $26 \pm 13 nm$[35] and they float freely in the membrane bilayer[36]. However, our results in Figs 5 and 6 cannot be explained by these features of the lipid rafts. If AChRs diffuse together with the lipid rafts, the difference in raft size is not enough to produce an exponential distribution of δ. This is because δ scales with the domain size a as $\ln(1/a)$ (ref. 5), which is essentially a constant for a moderately narrow distribution of raft size. As a result, the PDF $P(\Delta x')$ for lipid rafts of similar size should be of Gaussian form at the limit of long delay times, as they diffuse on the membrane at equilibrium with only a finite correlation time. In fact, this argument also applies to other models of membrane organization involving phase separation and critical fluctuations in membrane at equilibrium[37,38].

Another hypothesis is the picket-fence model[7,11,12], which envisions that the membrane is compartmentalized by cortical actin 'fences' and anchored transmembrane protein 'pickets.' For short times, the motion of membrane proteins and lipids is transiently confined in the corrals made of the protein pickets. Over long times, the proteins and lipids can hop among different corrals following a thermal activation process. Although this model of hop diffusion can qualitatively explain some previous SPT results, it contains two key assumptions that are inconsistent with the findings of the present experiment. First, the model assumes that the corrals are quasi-periodic with a size ranging from 32 to 230 nm depending on the cell type[7,11,12]. As mentioned above, it is difficult to produce an exponential $f(\delta)$ with motion confined by corrals with a narrow size distribution. Second, the model assumes that the hopping of membrane molecules among different corrals is made by thermal fluctuations, an equilibrium process with a finite correlation time which is unlikely to produce the non-Gaussian statistics shown in Fig. 5.

On the basis of the above experimental results, we propose a dynamic picket-fence model involving slow-active remodelling of the cortical actin network to explain the observed dynamic heterogeneity. In this model, we postulate that the anchored transmembrane proteins (both immobile and transiently confined proteins) have a dominant role in determining the diffusion dynamics of other (mobile) membrane molecules. We find that 36% of the AChRs, on average, are immobile during the experimental observation time (10–15 min). For other transmembrane proteins with stronger interactions with the cortical actin network, this ratio may be even larger. Due to the abundance of membrane proteins[2], the anchored proteins can form a continuous random network, partitioning the membrane into domains (corrals) of various sizes. Within each corral, the motion of the membrane molecules is strongly hindered by the rigid boundary of the protein network, giving rise to a local diffusion coefficient δ, which is strongly influenced by the size of the corral. Because the protein network on the membrane is anchored to the underlying cortical actin network, the two networks should share the same topological structure and dynamics. Without external stimulations, the protein network on the membrane will be randomly orientated having a large variety of meshes (corrals) of different sizes[39]. For short times, the mobile proteins and lipids on the membrane can move within the corrals, and over long times they also move between different corrals as the network remodels.

Our hypothesis has important biological implications, as it provides a mechanism of membrane organization for live cells to actively control the membrane fluidity and regulate the molecular transport on the membrane. It has specific predictions that can be tested in future experiments. First, the membrane protein network is not permanently stationary, as this would provide permanent barriers inhibiting the mobile proteins and lipids from moving over long distances, which is inconsistent with the measurements shown in the inset of Fig. 3. Although thermal fluctuations and ligand-binding equilibrium may provide some mobility for the protein network, these are Gaussian-like agitations and cannot produce the exponential (non-Gaussian) PDF as shown in Fig. 5. Under the dynamic picket-fence model, the dynamics of the protein network (and hence the long-time diffusion of the mobile proteins and lipids) is determined by the dynamics of the underlying cortical actin network, which is under constant active remodelling[40–43]. The slow-active remodelling of the cortical network (and hence the protein network) is caused by the activity of molecular motors (for example, myosin II motors) and other non-equilibrium cellular processes[44,45], and thus is capable of producing fluctuations with a long correlation time, to which the central limit theorem does not apply.

Second, the long-time non-Gaussian statistics shown in Fig. 5 should be a universal behaviour for all mobile molecules on the membrane including lipids and lipid-tethered proteins on the outer leaflet of the membrane, which do not have direct interactions with the underlying cortical actin network. In a recent experiment (W. He et al., manuscript in preparation), we studied the lateral motion of ganglioside GM1, which is a glycosphingolipid residing on the outer leaflet of the Xenopus muscle-cell membrane. The GM1s was found to have a similar non-Gausian behaviour as that of the AChRs. Finally, because the non-Gaussian dynamics of the membrane molecules is regulated by the active remodelling of the cortical actin network, it will change sensitively with the dynamics of the cortical network. Various drug manipulations of the actin network, such as

depletion of adenosine triphosphate and inhibition of myosin II motors, may be used to further test this prediction.

Methods

Cell culture. The AChR is a cation-selective, ligand-gated ion channel and consists of five subunits with diameter $d \simeq 7$ nm. It is an integral membrane protein that responds to the binding of acetylcholine, which is a neurotransmitter. The AChR spans the membrane of muscle cells with most of its mass in the extracellular space[46]. Xenopus muscle cells are dissected from myotomes of the fertilized Xenopus embryos developed at the stages 20–22, following the protocol described in ref. 47. The dissected muscle cells are seeded on the glass cover slips coated with Entactin, Collagen-IV, and Laminin (ECL, purchased from Upstate Co.), which are immersed in a culture medium composed of 88% Steinberg's solution, 10% L-15 medium (purchased from Leibovitz Co.), 1% foetal bovine serum and 1% penicillin/streptomycin/gentamincin[47]. The muscle-cell cultures are maintained at 23 °C and can be stored for 3 weeks if they are not contaminated.

QD labelling. To track the AChRs on a live muscle cell membrane, the individual AChRs are labelled by bright and photostable fluorescent QDs[8,17]. This is achieved by first labelling the AChRs with biotinylated α-bungarotoxin (biotin-BTX, purchased from Invitrogen Co.) for 10 min. The cells are then washed with the culture medium three times (5 min each). The concentration of biotin-BTX applied to the cells is adjusted according to the final labelling density of the QDs required. Typically, for a fast movie recording (for example, 80 and 5 fps), 0.5 nM biotin-BTX is used. For a slow movie recording (for example, 0.33 fps), 0.25 nM biotin-BTX is used. Lower QDs concentration is used to reduce tracking ambiguities between the consecutive images of the QDs. After repeated washing to remove unbounded biotin-BTX, 2.5 nM streptavidin-conjugated Qdot 655 solution (purchased from Invitrogen Co.) is added to the cells for 10 min after which the cells are washed with the culture medium three times (5 min each). The entire staining process requires ~ 1.5 h. Xenopus muscle cells in the primary culture present a large area for optical observation, typically 0.05 mm^2 on the bottom membrane and 0.002 mm^2 on the top apical membrane. QD-labelled AChRs are abundant on the membrane of the quiescent muscle cells. This is true even for sparsely labelled samples to avoid trajectory entanglement. In the experiment, several hundreds of AChRs are tracked concurrently. For a typical bottom membrane tracking at 5 fps, 200 QDs are labelled.

Other experimental details about the optical imaging and SPT are given in Supplementary Methods.

References

1. Singer, S. J. & Nicolson, G. L. The fluid mosaic model of the structure of cell membranes. *Science* **175**, 720–731 (1972).
2. Engelman, D. M. Membranes are more mosaic than fluid. *Nature* **438**, 578–580 (2005).
3. Lingwood, D. & Simons, D. Lipid rafts as a membrane-organizing principle. *Science* **327**, 46–50 (2010).
4. Munro, S. Lipid rafts: elusive or illusive? *Cell* **115**, 377–388 (2003).
5. Saffman, P. G. & Delbrück, M. Brownian motion in biological membranes. *Proc. Natl Acad. Sci. USA* **72**, 3111–3113 (1975).
6. Jacobson, K., Sheets, E. D. & Simson, R. Revisiting the fluid mosaic model of membranes. *Science* **268**, 1441–1442 (1995).
7. Kusumi, A., Umemura, Y., Morone, N. & Fujiwara, T. in *Anomalous Transport: Foundations and Applications* (eds Klages, R., Radons, G. & Sokolov, I. M.) ch19, 545–574 (Wiley-VCH Verlag GmbH & Co. KGaA, 2008).
8. Bannai, H., Lévi, S., Schweizer, C., Dahan, M. & Triller, A. Imaging the lateral diffusion of membrane molecules with quantum dots. *Nat. Protoc.* **1**, 2628–2634 (2006).
9. Gelles, J., Schnapp, B. J. & Sheetz, M. P. Tracking kinesin-driven movements with nanometre-scale precision. *Nature* **331**, 450–453 (1988).
10. Simons, K. & Sampaio, J. L. Membrane organization and lipid rafts. *Cold Spring Harb. Perspect. Biol.* **3**, a004697 (2011).
11. Ritchie, K. et al. Detection of non-Brownian diffusion in the cell membrane in single molecule tracking. *Biophys. J.* **88**, 2266–2277 (2005).
12. Kraft, M. L. Plasma membrane organization and function: moving past lipid rafts. *Mol. Biol. Cell* **24**, 2765–2768 (2013).
13. Crocker, J. C. et al. Two-point microrheology of inhomogeneous soft materials. *Phys. Rev. Lett.* **85**, 888–891 (2000).
14. Hoffman, B. D., Massiera, G., Van Citters, K. M. & Crocker, J. C. The consensus mechanics of cultured mammalian cells. *Proc. Natl Acad. Sci. USA* **103**, 10259–10264 (2006).
15. Höfling, F. & Franosch, T. Anomalous transport in the crowded world of biological cells. *Rep. Prog. Phys.* **76**, 046602 (2013).

16. Geng, L., Zhang, H. L. & Peng, H. B. The formation of acetylcholine receptor clusters visualized with quantum dots. *BMC Neurosci.* **10,** 80 (2009).

17. Geng, L. *Visualization of nicotinic acetylcholine receptor trafficking with quantum dots in* Xenopus *muscle cells.* PhD Thesis, HKUST (2006).

18. Alcor, D., Gouzer, G. & Triller, A. Single-particle tracking methods for the study of membrane receptors dynamics. *Eur. J. Neurosci.* **30,** 987–997 (2009).

19. Walck, C. *Internal Report SUFPFY/9601* (University of Stockholm, 2007).

20. Weigel, A. V., Simon, B., Tamkun, M. M. & Krapfa, D. Ergodic and nonergodic processes coexist in the plasma membrane as observed by single-molecule tracking. *Proc. Natl Acad. Sci. USA* **108,** 6438–6443 (2011).

21. Golding, I. & Cox, E. C. Physical nature of bacterial cytoplasm. *Phys. Rev. Lett.* **96,** 098102 (2006).

22. Jeon, J. H. *et al. In vivo* anomalous diffusion and weak ergodicity breaking of lipid granules. *Phys. Rev. Lett.* **106,** 048103 (2011).

23. Barkai, E., Garini, Y. & Metzler, R. Strange kinetics of single molecules in living cells. *Phys. Today* **65,** 29–35 (2012).

24. Meroz, Y. & Sokolov, I. M. A toolbox for determining subdiffusive mechanisms. *Phys. Rep.* **573,** 1–29 (2015).

25. Saxton, M. J. A biological interpretation of transient anomalous subdiffusion. I. Qualitative model. *Biophys. J.* **92,** 1178–1191 (2007).

26. Ghosh, A., Chikkadi, V., Schall, P. & Bonn, D. Connecting structural relaxation with the low frequency modes in a hard-sphere colloidal glass. *Phys. Rev. Lett.* **107,** 188303 (2011).

27. Hunter, G. L. & Weeks, E. R. The physics of the colloidal glass transition. *Rep. Prog. Phys.* **75,** 066501 (2012).

28. Hanes, R. D., Dalle-Ferrier, C., Schmiedeberg, M., Jenkinsa, M. C. & Egelhaaf, S. U. Colloids in one dimensional random energy landscapes. *Soft Matter* **8,** 2714–2723 (2012).

29. Smith, P. R., Morrison, I. E., Wilson, K. M., Fernández, N. & Cherry, R. J. Anomalous diffusion of major histocompatibility complex class I molecules on heLa cells determined by single particle tracking. *Biophys. J.* **76,** 3331–3344 (1999).

30. Martin, D. S., Forstner, M. B. & Käs, J. A. Apparent subdiffusion inherent to single particle tracking. *Biophys. J.* **83,** 2109 (2002).

31. Soula, H., Caré, B., Beslon, G. & Berry, H. Anomalous versus slowed-down Brownian diffusion in the ligand-binding equilibrium. *Biophys. J.* **105,** 2064–2073 (2013).

32. Wong, I. Y. *et al.* Anomalous diffusion probes microstructure dynamics of entangled F-actin networks. *Phys. Rev. Lett.* **92,** 178101 (2004).

33. Bauer, M., Valiullin, R., Radons, G. & Kärger, J. How to compare diffusion processes assessed by single-particle tracking and pulsed field gradient nuclear magnetic resonance. *J. Chem. Phys.* **135,** 144118 (2011).

34. Heidernätsch, M., Bauer, M. & Radons, G. Characterizing N-dimensional anisotrpic Brownian motion by the distribution of diffusivities. *J. Chem. Phys.* **139,** 184105 (2013).

35. Pralle, A., Keller, P., Florin, E. L., Simons, K. & Hörber, J. K. Sphingolipid-cholesterol rafts diffuse as small entities in the plasma membrane of mammalian cells. *J. Cell Biol.* **148,** 997–1008 (2000).

36. Simons, K. & Ehehalt, R. Cholesterol, lipid rafts, and disease. *J. Clin. Invest.* **110,** 597–603 (2002).

37. Veatch, S. L. *et al.* Critical fluctuations in plasma membrane vesicles. *ACS Chem. Biol.* **3,** 287–293 (2008).

38. Heberle, F. A. *et al.* Bilayer thickness mismatch controls domain size in model membranes. *J. Am. Chem. Soc.* **135,** 6853–6859 (2013).

39. Novikov, D. S., Fieremans, E., Jensen, J. H. & Helpern, J. A. Random walks with barriers. *Nat. Phys.* **7,** 508–514 (2011).

40. Gowrishankar, K. *et al.* Active remodeling of cortical actin regulates spatiotemporal organization of cell surface molecules. *Cell* **149,** 1353–1367 (2012).

41. Luo, W. W. *et al.* Analysis of the local organization and dynamics of cellular actin networks. *J. Cell. Biol.* **202,** 10571073 (2013).

42. Guo, M. *et al.* Probing the stochastic, motor-driven properties of the cytoplasm using force spectrum microscopy. *Cell* **158,** 822–832 (2014).

43. Parry, B. R. *et al.* The Bacterial cytoplasm has glass-like properties and is fluidized by metabolic activity. *Cell* **156,** 1–12 (2014).

44. Brangwynne, C. P., Koenderink, G. H., MacKintosh, F. C. & Weitz, D. A. Cytoplasmic diffusion: molecular motors mix it up. *J. Cell. Biol.* **183,** 583–587 (2008).

45. Prost, J., Jülicher, F. & Joanny, J. F. Active gel physics. *Nat. Phys* **11,** 111–117 (2015).

46. Unwin, N. Neurotransmitter action: opening of ligand-gated ion channels. *Cell Suppl.* **72** Suppl 31–41 (1993).

47. Peng, H. B., Baker, L. P. & Chen, Q. Tissue culture of *Xenopus* neurons and muscle cells as a model for studying synaptic induction. *Methods Cell Biol.* **36,** 511–526 (1991).

Acknowledgements

We thank X.-G. Ma for helpful discussions. This work was supported by the Hong Kong RGC under Grant Nos. HKUST-604310 and 16305214.

Author contributions

H.B.P. and P.T. designed and jointly supervised research; W.H. performed research; H.S. and L.G. developed new tools for imaging processing, SPT and QD labelling; W.H., Y.S. and B.J.A. analyzed data; and W.H., Y.S. and P.T. wrote the paper with further revisions from B.J.A. and H.B.P.

Additional information

20

Shaping metallic glasses by electromagnetic pulsing

Georg Kaltenboeck[1], Marios D. Demetriou[1], Scott Roberts[1] & William L. Johnson[1]

With damage tolerance rivalling advanced engineering alloys and thermoplastic forming capabilities analogous to conventional plastics, metallic glasses are emerging as a modern engineering material. Here, we take advantage of their unique electrical and rheological properties along with the classic Lorentz force concept to demonstrate that electromagnetic coupling of electric current and a magnetic field can thermoplastically shape a metallic glass without conventional heating sources or applied mechanical forces. Specifically, we identify a process window where application of an electric current pulse in the presence of a normally directed magnetic field can ohmically heat a metallic glass to a softened state, while simultaneously inducing a large enough magnetic body force to plastically shape it. The heating and shaping is performed on millisecond timescales, effectively bypassing crystallization producing fully amorphous-shaped parts. This electromagnetic forming approach lays the groundwork for a versatile, time- and energy-efficient manufacturing platform for ultrastrong metals.

[1] Keck Engineering Laboratories, California Institute of Technology, Pasadena, California 91125, USA. Correspondence and requests for materials should be addressed to M.D. (email: marios@caltech.edu).

The concept of a Lorentz force generated on a current-carrying conductor exposed to a magnetic field dates back to the nineteenth century work of Faraday and Maxwell on electromagnetism[1]. In this concept, moving point charges comprising the electric current experience Lorentz forces, which consist of electric and magnetic force components. The magnetic point forces combine to produce a magnetic body force, often referred to as the Laplace force, acting on the current-carrying conductor. If the conductor is a metallic glass, the Laplace force provides for innovative methods of forming. Owing to unique electrical resistivities, metallic glasses can be rapidly and uniformly heated when electrical energy is dissipated in them[2-4]. Combining ohmic dissipation with the application of Laplace force creates a powerful platform to process metallic glasses.

While metallic glasses are generally known for their attractive mechanical properties[5,6], perhaps their most promising attribute is their potential for 'thermoplastic' processing[7-14]. By virtue of being glasses, they can be softened to viscous liquid states above the glass transition where viscoplastic shaping can be carried out in a manner similar to that applied to process conventional thermoplastics. This prospect paved the way to nano-fabrication and nano-moulding, revealing a remarkable ability to replicate nano-features down to 10 nm length scales[10,11]. Unfortunately, this potential for thermoplastic forming is practically limited by the rapidly intervening crystallization of the relaxed 'supercooled' liquid. Electrical discharge heating has recently emerged as an effective means to overcome this limitation[2-4]. It enables rapid and spatially uniform heating to low-viscosity states considered to be optimal for thermoplastic shaping, over timescales sufficiently short to bypass crystallization.

In this work, we demonstrate that subjecting the metallic glass to an intense electric current pulse directed normal to an applied magnetic field can generate Laplace forces sufficiently large to perform thermoplastic shaping operations thereby producing high-quality net-shaped metallic glass articles.

Electromagnetic forming of conventional (that is, crystalline) metals has been explored since the late 1950s (refs 15,16). In the most widely used approach, an induction coil excited by a current pulse generates a transient magnetic field that induces eddy currents in a workpiece according to Faraday's law of induction. The coupling of the eddy currents and the magnetic field produces a repulsive Lorentz force, in accordance with Lenz's law, which accelerates the workpiece away from the restrained coil and against a tool causing it to form. This approach results in high workpiece velocities (typically $> 100\,\mathrm{m\,s^{-1}}$) and strain rates on the order of $10^4\,\mathrm{s^{-1}}$ during formation[16,17]. Although some heat may be generated in the sample by induction, the forming process takes place entirely in the solid (that is, the crystalline) state. Generally, the generated Lorentz forces must be high enough to produce an equivalent impact pressure (typically hundreds of MPa) that exceeds the material yield strength, thereby enabling large plastic deformation. The metallic sample must therefore have high electrical conductivity, low yield strength and also be of limited thickness. Because of these requirements, electromagnetic forming of metals has so far been limited mostly to aluminium and copper in thin sheet and tube geometries, while stronger, less conductive metals such as steels have been largely excluded.

Compared with crystalline metals, amorphous metals demonstrate considerably higher room-temperature yield strengths (1–2 orders of magnitude higher than copper and aluminium)[5,6]. However, in their relaxed viscous state above the glass transition, the flow stresses are much lower (on the order of tens of MPa or less at temperatures substantially higher than the glass transition)[18-20], and the associated viscosities within the range where conventional plastics are typically processed (that is, 10^0–$10^4\,\mathrm{Pa\,s}$)[21,22]. Processing a metallic glass in this optimum rheological window is limited by crystallization of the supercooled liquid, which typically limits the available processing time to under a second. Hence, a rapid forming approach for metallic glasses that utilizes an intense electromagnetic pulse to thermoplastically shape a bulk metallic glass in its relaxed viscous state would be attractive, as it would require much lower forming stresses than conventional aluminium and copper while final metallic glass parts would be considerably stronger.

Another fundamental distinction between crystalline and amorphous metals is their different electrical resistivities. Most amorphous metals exhibit electrical resistivities in the range of 150–200 $\mu\Omega$ cm (ref. 23), which are much larger than those of crystalline metals (typically 1–50 $\mu\Omega$ cm). Furthermore, unlike crystalline metals, the electrical resistivity of amorphous metals is nearly constant or even decreasing with temperature[24]. The large and temperature independent electrical resistivity hinders generation of large magnetic pressures during conventional electromagnetic forming; however, it enables efficient and stable volumetric heating by uniform ohmic dissipation. Indeed, it has been shown that ohmic heating of metallic glasses can be both rapid and uniform, enabling thermoplastic processing at temperatures far above the glass transition[2-4]. In this work, we demonstrate that the electrical and rheological properties of metallic glasses are such that a window exists for an electromagnetic forming process. Specifically, we show that by coupling a uniform electric current with an applied static magnetic field, one can simultaneously bring a metallic glass to a low-viscosity state while exploiting the accompanying Laplace force to form the sample to a net shape, all while avoiding crystallization.

Results

Identifying a process window. To process metallic glasses thermoplastically by electromagnetic forming, it is essential that the processes of ohmic heating and magnetic forming are independently controlled. This requires that the electric current, which travels through the metallic glass dissipating electrical energy and producing volumetric heating, be controlled independently of the magnetic field, which interacts with the electric current to generate the magnetic force. This is achieved by placing the metallic glass in series with an electric current source directed transverse to an independently applied magnetic field. Such configuration is shown in Fig. 1, where a metallic glass sample is connected in series to a capacitor via electrodes while situated transverse to a magnetic field generated by two permanent magnets. Application of an intense electric current pulse ohmically heats the metallic glass rapidly and uniformly to a predetermined temperature in the supercooled liquid region, while the coupling between electric current and magnetic field generates a force pulse on the sample urging it against a permanent die tool, where it forms and subsequently cools and revitrifies. Here, we show that typical currents produced by discharge of a conventional capacitor and typical magnetic field strengths generated by conventional permanent magnets are adequate to rapidly and uniformly heat a bulk metallic glass sample to a low-viscosity state and thermoplastically shape it against a die tool before the intervention of crystallization.

The Laplace force on a current-carrying conductor exposed to a magnetic field is given by $\vec{F} = I\vec{l} \times \vec{B}$, where I is the current travelling through the conductor, $\vec{B}$ is the magnetic field, and $\vec{l}$ is a conductor length vector along the current direction that traverses the magnetic field. When the magnetic field is normal to $\vec{l}$, the magnitude of the Laplace force reaches a maximum given by

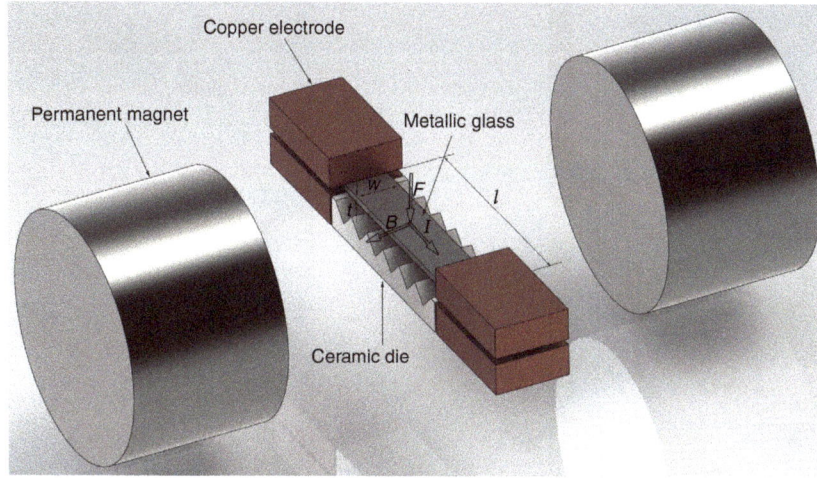

Figure 1 | Schematic representation of the electromagnetic forming configuration. The schematic presents a strip of metallic glass having length l, width w and thickness t, subject to traversing electrical and magnetic fields. The strip is in contact with copper electrodes delivering a current I along l, and normal to a pair of permanents magnets inducing a magnetic field B. A Laplace force F is shown to be generated on the strip normal to the electric and magnetic fields (according to a 'left-hand rule'). A ceramic die is also shown placed at the side of the strip opposite to the direction of F.

$F = BIl$, while any non-parallel configuration of $\vec{B}$ and $\vec{l}$ produces a fraction of this maximum value. For a typical metallic glass, one can show that the time-average current I discharged on a time constant τ required to ohmically heat a sample to a temperature in the supercooled liquid region where the viscosity is in the order of $\sim 10^2\,\mathrm{Pa\,s}$ is approximately (see Methods)[23,25,26]:

$$I \approx Cwt/\sqrt{\tau} \qquad (1)$$

where t is the sample thickness in the direction of the Laplace force, and C is a constant involving the material thermal and electrical properties, which for most metallic glasses ranges between 2 and $4 \times 10^7\,\mathrm{A}\sqrt{\mathrm{s}}\,\mathrm{m}^{-2}$ (see Methods)[27–30]. The magnitude of the applied Laplace force is then

$$F \approx CBwtl/\sqrt{\tau} \qquad (2)$$

where w is the conductor projected width along the direction of the magnetic field. The pressure P exerted by the applied Laplace force is

$$P \approx CBt/\sqrt{\tau} \qquad (3)$$

The Laplace force according to equation 2 appears dependent on the sample volume; its magnitude, however, can be independently controlled through B and τ. Pressure on the other hand, which is the fundamental parameter determining the shaping capacity, scales with the sample thickness and is likewise controllable through B and τ.

Equations 1–3 may be thought of as semi-quantitative relations allowing one to identify a process window for implementing this approach with a metallic glass. As an example, a metallic glass sheet feedstock undergoing electromagnetic forming having $w = l = 100\,\mathrm{mm}$ and $t = 1\,\mathrm{mm}$ may be considered. Using equations 1–3 and assuming $B \sim 1\,\mathrm{T}$, typical of a set of conventional permanent magnets, and $\tau \sim 1\,\mathrm{ms}$, typical of a capacitive discharge circuit, one may estimate $I \sim 10^5\,\mathrm{A}$, $F \sim 10\,\mathrm{kN}$ and $P \sim 10^5\,\mathrm{Pa}$ (that is, $P \sim 1\,\mathrm{atm}$). These are of the order of forces and pressures applied in typical blow moulding processes for plastics performed at viscosities in the range of 10^0–$10^4\,\mathrm{Pa\,s}$ (refs 21,22), and should be sufficient to induce substantial strain rates in a softened metallic glass heated high enough into the supercooled liquid region[18–20]. As the metallic glass would be ohmically heated to within this viscosity regime and the associated electromagnetic pressure would be on the order of

typical blow moulding pressures, the strains produced would likewise be comparable to those achieved in thermoplastic blow moulding of plastics. Hence, in terms of general forming capacity, electromagnetic forming of a metallic glass sheet would be akin to thermoplastic blow moulding. It is also important to note that the Laplace force will always be applied normal to the sheet (in the plane normal to the magnetic field). As such, the forming pressure is effectively 'hydrostatic', thereby mimicking the application of gas pressure in thermoplastic blow moulding.

Implementing electromagnetic forming. Here we demonstrate this concept by subjecting $Zr_{35}Ti_{30}Cu_{7.5}Be_{27.5}$ metallic glass strips with $l = 33\,\mathrm{mm}$, $w = 7\,\mathrm{mm}$ and $t = 1.0\,\mathrm{mm}$ to traversing electric and magnetic fields to electromagnetically form them against permanent die tools with semicircular corrugations, as illustrated in Fig. 1 (see Methods). A strip is placed between two FeNdB permanent magnets with its length l oriented normal to the magnetic field, while a capacitor with millisecond time constant is discharged to deliver a rapid current pulse to the strip along l through the contacting electrodes. A ceramic die is placed alongside the strip where deformation would be induced by the Laplace force (that is normal to the electric and magnetic fields according to a 'left-hand rule'). A capacitive discharge circuit with $0.264\,\mathrm{F}$ capacitance is used, and the applied voltages ranged between 68 and 71 V. A measured magnetic field of $\sim 0.275\,\mathrm{T}$ is produced by the permanent magnets at the location of the strip. The millisecond current pulse generated by the discharging capacitor rapidly heats the metallic glass in open air to a viscous state conducive for thermoplastic forming, while the Laplace force generated by the electromagnetic interaction between electrical and magnetic fields drives the softened metallic glass against a permanent die to shape it and simultaneously cool and revitrify the sample by thermal conduction to the die.

The evolution of electromagnetic forming against a die with a single semicircular corrugation is captured by a thermal imaging camera and is presented in Fig. 2. As seen in Fig. 2, the strip is heated rapidly and uniformly attaining a process temperature in the range of 500–550 °C in 1–2 ms. In this temperature range, the viscosity of $Zr_{35}Ti_{30}Cu_{7.5}Be_{27.5}$ is between 10 and 100 Pa s (ref. 8). As the heating appears to saturate, deformation of the strip initiates owing to the generated Laplace force. The plastically

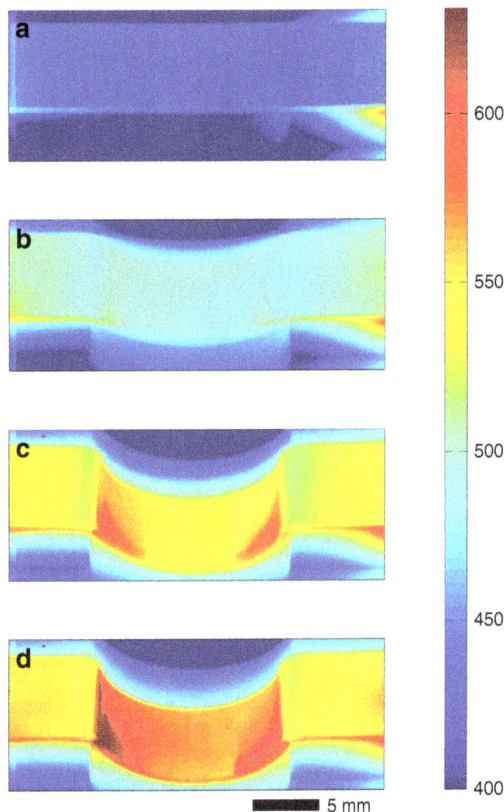

Figure 2 | Time evolution of the temperature distribution during electromagnetic forming. (**a**) 0 ms; (**b**) 1 ms; (**c**) 2 ms; and (**d**) 3 ms. An amorphous $Zr_{35}Ti_{30}Cu_{7.5}Be_{27.5}$ strip is undergoing heating and deformation by electromagnetic forming, as recorded by an infrared thermal imaging camera. Deformation is terminated at about 3 ms as the strip completely forms against the die (not visible by thermal imaging).

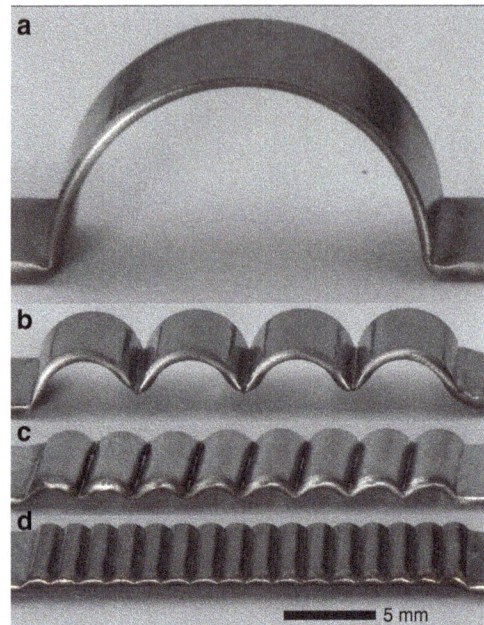

Figure 3 | Electromagnetically formed corrugated amorphous strips. Amorphous $Zr_{35}Ti_{30}Cu_{7.5}Be_{27.5}$ strips are formed electromagnetically against dies of progressively finer semicircular corrugations. Images **a–d** show strips formed with 1, 4, 8 and 16 semicircular corrugations. Image **a** presents the formed strip shown in the thermographs of Fig. 2.

deforming strip is pressed against the ceramic die at room temperature (not visible by thermal imaging) forming a semicircular bow. The heating and forming process are completed over a time of 3–4 ms, followed by cooling and revitrification occurring over a longer time (not shown). Crystallization is entirely avoided during the ultrarapid heating and forming processes, as well as during cooling; the amorphous structure of the formed strip is verified by X-ray diffraction. The final formed amorphous strip is shown in Fig. 3a. Figure 3b–d shows three more amorphous strips formed with 4, 8 and 16 semicircular corrugations produced using the same process conditions. The formed strips demonstrate fine replication of the progressively finer corrugated dies, which is a consequence of a low and fairly uniform viscosity, and also of a uniformly and hydrostatically applied force.

Analysis of the process parameters. In Fig. 4, we plot the time-dependent sample temperature, as monitored by an infrared pyrometer, the current through the sample as measured by a Rogowski coil, and the applied magnetic pressure as estimated by $P = BI/w$, during the heating and deformation of the strip shown in Fig. 2 (see Methods). The time interval of the 'forming region', where deformation of the strip initiates once a low enough viscosity is reached under an applied magnetic force and terminates when the strip is in full contact with the die, is indicated in all three plots. In Fig. 4a, the temperature is seen to rise rapidly crossing the dynamic glass transition at about 475 °C (about 170 °C above the calorimetric glass transition measured at

20 K min^{-1}) at about 2 ms. The temperature continues to rise in the deformation regime between 500 and 550 °C but in a somewhat discontinuous manner, as the strain developing in the strip increases its electrical resistance which contributes to incrementally more efficient ohmic heating. The temperature equilibrates to about 600 °C at about 6 ms. The current in Fig. 4b is shown to rapidly peak at slightly above 4,000 A in just less than 1 ms, then slowly declines towards zero at large timescales. The magnetic pressure as calculated from $P = BI/w$ (by assuming normal electric and magnetic fields) is plotted on the right-hand axis in Fig. 4b. With these assumptions, P is simply a linear superposition of I. The peak pressure reached in just less than 1 ms is about 170 kPa, while the pressure within the forming region declines gradually from 120 to 60 kPa N. These values are well within the range of pressures commonly used in thermoplastic blow moulding of plastics, as discussed above.

Discussion

In the simple configuration presented here, where the current travelling through a sample is directed normal to a magnetic field induced by two permanent magnets, the magnitude of the magnetic force can be independently varied by controlling the strength of the magnetic field. But since the magnetic field in this configuration is constant, the timing of the magnetic force is solely determined by the time-dependent current, and thus is not independently controllable. As shown in Fig. 4, being proportional to the current, the force peaks before the metallic glass reaches the least viscous supercooled state. Even though the applied force is sufficient for thermoplastic forming, the timing of the force application is not 'optimal'. More 'optimal' timing of the process, where a constant magnetic force of a desired magnitude is applied when a desirable viscous state is reached, can be achieved by more complex electric/magnetic configurations. For instance, configurations that involve two or more successive current pulses with appropriately chosen rise times and

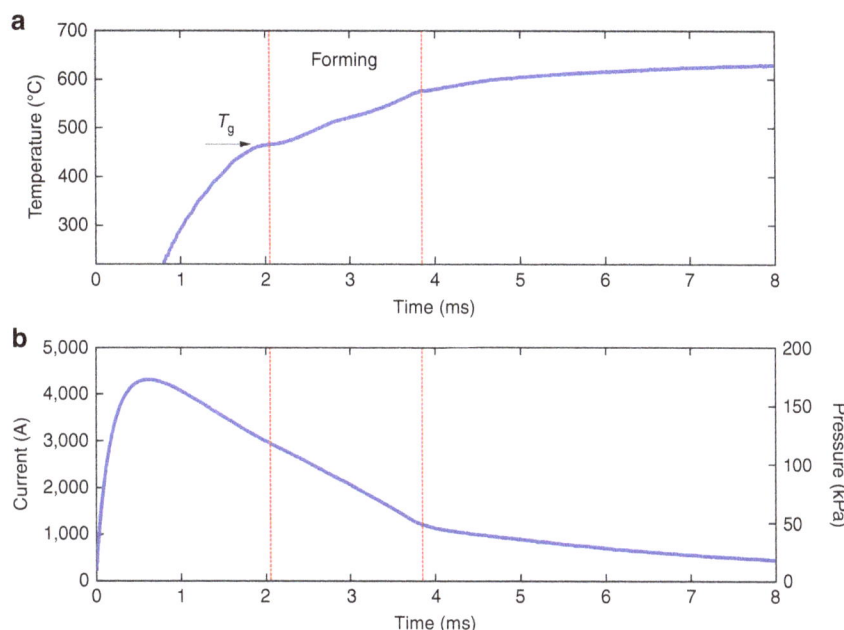

Figure 4 | Time-dependent thermal and electrical response. (**a**) Time-dependent sample temperature, as monitored by an infrared pyrometer; (**b**) time-dependent current through the sample as measured by a Rogowiski coil (left-hand axis) and time-dependent magnetic pressure estimated by $P = BI/w$ (right-hand axis), during ohmic heating and forming of the amorphous $Zr_{35}Ti_{30}Cu_{7.5}Be_{27.5}$ strip presented in Fig. 2. The 'forming' time interval is indicated by vertical lines in **a,b** and the dynamic glass-transition temperature T_g is indicated by an arrow in **a**.

amplitudes can achieve sequencing of the heating and forming process. Furthermore, configurations that utilize electromagnets (such as Helmholtz coils) instead of permanent magnets, where magnetic pulses with desirable profiles can be generated electrically, may achieve even more effective sequencing of the two processes.

The discussion and example above are focused on the 'forging' or 'stamping' of a strip or sheet using an operation mode akin to thermoplastic blow moulding. This is the most natural and straight-forward implementation of this forming approach. However, the general concept introduced here could in principle be configured to perform other thermoplastic shaping operations involving samples in bulk geometries. For example, using a different feedstock geometry and a different magnetic field configuration, one may be able to operate in an 'injection moulding' or 'calendaring' (sheet extrusion) mode. Specifically, using a suitable configuration of multiple permanent magnets or electromagnets disposed at different angles relative to the electric field axis, a distribution of magnetic forces can be generated on the feedstock that are capable of guiding it through a mould runner and gate and into a mould cavity of a desirable shape, or between rollers to shape it into a sheet. As feedstock, one may use a bulk prismatic sample having $l = 100$ mm, $w = 10$ mm and $t = 10$ mm. Assuming $B \sim 1$ T and ~ 1 ms, using equations 1–3 one may then estimate the current required to ohmically heat to a viscous supercooled state as $I \sim 10^5$ A, the magnetic force $F \sim 10$ kN, and the pressure $P \sim 10^6$ Pa (that is, ~ 1 MPa). While I and F are on the same order as in the sheet forming process, a pressure on the order of ~ 1 MPa is near the lower bound of pressures used in thermoplastic injection moulding or calendaring of conventional plastics (refs 21,22). As a t of 10 mm is near the upper bound of the achievable casting thickness of typical metallic glass formers and a B of 1 T is near the upper limit for conventional permanent magnets, the window for an injection moulding or calendaring process may be quite narrow.

The dependence of the Laplace force and pressure on the sample dimensions along with limitations in magnetic field strength may be seen as a drawback of the present approach when compared against conventional mechanical technologies (for example, utilizing hydraulic presses, pneumatic drives, electrical motors and so on), where larger forces and pressures are attainable. But the present concept offers unique advantages over conventional mechanical forming processes. As discussed above, the Laplace force is applied normal to the deforming sample leading to pressures that are effectively hydrostatic, unlike conventional methods where forces are generally unidirectional. Furthermore, spatially varying magnetic fields may be applied to produce varying Laplace force directions in different regions of the sample permitting complex shaping operations. From a manufacturing perspective, the absence of a working medium (that is, the lack of presses, motors, drives and massive fixtures) leads to an essentially frictionless process with minimal wear, and provides ease of automation. Moreover, because only the sample is accelerated during the process, large strains can be achieved in timescales significantly shorter than typical mechanical forming processes (that is, in milliseconds rather than tens of milliseconds; see Figs 2 and 4) such that process cycle times may be substantially reduced. Despite all these technical advantages, the net economic advantages over the incumbent technology and the overall commercial viability of this approach would be difficult to gauge at this stage.

In summary, using a simple electromagnetic setup comprising a capacitor bank and a pair of permanent magnets to subject a metallic glass to traversing electric and magnetic fields, we demonstrate that the metallic glass can be formed thermoplastically in a millisecond timescale in open air in the absence of any conventional heating source or any applied mechanical force by coupling ohmic heating and magnetic forming. This simple demonstration lays the foundation for a time and energy efficient all-electronic manufacturing platform for amorphous metals that could rival the simplicity and economics of plastics manufacturing.

Table 1 | Data for five metallic glass compositions for estimating constant C using Equation (10).

Metallic glass composition	ΔT_g (K)	v_m (m³ mol⁻¹)	c_p (J m⁻³ K⁻¹)	C (A $\sqrt{s}$ m⁻²)	Refs
$Au_{49}Cu_{26.9}Ag_{5.5}Pd_{2.3}Si_{16.3}$	103	9.23×10^{-6}	3.79×10^6	1.83×10^7	27
$Pt_{57.5}Cu_{14.7}Ni_{5.3}P_{22.5}$	210	8.75×10^{-6}	4.00×10^6	2.68×10^7	28
$Pd_{40}Ni_{10}Cu_{30}P_{20}$	295	7.92×10^{-6}	4.42×10^6	3.34×10^7	29
$Zr_{41.2}Ti_{13.8}Cu_{12.5}Ni_{10}Be_{22.5}$	320	9.95×10^{-6}	3.52×10^6	3.11×10^7	29
$Fe_{70}Ni_5Mo_5C_5B_{2.5}P_{12.5}$	398	6.89×10^{-6}	5.08×10^6	4.16×10^7	30

Methods

Sample preparation. Amorphous $Zr_{35}Ti_{30}Cu_{7.5}Be_{27.5}$ strips 1 mm in thickness, 7 mm in width and 4 cm in length were prepared by arc-melting the elemental constituents in a water-cooled copper hearth under a titanium-gettered argon atmosphere, followed by suction casting in a copper mould. The strips were ground flat and parallel, and their amorphous structure was verified by X-ray diffraction. Corrugated dies used for the shaping process were machined from MACOR ceramic.

Electromagnetic setup configuration. The electromagnetic forming setup comprises a capacitor bank with capacitance of 0.264 F rated for voltage of up to 100 V, a set of copper clamps connected to the capacitor bank via copper leads that hold the metallic glass sample and deliver the current discharge across it. Two NdFeB cylindrical magnets (3′ in diameter and 2′ in width) held on either side of the sample to create a magnetic field perpendicular to the current flow and parallel to the die's moulding surface. The setup operates in open air. The magnetic field strength in the vicinity of the sample is measured using a hall-probe Gaussmeter to be 0.275 T. Also, the time-dependent current through the sample during the electrical discharge was measured using a Rogowski coil wound around a current-carrying lead.

Thermal imaging. A high-speed infrared imaging camera (FLIR Corp., SC2500) with a spectral band from 0.9 to 1.7 mm outfitted with a band-pass filter allowing wavelengths from 1.5 to 1.9 mm was employed to record the evolution of the temperature distribution and deformation at frame rates in the range of 994–1,500 frames s⁻¹. An IMPAC IGA740-LO high-speed infrared pyrometer with a spectral band from 1.58 to 2.2 mm and a response time of 6 μs was also used to record temperature vs. time in a circular focal spot of 1 mm diameter in the section of the strip being formed. Both camera and pyrometer were calibrated simultaneously by tracking the melting of $Pd_{43}Ni_{10}Cu_{27}P_{20}$ alloy with a solidus temperature of 531 °C, which is within the temperature range considered in this work. Using this method, emissivities of 0.285 and 0.26 were found for the infrared camera and pyrometer, respectively. The emissivities are expected to be roughly representative of $Zr_{35}Ti_{30}Cu_{7.5}Be_{27.5}$ liquid in the temperature range of 450–600 °C, and the temperature error is expected to be within 20 °C.

Analysis of Ohmic heating. The electrical energy E dissipated to ohmically heat a prismatic metallic glass sample having length l along the direction of current flow, width w and thickness t from an initial temperature T_o (typically room temperature) to a process temperature T in the supercooled liquid region (that is, above the glass-transition temperature T_g) where the viscosity is on the order of 10^2 Pa s is given by:

$$E = wlt \int_{T_o}^{T} c_p dT \tag{4}$$

where c_p is the specific heat capacity of the material in J m⁻³ K⁻¹. Assuming that c_p of the glass and the supercooled liquid is a constant (that is, independent of temperature) as the material is heated through T_g, equation 4 can be approximated as

$$E = c_p wlt\Delta T \tag{5}$$

where ΔT is the temperature rise from T_o to T. For most metallic glasses,

$$\Delta T = 1.5\Delta T_g \tag{6}$$

where $\Delta T_g = T_g - T_o$ (ref. 25). If one approximates the time-dependence of a current pulse produced in an electrical discharge having a characteristic discharge time τ by an isosceles triangle, the electrical energy dissipated during the discharge can be estimated as:

$$E \approx I^2 R\tau \tag{7}$$

where I is the peak current attained at time τ, and R is the electrical resistance of the metallic glass sample, given by

$$R = \rho l/wt \tag{8}$$

where ρ is the electrical resistivity of the metallic glass. Substituting equations 6–8 into equation 4 and solving for I, one can obtain the electrical current required to ohmically heat a metallic glass sample a temperature where the viscosity is on the

order of 10^2 Pa s as follows:

$$I \approx Cwt/\sqrt{\tau} \tag{9}$$

The constant C has units of A $\sqrt{s}$ m⁻², and involves the material properties as follows:

$$C \approx \sqrt{1.5 \, c_p \Delta T_g / \rho} \tag{10}$$

Here we estimate the constant C for a wide variety of metallic glass compositions. The temperature-dependent resistivity of metallic glasses is known to vary over a narrow range, typically from 150 to 200 μΩ cm (ref. 23). For the sake of simplicity, we approximate the resistivity of all metallic glasses by a mean value of 175 μΩ cm (1.75×10^{-6} Ω m). Moreover, an average specific heat capacity between room temperature and $1.5\Delta T_g$ will be assumed for c_p. In this temperature range, the specific heat capacity in J mol⁻¹ K⁻¹ for most metallic glass compositions as they are heated through the glass transition varies between 25 and 45 J/mol⁻¹ K⁻¹; here we assume a mean value of $\sim$35 J mol⁻¹ K⁻¹ (ref. 26). But the specific heat capacity in J m⁻³ K⁻¹, as used in equations 4 and 10, can vary substantially between metallic glass compositions because of the variation in the molar volume between compositions. Here we estimate the specific heat capacity in J m⁻³ K⁻¹ for each metallic glass composition as the ratio between 35 J mol⁻¹ K⁻¹ and molar volume. Lastly, ΔT_g, estimated here as $\Delta T_g = T_g - 298$ K, also varies considerably between metallic glass compositions.

In Table 1 we present the values for ΔT_g, molar volume v_m and c_p in J m⁻³ K⁻¹ (estimated as the ratio between 35 J mol⁻¹ K⁻¹ and v_m) for a Au-based, Pt-based, Pd-based, Zr-based and Fe-based metallic glass to estimate the constant C corresponding to each composition using equation 10 (refs 27–30). As seen, C for this wide variety of metallic glass compositions is in the range of $\sim$2–4 $\times$ 10⁷ A $\sqrt{s}$ m⁻².

References

1. Darrigol, O. *Electrodynamics from Ampere to Einstein* 126–131; 139–144 (Oxford University Press, 2000).
2. Johnson, W. L. *et al.* Beating crystallization in glass-forming metals by millisecond heating and processing. *Science* **332**, 828–833 (2011).
3. Liu, X. *et al.* Description of millisecond ohmic heating and forming of metallic glasses. *Acta Mater.* **61**, 3060–3067 (2013).
4. Kaltenboeck, G. *et al.* Accessing thermoplastic processing windows in metallic glasses using rapid capacitive discharge. *Sci. Rep.* **4**, 06441 (2014).
5. Ashby, M. F. & Greer, A. L. Metallic glasses as structural materials. *Scripta Mater.* **54**, 321–326 (2006).
6. Demetriou, M. D. *et al.* A damage tolerant glass. *Nat. Mater.* **10**, 123–128 (2011).
7. Zhang, B. *et al.* Amorphous metal plastic. *Phys. Rev. Lett.* **94**, 205502 (2005).
8. Duan, G. *et al.* Bulk metallic glass with benchmark thermoplastic processability. *Adv. Mater.* **19**, 4272–2475 (2007).
9. Schroers, J. *et al.* Blow molding of bulk metallic glasses. *Scripta Mater.* **57**, 341–344 (2007).
10. Sharma, P *et al.* Nanofabrication with metallic glass—an exotic material for nano-electromechanical systems. *Nanotechnology* **18**, 035302 (2007).
11. Kumar, G., Tang, H. X. & Schroers, J. Nanomolding with amorphous metals. *Nature* **457**, 868–872 (2009).
12. Wiest, A. *et al.* Injection molding metallic glass. *Scripta Mater.* **60**, 160–163 (2009).
13. Schroers, J. Processing of bulk metallic glasses. *Adv. Mater.* **22**, 1566–1597 (2010).
14. Schroers, J. Thermoplastic blowmolding of metals. *Materials Today* **14**, 14–19 (2011).
15. Harvey, G. W. & Brower, D. F. Metal forming device and method. US-Patent 2,976,907 (1958).
16. Wilson, F. W. *High Velocity Forming of Metals* (ASTM, Prentice Hall, 1964).
17. Psyk, V. *et al.* Electromagnetic forming-a review. *J. Mater. Process. Technol.* **211**, 787–829 (2011).
18. Lu, J., Ravichandran, G. & Johnson, W. L. Deformation behavior of the $Zr_{41.2}Ti_{13.8}Cu_{12.5}Ni_{10}Be_{22.5}$ bulk metallic glass over a wide range of strain-rates and temperatures. *Acta Mater.* **51**, 3429–3443 (2003).

19. Johnson, W. L. & Demetriou, M. D. Deformation and flow in bulk metallic glasses and deeply undercooled glass forming liquids—a self-consistent dynamic free volume model. *Intermetallics* **10,** 1039–1046 (2002).

20. Demetriou, M. D. & Johnson, W. L. A free volume model for the rheology of undercooled $Zr_{41.2}Ti_{13.8}Cu_{12.5}Ni_{10}Be_{22.5}$ liquid. *Scripta Mater.* **52,** 833–837 (2005).

21. Han, D. C. *Rheology and Processing of Polymeric Materials: Polymer Processing* (Oxford University Press, 2007).

22. Dealy, J. M. & Wissbrun, K. F. *Melt Rheology and its Role in Plastics Processing: Theory and Applications* (Kluwer Academic Publishers, 1990).

23. Tschumi, A. Electrical resistivity and Hall coefficient of glassy and liquid alloys. *J. Non-Cryst. Solids* **61-62,** 1091–1096 (1984).

24. Nagel, S. R. Temperature dependence of the resistivity in metallic glasses. *Phys. Rev. B* **16,** 1694–1698 (1977).

25. Johnson, W. L. *et al.* Rheology and ultrasonic properties of metallic glass-forming liquids: a potential energy landscape perspective. *MRS Bulletin* **32,** 644–650 (2007).

26. Wei, S. *et al.* The impact of fragility on the calorimetric glass transition of bulk metallic glasses. *Intermetallics* **55,** 138–144 (2014).

27. Schroers, J. *et al.* Gold base bulk metallic glass. *Appl. Phys. Lett.* **87,** 061912 (2005).

28. Schroers, J. & Johnson, W. L. Highly processable bulk metallic glass forming-alloys in the Pt-Co-Ni-Cu-P system. *Appl. Phys. Lett.* **84,** 3666 (2004).

29. Johnson, W. L. & Samwer, K. A universal criterion for plastic yielding of metallic glasses with a $(T/T_g)^{2/3}$ temperature dependence. *Phys. Rev. Lett.* **95,** 195501 (2005).

30. Demetriou, M. D. *et al.* Glassy steel optimized for glass-forming ability and toughness. *Appl. Phys. Lett.* **92,** 161910 (2008).

Acknowledgements

We gratefully acknowledge partial support by the II-VI Foundation and Glassimetal Technology, and useful discussions with Konrad Samwer and Joseph P. Schramm.

Author contributions

G.K. and S.R. conducted the electromagnetic forming experiments and imaging analysis, G.K., M.D.D. and W.L.J. wrote the main manuscript.

Additional information

Competing financial interests: M.D.D. and W.L.J. are co-founders of Glassimetal Technology, Inc.

21

Large elasto-optic effect and reversible electrochromism in multiferroic BiFeO₃

D. Sando[1,*,†], Yurong Yang[2,*], E. Bousquet[3], C. Carrétéro[1], V. Garcia[1], S. Fusil[1], D. Dolfi[4], A. Barthélémy[1], Ph. Ghosez[3], L. Bellaiche[2] & M. Bibes[1]

The control of optical fields is usually achieved through the electro-optic or acousto-optic effect in single-crystal ferroelectric or polar compounds such as LiNbO₃ or quartz. In recent years, tremendous progress has been made in ferroelectric oxide thin film technology—a field which is now a strong driving force in areas such as electronics, spintronics and photovoltaics. Here, we apply epitaxial strain engineering to tune the optical response of BiFeO₃ thin films, and find a very large variation of the optical index with strain, corresponding to an effective elasto-optic coefficient larger than that of quartz. We observe a concomitant strain-driven variation in light absorption—reminiscent of piezochromism—which we show can be manipulated by an electric field. This constitutes an electrochromic effect that is reversible, remanent and not driven by defects. These findings broaden the potential of multiferroics towards photonics and thin film acousto-optic devices, and suggest exciting device opportunities arising from the coupling of ferroic, piezoelectric and optical responses.

[1] Unité Mixte de Physique, CNRS, Thales, Univ. Paris-Sud, Université Paris-Saclay, 91767 Palaiseau, France. [2] Department of Physics and Institute for Nanoscience and Engineering, University of Arkansas, Fayetteville, Arkansas 72701, USA. [3] Theoretical Materials Physics, Université de Liège, B-5, B-4000 Sart-Tilman, Belgium. [4] Thales Research and Technology France, 1 Avenue Augustin Fresnel, 91767 Palaiseau, France. * These authors contributed equally to this work. † Present address: School of Materials Science and Engineering, University of New South Wales, Sydney 2052, Australia. Correspondence and requests for materials should be addressed to D.S. (email: daniel.sando@unsw.edu.au) or to M.B. (email: manuel.bibes@thalesgroup.com).

Bismuth ferrite (BiFeO$_3$—BFO) is multiferroic at room temperature with strong ferroelectric polarisation[1] and G-type antiferromagnetic ordering with a cycloidal modulation of the Fe spins[2]. Most research on this material has been driven by the prospect of electrically controlled spintronic devices[3]. More recently, however, BFO has revealed further remarkable multifunctional properties. Notable discoveries include conductive domain walls[4], a strain-driven morphotropic phase boundary[5] and a specific magnonic response that can be tuned by epitaxial strain[6] or electric field[7]. Moreover, with a bandgap (~ 2.7 eV) in the visible[8], large birefringence[9] (0.25–0.3), a strong photovoltaic effect[10] and sizeable linear electro-optic coefficients[11], BFO is garnering interest in photonics and plasmonics[12].

Most of these physical properties are intimately linked to structural parameters, and may thus be tuned in thin films by epitaxial strain. Strain engineering[13] is a powerful tool through which, for instance, ferroelectricity is strongly enhanced in BaTiO$_3$ (ref. 14), or induced in otherwise non-ferroelectric materials such as SrTiO$_3$ (ref. 15). In BFO, two structural instabilities are sensitive to epitaxial strain: the polar distortion—responsible for the ferroelectricity—and antiferrodistortive (FeO$_6$ octahedra) rotations. In strained BFO films, the competition between both instabilities and their coupling to ferroic order parameters yields rich phase diagrams, revealing new structural, ferroelectric and magnetic phases[6,16], as well as large variations in the ferroelectric Curie temperature[17] and the spin direction[6].

Here, we present a combined experimental and theoretical study demonstrating that strain induces a very large change in the refractive index of BFO, which corresponds to an effective elasto-optic coefficient larger than in any ferroelectric, and larger than that of quartz[18]. This effect is accompanied by a shift of the optical bandgap, reminiscent of pressure-induced changes in light-absorption[19], a phenomenon known as piezochromism in other materials systems[20]. The trends in the optical properties as a function of strain are well reproduced by our first-principles calculations, and we are able to clarify precisely why the optical bandgap of tetragonal-like BFO is larger than that of the rhombohedral-like phase. Finally, we show how an electric field can be used to toggle between two strain states with different light absorption, corresponding to an electrochromic effect that is intrinsic, reversible and non-volatile.

Results

Sample preparation and structural characterization.
Fully strained BFO thin films were grown using pulsed laser deposition on (001)-oriented substrates (in pseudocubic notation, which we use throughout this paper) spanning a broad range of lattice mismatch (from -7.0% to $+1.0\%$; Methods section). At low strain—compressive or tensile—the films crystallize in the so-called R-like phase of BFO, derived from the bulk rhombohedral (R3c) phase. At high compressive strain ($\leq 4\%$), the films grow in the T-like phase[16] with a large tetragonality ratio $c/a \approx 1.26$ (cf. Fig. 1b). Reciprocal space maps around the (113)$_{pc}$ or (223)$_{pc}$ reflections (Supplementary Fig. 1) reveal that all our BFO films possess a monoclinic structure (M$_A$ or M$_B$ for R-like, M$_C$ for T-like[12,21]), and further scans (not shown) indicate the presence of two structural domain variants (see the sketch in Fig. 1a). The in-plane and out-of-plane pseudocubic lattice parameters are presented in Fig. 1b. The in-plane parameter shows a monotonic decrease with compressive strain, while the out-of-plane parameter concomitantly increases, albeit with a sharp jump at $\sim -3.5\%$ corresponding to the structural transition between the R-like and T-like phases.

First-principles calculations. To explore the effect of strain on the optical properties of BFO thin films, we performed first-principles calculations, using the Heyd–Scuseria–Ernzerhof (HSE) hybrid functional (see Methods section for details). In the following, we denote the electronic bandgap as the energy difference between the valence band maximum (VBM) and the conduction band minimum (CBM), while the optical bandgap corresponds to the extrapolation of the linear region of the Tauc plot (cf. Fig. 2c); theoretically, the optical bandgap is computed from the complex dielectric function.

Figure 1c,d show the computed electronic density of states for the R-like phase of BFO (at 0, $+2$ and -3% strain) and for the T-like phase (-5 and -7%). The insets of Fig. 1c,d show that the electronic bandgap is lower for the T-like phase than for the R-like phase, particularly for -7% strain, consistent with previous studies[22,23]. In the R-like phase, both compressive and tensile strains yield an increase of the electronic bandgap, similar to the situation in SrTiO$_3$ (ref. 24).

The partial density of states of Fig. 1e-l shows that the VBM mostly consists of O 2p orbitals for any considered misfit strain, and the CBM mainly comprises Fe d_{xy}, d_{xz}, d_{yz} orbitals for both the R-like and T-like phases. In R-like BFO, strain-induced changes in the FeO$_6$ octahedra rotations and the polar modes conspire[24] to slightly lift the degeneracy of the Fe 3d orbitals, but the nature of the electronic states at the VBM and CBM is globally preserved. In contrast, the pyramidal coordination of the FeO$_5$ unit in highly elongated T-like BFO yields a large splitting of the 3d states with the d_{xy} orbital sitting 300 meV lower in energy (Fig. 1h). Figure 1l shows that this d_{xy} state is weakly hybridized with O states and the states near the CBM in the T-like phase have very little O 2p character. This suggests that optical transitions from the VBM to those states should be very weak, and that the main optical transitions in the T-like phase should occur from the VBM to d_{xz} and d_{yz} states that lie in energy ~ 300 meV above the CBM. In other words, the optical bandgap should be higher in T-like BFO than in R-like BFO despite the opposite trend in the electronic bandgap. This presumption is confirmed by the energy dependence of the extinction coefficient derived from our calculations (Fig. 2b): the absorption edge appears at least 200 meV higher in the T-like phase than in the R-like phase.

Optical characterization. Figure 2a presents the experimental energy dependence of the extinction coefficient, extracted from spectroscopic ellipsometry measurements (Supplementary Fig. 2). These data confirm the theoretical prediction of a larger optical bandgap for the T-like phase, and are consistent with previous studies. The agreement between the experimental and calculated extinction coefficient curves (Fig. 2a,b) is very good, particularly for the onset of absorption. The corresponding experimental Tauc plots (cf. Methods section) for these three samples are shown in Fig. 2c, indicating that the optical bandgap for T-like BFO is 3.02 eV, while for R-like BFO, a compressive strain of 2.6% induces an increase in the bandgap from 2.76 to 2.80 eV. Figure 2d summarizes the strain dependence of the experimental and calculated optical bandgap. In the R-like phase, both compressive and tensile strains induce an increase of the optical bandgap, and the T-like polymorph exhibits an optical bandgap ~ 0.25 eV larger than the R-like phase, consistent with previous reports[25].

Our observation of a strain-induced change in optical bandgap and thus optical absorption is reminiscent of an effect called piezochromism[20], which corresponds to changes in light absorption driven by hydrostatic pressure. Piezochromic effects have been identified in several organic compounds, but for

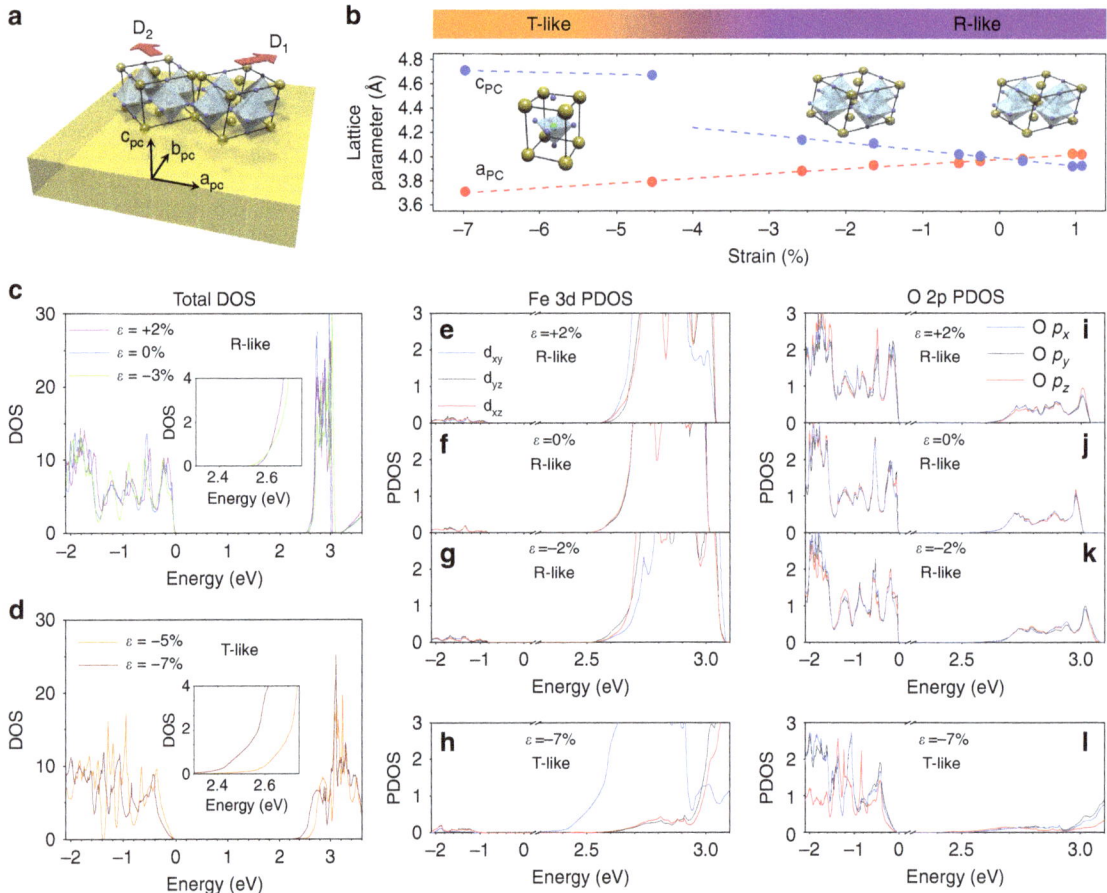

Figure 1 | Electronic structure of strained BiFeO₃ thin films. (a) Sketch of the two structural variants present in our monoclinic (M_A or M_B) R-like BiFeO₃ films. The red arrows indicate the direction of the monoclinic distortion for the two variants D₁ and D₂. **(b)** In-plane and out-of-plane lattice parameters of our strained BFO films. Total density of state (DOS) for R-like **(c)** and T-like phases **(d)**. The insets show the DOS near the CBM. Partial density of states (PDOS) of iron 3d **(e-h)** and oxygen **(i-l)** states for R-like and T-like BFO. Note the break between 0.2 and 2.3 eV in the horizontal axes in **(e-l)**. For all panels, only the spin-up channel states are shown; the spin-down channel states are the same as the spin up due to the antiferromagnetic order.

inorganics mainly in the CuMoO₄ family[26,27]. In this compound, the application of hydrostatic pressure triggers a first order transition between two polytypes having different optical absorption spectra due to changes in the oxygen cage surrounding the Cu ions[26]. Interestingly, in both CuMoO₄ and BiFeO₃ the absorption is stronger when the transition metal cation is in an octahedral oxygen environment, which suggests a possible trend, and strategies for engineering piezochromic effects in other perovskites.

In bulk BFO, the bandgap is known to decrease with pressure[19], particularly below 3.4 GPa and at the structural phase transition near 9.5 GPa. This corresponds to a piezochromic effect of amplitude 0.058 eV GPa⁻¹ at low pressure, and of 0.027 eV GPa⁻¹ on average between ambient pressure and 18 GPa (ref. 19). In our films, from the strain values and Young's modulus[28], we estimate the amplitude of the piezochromic effect in BiFeO₃ at ∼0.12 eV GPa⁻¹. Importantly, working with thin films may be advantageous for several applications[29]. For instance, the thin film geometry allows the application of large electric fields to toggle between two optical polytypes, thereby producing potentially high-speed electrochromic effects.

We have explored this possibility in BFO thin films with coexisting R-like and T-like regions[5]. We applied an electric field to transform the mixed R − T BFO into nominally pure T-like BFO over 10 × 10 μm² regions (cf. Fig. 3a) and probed the local optical transmission in and out of this area using a conventional optical microscope. Figure 3b shows a transmission image with a

dielectric filter (bandwidth 10 nm) centred at 420 nm inserted between the white light source and the sample. Clearly, a 10 × 10 μm² square with a higher intensity than the background is visible in the image. Remarkably, this effect is reversible: applying a voltage with the opposite polarity restores a mixed R + T state (cf. Fig. 3c), which restores a stronger optical absorption, see Fig. 3d. This contrast is stable for several weeks.

We have acquired similar optical images using various dielectric filters, recorded the transmitted intensity in and out of the 10 × 10 μm² square and calculated the contrast difference as a function of wavelength. The contrast is maximal between 420 and 450 nm, see Fig. 3e. This dependence agrees very well with the expected contrast difference, calculated from the extinction coefficients of pure R-like BiFeO₃ and pure T-like BiFeO₃ films of Fig. 2a. This confirms that the contrast in the optical images is indeed due to the intrinsic modulation of the optical bandgap induced by the electrical poling, rather than by defect-mediated processes as in Ca-doped BFO (ref. 30) or WO₃ (ref. 20).

Finally, we focus on the influence of strain on the real part of the complex refractive index n. In Fig. 4a, we highlight representative results of the variation of n with wavelength for R-like BFO that is weakly strained (on SmScO₃, SSO), strongly compressively strained (on (La,Sr)(Al,Ta)O₃ (LSAT)), and T-like BFO (on LaAlO₃ (LAO)). Below the optical bandgap (that is, for wavelengths longer than ∼460 nm), the refractive index systematically decreases with increasing strain. This is also visible in

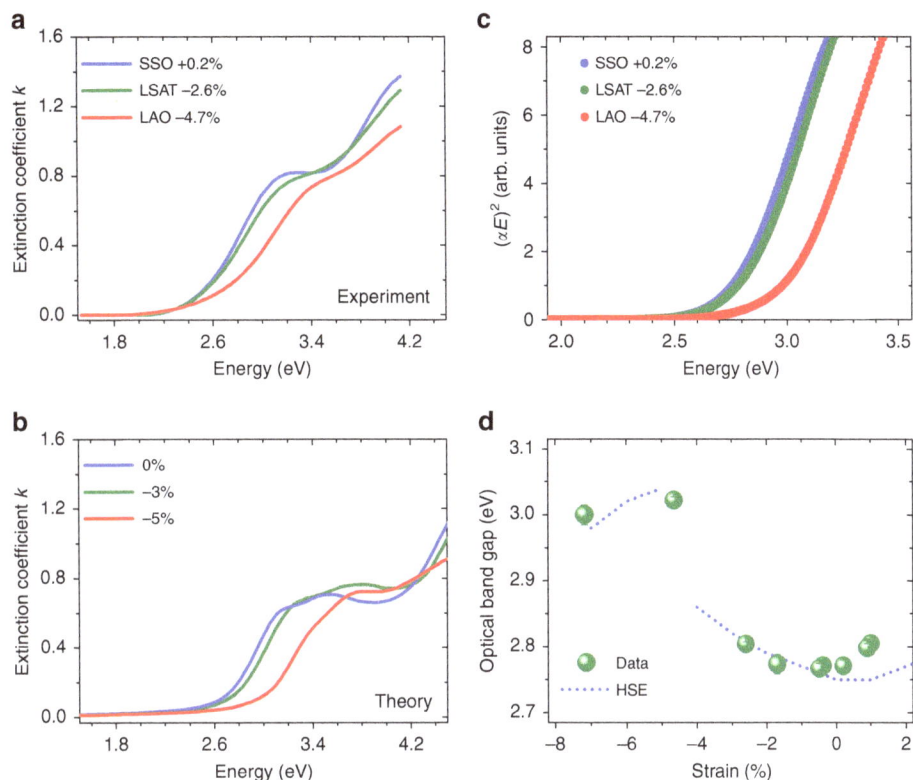

Figure 2 | Optical absorption properties of strained BiFeO₃ films. (**a**) Measured extinction coefficient for three representative strain levels. (**b**) Calculated extinction coefficient for strain levels comparable to those displayed in **a**. (**c**) Tauc plots generated from measurements for representative samples. (**d**) Summary of optical bandgap versus strain results, comparing theory and experiment. The error bars were determined by generating the dispersion laws using the upper and lower bounds of the Tauc–Lorentz oscillator parameters (from their uncertainties) and finding the resultant maximum variation in the bandgap.

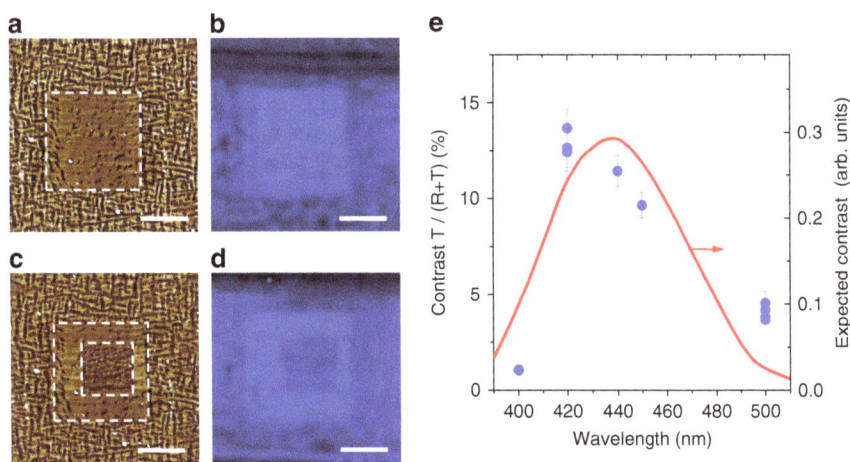

Figure 3 | Electrochromism in BiFeO₃ thin films. (**a**) Topography image after poling a $10 \times 10\,\mu m^2$ square, locally transforming R + T BFO into T-like BFO. (**b**) Transmission optical image acquired in the same region with a dielectric filter centred at 420 nm (bandwidth 10 nm). (**c**) Topography image of the same area after poling a $5 \times 5\,\mu m^2$ region with an opposite voltage, restoring the R + T structure. (**d**) Transmission optical image with a 420-nm filter. The horizontal dark features are due to twin boundaries in the LaAlO₃ substrate. All white scale bars are 5 μm. (**e**) Blue symbols: normalized difference in transmitted light in (T) and out (R + T) of the square in **a** with dielectric filters centred at different wavelengths. Red line: expected contrast calculated from the transmission of pure R-like and T-like films and the transmission function of the dielectric filters. The error bars in **e** are derived from the s.d. of the image pixel values in zones in and out of the T and (R + T) regions.

Fig. 4b, which displays the strain dependence of n at various wavelengths for all samples. The strain-induced change in refractive index measured at 633 nm is reproduced in Fig. 4c

and compared with first-principles calculations. The refractive index is higher in the R-like phase and globally decreases with strain, both compressive and tensile.

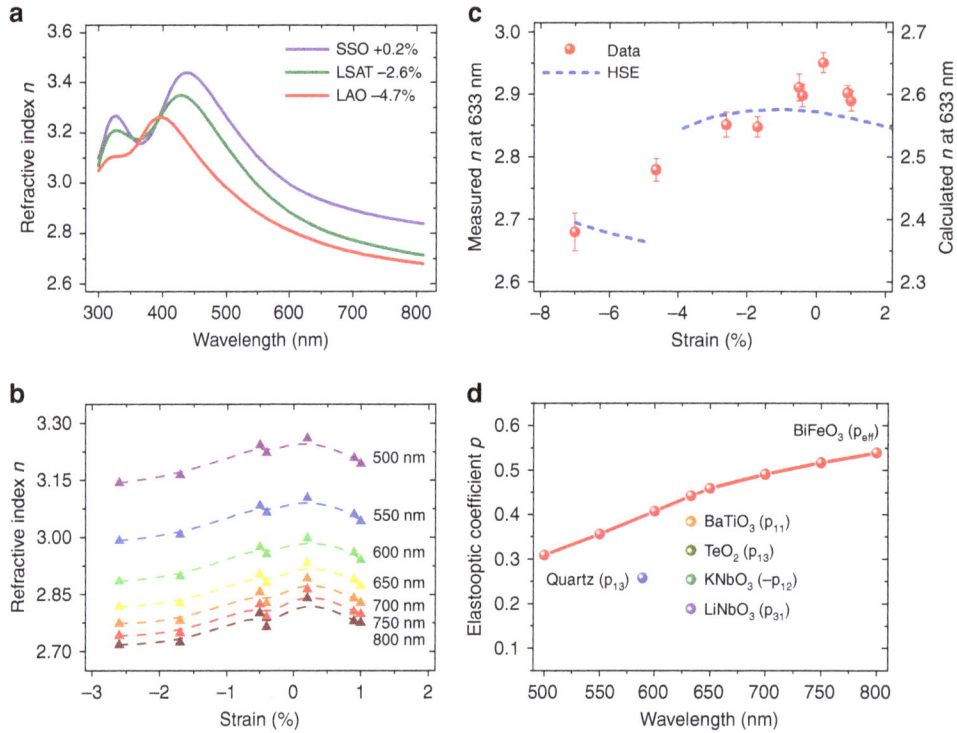

Figure 4 | Optical refractive index and elasto-optic coefficients in strained BiFeO₃ films. (**a**) Measured refractive index n as a function of wavelength for various strain levels. (**b**) Measured refractive index as a function of strain for various wavelengths, for the R-like phase only. The lines serve as guides to the eye. (**c**) Measured and calculated refractive index at 633 nm as a function of strain. The error bars were determined by generating the dispersion laws using the upper and lower bounds of the Tauc–Lorentz oscillator parameters (from their uncertainties) and finding the resultant maximum variation in the refractive index. (**d**) Effective elasto-optic coefficient of BFO as a function of wavelength. Representative reported largest elasto-optic coefficients of various other crystalline materials are plotted for comparison.

Discussion

The experimental results of Fig. 4 indicate that BFO exhibits a strong elasto-optic effect (change in refractive index on physical strain). Taking the slope of the change in $1/n^2$ with strain (Supplementary Fig. 3) for the weakly compressively strained BFO samples at various wavelengths larger than the BFO bandgap (at $\sim 460\,$nm), we obtain an effective elasto-optic coefficient for BFO, as shown in Fig. 4d (Supplementary Note 1). In this figure we also plot the elasto-optic tensor element with the largest magnitude for various elasto-optic media[18,31]. The results suggest that BFO should be a robust elasto-optic medium and, more specifically, that BFO has an effective elasto-optic coefficient at least twice as large as LiNbO₃.

Combined with its relatively low Young's modulus[28] and sound speed[32], the large elasto-optic coefficient of BFO yields an acousto-optic figure-of-merit[18] $M = \frac{n^6 p^2}{E v_{ac}}$ as large as $M = 365 \times 10^{-15}\,$s³ kg^{-1}, a value much larger than for any other material for longitudinal acoustic waves (cf. in TeO₂, $M = 23 \times 10^{-15}\,$s³ kg^{-1} and in LiNbO₃ $M = 1.8 \times 10^{-15}\,$s³ kg^{-1}, ref. 18). Importantly, as BFO may easily be grown[33] on isostructural perovskites with giant piezoelectric responses (such as PMN-PT or PZN-PT)[34], this huge figure-of-merit opens the way towards thin film acousto-optical components[35] with potential performance orders of magnitude greater than those currently based on single crystals. This would extend the potential of BFO to devices to deflect or modulate optical fields, and towards the emerging field of optomechanics, from back-action and laser cooling, to highly integrated sensors and frequency references[36].

More generally, our work has implications for the design of multifunctional devices exploiting the magnetic, ferroelectric or piezoelectric response of BiFeO₃ in conjunction with these unique optical properties. Importantly, the mechanisms that we identify to modulate bandgap and absorption are not specific to BiFeO₃ and can be transposed to many perovskites. Strain-induced elasto-optic and piezochromic effects even larger than in BiFeO₃, possibly by one order of magnitude or more, could be awaiting discovery in other oxide materials, particularly in Mott insulators[37] in which the bandgap falls between $3d$ states and is lower than in BiFeO₃. Giant, electric field-controllable optical absorption in the visible range could thus be exploited, opening the way towards devices harvesting both mechanical and solar energies.

Methods

BiFeO₃ thin film growth and structural characterization. Single phase R-like or T-like films of BFO were prepared by pulsed laser deposition (using the conditions of ref. 38) on the following single crystal substrates: YAlO₃, LAO, LSAT, SrTiO₃, DyScO₃, TbScO₃, SmScO₃, NdScO₃ and PrScO₃. The scandates and YAlO₃ were (110)-oriented (orthorhombic notation) while cubic SrTiO₃, LSAT and rhombohedral LAO were (001)-oriented. The nominal biaxial strain induced by these substrates ranges from -7.0% (compressive) to $+1.0\%$ (tensile). The thickness of the films was 50–70 nm as determined by X-Ray reflectometry, and confirmed by spectroscopic ellipsometry measurements. High-angle X-Ray diffraction 2θ–θ scans, collected with a Panalytical Empyrean diffractometer using CuK$_{\alpha 1}$ radiation, indicated that the films were epitaxial and grew in a single phase. Mixed R + T phase BFO (nominal thickness 100 nm) were grown using a KrF excimer laser at 540 °C and 0.36 mbar on (001)-oriented LAO substrates after the growth of a 10-nm-thick LaNiO₃ bottom electrode at 640 °C and 0.36 mbar.

Optical characterization. The films were characterized using spectroscopic ellipsometry with a UVISEL spectral-scanning near infrared spectroscopic phase-modulated ellipsometer from HORIBA Jobin-Yvon. The incidence angle was 70° and the wavelength range was 300–840 nm (0.62–4.13 eV). This range was imposed by the ellipsometer (maximum energy ~ 4.13 eV), while absorption peaks arising from colour centres in the scandate substrates limited the maximum wavelength to 840 nm. These boundaries do not adversely affect the present analysis since the optical bandgap of BFO is well within the explored spectral range. The raw ellipsometry data were fitted to a multilayer model consisting a semi-infinite substrate, BFO layer and roughness layer implemented by the Bruggeman approximation with a void and BFO mixture. The dispersion law of the BFO layer was described by three Tauc–Lorentz oscillators[8], where the central energies correspond to charge transfer transitions. An example of a typical fit (in this case for BFO on NdScO$_3$) is shown in Supplementary Fig. 2b. The complex dispersion law ($\tilde{n} = n + ik$) of BFO was determined for each sample, and for all fits the mean square error, χ^2, was <2 (as indicated in Supplementary Table 1). To extract the bandgap from the dispersion laws, Tauc plots of $(\alpha E)^2$ versus E were constructed, and the linear region was extrapolated to the E axis (Fig. 2c), yielding the gap value. For each sample, ellipsometric data were collected and the dispersion law and bandgap calculated a minimum of four times, and the results averaged. The error bars displayed in Figs 2d and 4c were determined by generating the dispersion laws using the upper and lower bounds of the Tauc–Lorentz oscillator parameters (from their uncertainties) and finding the resultant maximum variation in the bandgap and refractive index.

First-principles calculations. Calculations were performed within density-functional theory, as implemented in the Vienna *ab initio* simulation package[39,40]. An energy cutoff of 550 eV was used, and the set of projector-augmented wave potentials was employed to describe the electron–ion interaction. We considered the following valence electron configuration: $5d^{10}6s^26p^3$ for Bi, $3p^63d^64s^2$ for Fe and $2s^22p^4$ for O. Supercells containing 20 atoms were used, and G-type antiferromagnetism was adopted. Electronic relaxations converged within 10^{-6} eV and ionic relaxation was performed until the residual force was <1 meV Å^{-1}. We used the PBEsol + U functional[41] (selecting $U = 4$ eV for the Fe ions) to relax the structures, and used both this PBEsol + U functional and the HSE hybrid functional[42] to calculate physical properties such as electronic structure and the dielectric function. These two methods yielded very similar results (hence we only report results for HSE in Figs 1, 2 and 4), with the exception that the PBEsol + U functional underestimated the electronic bandgap by 0.4 eV, while HSE overestimated this bandgap by 0.8 eV. The imaginary part of the dielectric tensor was obtained via

$$\varepsilon''_{\alpha\beta}(\omega) = \frac{4\pi^2 e^2}{\Omega}\lim_{q \to 0}\frac{1}{q^2}\sum_{c,v,\mathbf{k}}2\omega_{\mathbf{k}}\delta(\epsilon_{c\mathbf{k}} - \epsilon_{v\mathbf{k}} - \omega) \times \langle u_{c\mathbf{k}+\mathbf{e}_\alpha q} \mid u_{v\mathbf{k}}\rangle\langle u_{c\mathbf{k}+e_\beta q} \mid u_{v\mathbf{k}}\rangle^*,$$

(1)

where the indices c and v refer to conduction and valence band states, respectively, $u_{c\mathbf{k}}$ is the cell periodic part of the orbitals at the k-point $\mathbf{k}$, and $\mathbf{e}_\alpha$ is a unit vector along the α Cartesian direction[43]. Finally, the real part of the dielectric tensor $\varepsilon'_{\alpha\beta}$ was obtained through the Kramers–Kronig transformation $\varepsilon'_{\alpha\beta}(\omega) = 1 + \frac{2}{\pi}P\int_0^\infty \frac{\varepsilon''_{\alpha\beta}(\omega')\omega'}{\omega'^2 - \omega^2}d\omega'$, where P denotes the principal value. We then obtained the extinction coefficient k and refractive index n by $\tilde{\varepsilon} = \varepsilon' + i\varepsilon'' = (n + ik)^2$. Note that local field effects were neglected in our calculations. The optical bandgap determined from the calculated dielectric function was seen to overestimate the experiment by 0.8 eV; therefore, in all figures in this manuscript, the conduction band has been systematically shifted by 0.8 eV with respect to the VBM to reflect this scissors correction.

For the refractive index we find a systematic quantitative difference of ~ 0.3 between experiment and theory, which can be understood by the fact that first-principles calculations consider defect-free samples, neglect local field and temperature effects, and only incorporate the average between the different components of the dielectric function tensor (which is, additionally, a quantity rather difficult to simulate precisely by *ab initio* methods).

References

1. Lebeugle, D. et al. Room-temperature coexistence of large electric polarization and magnetic order in BiFeO$_3$ single crystals. *Phys. Rev. B* **76**, 024116 (2007).
2. Sosnowska, I., Peterlin-Neumaier, T. & Steichele, E. Spiral magnetic ordering in bismuth ferrite. *J. Phys. C Solid State Phys.* **15**, 4835–4846 (1982).
3. Heron, J. T. et al. Deterministic switching of ferromagnetism at room temperature using an electric field. *Nature* **516**, 370–373 (2014).
4. Seidel, J. et al. Conduction at domain walls in oxide multiferroics. *Nat. Mater.* **8**, 229–234 (2009).
5. Zeches, R. J. et al. A strain-driven morphotropic phase boundary in BiFeO$_3$. *Science* **326**, 977–980 (2009).
6. Sando, D. et al. Crafting the magnonic and spintronic response of BiFeO$_3$ films by epitaxial strain. *Nat. Mater.* **12**, 641–646 (2013).
7. Rovillain, P. et al. Electric-field control of spin waves at room temperature in multiferroic BiFeO$_3$. *Nat. Mater.* **9**, 975–979 (2010).
8. Allibe, J. et al. Optical properties of integrated multiferroic BiFeO$_3$ thin films for microwave applications. *Appl. Phys. Lett.* **96**, 182902 (2010).
9. Rivera, J.-P. & Schmid, H. On the birefringence of magnetoelectric BiFeO$_3$. *Ferroelectrics* **204**, 23–33 (1997).
10. Choi, T., Lee, S., Choi, Y. J., Kiryukhin, V. & Cheong, S.-W. Switchable ferroelectric diode and photovoltaic effect in BiFeO$_3$. *Science* **324**, 63–66 (2009).
11. Sando, D. et al. Linear electro-optic effect in multiferroic BiFeO$_3$ thin films. *Phys. Rev. B* **89**, 195106 (2014).
12. Sando, D., Barthélémy, A. & Bibes, M. BiFeO$_3$ epitaxial thin films and devices: past, present and future. *J. Phys. Condens. Matter* **26**, 473201 (2014).
13. Schlom, D. G. et al. Elastic strain engineering of ferroic oxides. *MRS Bull.* **39**, 118–130 (2014).
14. Choi, K. J. et al. Enhancement of ferroelectricity in strained BaTiO$_3$ thin films. *Science* **306**, 1005–1008 (2004).
15. Haeni, J. H. et al. Room-temperature ferroelectricity in strained SrTiO$_3$. *Nature* **430**, 758–761 (2004).
16. Béa, H. et al. Evidence for room-temperature multiferroicity in a compound with a giant axial ratio. *Phys. Rev. Lett.* **102**, 217603 (2009).
17. Infante, I. C. et al. Bridging multiferroic phase transitions by epitaxial strain in BiFeO$_3$. *Phys. Rev. Lett.* **105**, 057601 (2010).
18. Dieulesaint, E. & Royer, D. *Elastic Waves in Solids II* (Springer-Verlag Berlin Heidelberg, 2000).
19. Gómez-Salces, S. et al. Effect of pressure on the band gap and the local FeO$_6$ environment in BiFeO$_3$. *Phys. Rev. B* **85**, 144109 (2012).
20. Bamfield, P & Hutchings, M. G. *Chromic Phenomena: Technological Applications of Colour Chemistry.* 2nd edn (Royal Society of Chemistry, 2010).
21. Vanderbilt, D. & Cohen, M. H. Monoclinic and triclinic phases in higher-order Devonshire theory. *Phys. Rev. B* **63**, 094108 (2001).
22. Ju, S. & Cai, T. Ab initio study of ferroelectric and nonlinear optical performance in BiFeO$_3$ ultrathin films. *Appl. Phys. Lett.* **95**, 112506 (2009).
23. Dong, H., Liu, H. & Wang, S. Optical anisotropy and blue-shift phenomenon in tetragonal BiFeO$_3$. *J. Phys. D Appl. Phys.* **46**, 135102 (2013).
24. Berger, R. F., Fennie, C. J. & Neaton, J. B. Band gap and edge engineering via ferroic distortion and anisotropic strain: the case of SrTiO$_3$. *Phys. Rev. Lett.* **107**, 146804 (2011).
25. Chen, P. et al. Optical properties of quasi-tetragonal BiFeO$_3$ thin films. *Appl. Phys. Lett.* **96**, 131907 (2010).
26. Rodríguez, F., Hernández, D., Garcia-Jaca, J., Ehrenberg, H. & Weitzel, H. Optical study of the piezochromic transition in CuMoO$_4$ by pressure spectroscopy. *Phys. Rev. B* **61**, 16497–16501 (2000).
27. Gaudon, M. et al. Unprecedented 'one-finger-push'-induced phase transition with a drastic color change in an inorganic material. *Adv. Mater.* **19**, 3517–3519 (2007).
28. Redfern, S. A. T., Wang, C., Hong, J. W., Catalan, G. & Scott, J. F. Elastic and electrical anomalies at low-temperature phase transitions in BiFeO$_3$. *J. Phys. Condens. Matter* **20**, 452205 (2008).
29. Greenberg, C. B. Optically switchable thin films: a review. *Thin Solid Films* **251**, 81–93 (1994).
30. Seidel, J. et al. Prominent electrochromism through vacancy-order melting in a complex oxide. *Nat. Commun.* **3**, 799 (2012).
31. Weber, M. J. *Handbook of Optical Materials* (CRC Press, 2003).
32. Smirnova, E. P. et al. Acoustic properties of multiferroic BiFeO$_3$ over the temperature range 4.2–830 K. *Eur. Phys. J. B* **83**, 39–45 (2011).
33. Biegalski, M. D. et al. Strong strain dependence of ferroelectric coercivity in a BiFeO$_3$ film. *Appl. Phys. Lett.* **98**, 142902 (2011).
34. Park, S.-E. & Shrout, T. R. Ultrahigh strain and piezoelectric behavior in relaxor based ferroelectric single crystals. *J. Appl. Phys.* **82**, 1804–1811 (1997).
35. Lean, E. G. H., White, J. M. & Wilkinson, C. D. W. Thin-film acoustooptic devices. *Proc. IEEE* **64**, 779–788 (1976).
36. Aspelmeyer, M., Kippenberg, T. J. & Marquardt, F. Cavity optomechanics. *Rev. Mod. Phys.* **86**, 1391–1452 (2014).
37. Imada, M., Fujimori, A. & Tokura, Y. Metal-insulator transitions. *Rev. Mod. Phys.* **70**, 1039–1263 (1998).
38. Béa, H et al. Influence of parasitic phases on the properties of BiFeO$_3$ epitaxial thin films. *Appl. Phys. Lett.* **87**, 072508 (2005).
39. Kresse, G. & Furthmüller, J. Efficient iterative schemes for ab initio total-energy calculations using a plane-wave basis set. *Phys. Rev. B* **90**, 11169–11186 (1996).
40. Kresse, G. & Hafner, J. Ab initio molecular dynamics for liquid metals. *Phys. Rev. B* **47**, 558–561 (1993).
41. Perdew, J. et al. Restoring the density-gradient expansion for exchange in solids and surfaces. *Phys. Rev. Lett.* **100**, 136406 (2008).
42. Krukau, A. V., Vydrov, O. A., Izmaylov, A. F. & Scuseria, G. E. Influence of the exchange screening parameter on the performance of screened hybrid functionals. *J. Chem. Phys.* **125**, 224106 (2006).
43. Gajdoš, M., Hummer, K., Kresse, G., Furthmüller, J. & Bechstedt, F. Linear optical properties in the projector-augmented wave methodology. *Phys. Rev. B* **73**, 045112 (2006).

Acknowledgements

This work was supported by the French Research Agency (ANR) projects 'Méloïc', 'Nomilops' and 'Multidolls,' the European Research Council Advanced Grant 'FEMMES'

(Contract No. 267579) and the European Research Council Consolidator Grant 'MINT' (Contract No. 615759). Y.Y. and L.B. thank the financial support of ONR Grant No N00014-12-1-1034 and DARPA Grant No. HR0011-15-2-0038 (Matrix program), and the Arkansas High Performance Computer Center for the use of its supercomputers. Ph.G. acknowledges a Research Professorship from the Francqui Foundation, financial support of the ARC project 'AIMED' and F.R.S.-FNRS PDR project ' HiT4FiT' as well as access to Céci-HPC facilities funded by F.R.S.-FNRS (Grant No 2.5020.1) and the Tier-1 supercomputer of the Fédération Wallonie- Bruxelles funded by the Walloon Region (Grant No 1117545). This work was also supported (E.B.) by F.R.S.-FNRS Belgium, and calculations were partly performed within the PRACE projects TheoMoMuLaM and TheDeNoMo. We thank J.-L. Reverchon for assistance with the ellipsometry measurements.

Author contributions

M.B. conceived and supervised the study with the help of A.B. and D.D. D.S. and C.C. grew the samples and characterized them with X-ray diffraction. D.S. performed ellipsometry measurements and analysed the data. V.G., S.F. and M.B. characterized the electrochromic response. Y.Y, E.B., Ph.G. and L.B. performed first-principles calculations. D.S. and M.B. wrote the manuscript with input from all authors.

Additional information

22

Mapping the electrostatic force field of single molecules from high-resolution scanning probe images

Prokop Hapala[1], Martin Švec[1], Oleksandr Stetsovych[1], Nadine J. van der Heijden[2], Martin Ondráček[1], Joost van der Lit[2], Pingo Mutombo[1], Ingmar Swart[2] & Pavel Jelínek[1]

How electronic charge is distributed over a molecule determines to a large extent its chemical properties. Here, we demonstrate how the electrostatic force field, originating from the inhomogeneous charge distribution in a molecule, can be measured with submolecular resolution. We exploit the fact that distortions typically observed in high-resolution atomic force microscopy images are for a significant part caused by the electrostatic force acting between charges of the tip and the molecule of interest. By finding a geometrical transformation between two high-resolution AFM images acquired with two different tips, the electrostatic force field or potential over individual molecules and self-assemblies thereof can be reconstructed with submolecular resolution.

[1] Department of Thin Films and Nanostructures, Institute of Physics, Academy of Sciences of the Czech Republic, v.v.i., Cukrovarnická 10, 162 00 Prague, Czech Republic. [2] Department of Chemistry, Condensed Matter and Interfaces, Debye Institute for Nanomaterials Science, Utrecht University, PO Box 80 000, 3508 TA Utrecht, The Netherlands. Correspondence and requests for materials should be addressed to P.J. (email: jelinekp@fzu.cz) or to I.S. (email: i.swart@uu.nl).

Scanning probe techniques routinely provide detailed information on the electronic and geometric structure of molecules. For example, the frontier molecular orbitals[1], the chemical structure of molecules[2-4] and bond orders[5] can be imaged. The possibility to image molecules[6] and atomic clusters[7] with nearly atomic resolution, also at elevated temperatures[8-11] provided a great stimulus for surface science[12-17].

From the perspective of chemistry, the capability to measure the charge distribution of a molecule is extremely useful as this property determines the chemical reactivity of a molecule, as well as many other molecular properties[18]. However, imaging the charge distribution within a single molecule remains a challenge. Thus far, kelvin probe force microscopy (KPFM)[19] is the only technique able to measure a quantity that is related to the charge distribution of an individual molecule[20]: the local contact potential difference[21]. The acquisition and unambiguous interpretation of KPFM data on the atomic[22,23] and submolecular level is a non-trivial task[20,24]. One of the primary difficulties is that there is no clear definition to which physical quantity (electrostatic potential, field or surface dipole and so on) the detected signal should be compared (see, for example, discussion in ref. 25). Furthermore, at the typical tip-sample distances required to obtain submolecular resolution in atomic force microscopy (AFM) images, the measured KPFM signal is governed by the complex interplay of local electrostatic fields of tip and sample, their mutual polarization[26], mechanical distortions and the conductance due to overlap of molecular orbitals[27]. In this regime, the usual interpretation of KPFM data is longer valid[28].

Very recently, two alternative techniques, the Scanning Quantum Dot Microscopy[29] and the kelvin probe force spectroscopy[28], were introduced. Both methods partially solve the deficiencies of the KPFM method discussed above. Namely, Scanning Quantum Dot Microscopy is able to provide a quantitative analysis of the electrostatic potential, but only in the far distance regime, limiting the spatial resolution. The kelvin probe force spectroscopy method provides high spatial resolution but suffers from the same drawback of ambiguous definition of the observable as KPFM (ref. 28).

As the charge distribution is to be imaged with high-resolution resolution, the use of chemically passivated tips is essential[30-32]. Several different types of forces and processes have been identified to be important for the contrast in AFM images acquired with such tips. These include the Pauli, van der Waals and electrostatic forces, as well as the flexibility of the functionalized tip[2,5,33-39]. The latter is especially important to understand the distortions in the appearance of molecules in submolecular resolution images[34,36,40-45].

Here we will show that the electrostatic forces acting between probe and an inspected molecule can significantly affect the submolecular contrast. Furthermore, we will show that distortions of the high-resolution images induced by the electrostatic force can be used to map the electrostatic potential of the molecule with submolecular resolution.

Results

General considerations. To illustrate the central idea behind the method proposed here, we consider imaging a neutral molecule with an inhomogeneous charge distribution with a tip terminated by a positively charged flexible probe particle. The probe particle is attracted to regions of excess electron density, whereas it is repelled from regions that have a positive charge. Consequently, positively/negatively charged areas will appear smaller/larger than they really are with such a tip, as illustrated in Fig. 1. The opposite tendency is true for negatively charged tips. Hence, the distortions in submolecular resolution AFM images acquired with charged

tips carry information on the charge distribution within the molecule. Here, we demonstrate how these image distortions can be used to determine the spatial distribution of the electrostatic field above molecules with submolecular resolution. The technique is applied to reconstruct the local electrostatic field of both individual molecules and self-assembled monolayers.

First, let us briefly discuss the origin and character of the apparent bonds or sharp edges in high-resolution AFM/STM images. At close tip-sample distances, the repulsive Pauli interaction induces a significant lateral deflection of the probe particle. There is a discontinuity in the deflection above saddle points of the energy landscape (Fig. 2a,b). The saddle points (sharp edges) are typically present over atoms or bonds at a tip-sample distance where the Pauli repulsion fully compensates the attractive forces. Consequently, the trajectory of the probe particle is split into branches, giving rise to sharp edges[35,38,43]. Hence, the apparent bonds correspond to saddle points of the potential

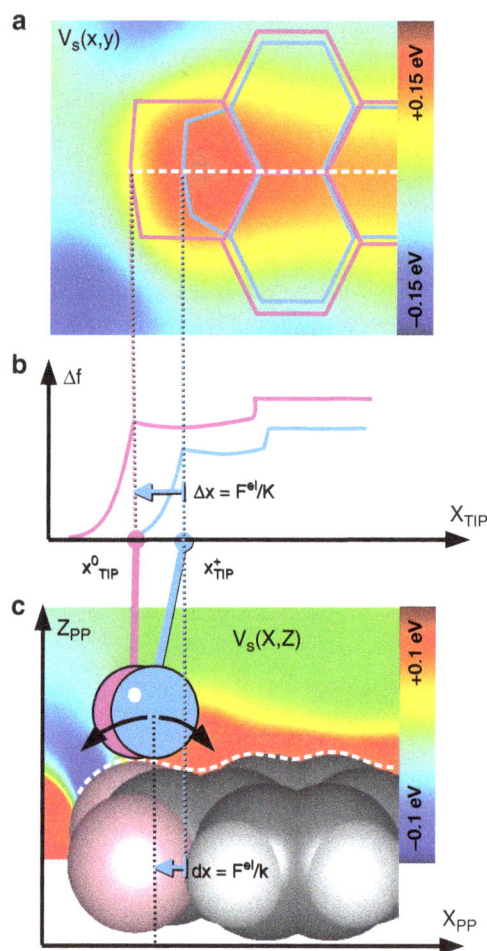

Figure 1 | Schematic view of the impact of the electrostatic forces on the high-resolution AFM images. (**a**) Blue and pink lines represent the positions of sharp edges observed in high-resolution AFM images acquired using positively charged and neutral tips, respectively. Corresponding x,y cut-plane of the Hartree potential $V_S(x, y)$ above the molecule is displayed in the background. (**b**) Variation of the frequency shift Δf as function of the tip positions x_{TIP} for the neutral x^0_{TIP} and positively charged x^+_{TIP} probe particle. (**c**) The lateral relaxation Δx of the probe particle with (blue) and without (pink) an effective charge on the tip is different above a molecule due to the presence of the electrostatic force. The electrostatic force originates from the interaction between the electrostatic potential of the molecule (V_S) on the surface and the effective charge on the probe particle at its given position x_{PP}.

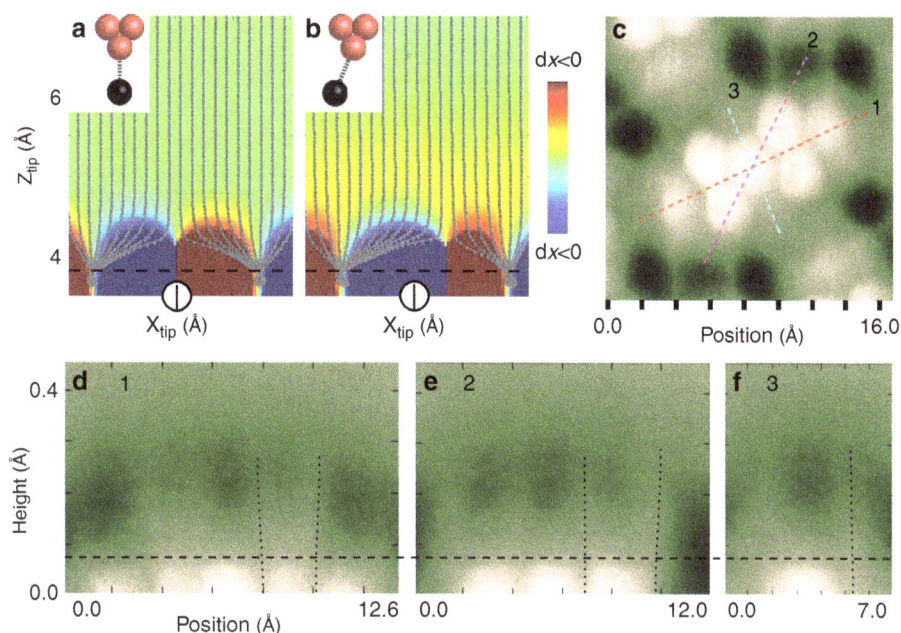

Figure 2 | Height dependence of the position of the sharp edges in AFM images. (a) Simulated deflections of the probe particle as a function of tip position. Grey lines represent the additional deflection of the probe particle with respect to its optimal configuration in far distance. The deflection of the probe particle to the left and right is indicated by blue and red, respectively. **(b)** Same as **a** but now in the presence of an additional constant lateral force (see Supplementary Fig. 2 and Supplementary Note 2 for more details). **(c)** Constant-height AFM image of PTCDA on Ag(111). Dashed lines with numbers 1,2,3 denote positions of different vertical cut planes. **(d–f)** Evolution of Δf with height along the lines shown in **c**. The black dotted lines guide the eye to see relevant edges. The horizontal dashed line indicates the approximate z-distance of contrast inversion.

energy surface experienced by the probe particle at a certain tip–sample distance.

Figure 2a shows the simulated deflections of the probe particle on tip approach over a one-dimensional chain of atoms separated by 2.9 Å (corresponding to the width of a typical benzene ring). The lateral deflection of the probe particle to the left and right is depicted in blue and red, respectively. Note that the trajectories of the probe particle are split into two branches. The deflection depends non-linearly on the tip height. However, the position of the sharp discontinuous boundary between bending left and right, that is, between blue and red regions, does not depend on the tip–sample distance. This is in agreement with our experimental observation that at close tip–sample distances the apparent position of various sharp edges in AFM images of a perylene-3,4,9,10-tetracarboxylic dianhydride (PTCDA) molecules on Ag(111) does not change with distance (Fig. 2c–f). This finding can be rationalized by the fact that the position of sharp edges is determined by the distance where the bifurcation of the probe particle trajectory on tip approach happens. Consequently, while the lateral distortions of the probe particle may be large beyond this point, the lateral apparent position of the sharp edge remains constant.

The total lateral force $\mathbf{F}_{tot}$ is the sum of van der Waals ($\mathbf{F}_{vdW}$), Pauli ($\mathbf{F}_{Pauli}$) and optionally electrostatic ($\mathbf{F}_{el}$) forces. The presence of the lateral force ($\mathbf{F}_{tot}$) induces a lateral deflection ($d\mathbf{x}$) of the flexible probe particle with respect to the tip position ($\mathbf{x}_{tip}$), see Fig. 2b. As long as $d\mathbf{x}$ is small, it is linearly proportional to $\mathbf{F}_{tot}$ acting on the probe particle, according to Hooke's law: $d\mathbf{x} = \mathbf{F}_{tot}/K$ (refs 34,39), where K is the lateral bending stiffness of the bond between the probe particle and the tip. Variation of the lateral electrostatic force $\mathbf{F}_{el}$ causes a shift of the characteristic feature at a different lateral tip position (indicated by $\mathbf{x}_{tip}^{+}$) with respect to the position with the absence of the electrostatic lateral force ($\mathbf{x}_{tip}^{0}$) as shown in Fig. 1. Consequently, the positions of the apparent bonds ($\mathbf{x}_{tip}$) in high-resolution AFM images do not correspond to their actual positions on the surface.

From the above, it is clear that the apparent shift of the characteristic features ($\Delta\mathbf{x}$) in AFM/STM images, therefore, carries information about the lateral forces $\mathbf{F}_{tot}$ with atomic resolution. The apparent shift is linearly related to the deflection of the probe particle from the tip base: $\Delta\mathbf{x} = \gamma d\mathbf{x} = \gamma(\mathbf{x}_{tip} - \mathbf{x}_{PP})$, see Fig. 1. Here, $\mathbf{x}_{PP}$ denotes the actual position of the probe particle and $\gamma \approx 2$. A detailed quantitative analysis including the definition of the γ coefficient can be found in Supplementary Note 1. In the following discussion, we will express everything in an experimentally observable $\Delta\mathbf{x}$ instead of in $d\mathbf{x}$. In this notation, $\Delta\mathbf{x}$ is linearly proportional to the lateral force: $\Delta\mathbf{x} = \mathbf{F}_{tot}/k$, where $k = K/\gamma$ is an effective lateral stiffness.

In our analysis of the electrostatic field, we will use the differences in the apparent positions of sharp contours recorded with two different tips. Specifically, we extract and compare the apparent positions of the same contour feature (for example, a particular atom vertex or bond edge) from two high-resolution AFM images obtained with different tips or scanning conditions (labelled tip A and tip B) at approximately the same tip–sample distances. The apparent position of features acquired with the different tips are indicated by $\mathbf{x}_{tip,A}$ and $\mathbf{x}_{tip,B}$, respectively. In the following, we are interested in the relative difference of the apparent positions $\delta\mathbf{x} = \mathbf{x}_{tip,A} - \mathbf{x}_{tip,B}$. Because we measure the same object on the surface, the real position of any atom or bond that corresponds to a particular contour feature is the same for both images. Therefore, $\delta\mathbf{x}$ can be expressed as the difference between the image distortions in the two images of the same object:

$$\delta\mathbf{x} = \mathbf{x}_{tip,A} - \mathbf{x}_{tip,B} \qquad (1)$$

$$= \Delta\mathbf{x}_A - \Delta\mathbf{x}_B \qquad (2)$$

$$= \mathbf{F}_A/k_A - \mathbf{F}_B/k_B. \qquad (3)$$

Here, $\mathbf{F}_A$ and $\mathbf{F}_B$ are the total lateral force acting on the probe particle in case A and B, respectively. As shown in the Supplementary Note 1, only the van der Waals and electrostatic force components of the total force contribute to the apparent lateral distortions $\Delta\mathbf{x}$. The reason why $\mathbf{F}_{Pauli}$ can be ignored is connected with the rather abrupt onset of the Pauli repulsion, as we explain in Supplementary Fig. 1 using our hard-sphere model. The distortion then depends linearly on the lateral van der Waals and electrostatic forces (Supplementary Note 2 and Supplementary Fig. 2). In the following, we therefore use the following expression for the differences in the lateral distortions:

$$\delta\mathbf{x} = \frac{\mathbf{F}_{vdW,A} + \mathbf{F}_{el,A}}{k_A} - \frac{\mathbf{F}_{vdW,B} + \mathbf{F}_{el,B}}{k_B}. \tag{4}$$

In this general form, there is unfortunately no clear way how to attribute partial relative distortions to each force component, for example, to determine the lateral electrostatic force field $\mathbf{F}_{el}$ component only from the high-resolution images. However, under certain assumptions and/or with the help of numerical simulations this problem can be circumvented.

We will now discuss an approach to extract the lateral electrostatic force field component from equation (4). A detailed discussion is given in Supplementary Note 3. In general, the effective stiffness (k) and charge (Q) of the tips, as well as the van der Waals contributions are different for each tip. The charge of the probe particle depends on the configuration and chemical nature of the metal apex[46], as well as how the Xe/CO is coordinated. Hence, even tips terminated with the same species can have a different charge. First, we will consider the simplest case: two high-resolution images are acquired with (nearly) identical tips, for example, Xe-terminated metallic tips, differing only in their effective charge. In principle, they can have different dipoles instead of effective charges, but this would not change the conclusion drawn later (Supplementary Note 4 and Supplementary Fig. 3). In this case, the following approximations hold: (i) the effective lateral stiffness k of both tips is identical or very similar (that is, $k_A \approx k_B = k$); (ii) the lateral components of the van der Waals forces for tip A and B are also almost identical at a given tip-sample distance. Since both tips are used to image the same object, the surface electrostatic field ($\mathbf{E}_S$) must be the same in both images. Under these approximations, equation (4) simplifies to:

$$\begin{aligned} \delta\mathbf{x} &= \frac{\mathbf{F}_{el,A} + \mathbf{F}_{el,B}}{k} \\ &= \frac{Q_A - Q_B}{k}\mathbf{E}_S, \end{aligned} \tag{5}$$

where Q_A, Q_B are the effective charges of tips A and B, respectively. This equation shows that we can obtain quantitative information about $\mathbf{E}_S$ directly from the difference in the image distortions. The only parameters that are needed are the effective lateral stiffness k and the difference between the effective charges Q_A and Q_B. These can be estimated for each tip from a direct comparison between experimental and simulated high-resolution AFM/STM images[35,47] (for more details see Supplementary Method: correlating the experimental and theoretical data sets to obtain the probe characteristics K and Q and Supplementary Fig. 4). Alternatively, k can be obtained directly from experimental measurements[48].

Next, we consider a more general case where the van der Waals contribution for the two tips is significantly different but one of the tips is neutral, that is, $Q_B \approx 0$. This would correspond to the situation where macroscopically different tips, possibly with different tip termination, are used. In this case, we obtain the following relation for the surface electrostatic field $\mathbf{E}_S$ (for details see Supplementary Note 3):

$$\mathbf{E}_S = \frac{k_A(\delta\mathbf{x} - \delta\mathbf{x}_{vdW})}{Q_A}, \tag{6}$$

where $\delta\mathbf{x}_{vdW} = \mathbf{F}_{vdW,A}/k_A - \mathbf{F}_{vdW,B}/k_B$. The differences of the van der Waals deformation field components ($\delta\mathbf{x}_{vdW}$) can be

estimated from numerical simulations, as discussed later. We note here that the effect of the van der Waals contribution is indispensable only at the periphery of the molecule. Although it may be difficult to extract absolute values in this scenario, the overall shape of the electrostatic field is preserved.

To test our approach to measure the electrostatic field with submolecular resolution, we performed two different sets of experiments. First, we studied a densely packed self-assembled layer of PTCDA on Ag(111) with two differently charged but otherwise similar Xe-terminated tips. Second, we studied individual 1,5,9-trioxo-13-azatriangulene (TOAT) molecules on Cu(111) with a neutral CO terminated tip and a positively charged Xe tip.

Molecular layers of PTCDA on Ag(111). In the case of densely packed self-assembled molecular layers, the van der Waals force component varies slowly. In addition, the effective stiffness k for different tips typically has similar values (see later). Therefore, the term $\delta\mathbf{x}_{vdW}$ in equation (6) can be neglected and the surface electrostatic field is given by $\delta\mathbf{E}_S = k_A\delta\mathbf{x}/Q_A$.

Figure 3a,b show two constant-height AFM images of a self-assembled monolayer of PTCDA on Ag(111) acquired with a neutral and positively charged Xe tips at the same tip-sample distance (for more details see Methods section). Note that the apparent size of the anhydride groups, indicated by the green circles, is different in the two images. We attribute these differences to a repulsive electrostatic interaction between the positively charged Xe tip and the positively charged anhydride groups[49]. This assignment is supported by a very good agreement between experimental and simulated AFM images of PTCDA/Ag(111) with different effective charges on the Xe tip (Fig. 3c,d). The determined values for k_A, k_B, Q_A and Q_B are: 0.16 Nm^{-1}, 0.20 Nm^{-1}, 0.0 e and $+0.3$e, respectively, as shown in Supplementary Fig. 4.

The abundant presence of sharp features in the AFM images allows us to use an automatic computer algorithm to determine the differences in the image distortions, that is, $\delta\mathbf{x}$. First, two AFM images are brought into register. Subsequently, the distortion field is found by comparing the corresponding sharp features between the two images. The algorithm is based on matching small regions of the two images. The image is divided into small tiles, in our case regularly distributed circular areas with a diameter of the characteristic image feature (for example, C–C bonds, carbon hexagons and so on). These circular areas are each matched to the other image by moving them laterally, searching for maximum correlation. The resulting shift vectors represent a good approximation of the distortion between the two images, and serve as input for the electrostatic potential determination. The green grids plotted in Fig. 3e,f visualize the determined deformation.

As argued above, the as-obtained deformation field (grey arrows in Fig. 3g) is linearly proportional to the lateral electrostatic field above the molecular layer, with proportionality constant k_A/Q_A. The electrostatic potential obtained from the experimental images is shown in Fig. 3g and is in very good agreement with the electrostatic potential as calculated from density functional theory (DFT; Fig. 3h). The absolute magnitude $+0.04$ to -0.04 eV of the electrostatic field as determined experimentally is approximately three times smaller than estimated from DFT calculations. This discrepancy can be attributed to several effects, such as uncertainties in the absolute tip-sample distance where the electrostatic potential is measured; in the values of the effective charge Q and lateral stiffness k; in the finite oscillation amplitude and so on. We will address the limits of the method and their possible solutions later (see Discussion).

From the correlation analysis of the experimental and theoretical AFM images of PTCDA (Supplementary Fig. 5), we can estimate the uncertainties in Q and k. For the neutral tip, the maximum correlation is well defined within $\pm 0.05e$ and $\pm 0.08\,Nm^{-1}$. However, for the positively charged tip there are multiple Q-k combinations that provide a similar correlation between experiment and theory. As the scaling term of the vector field is the ratio Q/k, we can estimate the systematic error from its variation. By selecting the different favourable Q,k pairs, we find a systematic error of approximately 20%. It is important to note that this uncertainty only affects the absolute values, that is, the relative variation of the electrostatic potential is correct.

Single TOAT molecule on Cu(111). As the second example, we studied individual TOAT molecules, since they have a highly non-homogeneous charge distribution. The central N atom donates an electron to the delocalized π-system and is thus positively charged. In contrast, the three ketone groups at the edge of the molecule withdraw electron density from the ring system and therefore have a partial negative charge. High-pass filtered constant-height AFM images of a TOAT molecule on Cu(111) acquired with CO and Xe tips are shown in Fig. 4a,b, respectively. There are significant differences between images acquired with the two different tip terminations. The central region of the molecule appears smaller while the peripheral benzene rings are elongated for images acquired with a Xe-terminated tip compared to images obtained with a CO tip. Again, this effect is attributed to a repulsive interaction between the positively charged Xe tip and the positively charged central area

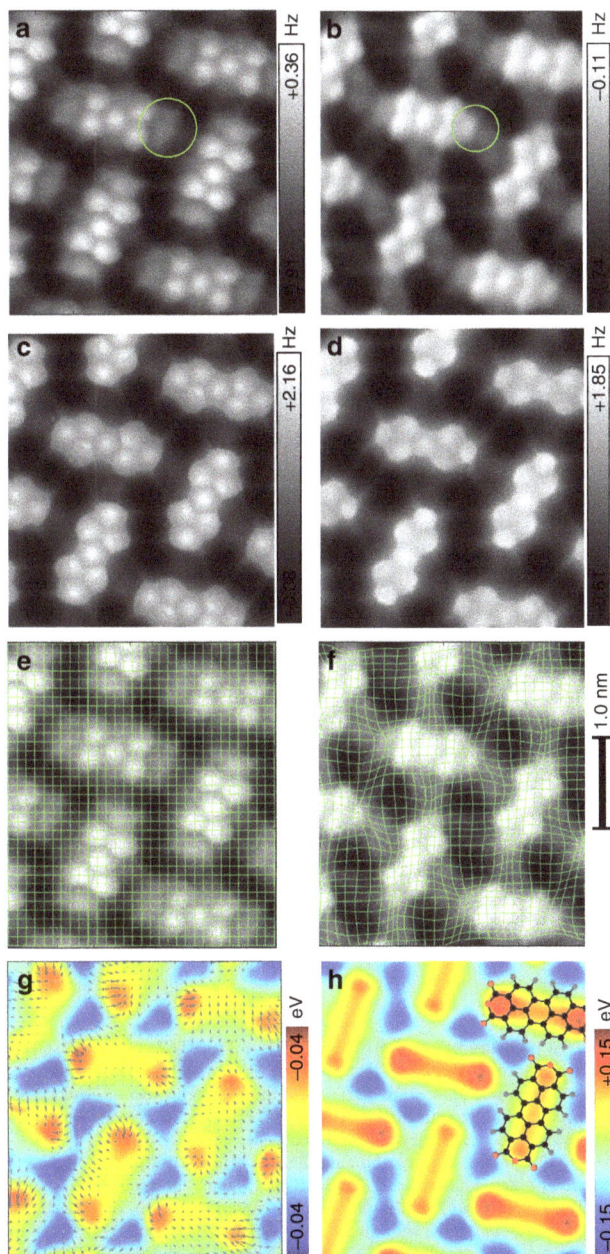

Figure 3 | Determining the electrostatic field above a close-packed PTCDA layer. (**a,b**) experimental high-resolution AFM images of a self-assembled monolayer of PTCDA deposited on Ag(111) obtained with two different Xe tips. (**c**) simulated AFM image using an effective charge $Q = 0.0e$ and the effective lateral stiffness $k = 0.16\,Nm^{-1}$. (**d**) same as **c** but with $Q = +0.3e$ and $k = 0.20\,Nm^{-1}$. (**e,f**) the experimental images superimposed with a deformation grid defined by comparing the corresponding sharp features between the two images in **a** and **b**. (**g**) electrostatic potential calculated from the deformation field (grey arrows). (**h**) calculated Hartree potential from DFT simulations 3.0 Å above the molecular layer.

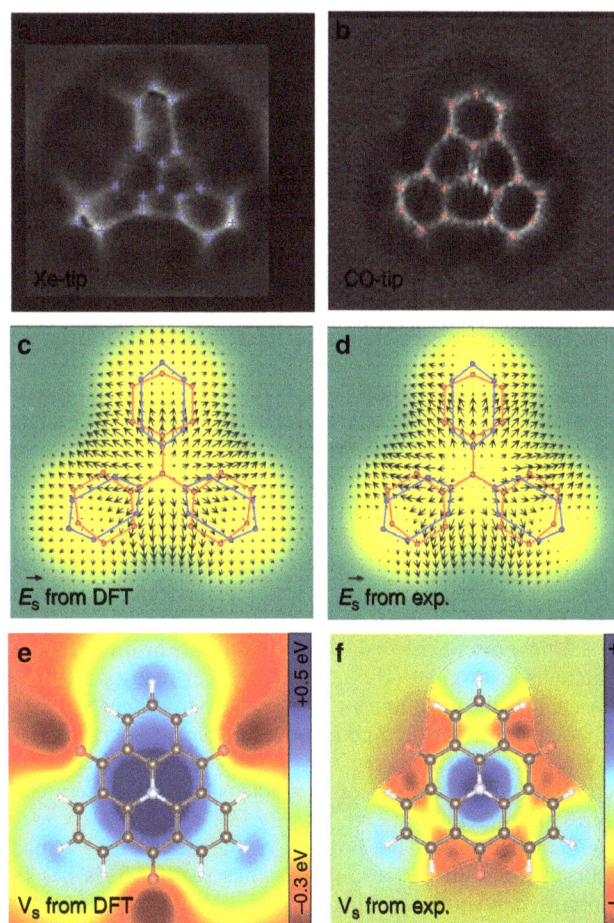

Figure 4 | Determining the electrostatic field above an individual molecule. (**a**) High-pass filtered constant-height AFM images of a TOAT molecule on Cu(111) acquired with a Xe tip. Crosses indicate characteristic vertices. (**b**) Same as **a** but measured with a CO tip. (**c**) electrostatic force field calculated from DFT. (**d**) experimentally determined electrostatic force field obtained after subtraction of the van der Waals component from the deformation field obtained from the images shown in **a** and **b**. (**e**) calculated Hartree potential; (**f**) electrostatic potential calculated from the experimental deformation field shown in **d**.

of the molecule. This effect is reproduced by our simulations for a Xe tip with an effective charge $Q = +0.3$ e and lateral stiffness $k = 0.24\,\text{Nm}^{-1}$. Similarly, we found the best match between experimental and theoretical AFM images acquired with the CO tip with an effective charge $Q = 0.0$e and lateral stiffness $k = 0.24\,\text{Nm}^{-1}$ (details can be found in Supplementary Fig. 5). For individual molecules, positions of vertices were determined manually (blue and red markers in Fig. 4a,b). The deformation field can be obtained by alignment of corresponding vertices ($\delta\mathbf{x}$) in the two AFM images using interpolation by radial basis functions and exploiting the threefold rotational symmetry of the TOAT molecule (see Supplementary Method: estimation of the image distortion from high-resolution AFM images of TOAT molecule and Supplementary Fig. 6). The obtained deformation field shown in Supplementary Fig. 7 is directly proportional to total lateral force field.

Here, the estimated lateral force field also contains the van der Waals force component, which can be determined with the help of numerical modelling[34]. Thus the electrostatic field $\mathbf{E}_S$ can be reconstructed from equation (6) using the fitted lateral stiffness (k_A, k_B) and differential van der Waals deformation field $\delta\mathbf{x}_{vdW}$ (details of how $\delta\mathbf{x}_{vdW}$ was subtracted are provided in the Supplementary Method: subtraction of van der Waals component from distortion field on TOAT). Figure 4d shows the as determined electrostatic field while its calculated counterpart is given in Fig. 4c. The agreement between theory and experiment above the molecule is again very good. Note that the method cannot provide resolution outside of the molecule (green area in Fig. 4c,d), due to the lack of sharp features in this region. Hence we nullified the obtained electrostatic field in this area.

For the TOAT molecules, we obtained the experimental electrostatic potential V_S by determining the derivative of the electrostatic field $\mathbf{E}_S$. The resulting electrostatic potential, shown in Fig. 4f, matches the calculated electrostatic field over the molecule including a complex charge distribution on the benzene lobes. The charge distribution near the oxygen atoms can not be described properly due to the lack of sharp features in this area. In the TOAT case, our method cannot reliably quantify the absolute magnitude of the electrostatic potential due to uncertainties associated with subtraction of the vdW force field and the absence of sharp features outside the molecule. We decided not to provide a quantitative comparison of the electrostatic potential V_S to avoid an over-interpretation of our method.

Discussion

We will now discuss several important aspects of the method to facilitate its assessment. The resolution of our method is directly connected to the requirement of having sharp edges in images originating from a saddle point of the potential energy surface. Therefore, the method can only be used to determine the electrostatic field at close tip-sample distances where such features are present.

As discussed above, the position of the sharp edges remains practically constant in the close distance regime. This has two important consequences. First, the position of sharp edges is determined by a bifurcation (cusp) in the probe particle trajectories upon tip approach. The deflection of the probe particle beyond this branching point does not further affect the apparent position of the edges in the images. Consequently, only the value of the lateral spring constant k at the tip-sample distance where the branching occurs will influence the results. Therefore, variations of the lateral stiffness k with tip-sample distance[39] do not affect the analysis. Second, our method can only map the electrostatic field at the height where the trajectory of the probe particle branches.

It is important to note that sharp edges are visible in both simultaneously acquired AFM and STM channels, as shown in Supplementary Fig. 8. In addition, sharp features are also present in high-resolution STM (ref. 3) and IETS-STM (ref. 4) images. Hence, in principle, our method can also be applied to such data.

The possibility to extract quantitative information depends critically on several factors. First, uncertainties in the values of k and Q could potentially be eliminated by acquiring two images with the same functionalized tip, the charge of which can be effectively modified by other means (for example, by some oxidation/reduction process of the moiety attached to the tip). A search for new functionalized tips with the possibility to modulate an effective charge without loosing its mechanical stability is the subject of current investigations. Alternatively, one can try to reduce the uncertainties in k and Q by, for example, using more sophisticated algorithms for image analysis. The simulation of the electrostatic interaction can be further improved by implementation of a more realistic charge distribution on the probe particle using dipole/quadrupole or even the charge distribution obtained directly from $ab\ initio$ calculations[50].

In conclusion, we showed that the electrostatic interaction between the probe and a molecule on the surface affects distortions in high-resolution images. In particular, the electrostatic field originating from polar molecules can be mapped with high resolution by analysing the differences in the distortions in images acquired with differently charged tips. The arguments and results presented above demonstrate the background, advantages and limitations of the method to probe the electrostatic potential of molecules with submolecular spatial resolution. The main advantages of this method are the clear relation between the physical observables and the electrostatic field, the high spatial resolution and its applicability to STM and AFM images. In addition, it offers the prospect of extracting quantitative information. Here, we applied the method for molecules, but it can be easily extended to surfaces and surface defects (for example, impurities, vacancies, subsurface defects and so on). As such, it constitutes a valuable complementary tool to existing techniques.

Finally we would like to stress that the general idea behind the technique can be applied to any lateral force acting on the last atom of the tip (the probe particle). Consequently, new potential applications can be envisaged, such as imaging the electrostatic field of the probe itself or that of excited molecules. In addition, it may be possible to map molecular magnetic field as well.

Methods

AFM/STM measurements of PTCDA/Ag(111). The PTCDA on Ag(111) experiments were carried out with a Specs LT STM/AFM with a commercially available Kolibri sensor, operating at $\sim 1.2\,\text{K}$ in ultra-high vacuum. Kolibri sensor parameters used in experiment are: $f_0 \approx 985,387\,\text{Hz}$, $Q \approx 230,000$ and $A \approx 70\,\text{pm}$.

The Ag surface was cleaned by repeated cycles of sputtering (Ar$^+$, $p\text{Ar} \approx 5 \times 10^{-6}\,\text{mbar}$, 10 min) and annealing ($\approx 800\,\text{K}$, 5 min). PTCDA was evaporated in ultra-high vacuum ($P < 1.5 \times 10^{-9}\,\text{mbar}$) for 4 min from a crucible thermally heated to $\approx 673\,\text{K}$. Evaporation was performed $\approx 10\,\text{min}$ after the final annealing of the Ag sample with no post evaporation annealing. Xe (99.99% purity) was deposited on the cold sample (T < 10 K) by opening shutters for $\approx 14\,\text{s}$ to $p_{Xe} = 5 \times 10^{-7}\,\text{mbar}$. The tip was functionalized in two steps. First, a metal terminated tip was obtained by few nm dipping of the sensor into the clean Ag surface with a $\approx 2\,\text{V}$ bias pulse. Second, Xe-terminated tip was obtained by spontaneous picking up a Xe atom from a Xe island by the metal terminated tip, while scanning in STM mode (0.1 V, 10 pA).

The acquisition of the three-dimensional (3D) force maps was done automatically, by measuring a sequence of constant-height images and changing the tip-sample separation in between the subsequent images. Apart from the frequency shift, tunnelling current, dissipation and also the amplitude channels have been recorded simultaneously. The step in z was chosen to be in the order of picometres and positive, that is, increasing the tip-sample distance.

Images acquired with different tips were aligned vertically using the following procedure. First, for each tip a data cube with simulated 3D frequency shift values is generated for a particular set of k and Q values. The offsets in z-distance of the experimental and theoretical data sets are then determined by aligning the z-position of the frequency shift minimum for the centres of the molecules. Once this information is available for each tip, images corresponding to approximately the same tip-sample distance can be selected.

AFM/STM measurements of TOAT/Cu(111).

Individual TOAT molecules on Cu(111) with a neutral CO terminated tip and a positively charged Xe tip were imaged using a Scienta-Omicron LT STM/AFM with a commercially available Qplus sensor, operating at ~ 4.6 K in ultrahigh vacuum with an average pressure of 5×10^{-10} mbar. The baked qPlus sensor (3 h at 120 °C) had a quality factor of $Q = 30{,}000$, a resonance frequency of $f_0 = 25{,}634$ Hz and a peak-to-peak oscillation amplitude of approximately 2 Å.

A Cu(111) crystal surface was cleaned with several sputter and anneal cycles before inserting it in the microscope head. The TOAT molecules were thermally evaporated onto the cold surface using an e-beam evaporator (Focus GmbH). For STM imaging, the bias voltage was applied to the sample. After approaching the tip to the surface, an atomically sharp metal tip was prepared by controlled crashes into the copper surface and bias pulses. Each chemically passivated tip was prepared by subsequent pick-up of either a Xe atom or CO molecule[2,51-53]. After a free-lying TOAT molecule was located on the surface, the tip was left in tunnelling contact ($I = 10$ pA at $V = 0.1$ V) and allowed to relax for 12 h to minimize drift and piezo-creep. All AFM images were acquired in constant-height mode. After each AFM image, the STM feedback loop was enabled for 2 s to further minimize tip-sample drift. A complete stack of images resulting in a 3D force grid took ~ 13 h to acquire.

DFT calculation of PTCDA/Ag(111).

We used a pre-optimized herringbone structure of PTCDA molecules on Ag(111) surface[54] consisting of two molecules in the unit cell and a slab of 3 Ag layers (99 Ag atoms). The Hartree potential used for generating the theoretical electrostatic force field[35] was obtained from self-consistent total energy DFT using the Vienna ab initio simulation package[55] with generalized gradient approximation based functional PW91 (ref. 56) and projector augmented-wave method[57]. Plane wave basis set was chosen with $E_{cut} = 396$ eV.

DFT calculation of TOAT/Cu(111).

Total energy DFT calculations were performed using the FHI-aims code[58]. We used a 6×6 supercell made of four Cu layers to describe the Cu(111) surface. The TOAT molecule was placed on the surface with the N atom in a top position. This position was chosen based on the experimental findings. All the atoms except the two bottom Cu layers were relaxed until the remaining atomic forces and the total energy were below 10^{-2} eV Å^{-1} and 10^{-5} eV, respectively. A Monkhorst-Pack grid of $2 \times 2 \times 1$ was used for the integration in the Brillouin zone. All the calculations were carried out at the GGA-PBE level[59] including the Tkatchenko-Scheffler treatment[60] of the van der Waals interaction. The use of van der Waals interactions was necessary to correctly describe the interaction between the molecule and surface. The basis set, pseudo-potentials, integration grids and Hartree potential accuracy were specified using the 'tight' settings. For species like H, O, N the basis set level was set to 'tier 2' while for Cu a first tier was used. Note that a 'tier' represents a single set of radial functions added to the minimal basis to effectively describe the chemical bond.

AFM simulations.

To calculate high-resolution AFM images we used a home built AFM simulation toolkit[34,35] (avaible opensource at https://github.com/ProkopHapala/ProbeParticleModel; see also webpage http://nanosurf.fzu.cz/ppr/). We used default parameters of pairwise LJ potentials[34]. The optimized structures and corresponding surface Hartree potentials were obtained from fully relaxed total energy DFT simulations of the system, see above. The effective tip charge Q and lateral stiffness k of probe particle are a free input of the model. The positions of the $\Delta f(z)$ minima in the centres of the molecules were taken as the reference points in Z, for both the experiment and theory.

Processing of experimental data and matching to simulation.

Iterative algorithm for registration of experimental AFM/STM images using linear correlation is described in the Supplementary Method: data set registration procedure and Supplementary Fig. 9. The estimation of the effective stiffness K and effective charge Q is described in the Supplementary Method: correlating the experimental and theoretical data sets to obtain the probe characteristics K and Q and Supplementary Fig. 4 for PTCDA and Supplementary Fig. 5 for TOAT. The evaluation of the electrostatic potential by fitting its derivatives to image distortions is described in the Supplementary Method: evaluation of the electrostatic potential from the distortion vector field.

References

1. Repp, J., Meyer, G., Stojkovic, S. M., Gourdon, A. & Joachim, C. Molecules on insulating films: scanning-tunneling microscopy imaging of individual molecular orbitals. *Phys. Rev. Lett.* **94**, 026803 (2005).
2. Gross, L., Mohn, F., Moll, N., Liljeroth, P. & Meyer, G. The chemical structure of a molecule resolved by atomic force microscopy. *Science* **325**, 1110–1114 (2009).
3. Temirov, R., Soubatch, S., Neucheva, O., Lassise, A. C. & Tautz, F. S. A novel method achieving ultra-high geometrical resolution in scanning tunnelling microscopy. *New J. Phys.* **10**, 053012 (2008).
4. Chiang, C. l., Xu, C., Han, Z. & Ho, W. Real-space imaging of molecular structure and chemical bonding by single-molecule inelastic tunneling probe. *Science* **344**, 885–888 (2014).
5. Gross, L. *et al.* Bond-order discrimination by atomic force microscopy. *Science* **337**, 1326–1329 (2012).
6. Gross, L. Recent advances in submolecular resolution with scanning probe microscopy. *Nat. Chem.* **3**, 273–278 (2011).
7. Emmrich, M. *et al.* Subatomic resolution force microscopy reveals internal structure and adsorption sites of small iron clusters. *Science* **348**, 308–311 (2015).
8. Sweetman, A. *et al.* Intramolecular bonds resolved on a semiconductor surface. *Phys. Rev. B* **90**, 165425 (2014).
9. Iwata, K. *et al.* Chemical structure imaging of a single molecule by atomic force microscopy at room temperature. *Nat. Commun.* **6**, 7766 (2015).
10. Moreno, C., Stetsovych, O., Shimizu, T. K. & Custance, O. Imaging three-dimensional surface objects with submolecular resolution by atomic force microscopy. *Nano Lett.* **15**, 2257–2262 (2015).
11. Huber, F. *et al.* Intramolecular force contrast and dynamic current-distance measurements at room temperature. *Phys. Rev. Lett.* **115**, 066101 (2015).
12. de Oteyza, D. G. *et al.* Direct imaging of covalent bond structure in single-molecule chemical reactions. *Science* **340**, 1434–1437 (2013).
13. Pavlicek, N. *et al.* Atomic force microscopy reveals bistable configurations of dibenzo[a,h]thianthrene and their interconversion pathway. *Phys. Rev. Lett.* **108**, 08610 (2012).
14. Kawai, S. *et al.* Obtaining detailed structural information about supramolecular systems on surfaces by combining high-resolution force microscopy with ab initio calculations. *ACS Nano* **7**, 9098–9105 (2013).
15. Schuler, B., Meyer, G., Peña, D., Mullins, O. C. & Gross, L. Unraveling the molecular structures of asphaltenes by atomic force microscopy. *J. Am. Chem. Soc.* **137**, 9870–9876 (2015).
16. Pavlicek, N. *et al.* On-surface generation and imaging of arynes by atomic force microscopy. *Nat. Chem.* **7**, 623–628 (2015).
17. Dienel, T. *et al.* Resolving atomic connectivity in graphene nanostructure junctions. *Nano Lett.* **15**, 5185–5190 (2015).
18. Clayden, J., Greeves, N. & Warren, S. *Organic Chemistry* (Oxford University Press, 2001).
19. Nonnenmacher, M., O'Boyle, M. P. & Wickramasinghe, H. K. Kelvin probe force microscopy. *Appl. Phys. Lett.* **58**, 2921 (1991).
20. Mohn, F., Gross, L., Moll, N. & Meyer, G. Imaging the charge distribution within a single molecule. *Nat. Nanotechnol.* **7**, 227–231 (2012).
21. Baier, R., Leenderts, C., Lux-Steiner, M. C. & Sadewasser, S. Toward quantitative Kelvin probe force microscopy of nanoscale potential distributions. *Phys. Rev. B* **85**, 165436 (2012).
22. Enevoldsen, G. H., Glatzel, T., Christensen, M. C., Lauritsen, J. V. & Besenbacher, F. Atomic scale kelvin probe force microscopy studies of the surface potential variations on the TiO2(110) surface. *Phys. Rev. Lett.* **100**, 236104 (2008).
23. Sadewasser, S. *et al.* New insights on atomic-resolution frequency-modulation kelvin-probe force-microscopy imaging of semiconductors. *Phys. Rev. Lett.* **103**, 266103 (2009).
24. Schuler, B. *et al.* Contrast formation in Kelvin probe force microscopy of single π-conjugated molecules. *Nano Lett.* **14**, 3342–3346 (2014).
25. Neff, J. L. & Rahe, P. Insights into Kelvin probe force microscopy data of insulator-supported molecules. *Phys. Rev. B* **91**, 085424 (2015).
26. Corso, M. *et al.* Charge redistribution and transport in molecular contacts. *Phys. Rev. Lett.* **115**, 136101 (2015).
27. Weymouth, A., Wutscher, T., Welker, J., Hofmann, T. & Giessibl, F. Phantom force induced by tunneling current: a characterization on Si(111). *Phys. Rev. Lett.* **106**, 226801 (2011).
28. Albrecht, F. *et al.* Probing charges on the atomic scale by means of atomic force microscopy. *Phys. Rev. Lett.* **115**, 076101 (2015).
29. Wagner, C. *et al.* Scanning quantum dot microscopy. *Phys. Rev. Lett.* **115**, 026101 (2015).
30. Bartels, L. *et al.* Dynamics of electron-induced manipulation of individual CO molecules on Cu(111). *Phys. Rev. Lett.* **80**, 2004 (1998).
31. Welker, J. & Giessibl, F. J. Revealing the angular symmetry of chemical bonds by atomic force microscopy. *Science* **336**, 444–449 (2012).
32. Mohn, F., Schuler, B., Gross, L. & Meyer, G. Different tips for high-resolution atomic force microscopy and scanning tunneling microscopy of single molecules. *Appl. Phys. Lett.* **102**, 073109 (2013).

33. Moll, N., Gross, L., Mohn, F., Curioni, A. & Meyer, G. The mechanisms underlying the enhanced resolution of atomic force microscopy with functionalized tips. *New J. Phys.* **12**, 125020 (2010).

34. Hapala, P. *et al.* The mechanism of high-resolution STM/AFM imaging with functionalized tips. *Phys. Rev. B* **90**, 085421 (2014).

35. Hapala, P., Temirov, R., Tautz, F. S. & Jelinek, P. Origin of high-resolution IETS-STM images of organic molecules with functionalized tips. *Phys. Rev. Lett.* **113**, 226101 (2014).

36. Guo, C., Xin, X., Van Hove, M. A., Ren, X. & Zhao, Y. Origin of the contrast interpreted as intermolecular and intramolecular bonds in atomic force microscopy images. *J. Phys. Chem. C* **119**, 14195–14200 (2015).

37. van der Lit, J., Di Cicco, F., Hapala, P., Jelinek, P. & Swart, I. Submolecular resolution imaging of molecules by atomic force microscopy: the influence of the electrostatic force. *Phys. Rev. Lett.* **105**, 096102 (2016).

38. Boneschanscher, M. P., Hämäläinen, S. K., Liljeroth, P. & Swart, I. Sample corrugation affects the apparent bond lengths in atomic force microscopy. *ACS Nano* **8**, 3006–3014 (2014).

39. Neu, M. *et al.* Image correction for atomic force microscopy images with functionalized tips. *Phys. Rev. B* **89**, 205407 (2014).

40. Zhang, J. *et al.* Real-space identification of intermolecular bonding with atomic force microscopy. *Science* **342**, 611–614 (2013).

41. Sweetman, A. M. *et al.* Mapping the force field of a hydrogen-bonded assembly. *Nat. Commun.* **5**, 3931 (2014).

42. Moll, N. *et al.* Image distortions of a partially fluorinated hydrocarbon molecule in atomic force microscopy with carbon monoxide terminated tips. *Nano Lett.* **14**, 6127–6131 (2014).

43. Hämäläinen, S. K. *et al.* Intermolecular contrast in atomic force microscopy images without intermolecular bonds. *Phys. Rev. Lett.* **113**, 186102 (2014).

44. Kawai, S. *et al.* Extended halogen bonding between fully fluorinated aromatic molecules. *ACS Nano* **9**, 2574–2583 (2015).

45. Jarvis, S. P. *et al.* Intermolecular artifacts in probe microscope images of C60 assemblies. *Phys. Rev. B* **92**, 241405 (2015).

46. Gao, D. Z. *et al.* Using metallic noncontact atomic force microscope tips for imaging insulators and polar molecules: tip characterization and imaging mechanisms. *ACS Nano* **8**, 5339–5351 (2014).

47. Sun, Z., Boneschanscher, M., Swart, I., Vanmaekelbergh, D. & Liljeroth, P. Quantitative atomic force microscopy with carbon monoxide terminated tips. *Phys. Rev. Lett.* **106**, 046104 (2011).

48. Weymouth, A. J., Hofmann, T. & Giessibl, F. J. Quantifying molecular stiffness and interaction with lateral force microscopy. *Science* **343**, 1120–1122 (2014).

49. Gross, L. *et al.* Investigating atomic contrast in atomic force microscopy and Kelvin probe force microscopy on ionic systems using functionalized tips. *Phys. Rev. B* **90**, 155455 (2014).

50. Ellner, M. *et al.* The electric field of CO tips and its relevance for atomic force microscopy. *Nano Lett.* **16**, 1974–1980 (2016).

51. Eigler, D. M., Lutz, C. P. & Rudge, W. E. An atomic switch realized with the scanning tunnelling microscope. *Nature* **352**, 600–603 (1991).

52. Bartels, L., Meyer, G. & Rieder, K.-H. Controlled vertical manipulation of single CO molecules with the scanning tunneling microscope: a route to chemical contrast. *Appl. Phys. Lett.* **71**, 213–215 (1997).

53. Giessibl, F. Advances in atomic force microscopy. *Rev. Mod. Phys.* **75**, 949 (2003).

54. Rohlfing, M., Temirov, R. & Tautz, F. Adsorption structure and scanning tunneling data of a prototype organic-inorganic interface: PTCDA on Ag(111). *Phys. Rev. B* **76**, 115421 (2007).

55. Kresse, G. & Furthmüller, J. Efficient iterative schemes for ab initio total-energy calculations using a plane-wave basis set. *Phys. Rev. B* **54**, 11169 (1996).

56. Perdew, J. P. *et al.* Atoms, molecules, solids, and surfaces: applications of the generalized gradient approximation for exchange and correlation. *Phys. Rev. B* **46**, 6671–6687 (1992).

57. Kresse, G. & Joubert, D. From ultrasoft pseudopotentials to the projector augmented-wave method. *Phys. Rev. B* **59**, 1758–1775 (1999).

58. Blum, V. *et al.* Ab initio molecular simulations with numeric atom-centered orbitals. *Comput. Phys. Commun.* **180**, 2175–2196 (2009).

59. Perdew, J. P., Burke, K. & Ernzerhof, M. Generalized gradient approximation made simple. *Phys. Rev. Lett.* **77**, 3865–3868 (1996).

60. Tkatchenko, A. & Scheffler, M. Accurate molecular van der waals interactions from ground-state electron density and free-atom reference data. *Phys. Rev. Lett.* **102**, 073005 (2009).

Acknowledgements

This work was financially supported by grants from the Czech Science Foundation (grant no. 14-16963J), the Ministry of Education of the Czech Republic grant no. LM2015087 and the Netherlands Organization for Scientific Research (ECHO-STIP grant no. 717.013.003).

Author contributions

P.H., M.Š and P.J. conceived the method. I.S. and P.J. conceived the experiments. M.Š, O.S., J.L, N.J.v.d.H. and I.S. carried out the experiments and analysed the experimental data. P.H., M.O., P.M. and P.J. performed and analysed the total energy DFT and AFM calculations. P.H., I.S., M.Š and P.J. wrote the paper. All coauthors provided feedback on the manuscript.

Additional information

23

Frequency and bandwidth conversion of single photons in a room-temperature diamond quantum memory

Kent A.G. Fisher[1], Duncan G. England[2], Jean-Philippe W. MacLean[1], Philip J. Bustard[2], Kevin J. Resch[1] & Benjamin J. Sussman[2,3]

The spectral manipulation of photons is essential for linking components in a quantum network. Large frequency shifts are needed for conversion between optical and telecommunication frequencies, while smaller shifts are useful for frequency-multiplexing quantum systems, in the same way that wavelength division multiplexing is used in classical communications. Here we demonstrate frequency and bandwidth conversion of single photons in a room-temperature diamond quantum memory. Heralded 723.5 nm photons, with 4.1 nm bandwidth, are stored as optical phonons in the diamond via a Raman transition. Upon retrieval from the diamond memory, the spectral shape of the photons is determined by a tunable read pulse through the reverse Raman transition. We report central frequency tunability over 4.2 times the input bandwidth, and bandwidth modulation between 0.5 and 1.9 times the input bandwidth. Our results demonstrate the potential for diamond, and Raman memories in general, as an integrated platform for photon storage and spectral conversion.

[1] Institute for Quantum Computing and Department of Physics and Astronomy, University of Waterloo, 200 University Avenue West, Waterloo, Ontario, Canada N2L 3G1. [2] National Research Council of Canada, 100 Sussex Drive, Ottawa, Ontario, Canada K1A 0R6. [3] Department of Physics, University of Ottawa, 150 Louis Pasteur, Ottawa, Ontario, Canada K1N 6N5. Correspondence and requests for materials should be addressed to K.J.R. (email: kresch@uwaterloo.ca) or to B.J.S. (email: ben.sussman@nrc.ca).

The fragility of the quantum state is a challenge facing all quantum technologies. Great efforts have been undertaken to mitigate the deleterious effects of decoherence by isolating quantum systems, for example, by cryogenically cooling and isolating in vacuum. State-of-the-art decoherence times are now measured in hours[1]. An alternative approach is to build quantum technologies that execute on ultrafast timescales—as short as femtoseconds—such that operations can be completed before decoherence overwhelms unitarity. A shining example is the Raman quantum memory[2–5], which can absorb single photons of femtosecond duration and release them on demand several picoseconds later[6]. While picosecond storage times are not appropriate for conventional quantum memory applications such as long-distance communication, it has been suggested[2] that Raman quantum memories can find additional uses such as frequency and bandwidth conversion.

Controlling the spectral properties of single photons is essential for a wide array of emerging optical quantum technologies spanning quantum sensing[7], quantum computing[8] and quantum communications[9]. Essential components for these technologies include single-photon sources[10], quantum memories[11], waveguides[12] and detectors[13]. The ideal spectral operating parameters (wavelength and bandwidth) of these components are rarely similar; thus, frequency conversion and spectral control are key enabling steps for component hybridization[14]. Beyond hybridization, frequency conversion is an area of emerging interest in quantum optical processing. The frequency degree of freedom can be used along side conventional encoding in, for example, polarization, or time-bin, to build quantum states of higher dimensionality[15].

Spectral control is a mature field in ultrafast optics where phase- and amplitude-shaping of a THz-bandwidth pulse can be achieved using passive pulse-shaping elements in the Fourier plane[16]. Meanwhile, a range of nonlinear optical techniques[17] such as second harmonic generation (SHG), sum- and difference-frequency generation, four-wave mixing and Raman scattering are routinely employed to shift the frequency of laser pulses.

Extending these frequency conversion techniques into the quantum regime is a critical task for many quantum technologies but is made difficult by the low intensity of single photons, and the sensitivity of quantum states to loss and noise. Despite these challenges, quantum frequency conversion[18] has been demonstrated in a number of systems including waveguides in nonlinear crystals[19–24], photonic crystal fibres[25] and atomic vapour[26]. Similarly, photon bandwidth compression has been shown using chirped-pulse upconversion[27]. Full control over the spectral properties of single photons[28] has been proposed using second-[14,29] and third-order[30] optical nonlinearities.

Large frequency shifts, such as those achieved using sum- and difference-frequency generation, are desirable to convert photons to, and from, the telecommunication band. Meanwhile, smaller shifts can be useful for frequency-multiplexing in several closely spaced bins. This concept is widely used in classical fibre optics, where wavelength division multiplexing is employed to achieve data rates far beyond that which could be achieved with monochromatic light[31]. However, in quantum optics the utility of frequency multiplexing has only recently been explored[15,32–34]. In a frequency-multiplexed quantum architecture, it will be critical to build components that can add, drop, and manipulate different frequency bins: it has been proposed that frequency-selective quantum memories can perform this task[15,33]. As with classical wavelength division multiplexing, the frequency bins will likely be closely spaced so that small shifts around a central frequency will be required.

In this article, we demonstrate the use of a Raman quantum memory to perform quantum frequency conversion; we manipulate the spectral properties of THz-bandwidth photons using a memory in the optical phonon modes of diamond[6]. Crucially, the quantum properties of the photon must be maintained even while the carrier frequency and bandwidth are modified. In our demonstration, a signal photon is mapped into an optical phonon by the write pulse, and then retrieved with its spectral properties modified according to the properties of the read pulse.

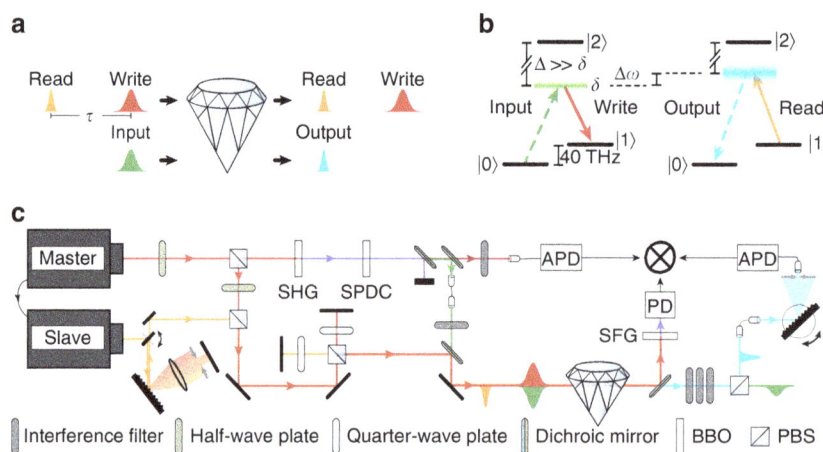

Figure 1 | Concept and experiment. (a) Input signal photons stored in the diamond by the strong write pulse can be retrieved with modified spectral properties upon output. The output spectrum is controlled by the spectrum of the read pulse. **(b)** Photons are Raman-absorbed to create optical phonons ($|1\rangle$), 40 THz above the ground state ($|0\rangle$). A read pulse of tunable wavelength and bandwidth retrieves the photon, determining its spectrum. Here, Δ is the detuning from the conduction band ($|2\rangle$), δ is the input photon bandwidth, $\Delta\omega$ is the detuning between input and output frequencies. **(c)** The master laser (red) is split between the write field and photon source. In the photon source, frequency-doubled laser light pumps SPDC and heralded input signal photons (green) are generated. Signal photons are Raman-absorbed into the optical phonon modes in the diamond via the write field. The slave laser (orange) emits the read field which retrieves the photon from the diamond after time τ. The output signal photon (blue) spectrum is measured on a monochromator. The SFG of read and write pulses triggers the experiment. Coincident detections of output, herald photons and SFG events are measured by a coincidence logic unit. SFG, second-harmonic generation; SPDC, spontaneous parametric downconversion; APD, avalanche photodiode; PD, photodiode; SFG, sum-frequency generation; BBO, β-barium borate; PBS, polarizing beamsplitter.

Results

Experiment. The frequency converter is based on a quantum memory[6] modelled by the Λ-level system shown in Fig. 1b, where an input signal photon (723.5 nm centre wavelength and bandwidth $\delta = 4.1$ nm full-width at half-maximum (FWHM)) and a strong write pulse (800, 5 nm FWHM) are in Raman resonance with the optical phonon band (frequency 40 THz). The large detuning of both fields from the conduction band (detuning $\Delta \approx 950$ THz) allows for the storage of high-bandwidth photons, while the memory exhibits a quantum-level noise floor even at room temperature[6]. The input signal photon is stored in the memory by Raman absorption with the write pulse, creating an optical phonon. After a delay τ, the read pulse annihilates the phonon and creates a modified output photon. By tuning the wavelength and bandwidth of the read pulse, we convert the wavelength of the input signal photon over a range of 17 nm as well as performing bandwidth compression to 2.2 nm and expansion to 7.6 nm (FWHM). The diamond memory is ideally suited to this task, offering low-noise frequency manipulation of THz-bandwidth quantum signals at a range of visible and near-infrared wavelengths in a robust room-temperature device[35].

The experimental setup is shown in Fig. 1c. The master laser for the experiment is a Ti:sapphire oscillator producing 44 nJ pulses at a repetition rate of 80 MHz and a central wavelength of 800 nm. This beam is split in two parts: the photon-source pump and the write field. In the photon source, the SHG of the laser light pumps collinear type-I spontaneous parametric down-conversion (SPDC) in a β-barium borate (BBO) crystal, generating photons in pairs, with one at 723.5 nm (input signal) and the other at 894.6 nm (herald). The herald photon is detected on an avalanche photodiode (APD), while the input is spatially and spectrally filtered and overlapped with the orthogonally polarized write pulse on a dichroic mirror. The input signal photon and write field are incident on the ⟨100⟩ face of the diamond and the input is Raman-absorbed.

The photon is retrieved from the diamond using a read pulse produced by a second Ti:sapphire laser (slave), whose repetition rate is locked to the master, but whose frequency and bandwidth can be independently modified. In this experiment we vary the

read field wavelength between 784 and 814 nm, and its bandwidth between 2.1 and 12.1 nm FWHM. To narrow the bandwidth of the read pulse, a folded-grating 4f-system[16] with a narrow slit is used, while in all other configurations the 4f line is removed. The read pulse is then overlapped with the write on a polarizing beamsplitter, arriving at the diamond a time τ after storage. The horizontally polarized read pulse retrieves a vertically polarized photon (output) from the diamond with spectral shape close to that of the read pulse, blue-shifted by the phonon frequency (40 THz).

The read and write pulses are separated from the signal photons after the diamond by a dichroic mirror; sum-frequency generation (SFG) of the pulses is detected on a fast-photodiode (PD) and used to confirm their successful overlap. Frequency-converted output photons are separated from any unstored input photons by a polarizing beamsplitter, coupled into a single-mode fibre and directed to a monochromator. The spectrally filtered output from the monochromator is coupled into a multi-mode fibre and detected on an APD. Coincident detections between output, herald and read–write SFG events are measured; the experiment is triggered by the joint detection of a herald photon and an SFG signal. (see Methods for further details.)

Frequency shifts. Frequency conversion of the signal photon is observed by tuning the slave laser wavelength. We vary this from 784 to 812 nm and measure the output photon wavelength using the monochromator, recording threefold coincidence events. The resulting output spectra, with the read pulse centred at 792 and 808 nm, are shown in Fig. 2a (hollow circles). The spectrum of each read pulse (solid lines), blue-shifted by the phonon frequency, is plotted alongside the relevant output photon spectrum to show how the photon spectrum is determined by the read pulse. We find the peaks of the output spectra to be 716 and 728 nm with bandwidths 3.3 and 3.5 nm, FWHM respectively, making the output spectrally distinguishable from the input (green).

Following retrieval, the time-correlations characteristic of SPDC photon-pairs are preserved. This is measured by scanning the electronic delay between the signal and herald detection

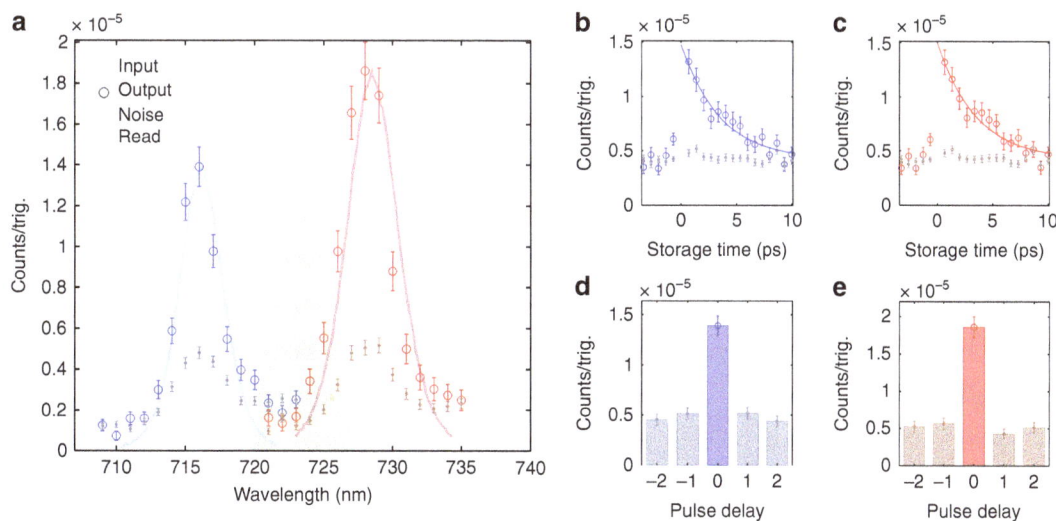

Figure 2 | Frequency conversion. (**a**) The measured blue- and red-shifted output photon spectra (hollow circles), and noise (dots), when the read beam is tuned to 792 and 808 nm, respectively. Corresponding read beam spectra, blue-shifted by the phonon frequency, are shown (solid lines) for reference along with the input photon spectrum (green). (**b**) Retrieved blue- and (**c**) Red-shifted signal (hollow circles), and noise (dots), as read–write delay is scanned. An exponential fit gives a phonon lifetime of 3.5 ps. (**d**) Coincidence detection events between blue- and (**e**) red-shifted output and herald photons while scanning the electronic delay between them, as measured at the peak of the spectrum. (**a-e**) Error bars show 1 standard deviation calculated assuming Poissonian noise.

events in steps of 12.5 ns (the time between adjacent oscillator pulses) and counting coincident detections. Results are shown in Fig. 2d,e for blue- and red-shifted cases, respectively. We quantify this using the two-mode intensity cross-correlation function between output signal and herald fields given by $g_{s,h}^{(2)} = P_{s,h}/(P_s P_h)$. Here, $P_s (P_h)$ is the probability of detecting a photon in the signal (herald) mode, and $P_{s,h}$ is the probability of measuring a joint detection. A measurement of $g_{s,h}^{(2)} > 2$ indicates non-classical correlation[36,37] (see Methods), whereas uncorrelated photon detections, for example, from noise, give $g_{s,h}^{(2)} = 1$. We calculate the values of $g_{s,h}^{(2)}$ at the peak of the blue- and red-shifted spectra to be 2.7 ± 0.2 and 3.4 ± 0.3, respectively.

Figure 2b,c shows the blue-(red-)shifted photon retrieval rate as a function of the optical delay τ between read and write pulses. An exponential function is fit to the data and we find a memory lifetime of 3.5 ps as expected from the lifetime of the optical phonon[35,38]. This exceeds 12 times the duration of the input photon (see Methods). Also plotted in Fig. 2a–c are the measured coincidences due to noise (dots), which are measured by taking the average of the ± 12.5 and ± 25 ns time-bins as shown in Fig. 2d,e. Noise comes from two processes: four-wave mixing[35]; and read pulses scattering from thermally populated phonons producing anti-Stokes light[6,35]. The latter can be reduced by an order of magnitude by cooling the diamond to $-60\,°C$.

Figure 3 shows $g_{s,h}^{(2)}$ measured at the peak of each output signal spectrum as the read wavelength is tuned over a 30 nm range. We find that blue- and red-shifted single photons maintain non-classical correlations over a 17 nm range. Since noise is uncorrelated with herald photons, we expect the noise to have cross-correlation $g^{(2)} = 1$. The $g_{s,h}^{(2)}$ will increase from 1 in proportion to the signal-to-noise ratio (see Methods for further details),

$$g_{s,h}^{(2)} \approx 1 + \frac{\eta_h \eta_{fc}(\Delta\omega)}{P_n}. \qquad (1)$$

Here $\eta_h = P_{s,h}/P_h = 0.13\%$ is the photon heralding efficiency in the signal arm including the monochromator, $P_n = 3.8 \times 10^{-6}$ is the probability of detecting a noise photon, and $\eta_{fc}(\Delta\omega)$ is the conversion efficiency as a function of frequency detuning, $\Delta\omega$, between input and output photons. The conversion efficiency

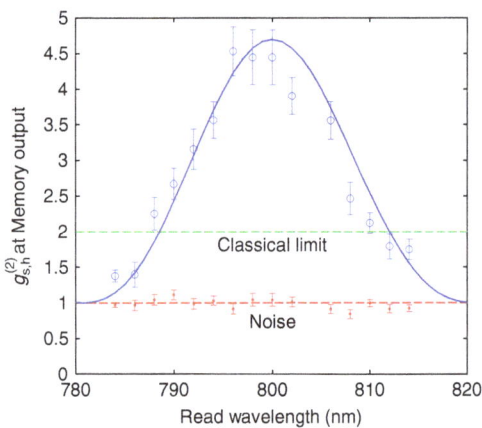

$\eta_{fc}(\Delta\omega) = \eta_{fc}(0) \times sinc^2(\Delta k L/2)$, where $L = 2.3$ mm is the length of the diamond along the propagation axis, and $\Delta\mathbf{k} = \mathbf{k}_i - \mathbf{k}_o + \mathbf{k}_r - \mathbf{k}_w$ is the phase mismatch between the input signal (i), output signal (o), read (r) and write (w) fields because of material dispersion in diamond[35]. The conversion efficiency was measured to be $\eta_{fc}(0) = 1.1\%$, inside the diamond at zero detuning. As the diamond is not anti-reflection coated, a 17% reflection loss occurs at each face.

Inserting experimental parameters into equation 1 returns $g_{s,h}^{(2)} \approx 1 + 3.7 \times sinc^2(\Delta k L/2)$, which is plotted along side data in Fig. 3 (solid line). The close agreement with experiment suggests that the limitation on frequency conversion comes primarily from phase-matching conditions. We then expect that the range of frequency conversion in diamond can be extended by modifying the phase-matching conditions. This could be achieved by shortening the diamond crystal, or by employing non-collinear beam geometries[39]. In the current configuration, the maximum conversion efficiency is limited to $\sim 1\%$ due to the efficiency of the quantum memory[6]. However, we note that this could be improved with increased intensity in read and write pulses, or by increasing the Raman coupling, for example, by the use of a waveguide.

Bandwidth manipulation. Bandwidth conversion is observed by tuning the slave laser bandwidth. With the read pulse wavelength centred at 801 nm, its bandwidth could be tuned from 12.1 to 2.1 nm FWHM using a slit in a grating 4f line. Figure 4a,b shows the resulting narrowed (expanded) output photon spectrum with the corresponding read pulse spectrum, blue-shifted by the phonon frequency, and the input signal photon spectrum for reference. The resulting narrowed and expanded photon bandwidths are 2.2 and 7.6 nm, FWHM, respectively. Figure 4c,d shows the conservation of timing correlations between bandwidth-narrowed (-expanded) photons and herald photons, respectively. We measure $g_{s,h}^{(2)} = 2.6 \pm 0.2$ in the narrowed bandwidth case, $g_{s,h}^{(2)} = 2.9 \pm 0.2$ in the expanded bandwidth case, showing that bandwidth-converted output light from the diamond maintains non-classical correlations with herald photons.

Discussion

We have demonstrated ultrafast quantum frequency manipulation by adjusting the central wavelength and spectrum, of THz-bandwidth heralded single photons. We achieve this spectral control using a modified Raman quantum memory in diamond. The single photons are written to the memory from one spectral mode, and recalled to another. Critically—and unlike frequency conversion based on, for example, amplification—the non-classical photon statistics in our demonstration were retained after spectral manipulation. Diamond therefore offers low-noise THz-bandwidth storage and frequency control of single photons on a single, robust, room-temperature platform. We have demonstrated frequency conversion of a single polarization; two memories could be used in parallel to convert a polarization qubit. Quantum memories for long-distance quantum communication typically demand long storage times at the expense of high bandwidth; this application leverages a high-bandwidth memory where long storage time is not relevant. We believe that this system could find use in a number of applications, including entangling two photon-pair sources of different colour; reducing the bandwidth of the ubiquitous ultrafast SPDC photon source, without losing photons (as occurs in passive filtering); broadening the bandwidth of an SPDC photon source to match a chosen material system (not possible with a passive filtering system); and increasing the dimensionality of quantum encoded information[15]. Ultimately, we believe that arbitrary optical

Figure 3 | Range of frequency conversion. Measured $g_{s,h}^{(2)}$ of frequency-shifted photons. Frequency conversion, tuning the read beam over a 30 nm range, is observed. Non-classical statistics, that is, $g_{s,h}^{(2)} > 2$, are maintained over a 18 nm range. The estimated $g_{s,h}^{(2)}$ (solid line) which depends on $sinc^2(\Delta k L/2)$ agrees well with experimental data suggesting that the range of frequency conversion is determined by phase-matching conditions. Error bars show 1 standard deviation calculated assuming Poissonian noise.

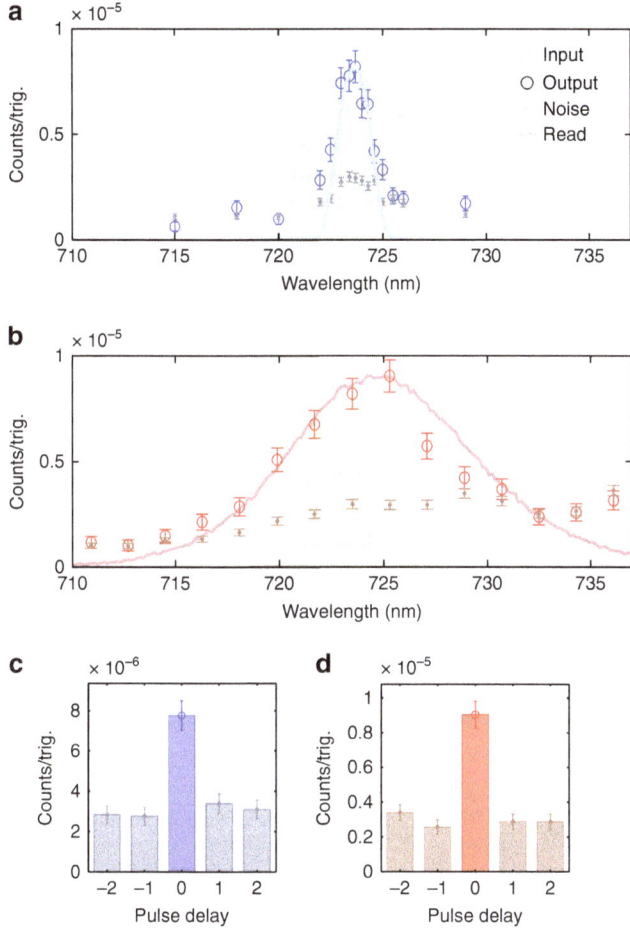

Figure 4 | Bandwidth conversion. (**a**) Narrowed output spectrum (hollow circles) and noise (dots) with read beam FWHM at 2.1 nm. (**b**) Expanded output photon spectrum (hollow circles) and noise (dots) with read beam FWHM at 12.1 nm. (**a**,**b**) Corresponding read beam spectra, blue-shifted by the phonon frequency, are shown (solid lines) for reference along with the input photon spectrum (green). (**c**) Coincidence detection events between bandwidth narrowed output photons, and (**d**) expanded output photons and heralds while scanning the electronic delay between them, as measured at the peak of the spectrum. (**a**–**d**) Error bars show 1 standard deviation calculated assuming Poissonian noise.

function generation[40]—with both classical and quantum light—and associated signal processing capabilities will be a platform for future technology development. We expect that the large-bandwidth nonlinear optical conversion of Raman-based quantum memories will find use in implementations of these generators.

Methods

Photon source. Laser light is frequency-doubled by type-I SHG in a 1 mm BBO crystal before pumping collinear type-I SPDC in a second 1 mm BBO crystal. Horizontally polarized photon pairs are generated at 894.6 and 723.5 nm. Remaining pump light is filtered out and photon pairs are separated by a 801 nm long-pass dichroic mirror. The 894.6 nm photon passes through a 5 nm interference filter, is coupled into a single-mode fibre, and detected on an APD. A detection heralds the presence of the 723.5 nm photon, which is spatially filtered in a single-mode fibre and spectrally filtered by an interference filter with bandwidth of 5 nm (FWHM). The input and write pulses are overlapped using a 750 nm shortpass dichroic mirror. The input and write pulses are focused into the diamond by an achromatic lens of focal length 6 cm.

Diamond. The diamond is a high-purity, low birefringence crystal grown by chemical vapour deposition by Element Six Ltd. The crystal is 2.3 mm long, cut along the ⟨100⟩ lattice direction, and polished on two sides.

Storage time. Absorption of the input photon by the diamond lattice is observed by an 18% dip in input-herald coincidences when the input photon and write field arrive at the diamond simultaneously. The duration of the input photon can be deconvolved from the width of the absorption dip, 346 fs. With write pulses 190 fs in duration, the input photon pulse duration time is $\sqrt{346^2 - 190^2} = 289$ fs, assuming transform-limited Gaussian pulses. The characteristic storage time of the diamond memory is 3.5 ps, found from an exponential fit to storage data, over 12 times the duration of the input pulse.

Laser locking. The repetition rate of the slave laser is locked to that of the master using a Spectra Physics Lok-to-Clock device. We send read and write beams through a cross-correlator (type-II SFG in a 1 mm BBO crystal) and detect the resulting signal on a PD confirming that the time difference between the two pulses is ≤ 200 fs. We measure a typical SFG signal rate of 2.5 MHz; we use this signal to trigger the experiment.

Monochromator. The monochromator (Acton SP2300) is comprised of a 1,200 g mm^{-1} grating between two 30 cm focal length spherical mirrors. The output is coupled to a multi-mode fibre (105 µm core). The apparatus has a spectral resolution of 1.1 nm and an overall efficiency of 10% at 723 nm.

Cross-correlation function. The cross-correlation function between the herald and frequency-converted light is given by $g_{s,h}^{(2)} = P_{s,h}/(P_s P_h)$. Classically, $g_{s,h}^{(2)}$ is upper-bounded by a Cauchy–Schwarz inequality[36,37] $g_{s,h}^{(2)} \leq \sqrt{g_{s,s}^{(2)} g_{h,h}^{(2)}}$. Here, the terms on the right-hand side are the intensity auto-correlation functions for the output signal and herald fields, which we assume, being produced by SPDC, follow thermal statistics and have $g_{s,s}^{(2)} = g_{h,h}^{(2)} = 2$. Adding any uncorrelated noise would strictly lower terms on the right-hand side towards 1. To model the effect of noise on this measurement, we assume that the signal is made up of a mixture of noise photons (detected with probability $P_n = 3.8 \times 10^{-6}$) and frequency-converted photons (probability P_γ), such that

$$P_s = P_\gamma + P_n \approx P_h \eta_h \eta_{fc}(\Delta\omega) + P_n, \tag{2}$$

$$P_{s,h} = P_{\gamma,h} + P_{n,h} \approx P_h \eta_h \eta_{fc}(\Delta\omega) + P_n P_h, \tag{3}$$

where $\eta_h = 1.3 \times 10^{-3}$ is the heralding efficiency which equates to the collection efficiency of the entire signal arm, including the monochromator, and $\eta_{fc}(\Delta\omega)$ is the efficiency of the quantum frequency conversion. This returns

$$g_{s,h}^{(2)} = \frac{\eta_h \eta_{fc}(\Delta\omega) + P_n}{P_h \eta_h \eta_{fc}(\Delta\omega) + P_n}, \tag{4}$$

from which equation 1 follows, given that $P_h \eta_h \eta_{fc}(\Delta\omega) \ll P_n$.

Background subtraction. When measured at the photon source the input and herald photon cross-correlation is $g_{in,h}^{(2)} = 164$. Because of imperfect polarization extinction the input photon can, with low probability, traverse the monochromator and be detected, thereby artificially inflating the measured $g_{s,h}^{(2)}$ of the converted output. For this reason we make a measurement with no read/write pulses present and subtract these counts from the output signal when read/write pulses are present to portray an accurate value of $g_{s,h}^{(2)}$. As an example, in Fig. 3, the peak count rate is 19×10^{-6} photons per pulse compared with a background of 3×10^{-6} photons per pulse.

References

1. Zhong, M. et al. Optically addressable nuclear spins in a solid with a six-hour coherence time. *Nature* **517**, 177–180 (2015).
2. Nunn, J. et al. Mapping broadband single-photon wave packets into an atomic memory. *Phys. Rev. A* **75**, 011401(R) (2007).
3. Reim, K. et al. Towards high-speed optical quantum memories. *Nat. Photon.* **4**, 218–221 (2010).
4. Bustard, P. J., Lausten, R., England, D. G. & Sussman, B. J. Toward quantum processing in molecules: A THz-bandwidth coherent memory for light. *Phys. Rev. Lett.* **111**, 083901 (2013).
5. Michelberger, P. S. et al. Interfacing GHz-bandwidth heralded single photons with a warm vapour Raman memory. *New J. Phys.* **17**, 043006 (2015).
6. England, D. G. et al. Storage and retrieval of THz-bandwidth single photons using a room-temperature diamond quantum memory. *Phys. Rev. Lett.* **114**, 053602 (2015).
7. Giovannetti, V., Lloyd, S. & Maccone, L. Quantum-enhanced measurements: beating the standard quantum limit. *Science* **306**, 1330–1336 (2004).
8. Knill, E., Laflamme, R. & Milburn, G. J. A scheme for efficient quantum computation with linear optics. *Nature* **409**, 46–52 (2001).
9. Duan, L.-M., Lukin, M. D., Cirac, J. I. & Zoller, P. Long-distance quantum communication with atomic ensembles and linear optics. *Nature* **414**, 413–418 (2001).

10. Kurtsiefer, C., Mayer, S., Zarda, P. & Weinfurter, H. Stable solid-state source of single photons. *Phys. Rev. Lett.* **85**, 290–293 (2000).
11. Kozhekin, A. E., Mølmer, K. & Polzik, E. Quantum memory for light. *Phys. Rev. A.* **62**, 033809 (2000).
12. Politi, A., Cryan, M. J., Rarity, J. G., Yu, S. & O'Brien, J. L. Silica-on-silicon waveguide quantum circuits. *Science* **320**, 646–649 (2008).
13. Lita, A. E., Miller, A. J. & Nam, S. W. Counting near-infrared single-photons with 95% efficiency. *Opt. Express* **16**, 3032–3040 (2008).
14. Kielpinski, D., Corney, J. F. & Wiseman, H. M. Quantum optical waveform conversion. *Phys. Rev. Lett.* **106**, 130501 (2011).
15. Humphreys, P. C. *et al.* Continuous-variable quantum computing in optical time-frequency modes using quantum memories. *Phys. Rev. Lett.* **113**, 130502 (2014).
16. Weiner, A. M. Femtosecond pulse shaping using spatial light modulators. *Rev. Sci. Instrum.* **71**, 1929–1960 (2000).
17. Boyd, R. W. *Nonlinear Optics* (Academic Press, 2008).
18. Kumar, P. Quantum frequency conversion. *Opt. Lett.* **15**, 1476 (1990).
19. Rakher, M. T., Ma, L., Slattery, O., Tang, X. & Srinivasan, K. Quantum transduction of telecommunications-band single photons from a quantum dot by frequency upconversion. *Nat. Photon.* **4**, 786–791 (2010).
20. Rakher, M. T. *et al.* Simultaneous wavelength translation and amplitude modulation of single photons from a quantum dot. *Phys. Rev. Lett.* **107**, 083602 (2011).
21. Tanzilli, S. *et al.* A photonic quantum information interface. *Nature* **437**, 116–120 (2005).
22. Ikuta, R. *et al.* Wide-band quantum interface for visible-to-telecommunication wavelength conversion. *Nat. Commun.* **2**, 537 (2011).
23. De Greve, K. *et al.* Quantum-dot spin-photon entanglement via frequency downconversion to telecom wavelength. *Nature* **491**, 421–425 (2012).
24. Guerreiro, T. *et al.* Nonlinear interaction between single photons. *Phys. Rev. Lett.* **113**, 173601 (2014).
25. McGuinness, H. J., Raymer, M. G., McKinstrie, C. J. & Radic, S. Quantum frequency translation of single-photon states in a photonic crystal fiber. *Phys. Rev. Lett.* **105**, 093604 (2010).
26. Dudin, Y. O. *et al.* Entanglement of light-shift compensated atomic spin waves with telecom light. *Phys. Rev. Lett.* **105**, 260502 (2010).
27. Lavoie, J., Donohue, J. M., Wright, L. G., Fedrizzi, A. & Resch, K. J. Spectral compression of single photons. *Nat. Photon.* **7**, 363–366 (2013).
28. Raymer, M. G. & Srinivasan, K. Manipulating the color and shape of single photons. *Phys. Today* **65**, 32–37 (2012).
29. Brecht, B., Eckstein, A., Christ, A., Suche, H. & Silberhorn, C. From quantum pulse gate to quantum pulse shaper—engineered frequency conversion in nonlinear optical waveguides. *New J. Phys.* **13**, 065029 (2011).
30. McKinstrie, C. J., Mejling, L., Raymer, M. G. & Rottwitt, K. Quantum-state-preserving optical frequency conversion and pulse reshaping by four-wave mixing. *Phys. Rev. A* **85**, 053829 (2012).
31. Brackett, C. A. Dense wavelength division multiplexing networks: principles and applications. *IEEE J. Sel. Areas Commun.* **8**, 948–964 (1990).
32. Sinclair, N. *et al.* Spectral multiplexing for scalable quantum photonics using an atomic frequency comb quantum memory and feed-forward control. *Phys. Rev. Lett.* **113**, 053603 (2014).
33. Campbell, G. T. *et al.* Configurable unitary transformations and linear logic gates using quantum memories. *Phys. Rev. Lett.* **113**, 063601 (2014).
34. Donohue, J. M., Lavoie, J. & Resch, K. J. Ultrafast time-division demultiplexing of polarization-entangled photons. *Phys. Rev. Lett.* **113**, 163602 (2014).
35. England, D., Bustard, P., Nunn, J., Lausten, R. & Sussman, B. J. From photons to phonons and back: A THz optical memory in diamond. *Phys. Rev. Lett.* **111**, 243601 (2013).
36. Loudon, R. *The Quantum Theory of Light* (Oxford University Press, 2004).
37. Clauser, J. F. Experimental distinction between the quantum and classical field-theoretic predictions for the photoelectric effect. *Phys. Rev. D* **9**, 853–860 (1974).
38. Lee, K. C. *et al.* Macroscopic non-classical states and terahertz quantum processing in room-temperature diamond. *Nat. Photon.* **6**, 41–44 (2012).
39. Eckbreth, A. C. BOXCARS: Crossed-beam phase-matched CARS generation in gases. *Appl. Phys. Lett.* **32**, 421–423 (1978).
40. Kowligy, A. S. *et al.* Quantum optical arbitrary waveform manipulation and measurement in real time. *Opt. Express* **22**, 27942–27957 (2014).

Acknowledgements

We thank Matthew Markham and Alastair Stacey of Element Six Ltd. for the diamond sample. We also thank Rune Lausten, Paul Hockett, John Donohue, Michael Mazurek and Khabat Heshami for fruitful discussions. Doug Moffatt and Denis Guay provided important technical assistance. This work was supported by the Natural Sciences and Engineering Research Council of Canada, Canada Research Chairs, Ontario Centres of Excellence, and the Ontario Ministry of Research and Innovation Early Researcher Award.

Author contributions

K.A.G.F, D.G.E. and J.-P.W.M. performed the experiment and analysed the data. All authors contributed to the final manuscript.

Additional information

Competing financial interests: The authors declare no competing financial interests.

24

Topographical pathways guide chemical microswimmers

Juliane Simmchen[1], Jaideep Katuri[1], William E. Uspal[1,2], Mihail N. Popescu[1,2], Mykola Tasinkevych[1,2] & Samuel Sánchez[1,3,4]

Achieving control over the directionality of active colloids is essential for their use in practical applications such as cargo carriers in microfluidic devices. So far, guidance of spherical Janus colloids was mainly realized using specially engineered magnetic multilayer coatings combined with external magnetic fields. Here we demonstrate that step-like submicrometre topographical features can be used as reliable docking and guiding platforms for chemically active spherical Janus colloids. For various topographic features (stripes, squares or circular posts), docking of the colloid at the feature edge is robust and reliable. Furthermore, the colloids move along the edges for significantly long times, which systematically increase with fuel concentration. The observed phenomenology is qualitatively captured by a simple continuum model of self-diffusiophoresis near confining boundaries, indicating that the chemical activity and associated hydrodynamic interactions with the nearby topography are the main physical ingredients behind the observed behaviour.

[1] Max-Planck-Institut für Intelligente Systeme, Heisenbergstrasse 3, D-70569 Stuttgart, Germany. [2] IV. Institut für Theoretische Physik, Universität Stuttgart, Pfaffenwaldring 57, D-70569 Stuttgart, Germany. [3] Institut de Bioenginyeria de Catalunya (IBEC), Baldiri I Reixac 10-12, 08028 Barcelona, Spain. [4] Institució Catalana de Recerca i Estudis Avancats (ICREA), Pg. Lluís Companys 23, 08010 Barcelona, Spain. Correspondence and requests for materials should be addressed to M.T. (email: miko@is.mpg.de) or to S.S. (email: sanchez@is.mpg.de and ssanchez@ibecbarcelona.eu).

Catalytically active micrometre-sized objects can self-propel by various mechanisms including bubble ejection, diffusio- and electrophoresis, when parts of their surface catalyse a chemical reaction in a surrounding liquid. In future, such chemically active micromotors may serve as autonomous carriers working within microfluidic devices to fulfill complex tasks[1-3]. However, to achieve this goal, it is essential to gain robust control over the directionality of particle motion. Although it has been more than a decade since motile chemically active colloids were first reported[4-7], this remains a challenging issue, in particular for the case of spherical particles.

Two main methods of guidance have been so far employed with varying degrees of success. The first one uses controlled spatial gradients of 'fuel' concentration. This approach suffers, however, from severe difficulties in creating and maintaining chemical gradients, and the spatial precision of guidance remains rather poor[8-12]. The second approach relies on the use of external magnetic fields in combination with particles with suitably designed magnetic coatings or inclusions[5,13]. This proved to be a very precise guidance mechanism, which could be employed straightforwardly for the case of tubular particles[14], but difficult to extend to the case of spherical colloids, where it requires sophisticated engineering of multilayer magnetic coatings[7,15-17]. In addition, individualized guidance of specific particles is difficult to achieve without complicated external apparatus and feedback loops[18]. The advantages of autonomous operation are thereby significantly hindered.

Although these methods are quite general in their applicability, we note here that the synthetic micromotors are, in general, density mismatched with the suspending medium and therefore tend to sediment and move near surfaces. Furthermore, even in situations in which sedimentation can be neglected (for example, in the case of neutrally buoyant swimmers), the presence of confining surfaces has profound consequences on swimming trajectories, as discussed below. Theoretical studies have shown that long-range hydrodynamic interactions (HIs) between microswimmers and nearby surfaces[19,20] can give rise to trapping at the walls or circular motion. Moreover, a theoretical study of a model active Janus colloid moving near a planar inert wall has revealed complex behaviour, including novel sliding and hovering steady states[21]. Experimentally, wall-bounded motion of active Janus particles was evidenced in the study by Bechinger et al.[22], whereas capture into orbital trajectories of active bimetallic rods by large spherical beads or of Janus colloids in colloidal crystals[23] has been reported in recent times. Capture of microswimmers by spherical obstacles via HIs has been modelled theoretically by Lauga et al.[24].

This intrinsic tendency of the active swimmers to operate near bounding surfaces motivated us to examine whether it can be further exploited to achieve directional guidance of chemically active microswimmers by endowing the wall with small height step-like topographical features, as shown in Fig. 1, which the particles can eventually exploit as pathways. Recently, Palacci et al.[25] have shown that shallow rectangular grooves can efficiently guide photocatalytic hematite swimmers that have size comparable with the width of the groove. Because of the strong lateral confinement, it is hard to discriminate between the different physical contributions that lead to particle guidance. Here we use a much less restrictive geometry—a shallow topographical step—and it is a priori not clear whether a self-phoretic swimmer can follow such features. We report experimental evidence that Janus microswimmers can follow step-like topographical features that are only a fraction of the particle radius in height. This is, in some sense, similar to the strategy employed in natural systems well below the microscale: within cells, protein motors such as myosin, kinesin

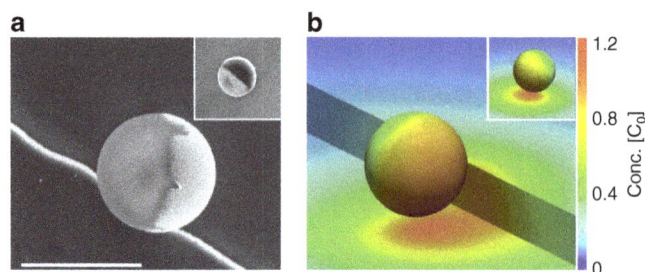

Figure 1 | Janus microswimmers near submicrometre steps. (**a**) Top view: scanning electron microscopy (SEM) image of a spherical Janus motor on a silicon substrate with a silicon step. The lighter part of the Janus particle corresponds to Pt, while the grey part is the SiO_2; scale bar, 2 µm. (**b**) Colour-coded steady-state distribution $c(\mathbf{r})$ of reaction products around a half covered Janus particle at an inert wall and a step with height $h_{step} = R$, where R is the particle radius. The colour map shows $c(\mathbf{r})$ at the surfaces of the particle and substrate, and is represented in units of c_0 defined in Methods. The insets show the particles on a flat surface.

and dynein use binding to microtubules to switch to directional motion[26,27]. The guidance of microswimmers through patterned device topography that we propose and demonstrate in this study may pave the way for new methods of self-propeller motion control based on patterned walls.

Results

Dynamics of Janus microswimmers at a planar wall. Janus particles are fabricated by vapour deposition of a thin layer of Pt (7 nm) on SiO_2 particles (diameters of ∼2 and 5 µm). For details on the fabrication, see Methods section. Scanning electron microscopy images of a Janus particle at the step edge are shown in Fig. 1a, whereas Fig. 1b illustrates numerically calculated steady-state distributions of the reaction products around a model half Pt-covered Janus sphere near a step.

Initially, the particles are introduced to the system with no H_2O_2 present and due to their weight they sediment near the bottom surface. After sedimentation, we find the particles uniformly distributed over the substrate and most of the 5-µm particles have their much denser (compared with the SiO_2 cores) Pt caps oriented downwards (Fig. 2a left), while smaller particles have a wider distribution of orientations. The particles are seen in the same focal plane of the microscope, which indicates that they are at similar vertical distances from the substrate. On addition of H_2O_2 to the system, we observe that the Pt caps of the microswimmers are oriented parallel to the substrate plane (see Fig. 2a right and Fig. 2b, and for the definition of the geometrical parameters see Supplementary Fig. 1). Following this reorientation, the microswimmers start moving parallel to the substrate in the direction away from the catalytic caps. In Fig. 2c we show snapshots from an optical microscopy video recording of Janus microswimmers in the vicinity of a step with $h_{step} = 800$ nm. Similar to the case depicted in Fig. 2a (left), in the absence of H_2O_2 the particles are oriented cap down. After addition of hydrogen peroxide, the particle caps turn away from the substrate and the particles start moving in random directions until some of them encounter a step; if the step is sufficiently tall (depending on the particle size) the particles stop, reorient and continue self-propelling along it. These observations confirm our hypothesis that the presence of a side step near the active microswimmers, even if small compared with the particle radius, has an influence on their orientation.

We show that this behaviour (alignment with both the wall and the step) is captured by a simple model of neutral

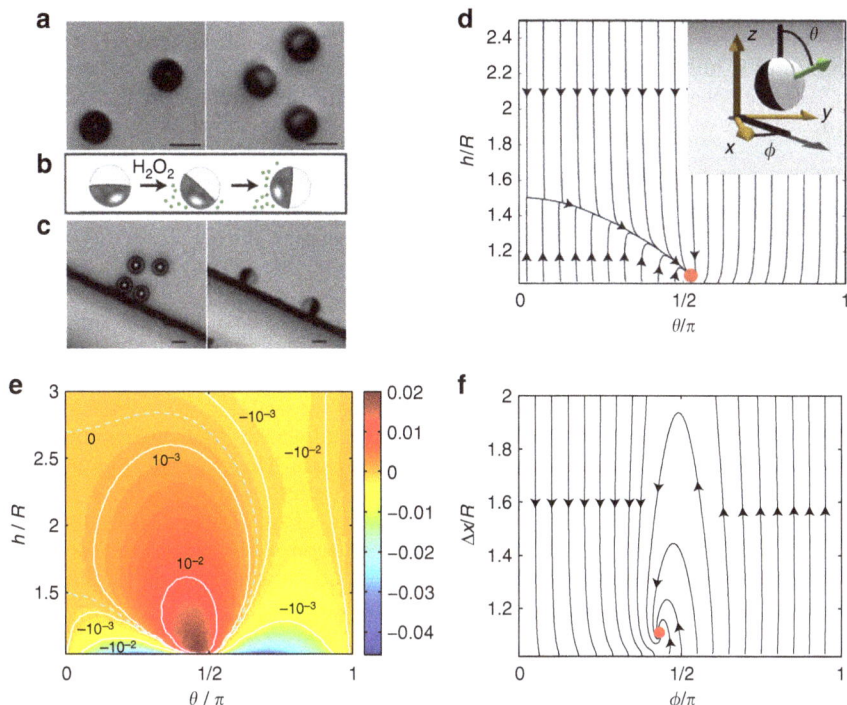

Figure 2 | Behaviour of Janus particles near planar surfaces. (**a**) Particles that sedimented to the bottom surface in a water suspension (inactive system) tend to align with their Pt caps facing downwards, which is more pronounced for larger particles. The Pt caps ($\rho = 21.45 \, \mathrm{g \, cm^{-3}}$), which are much denser than the silica parts of the particles, render them bottom heavy (in the image one sees transparent SiO_2 on top of heavily absorbing Pt). However, on addition of H_2O_2 (active system), the particles reorient their symmetry axis parallel to the bottom surface and can be seen as half-covered circles in the right micrograph in **a**, where dark semi-spheres correspond to the Pt cap and the SiO_2 parts that do not absorb the light appear lighter. (**b**) Schematic of a particle rotating from bottom-down configuration on peroxide addition. (**c**) Micrograph of Janus colloids ($R = 2.5 \, \mu m$) in the vicinity of the step with height $h_{step} = 800 \, nm$. In the absence of H_2O_2 (left image), the step (seen as a black line) has no influence on the orientation of the particles, (their caps are facing downwards, same as far from the step). On addition of fuel (right image), the particles orient with their symmetry axis parallel to both the bottom surface and the step. Scale bars, $5 \, \mu m$ (**a,c**). (**d**) The phase portrait for a bottom-heavy (see Supplementary Note 1) particle with $R = 2.5 \, \mu m$ at an infinite planar wall oriented with its normal parallel to the direction of gravity. The phase portrait is calculated at $b_{inert}/b_{cap} = 0.3$ and $b_w/b_{cap} = -0.2$, where b_{inert} and b_{cap} are the surface mobilities (see Methods for details) at the inert and catalytic faces of the particle, respectively, and b_w is the surface mobility at the wall (the phase portrait for $R = 1.0 \, \mu m$ is shown in Supplementary Fig. 2). The phase portrait indicates that a particle will rotate to its steady-state orientation $\theta = \theta_{eq} \approx 90°$ for all initial conditions. The inset represents a schematic diagram of the system: a Janus sphere of radius R is placed at distance h above an inert wall; θ describes the orientation of the particle's cap with respect to the wall normal. Δx is the step particle distance and ϕ is the cap orientation with respect to the step normal. (**e**) The rate of rotation $\dot{\theta} = -\Omega_x/\Omega_0$ of a particle with $R = 2.5 \, \mu m$ above a planar substrate, including contributions from activity, gravity and chemiosmotic flows on the substrate. This function is the sum of Fig. 3a,e,f. (**f**) Phase portrait similar to the one in **d** but in absence of gravity. All other parameters are as in **d**; this portrait is supposed to capture qualitatively the effect of the vertical step wall.

self-diffusiophoresis (see Methods section for details of the model), in which we assume that the activity of the Janus particle is captured by the release of a neutral solute (O_2 molecules) at a constant rate from its catalytic cap[28]. The resulting anisotropic solute distribution around the particle drives a surface flow in a thin layer surrounding the particle, leading to its directed motion[28,29]. The catalytically active particle has several types of interaction with a nearby impermeable wall. The particle drives long-range flows in the suspending solution. These flows are reflected from the wall, coupling back to the particle (HI). Second, the particle's self-generated solute gradient is modified by the presence of the wall. The wall-induced modification of the solute concentration field can contribute to translation and rotation of the particle ('phoretic interaction'). In particular, when the solute interacts more weakly with the inert region of the particle than with the catalytic cap and both interactions are repulsive (see Methods for details), the confinement and accumulation of solute near the substrate tends to drive rotation of the cap away from the substrate. On the other hand, the bottom heaviness of the particle, along with the HI of the particle with the substrate, tends to drive the rotation of the cap towards the

substrate. Finally, the inhomogeneous solute distribution along the wall induces a solute gradient-driven 'chemiosmotic' flow along the substrate. For repulsive solute–substrate interactions, this surface slip velocity is directed quasi-radially inward towards the particle, driving a particle-uplifting flow in the suspending solution and causing the particle cap to rotate away from the substrate. For attractive solute–substrate interactions, the opposite directions of flows apply. Numerical analysis of this model system shows that depending on the relative strengths of these interactions (that is, the parameters characterizing the surface chemistry of the particle and the wall), the various contributions to rotation discussed above may balance at a steady height h_{eq} and orientation $\theta_{eq} \approx 90°$, and that this steady state is robust and stable against perturbations in height and orientation. The particle cap orientation would therefore evolve to $\theta_{eq} \approx 90°$ (that is, the symmetry axis almost parallel to the substrate) from nearly all initial orientations, including a cap-down one. In Fig. 2d, a phase portrait shows the dynamical evolution of particle height and orientation, and the colour-coded rate of rotation $\dot{\theta} = -\Omega_x$ is depicted in Fig. 2e. The steady state (red dot) clearly has a large basin of attraction. We note that our numerical calculations were carried out for $h/R \geq 1.02$. Therefore, some trajectories in the region of the cap up ($\theta = 180°$) orientation encounter a numerical cutoff. However, based on the structure of the phase portrait, we expect such trajectories to roll towards $\theta \approx 90°$ after a close encounter with the wall.

If a particle as above would encounter now a second vertical side wall, numerical simulations for the same interaction parameters show (see Fig. 2f) that for this wall, for which gravity now plays no role, a similar sliding along the wall attractor emerges with $\phi_{eq} \approx 90°$, that is, with the particle oriented with its axis almost parallel to the vertical wall. The combination of the two sliding states thus aligns the axis of the particle along the edge formed by the two walls. It is noteworthy that although this second fixed point appears to have a smaller basin of attraction, it should capture the whole $\phi \leq \pi/2$ range. A particle on a trajectory that 'crashes' into the vertical wall would diffuse along the wall until it reaches the basin of attraction in the vicinity of $\phi_{eq} \approx 90°$. Although the argument is developed for the superposition of two infinite planar walls, we expect that similar features may occur for a vertical step with finite height. The above results will also hold for particles with the catalytic cap less dense than Pt. It is easy to see that in this case, while keeping all the other parameters of the system, such as geometry of the cap, activity and so on, fixed, a sliding fixed point along the bottom wall will also emerge. Moreover, we have checked via numerical simulations (results not shown) that the corresponding height and the orientation will lie between the values corresponding to the fixed points shown in Fig. 2d,f and thus the corresponding orientation will remain close to 90°.

Within our model, we can isolate and quantify the various wall-induced contributions to particle motion discussed above. The mathematical details of the decomposition are given in Supplementary Note 2. In Fig. 3b, we show the contribution to the rate of rotation $\dot{\theta}$ of the particle from HI with the wall as a function of particle height and orientation. HIs always rotate the particle cap towards the wall. Therefore, for the particular combination of parameters used in this work, HIs cannot by themselves produce a steady orientation $\theta_{eq} \approx 90°$. On the other hand, phoretic interactions always rotate the cap away from the wall, as described above (Fig. 3c). Therefore, the interplay of hydrodynamic and phoretic interactions can produce a curve with $\dot{\theta} = 0$ in the region of $\theta_{eq} \approx 90°$ (Fig. 3a). Moreover, the contributions of bottom heaviness (Fig. 3e) and chemiosmotic flow on the wall (Fig. 3f) to the angular velocity are comparable in magnitude with the contributions from HI and phoretic interactions. Therefore, for the parameters used in this work, all of these effects are important in determining the emergence and location of a 'sliding state' attractor. The surface chemistry parameters were chosen as providing the best fit to the experimental observations of the two sliding states (above a substrate and along a side wall).

As noted above, our model includes several types of interaction of the particle with the wall. However, many theoretical[24] and experimental[22,23] studies have sought to characterize the interaction of active particles and solid boundaries strictly in terms of effective HIs. It is therefore interesting to compare our full model against the best fit results from effective HI models. We consider two such approaches, the details of which are in Supplementary Note 3. Briefly, in the first approach, we use the

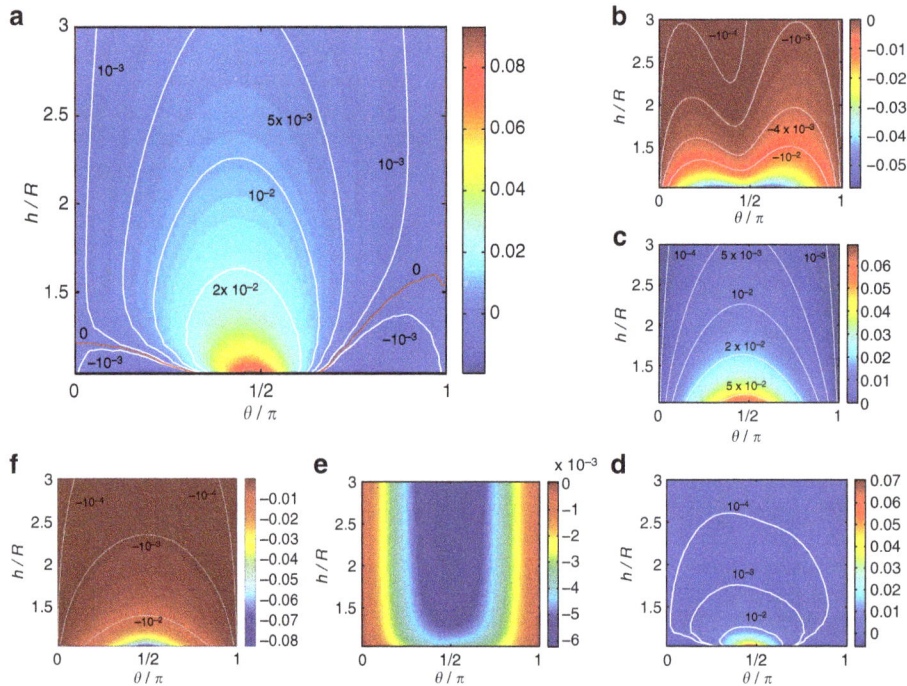

Figure 3 | Various contributions to particle angular velocity Ω_x. (a) Contribution from self diffusiophoresis $-\Omega_x^a/\Omega_0$ (see Supplementary Note 2 and equation (2) as a function of height h/R and orientation θ for half-covered Janus microswimmer and unequal surface mobilities $b_{inert}/b_{cap} = 0.3$. Throughout, white curves correspond to constant values of Ω_x^a. Note that, by definition, panel **a** is the sum of panels **b,c,** and **d**. (**b**), Contribution $-\Omega_x^{a,hi}/\Omega_0$ obtained by using the free space number density of solute distribution $c^{fs}(\mathbf{r})$ around the particle, i.e., neglecting the influence of the wall on the number density of solute, but including the influence of the wall on the hydrodynamic flow. (**c**) Contribution $-\Omega_x^{a,sol}/\Omega_0$ obtained by using the free space hydrodynamics stress tensor σ^{fs} in the dual Stokes problems employed in the reciprocal theorem, i.e., neglecting the effect of the wall on the hydrodynamics, but including the chemical effect. (**d**) Contribution $-\Omega_x^{a,\delta\delta}/\Omega_0$ due to higher order coupling between the two effects. (**e**) Contribution $-\Omega_x^{a,g}/\Omega_0$ to rate of rotation from the bottom-heaviness of the particle. (**f**) Chemio-osmotic contribution $-\Omega_x^{a,ws}/\Omega_0$ due to the activity-induced phoretic slip at the wall calculated at $b_w/b_{cap} = -0.2$.

classical 'squirmer' model and specify *a priori* the amplitude of the first two squirming modes. Higher-order modes are taken to have zero amplitude. In the second approach, we consider the 'effective squirmer' obtained within our model by neglecting phoretic and chemiosmotic effects. The 'effective squirmer' approach intrinsically covers a broad range of squirming mode amplitudes. In Table 1, we show that the results of the full model match the experimental observations significantly better than the best results of the two HI-only approaches.

Dynamics of Janus microswimmers at a rectangular step. We designed a system with microfabricated three-dimensional structures by patterning of photoresist through a circular or square mask, followed by e-beam deposition of the required material (Si or SiO_2 in our case), and the removal of the developed photoresist resulting in desired structures (for detailed information, see Methods). Depending on the use of positive or negative photoresist, we obtain patterns with posts or wells of different shapes (see Supplementary Fig. 5). The height of the features patterned on a substrate is tunable in a wide range; in this study, we have tested step heights h_{step} between 100 and 1,000 nm.

Characterization of particle trajectories approaching a step. Figure 4a shows snapshots of a typical trajectory of a microswimmer moving towards a step at almost perpendicular

direction. Once the particle hits the step (Fig. 4a third panel) it starts reorienting its axis (Fig. 4a fourth panel) towards the direction along the step (Fig. 4a fifth panel). We observe that in most cases the complete process of reorientation takes < 10 s, independent of the initial angle at which the particle approaches the step. Within the resolution of our experimental equipment, we do not observe any systematic deflection in the trajectory of the particle in the vicinity of the steps. Therefore, we conclude that if any long-range effective interaction exists between the particles and the steps, it must be very weak. This observation is reproduced by our numerical model: we calculate that the effects of a wall on the velocity of a particle are negligible when the particle is more than three radii away from the wall (Supplementary Fig. 4 and Supplementary Note 4). We attribute the suppression of the motion of the particles in the direction normal to the step upon collision to purely steric interactions.

In Fig. 4b we present the distribution of the reaction products around a microswimmer calculated numerically for particle positions and orientations approximately corresponding to those shown in the experimental micrographs in Fig. 4a. When the particle is far away from the step (Fig. 4b first and second panels), approaching it in a head-on direction, the generated concentration field confirms the expected mirror symmetry with respect to the plane defined by the motion axis and normal to the substrate. As the result of this symmetry, there are no activity-induced rotations and the particle stays on its head-on

Table 1 | Comparison of full model with two hydrodynamics-only models.

Model	Best fit parameters	h_{eq}/R, no gravity	θ_{eq}, no gravity	h_{eq}/R, with gravity	θ_{eq}, with gravity
Full model	$b_{inert}/b_{cap} = 0.3$, $b_{wall}/b_{cap} = -0.2$, $b_{cap} < 0$	1.11	77.9°	1.06	94.8°
Squirmer, first two squirming modes only	$B_2/B_1 = 0.3$	1.64	102°	Below 1.02	Around 45°
Effective squirmer	$b_{inert}/b_{cap} = -0.8$, $b_{cap} < 0$	1.063	69.7°	1.09	65.3°

For each model, we list the parameters that give the best fit to the experimental observations. For each model and set of best-fit parameters, we give the height and orientation of the particle when it is in a 'sliding state' above a planar wall in both the presence of gravity (corresponding to motion above a substrate) and the absence of gravity (corresponding to motion near a side wall). Experimentally, it is observed that $\theta_{eq} \approx 90°$ in both cases. Of the three models, the full model shows the best fit with these experimental observations. For the squirmer with only the first two squirming modes, there are clear signs of an attractor with h_{eq}/R below the numerical cutoff of $h/R = 1.02$ in the presence of gravity, but this attractor has θ_{eq} far from 90° (see also Supplementary Note 3 and Supplementary Table 1). The best-fit effective squirmer agrees moderately well with the experimental observations. However, the orientation of the sliding seems significantly different from the experiment and, as discussed in Supplementary Note 3 and Supplementary Table 2, the best-fit parameters correspond to an unrealistically large force dipole.

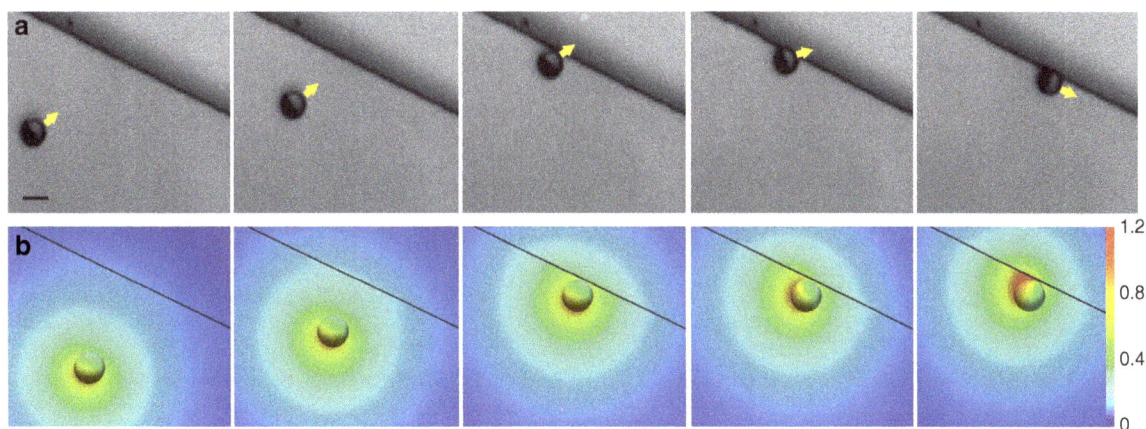

Figure 4 | Effect of 800 nm height step on the dynamics of a Janus microswimmer. (**a**) An active Janus particle approaching a step; after direct contact with the step, it reorients until its propulsion axis is parallel to the step. $R = 2.5$ μm, $h_{step} = 800$ nm, 2.5%vol. H_2O_2. (**b**) Numerically calculated steady-state distribution $c(\mathbf{r})$ of reaction products around half catalyst-covered Janus sphere as a function of the step distance and the cap orientation with respect to the step. The colour map shows $c(\mathbf{r})$ at the surfaces of the particle and substrate; $c(\mathbf{r})$ is in units of c_0 (see Methods). Scale bar, 5 μm.

track (up to Brownian rotational diffusion) towards the step. However, closer to the step, a head-on collision becomes unstable to small fluctuations of the propulsion axis, as any such fluctuation gets amplified by the buildup of asymmetric product distribution in the region between one side of the particle and the step (Fig. 4b fourth panel). This eventually leads to the reorientation of the motion axis parallel to the step (Fig. 4b fifth panel).

Effects of the step height on the capture efficiency. We observe that submicrometre steps are able to capture and guide particles as shown in Fig. 5 (insets). To evaluate the minimum height h_{step}^* that can still influence the trajectory of the particles, we fabricated a set of patterns with h_{step} varying in a range from 100 to 1,000 nm. The results for the two different particle sizes show that h_{step}^* decreases as the particles size increases. In Fig. 5 we summarize the responses of $R = 1\,\mu m$ and $R = 2.5\,\mu m$ active particles to variation of the step height. Both types of particles can swim over the step of 100 nm height. For $R = 2.5\,\mu m$ particles, steps of height 200 nm already ensure ~90% docking of particles upon collision with the step, while a significant fraction of the $R = 1\,\mu m$ particles managed to pass over the 200-nm-high step, and 400-nm-high steps were required for efficient docking. From Fig. 5 we infer that h_{step}^*/R is smaller for larger particles.

Having estimated the threshold values h_{step}^* for particle trapping, we now select steps of sufficient height to ensure full trapping upon collisions. Therefore, all the following experiments were carried out on 800 nm features, for which both 2.5 and 1 μm particles follow the step upon collision.

Guidance of microswimmers by low-height topographic steps. The step-like topography and particle alignment along step edges can be used to guide microswimmers, as it is shown in Fig. 6a. This corresponds to particles in a well structure with straight steps. Upon collisions, the particles align along the steps and follow them (Fig. 6a and Supplementary Movie 1). In the same

well-like structure, particles eventually encounter a corner and after spending some time adjusting their orientation can manoeuvre around the corner (Fig. 6b and Supplementary Movie 2). In the case of a post-like structure as displayed in Fig. 6c, particles also follow straight features but fail to reorient and manoeuvre around the 270° corner (see also Supplementary Movie 3). These findings suggest that certain critical value must exist for reflex angles between features above which guidance along the step is lost.

Active Janus microswimmers also follow circular trajectories around circular posts as shown in Fig. 7, which requires constant reorientation of the axis. In Fig. 7a–c, paths are shown where particles with $R = 1\,\mu m$ circle around posts with the diameter of 15, 40 and 60 μm, respectively, for more than 12 s (see Supplementary Movies 4–6). Supplementary Fig. 3 and Supplementary Movie 7 show an additional example of cycling motion with long retention times. We find that the retention time of microswimmers at the circular posts increases with increasing peroxide concentration, as displayed in Fig. 7d. At 1%, few particles completely circle around a whole post and, in most cases, the microswimmers detach from the post before a complete revolution (at lower peroxide concentrations, the particles hardly move and get easily stuck at the steps; thus, measurements were not considered). At 2% H_2O_2, the path length along the posts is increased and likewise in 3 and 5% H_2O_2, where many particles circle around posts multiple times. At even higher concentrations of H_2O_2, we observe vigorous formation of oxygen bubbles and occurrence of convective flows, and thus no reliable measurements could be performed above 5% H_2O_2.

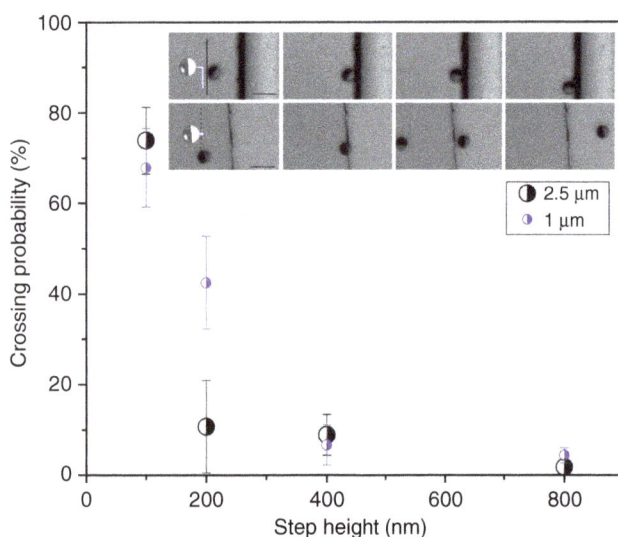

Figure 5 | Submicrometre steps as rectifiers of active particles' trajectories. A summary of the crossing behaviour of Janus SiO_2 microswimmers of different sizes at several values of h_{step}; the error bars are s.e.m. Inset: a sequence of micrographs showing a Janus particle with $R = 2.5\,\mu m$ approaching a step with $h_{step} = 800\,nm$, reorienting and then moving parallel to it (upper row). Micrograph sequence of a Janus particle, 2.5 μm, passing over a step, $h_{step} = 100\,nm$ (lower row). Scale bars, 10 μm (all).

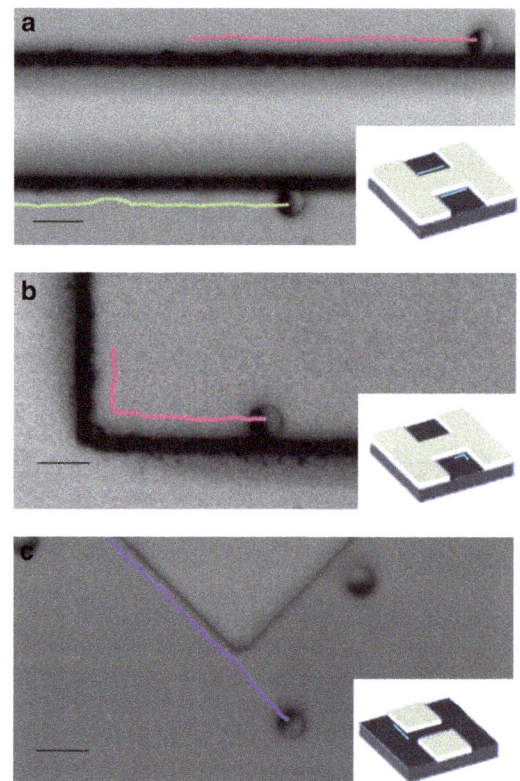

Figure 6 | Guidance of Janus microswimmers by step features. (a) A micrograph showing trajectories of two $R = 2.5\,\mu m$ Janus particles following a straight step. **(b)** A Janus particle tracked while manoeuvring around a 90° corner. **(c)** A Janus particle unable to follow a reflex angle of 270°. The insets show schematically the structures of wells **a,b** and posts **c**. The blue lines on the insets schematically indicate the position of Janus particles in actual experiments. Scale bars, 10 μm.

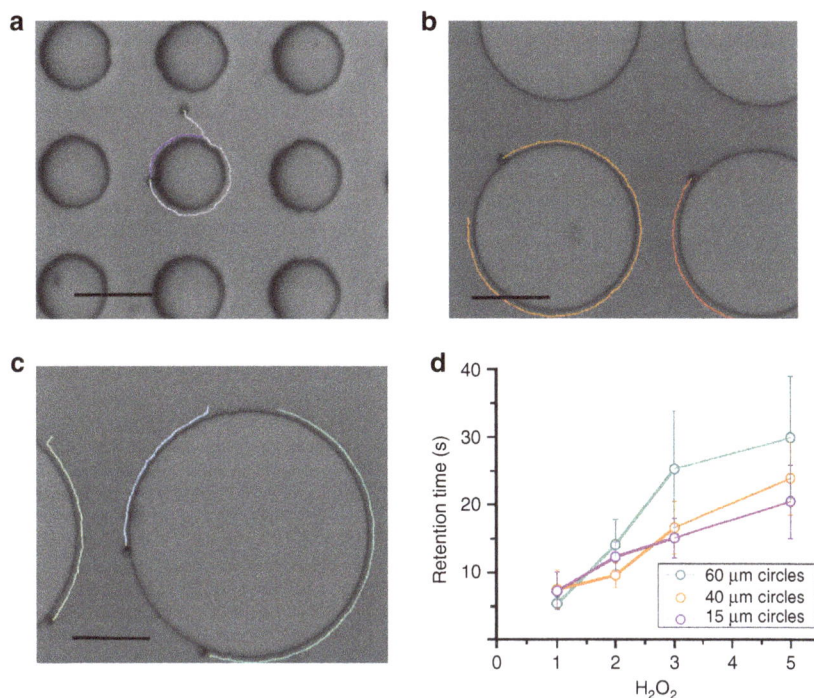

Figure 7 | Trapping and guiding active particles by circular posts. (**a–c**) Optical snapshots of $R = 1\,\mu m$ active particles moving for 12 s around circular posts of 15, 40 and 60 μm diameter d, respectively. Scale bar, 20 μm. (**d**) Average retention time as a function of peroxide concentration for $R = 1\,\mu m$. The average is determined for 15–20 trajectories per data point; the error bars are standard errors of the mean.

As the retention time increases with the concentration of H_2O_2, we conclude that it is the activity of the microswimmers that is directly responsible for the effective particle attraction to the posts and for the occurrence of the sliding attractor: it is the net result of the particle-step HI and confinement-induced modification of the distribution of the oxygen concentration. The strength of both effects depends on the fuel concentration: increased fuel concentration leads to a higher production rate of solute (that is, stronger phoretic and chemiosmotic interactions) and a higher self-propulsion velocity (that is, stronger HIs). The finite retention time is set by the competition between activity-induced effective attraction to the post side walls and rotational diffusion. Increased fuel concentration increases the strength of the first factor without affecting the second one and therefore increases the retention time. The robustness of the sliding state attractor is further discussed in Supplementary Note 5.

Discussion

We report experimental results showing the dynamics of chemically active Janus microswimmers at geometrically patterned substrates and a qualitative interpretation in terms of a minimal continuum model of self-diffusiophoresis of chemically active colloids. Employing a lithography-based method to fabricate submicrometre topographic features in the form of rectangular stripes, square posts, cylindrical posts or square wells on glass surface or silicon wafer, we demonstrate that the motion of chemically active Janus microswimmers can be restricted to proceed along these small height patterns for significant time intervals. Furthermore, the motion along the circumference of cylindrical posts reveals that the retention time increases with increasing H_2O_2 concentration. This allows us to unequivocally identify the particle's chemical activity, which modulates the distribution of the phoretic slip at the particle surface and thus the HI with the nearby topography, as playing a dominant role in the observed phenomenology.

We also show that a minimal continuum model of self-diffusiophoresis captures the qualitative features of the experimental observations if one accounts for the difference in material properties of the two parts of the colloid and for chemiosmotic flows induced at the wall. This latter aspect highlights the need for models that explicitly include chemical activity, without which a no-slip boundary condition would apply at the wall. The model exploited here allows us to understand the emergence of states of motion along the edges as a simultaneous attraction to two fixed-point attractors corresponding to steady sliding states along the bottom wall and along the vertical wall of the step.

The microstructuring method presented here avoids the use of any external fields and relies solely on the intrinsic properties of the system to control particle motion. The phenomenology reported here is, in some sense, a mesoscale analogue of the binding of motor proteins to microtubules, to switch to directional motion. However, in distinction to biological nanomotors, the Janus microswimmers bypass the binding and rather elegantly exploit an effective attraction that stems from the feedback between geometric confinement, and chemical and hydrodynamic activity. The results presented here open the possibility of robust guidance of particles along complex paths via minimal surface modifications, that is, by sculpting a pattern with the edge in the desired shape. This may have significant implications in designing new applications based on artificial swimmers. Finally, we consider that these findings will allow further developments by employing smart, chemically patterned walls, where features of the nearby surfaces (and thus the guiding of the microswimmers) can be turned on and off.

Methods

Sample preparation. Janus particles were obtained by drop casting of a suspension of spherical silica colloids (diameter of 2 or 5 μm, Sigma Aldrich) on an oxygen-plasma cleaned glass slide followed by slow evaporation of the solvent and subsequent placement in an e-beam system. High vacuum was applied and subsequently a monolayer of 7 nm Pt was evaporated, to guarantee catalytic properties. To release particles from the glass slides into deionized water, short ultrasound pulses were sufficient.

Photoresist patterns were prepared on 24 mm square glass slides or on silicon wafers. In case of positive photoresist, AR-P 3,510 was spin coated onto the cleaned substrate at 3,500 r.p.m. for 35 s, followed by a soft bake using a hotplate at 90 °C for 3 min and exposure to ultraviolet light with a Mask Aligner (400 nm) for 2 s. Patterns were developed in a 1:1 AR300-35:H_2O solution. In case of negative photoresist, a layer of adhesion promoter TI prime was spin coated on the substrate during 20 s at 3,500 r.p.m. After 2 min of drying at 120 °C, the negative photoresist was coated employing a programme of 35 s spinning at 4,500 r.p.m., followed by 5 min baking at 90°. The exposure was carried out with a Mask aligner for 2 s followed by 2 min on the hotplate at 120 °C. Finally, an additional exposure to 2 s ultraviolet light is applied and the patterns were developed in pure AZ726MIF. The steps were obtained by e-beam deposition of the desired material (SiO_2, Si) in the desired thickness. By dissolving the photoresist layer in acetone the pattern structures of the substrate are exposed; the whole process is illustrated in Supplementary Fig. 5.

Before experiment, the patterned substrates were cleaned by oxygen plasma. Experiments were performed directly on the substrates by adding equal volumes of particles in deionized water and diluted peroxide solutions. Videos were recorded with a Leica DFC 300G camera mounted to a Leica upright microscope at ~ 30 fps. Evaluation and tracking was performed using Fiji analysis software.

Tracking. Accurate tracking of Janus particles was performed automatically by a specially developed script in Python 2.7 using the OpenCV library. The position of the Janus swimmers at every frame is found by extracting the background, which erases the static posts from the image, leaving only the moving particles.

Theoretical modelling. We model particle motion within a continuum, neutral self-diffusiophoretic framework. A particle emits solute at a constant rate from its catalytic cap. The number density $c(\mathbf{r})$ of solute is quasi-static, where $\mathbf{r}$ is a position in the fluid. The solute field is governed by the Laplace equation $\nabla^2 c = 0$ and obeys the boundary conditions $-D\nabla c \cdot \mathbf{n} = \kappa$ on the catalytic cap and $-D\nabla c \cdot \mathbf{n} = 0$ on the inert face of the particle and the substrate, where κ is the rate of emission (uniform over the cap), D is the diffusion coefficient of oxygen and $\mathbf{n}$ is the local surface normal. Our model neglects the details of the catalytic reaction, which might involve the transport of charged intermediates[28,29]. Nevertheless, we expect this model to capture the gross effects of both near-wall confinement of the solute field and HI with nearby walls. The surface gradient of solute drives a surface flow ('slip velocity') $\mathbf{v}_s$ in a thin fluid layer surrounding the particle surface $\mathbf{v}_s(\mathbf{r}) = -b_s(\mathbf{r})\nabla_{\parallel} c$, where $\nabla_{\parallel} = (\mathbf{I} - \mathbf{n} \otimes \mathbf{n}) \cdot \nabla$. $\mathbf{I}$ is the 3×3 unit matrix.

The coefficient $b_s(\mathbf{r})$ of the slip velocity, the so-called 'surface mobility', is determined by the molecular interaction potential between the solute and the particle surface[30]. We allow $b_s(\mathbf{r})$ to differ between the inert and catalytic regions, but assume it is uniform in each region, that is, take $b_s = b_{inert}$ or $b_s = b_{cap}$. In addition, when we consider the effect of chemiosmotic flow on the substrate, we calculate a wall slip velocity $\mathbf{v}_w(\mathbf{r}_s) = -b_w\nabla_{\parallel} c$, where b_w is a constant. We always take the interaction between the solute and particle surface to be repulsive, that is, $b_s < 0$, so that the model is consistent with the observed motion of particles away from their caps.

The velocity $\mathbf{u}(\mathbf{r})$ in the fluid is governed by the Stokes equation $-\nabla p + \eta\nabla^2\mathbf{u} = 0$ and the incompressibility condition $\nabla \cdot \mathbf{u} = 0$, where $p(\mathbf{r})$ is the fluid pressure and η is the dynamic viscosity of the solution. The velocity obeys the boundary conditions $\mathbf{u} = \mathbf{v}_w(\mathbf{r})$ on the substrate and $\mathbf{u}(\mathbf{r}) = \mathbf{U}^a + \mathbf{\Omega}^a \times (\mathbf{r} - \mathbf{r}_O) + \mathbf{v}_s(\mathbf{r})$ on the particle surface, where $\mathbf{r}_O$ is the position of the particle centre and $\mathbf{U}^a$ and $\mathbf{\Omega}^a$ are the contributions of activity to the translational and rotational velocities of the particle, respectively. To obtain $\mathbf{U}^a$ and $\mathbf{\Omega}^a$ for a given position and orientation of the particle, we first solve for $c(\mathbf{r})$ numerically, using the boundary element method[31]. The slip velocities $\mathbf{v}_s$ and $\mathbf{v}_w$ are then calculated from $c(\mathbf{r})$. Inserting the slip velocities in the boundary conditions and requiring that the particle is force and torque free, we solve the Stokes equation numerically via the boundary element method, to obtain $\mathbf{U}^a$ and $\mathbf{\Omega}^a$ in terms of characteristic velocity scales $U_0 = |b_{cap}|\kappa/D$ and $\Omega_0 \stackrel{\text{def}}{=} U_0/R$. In addition, $c(\mathbf{r})$ is calculated in terms of a characteristic concentration $c_0 \stackrel{\text{def}}{=} \kappa R/D$.

When we include the effects of gravity, we adopt the geometrical model of Campbell and Ebbens[32], taking the Janus particle as having a platinum cap that smoothly varies in thickness between a maximum of 7 nm at the pole and zero thickness at the particle equator. The gravitational contributions to particle velocity, $\mathbf{U}^g$ and $\mathbf{\Omega}^g$, are calculated using standard methods (see Supplementary Note 1).

We obtain complete particle trajectories by numerically integrating $\mathbf{U} = \mathbf{U}^a + \mathbf{U}^g$ and $\mathbf{\Omega} = \mathbf{\Omega}^a + \mathbf{\Omega}^g$. Further details of the numerical method are given in ref. 21. We note that the assumption that the solute field is quasi-static is valid in the limit of small Peclet number $Pe = U_0 R/D$. We have neglected the inertia of the fluid, which is valid for small Reynolds number $Re = \rho U_0 R/\eta$, where ρ is the mass density of the solution. These dimensionless numbers are $Pe \approx 4 \times 10^{-3}$ and $Re \approx 10^{-5}$ for a 5-μm-diameter catalytic Janus particle that swims at 6 μm s^{-1} (ref. 33).

References

1. Wang, W., Duan, W. T., Ahmed, S., Mallouk, T. E. & Sen, A. Small power: autonomous nano- and micromotors propelled by self-generated gradients. *Nano Today* **8**, 531–554 (2013).
2. Sánchez, S., Soler, L. & Katuri, J. Chemically powered micro- and nanomotors. *Angew. Chem. Int.Ed.* **54**, 1414–1444 (2015).
3. Wang, J. *Nanomachines: Fundamentals and Applications* (Wiley-VCH, 2013).
4. Paxton, W. F. *et al.* Catalytic nanomotors: autonomous movement of striped nanorods. *J. Am. Chem. Soc.* **126**, 13424–13431 (2004).
5. Kline, T. R., Paxton, W. F., Mallouk, T. E. & Sen, A. Catalytic nanomotors: remote-controlled autonomous movement of striped metallic nanorods. *Angew. Chem. Int. Ed.* **44**, 744–746 (2005).
6. Fournier-Bidoz, S., Arsenault, A. C., Manners, I. & Ozin, G. A. Synthetic self-propelled nanorotors. *Chem. Commun.* 441–443 (2005).
7. Baraban, L. *et al.* Catalytic Janus motors on microfluidic chip: deterministic motion for targeted cargo delivery. *ACS Nano* **6**, 3383–3389 (2012).
8. Hong, Y., Blackman, N. M. K., Kopp, N. D., Sen, A. & Velegol, D. Chemotaxis of nonbiological colloidal rods. *Phys. Rev. Lett.* **99**, 178103 (2007).
9. Baraban, L., Harazim, S. M., Sanchez, S. & Schmidt, O. G. Chemotactic behavior of catalytic motors in microfluidic channels. *Angew. Chem. Int. Ed.* **52**, 5552–5556 (2013).
10. Dey, K. K., Bhandari, S., Bandyopadhyay, D., Basu, S. & Chattopadhyay, A. The pH taxis of an intelligent catalytic microbot. *Small* **9**, 1916–1920 (2013).
11. Saha, S., Golestanian, R. & Ramaswamy, S. Clusters, asters, and collective oscillations in chemotactic colloids. *Phys. Rev. E* **89**, 062316 (2014).
12. Peng, F., Tu, Y., van Hest, J. C. M. & Wilson, D. A. Self-guided supramolecular cargo-loaded nanomotors with chemotactic behavior towards cells. *Angew. Chem. Int.Ed.* **127**, 11828–11831 (2015).
13. Solovev, A. A., Sanchez, S., Pumera, M., Mei, Y. F. & Schmidt, O. G. Magnetic control of tubular catalytic microbots for the transport, assembly, and delivery of micro-objects. *Adv. Funct. Mater.* **20**, 2430–2435 (2010).
14. Khalil, I. S. M., Magdanz, V., Sanchez, S., Schmidt, O. G. & Misra, S. Three-dimensional closed-loop control of self-propelled microjets. *App. Phys. Lett.* **103**, 172404-1–172404-4 (2013).
15. Ulbrich, T. C. *et al.* Effect of magnetic coupling on the magnetization reversal in arrays of magnetic nanocaps. *Phys. Rev. B* **81**, 054421 (2010).
16. Albrecht, M. *et al.* Magnetic multilayers on nanospheres. *Nat. Mater.* **4**, 203–206 (2005).
17. Gunther, C. M. *et al.* Microscopic reversal behavior of magnetically capped nanospheres. *Phys. Rev. B* **81**, 064411 (2010).
18. Khalil, Islam S. M., Magdanz, V., Sanchez, S., Schmidt, O. G. & Misra, S. Precise localization and control of catalytic Janus micromotors using weak magnetic fields. *Int. J. Adv. Robot. Syst.* **12**, 2 (2015).
19. Evans, A. A. & Lauga, E. Propulsion by passive filaments and active flagella near boundaries. *Phys.Rev. E* **82**, 041915 (2010).
20. Spagnolie, S. E. & Lauga, E. Hydrodynamics of self-propulsion near a boundary: predictions and accuracy of far-field approximations. *J. Fluid Mech.* **700**, 105–147 (2012).
21. Uspal, W. E., Popescu, M. N., Dietrich, S. & Tasinkevych, M. Self-propulsion of a catalytically active particle near a planar wall: from reflection to sliding and hovering. *Soft Matter* **11**, 434–438 (2015).
22. Volpe, G., Buttinoni, I., Vogt, D., Kummerer, H. J. & Bechinger, C. Microswimmers in patterned environments. *Soft Matter* **7**, 8810–8815 (2011).
23. Brown, A. T. *et al.* Swimming in a crystal. *Soft Matter* **12**, 131–140 (2016).
24. Spagnolie, S. E., Moreno-Flores, G., Bartolo, D. & Lauga, E. Geometric capture and escape of a microswimmer colliding with an obstacle. *Soft Matter* **11**, 396–411 (2015).
25. Palacci, J., Sacanna, S., Vatchinsky, A., Chaikin, P. M. & Pine, D. J. Photoactivated colloidal dockers for cargo transportation. *J. Am. Chem. Soc.* **135**, 15978–15981 (2013).
26. Howard, J., Hudspeth, A. J. & Vale, R. D. Movement of microtubules by single kinesin molecules. *Nature* **342**, 154–158 (1989).
27. Howard, J. in *Physics of Bio-molecules and Cells. Physique des biomolécules et des cellules*. Vol. 75 Les Houches - Ecole d'Ete de Physique Theorique. (eds Flyvbjerg, F., Jülicher, F., Ormos, P. & David, F.) Ch. 2, 69–94 (Springer, 2002).
28. Golestanian, R., Liverpool, T. B. & Ajdari, A. Propulsion of a molecular machine by asymmetric distribution of reaction products. *Phys. Rev. Lett.* **94**, 220801 (2005).
29. Golestanian, R. Anomalous diffusion of symmetric and asymmetric active colloids. *Phys. Rev. Lett.* **102**, 188305 (2009).
30. Anderson, J. L. Colloid transport by interfacial forces. *Ann. Rev. Fluid Mech.* **21**, 61–99 (1989).
31. Pozrikidis, C. *A Practical Guide to Boundary Element Methods with the Software Library BEMLIB* (CRC Press, 2002).
32. Campbell, A. I. & Ebbens, S. J. Gravitaxis in spherical Janus swimming devices. *Langmuir* **29**, 14066–14073 (2013).

33. Popescu, M. N., Dietrich, S., Tasinkevych, M. & Ralston, J. Phoretic motion of spheroidal particles due to self-generated solute gradients. *Eur. Phys. J. E* **31**, 351–367 (2010).

Acknowledgements

We thank Albert Miguel Lopez for help with the automated tracking programme. W.E.U., M.T. and M.N.P. acknowledge financial support from the DFG grant number TA 959/1-1. S.S., J.S. and J.K. acknowledge the DFG grant number S.A 2525/1-1. The research also has received funding from the European Research Council under the European Union's Seventh Framework Programme (FP7/2007-2013)/ERC grant agreement 311529.

Author contributions

S.S. and J.S. designed the experiments. J.S. and J.K. performed the experiments and analysed the data. M.T. and W.E.U. performed numerical calculations. M.N.P. contributed to the theoretical analysis of the numerical results. J.S., W.E.U. and M.T. wrote the manuscript. All the authors discussed the results and commented on the manuscript.

Additional information

Competing financial interests: The authors declare no competing financial interests.

25

Magnetic-free non-reciprocity based on staggered commutation

Negar Reiskarimian[1] & Harish Krishnaswamy[1]

Lorentz reciprocity is a fundamental characteristic of the vast majority of electronic and photonic structures. However, non-reciprocal components such as isolators, circulators and gyrators enable new applications ranging from radio frequencies to optical frequencies, including full-duplex wireless communication and on-chip all-optical information processing. Such components today dominantly rely on the phenomenon of Faraday rotation in magneto-optic materials. However, they are typically bulky, expensive and not suitable for insertion in a conventional integrated circuit. Here we demonstrate magnetic-free linear passive non-reciprocity based on the concept of staggered commutation. Commutation is a form of parametric modulation with very high modulation ratio. We observe that staggered commutation enables time-reversal symmetry breaking within very small dimensions ($\lambda/1{,}250 \times \lambda/1{,}250$ in our device), resulting in a miniature radio-frequency circulator that exhibits reduced implementation complexity, very low loss, strong non-reciprocity, significantly enhanced linearity and real-time reconfigurability, and is integrated in a conventional complementary metal–oxide–semiconductor integrated circuit for the first time.

[1] Department of Electrical Engineering, Columbia University, 1300 South West Mudd, 500 West 120th Street, New York, New York 10027, USA. Correspondence and requests for materials should be addressed to H.K. (email: harish@ee.columbia.edu).

Reciprocity in electronics or, equivalently, the principle of time reversibility in optics is a fundamental property of any linear system or material described by symmetric and time-independent permittivity and permeability tensors[1]. Non-reciprocity, however, enables new applications that span radio frequencies (RF) to optical frequencies. Optical isolators are critical to on-chip all-optical information processing systems for the protection of lasers and amplifiers, and the mitigation of multipath reflections. RF circulators enable full-duplex wireless, an emerging wireless communication paradigm that has been historically considered impractical[2], where the transmitter and the receiver operate simultaneously on the same frequency band, potentially doubling network capacity at the physical layer while offering numerous other benefits at the network layer[3,4].

Non-reciprocal components today are almost exclusively realized through the magneto-optic Faraday effect (Fig. 1a–c)—the application of a magnetic field bias parallel to the direction of propagation rotates the polarization vector of light due to the different propagation velocities of left- and right-circularly polarized waves[5]. Despite significant research efforts in the optical[6,7] and RF domains[8–10], non-reciprocal components based on magneto-optic materials remain incompatible with complementary metal–oxide–semiconductor (CMOS) integrated circuit (IC) fabrication processes due to material incompatibilities and the need for a magnetic field bias, significantly restricting their impact.

As early as 50 years ago, magnetic-free non-reciprocity and circulators had been investigated at microwave frequencies through the use of the inherent non-reciprocal nature of active devices such as direct-current/voltage-biased transistors[11]. More recently, metamaterials with embedded active transistors have been explored at RF and microwave frequencies[12–14]. However, such approaches are fundamentally limited by the noise and

nonlinear distortion generated by the active transistors. Indeed, for non-reciprocal components used at the front end of RF communication systems (such as circulators), and for front-end components more broadly (such as filters and diplexers), passive approaches with superior linearity and noise performance are paramount[15,16]. Techniques for non-reciprocity leveraging nonlinearity[17–21] have also been proposed. These techniques, although potentially useful and extensively explored in optical applications, exhibit non-reciprocity over certain signal power levels only and have limited applicability to scenarios where linearity to the signal is required, such as RF communication. Lorentz reciprocity may also be broken by introducing time dependency into the material or system[1]. Techniques at optical frequencies based on electro-optic phase modulators have been investigated[22,23] but require complex electro-optic modulation networks. Approaches based on spatio-temporal parametric modulation of waveguides have been investigated[24–28]. In such approaches, a travelling-wave modulation of the waveguide's properties produces direction-dependent mode conversion of the desired signal. In the optical domain, the size of the structure is limited by the weak electro-optic or acousto-optic effect[24–26], which results in an extremely low modulation ratio (modulation ratio is defined as the ratio between the maximum and minimum values of the modulated parameter and extremely weak modulation corresponds to a modulation ratio of practically unity). At RF, varactors are able to achieve modulation ratios of around two to four, resulting in structures that are of the order of a wavelength[27]. A second disadvantage of these approaches is that the mode/frequency conversion is undesirable in some applications and necessitates the use of filters[24] or diplexers[27]. In recent times, non-reciprocity through spatio-temporal parametric modulation of resonant rings, resulting in angular momentum biasing[29], has been demonstrated in the acoustic domain[30] and at

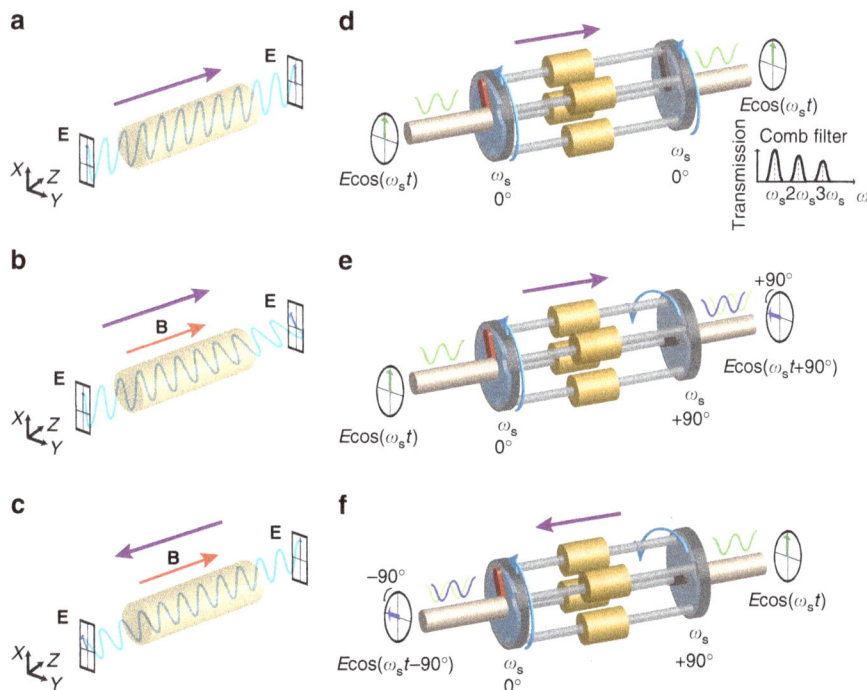

Figure 1 | Comparison between non-reciprocity induced by the magneto-optic Faraday effect and staggered commutation. (a) A wave propagating in a Faraday-active magneto-optic material experiences no Faraday rotation in the absence of magnetic bias. **(b)** In the presence of magnetic bias in the positive z direction, a wave travelling in the positive z direction experiences Faraday rotation due to the difference in the propagation velocities of right-handed and left-handed circularly polarized waves, whereas **(c)** a wave travelling in the negative z direction experiences an opposite rotation. **(d)** A wave propagating through a commutated network with no staggering experiences no phase shift. **(e)** Staggered commutation acts as a bias that breaks time reversibility, producing a phase shift for waves travelling from left to right and **(f)** an opposite phase shift for waves travelling from right to left.

RF[31], resulting in magnetic-free non-reciprocal circulators at or well below the wavelength scale and with low loss[32]. A key challenge with these spatio-temporal parametric modulation approaches in general is that the property that enables modulation (for instance, varactors[27,31] or opto-acoustic interactions[25]) often represents a nonlinearity to the signal itself, in particular when the modulation ratio is high, resulting in nonlinear distortion at higher signal levels. There has also been theoretical work on non-reciprocity and a microwave circulator based on parametric modulation of coupled resonators[33].

Here we introduce magnetic-free, linear and passive phase non-reciprocity based on staggered commutation and a highly-miniaturized RF circulator that embeds the phase-nonreciprocal component within a ring resonator. Commutation may be seen as a form of parametric modulation with very high modulation ratio (practically infinite in our prototype), making the phase-nonreciprocal component effectively a point parametric modulator. This enables its specific placement in the ring relative to the three circulator ports such that the signal across it is suppressed for excitations from one port, yielding significantly enhanced linearity to that port. Furthermore, the need for spatio-temporal modulation in the ring is eliminated, easing modulation complexity. Implementation complexity is further eased as no frequency conversion is seen at the circulator ports. In addition, when compared with prior art employing spatio-temporal parametric modulation of wave-guides[24–27] or electro-optic phase modulation[22,23], these concepts allow miniaturization of the unmodulated ring to deeply subwavelength scales ($\approx \lambda/80$ in our prototype). When compared with prior art employing angular-momentum biasing in parametrically modulated resonant rings[31,32], linearity enhancement is achieved due to the ability to suppress the signal across the point parametric modulator, although it is interesting to consider synergies between the approaches offering the dual benefits of both.

Results

Phase non-reciprocity through staggered commutation.

Commutated networks are a class of linear, periodically time-varying (LPTV) networks where the signal is periodically commutated through a bank of linear, time-invariant (LTI) networks (or media as in Fig. 1d). The first commutated networks relied on mechanical commutation through a rotating brush that periodically contacted a bank of capacitors[34,35] to realize narrow comb filters around harmonics of the commutation frequency. More recently, electronic commutation using passive transistor-based switches has resulted in high quality-factor (Q) comb filters, commonly called N-path filters, which operate at RF, exhibit significantly lower noise and higher linearity when compared with active RF filters, are compatible with conventional CMOS IC

technology and exhibit the potential to replace front-end off-chip surface acoustic wave filters for RF communication applications[15,36–40]. A key requirement is the availability of a high-quality switch with high ON/OFF transmission ratio (or modulation ratio when viewed as parametric modulation) and sufficient switching speed. Modern CMOS transistor switches boast ON/OFF conductance ratios as high as 1,000–100,000 (ref. 41), enough to be practically infinite from the perspective of commutated network operation, and switching speeds that enable such commutated networks to operate well into the RF frequency range and continuously improve with CMOS technology scaling. It should be emphasized that the transistors are used as reciprocal, highly linear, passive switches without direct-current bias. In other words, the switches cannot provide power gain to the desired signal by sourcing power from the modulating clock signals that control the commutation. These clocking signals that control the commutation are easily implemented in CMOS and the associated power consumption is due to the charging and discharging of parasitic capacitance in the clock path. This power consumption also reduces with CMOS technology scaling due to reducing parasitics.

An interesting property of commutated networks that we observe here is that staggering the commutation on either side of the bank of LTI networks results in time-reversal symmetry breaking and phase non-reciprocity (Fig. 1e,f). Waves travelling in the forward and reverse direction see a different ordering of the first and second commutating switches in time and experience opposite shifts in phase, essentially analogous to the magneto-optic Faraday effect. Here, the staggered commutation plays the role of the magnetic Faraday bias.

When the first and second set of switches are staggered by $+90°$, forward and reverse travelling waves at or near the commutation frequency experience phase shifts of $+90°$ and $-90°$, respectively. When a transmission line or waveguide of length $3\lambda/4$ is wrapped around the staggered commutated network, non-reciprocal wave propagation is achieved as waves may propagate in only one direction (Fig. 2a,b). In that direction, the $-270°$ phase delay through the $3\lambda/4$ ring adds with $-90°$ phase shift of the staggered commutated network to satisfy the boundary condition, enabling wave propagation. In the other direction, the $-270°$ phase delay adds with the $+90°$ phase shift of the staggered commutated network to prohibit wave propagation.

Unidirectional wave propagation not only requires phase non-reciprocity but also requires (near-)perfect transmission through the commutated network. In Fig. 3, we examine the requirements on the media across which commutation is being performed. We consider electrically short transmission-line media, in line with our goal of achieving a point parametric modulator. In the depicted simulations, eight-way commutation is considered, the transmission-line characteristic impedance (Z_{medium}), wave velocity (v) and length (l) are varied, and ideal sets of switches are

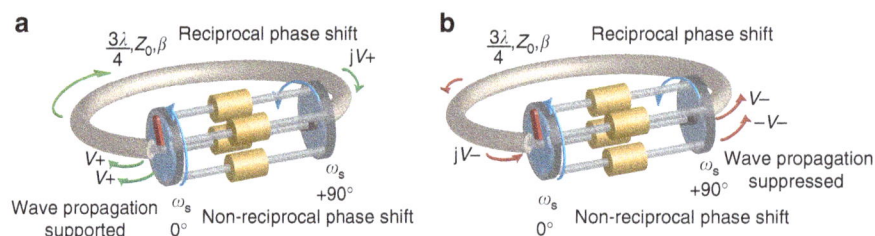

Figure 2 | Embedding the staggered commutated network within a $3\lambda/4$ transmission-line ring results in unidirectional wave propagation. (a) When the commutated network with $+90°$ staggering is embedded in a $3\lambda/4$ ring, then in one direction, the $-270°$ phase delay of the ring adds to the $-90°$ phase shift through the commutated network, enabling wave propagation. (**b**) In the other direction, the $-270°$ phase delay adds with the $+90°$ phase shift of the staggered commutated network, prohibiting wave propagation.

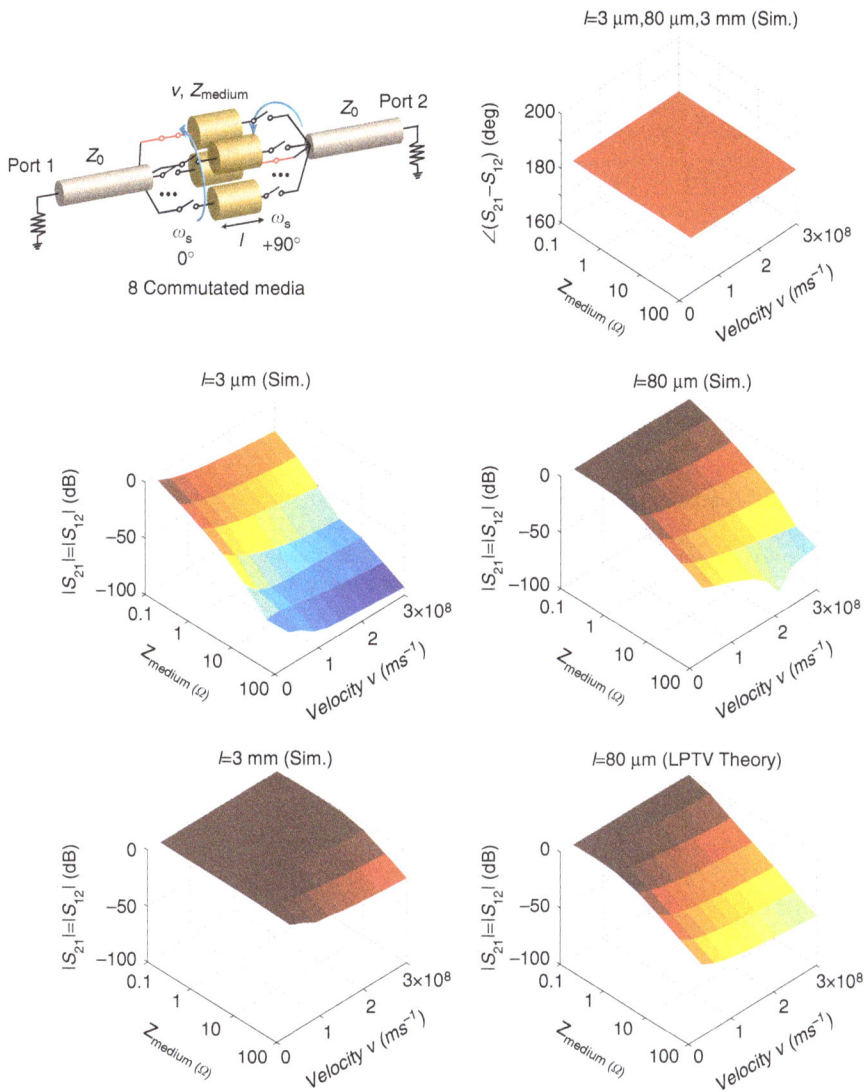

Figure 3 | Signal transmission in a staggered commutated network. Simulated S-parameters of an eight-way staggered commutated network are depicted assuming electrically short transmission line media of varied length l, characteristic impedance Z_{medium} and velocity v. Ideal switches are commutated at 750 MHz, with each switch active for 12.5% of the time period, and the reference impedance is assumed to be 50 Ω. Theoretical calculations based on the analytical formulation presented in Supplementary Note 1, where the electrically short transmission lines are approximated by their capacitance, are also shown and agree very well with simulations. The phases of S_{21} and S_{12} are always non-reciprocal and differ by 180° for 90° staggering. The magnitudes are always reciprocal; however, for substantial transmission, high l, low Z_{medium} and low v, or equivalently large capacitance, are required.

used for commutation at 750 MHz frequency, with each switch active for 12.5% of the time period. With +90° staggering, phase non-reciprocity is always observed (phase of the scattering or S-parameters S_{21} and S_{12} at the commutation frequency always have a 180° difference). The magnitudes of S_{21} and S_{12} at the commutation frequency are always reciprocal; however, low Z_{medium}, low velocity v and higher lengths l are required for significant signal transmission. In other words, the media must have a significant capacitance, similar to the (reciprocal) comb filter implementations described earlier, which use commutation across capacitor banks without staggering.

The need for capacitance may be intuitively understood by the fact with +90° staggering, no direct connection exists between the input and output at any instant assuming least four paths. Therefore, a certain amount of capacitance is required for the media to store sufficient energy for subsequent transmission. A formal analysis of a commutated network with transmission-line media based on LPTV network theory is challenging. In Supplementary Note 1, with the aid of Supplementary Figs 1

and 2, we have completed an analysis of a staggered commutated network with N capacitors and ideal switches. As electrically short transmission lines are being considered, the approximation with capacitors is expected to be accurate. Figure 3 also depicts the magnitudes of S_{21} and S_{12} at the commutation frequency based on the exact analytical solution in Supplementary Note 1 for the case of $l = 80\,\mu m$ and an excellent agreement is seen to the simulations.

The analytical formulation quantifies the capacitance condition for substantial transmission as $C >> \frac{1}{2\pi f_s Z_0}$ where f_s is the commutation frequency and Z_0 is the reference impedance. Under this condition, at f_s for +90° staggering, the S-parameters of the staggered commutated network reduce to

$$S(f_s) \approx \begin{bmatrix} \frac{N^2\left(1-\cos\left(\frac{2\pi}{N}\right)\right)}{2\pi^2} - 1 & \frac{N^2\left(1-\cos\left(\frac{2\pi}{N}\right)\right)}{2\pi^2}e^{-j\pi/2} \\ \frac{N^2\left(1-\cos\left(\frac{2\pi}{N}\right)\right)}{2\pi^2}e^{+j\pi/2} & \frac{N^2\left(1-\cos\left(\frac{2\pi}{N}\right)\right)}{2\pi^2} - 1 \end{bmatrix} \underset{N\to\infty}{\approx} \begin{bmatrix} 0 & e^{-j\pi/2} \\ e^{+j\pi/2} & 0 \end{bmatrix},$$

$$(1)$$

clearly showing the non-reciprocity in the phase of S_{21} and S_{12}. The (reciprocal) magnitudes of S_{21} and S_{12} approach unity as $N\to\infty$ but even $N=4$ or $N=8$ are sufficient to achieve low loss, resulting in -1.8 and -0.45 dB transmission, respectively. The magnitudes of S_{11} and S_{22} approach 0 for $N\to\infty$, implying perfect impedance matching.

A frequency up-/down-conversion and filtering-based explanation. The behaviour may also be explained by viewing each set of commutating switches as an in-phase and quadrature reciprocal modulator that performs a frequency up-conversion and down-conversion, similar to the approach described in prior literature[33]. Indeed, such commutating transistor-based switches are regularly used as frequency converters in RF and microwave integrated electronics[42]. Figure 4 depicts a graphical view of signal propagation through the staggered commutated network in the forward and the reverse directions. In each direction, a sinusoidal input signal, $\cos(\omega_{in}t)$, is assumed at a frequency ω_{in} near the commutation frequency ω_s. Each set of switches is modelled as an in-phase and quadrature reciprocal modulator that multiplies the input signal with cosine and sine versions of the pump signal at ω_s. The second set of pump signals (2 and 4) are assumed to lead the first set (1 and 3) by $+90°$, representing the staggering. The capacitances of the media are assumed to effectively form a

low-pass filter in conjunction with the source and load impedances. The frequency translation of this low-pass filtering to RF by the commutating switches is the basis for the use of unstaggered commutated networks to realize (reciprocal) comb filters[15,36–40]. This low-pass filter effect substantially attenuates the up-converted signal at $\omega_s + \omega_{in}$ and $-\omega_s - \omega_{in}$ after the first modulation. It is seen that this leads to signal transmission with non-reciprocal phase response ($+90°/-90°$) in the forward and reverse directions. It should be noted that although the up-converted frequency components are filtered away, this does not result in any power loss for the desired signal as the commutated media are purely capacitive. Consequently, the staggered commutated network is lossless for $N\to\infty$, as discussed earlier (for finite N, there is loss due to harmonic conversion, albeit small even for $N=4$ or $N=8$ as discussed earlier). The reader may also verify that in the absence of low-pass filtering, all frequency components at the outputs cancel and the transmission of the structure in both directions is identically zero, which agrees with the theory and simulations in Fig. 3 that show very weak tranmission for high Z_{medium}, low l and high v (that is, low capacitance in the commutated media), and our expectation given a lack of a direct connection between the input and the output.

A comparison may be made with multi-arm optical isolators based on tandem electro-optic phase modulation[22], where back-to-back phase modulation with quarter-wavelength optical

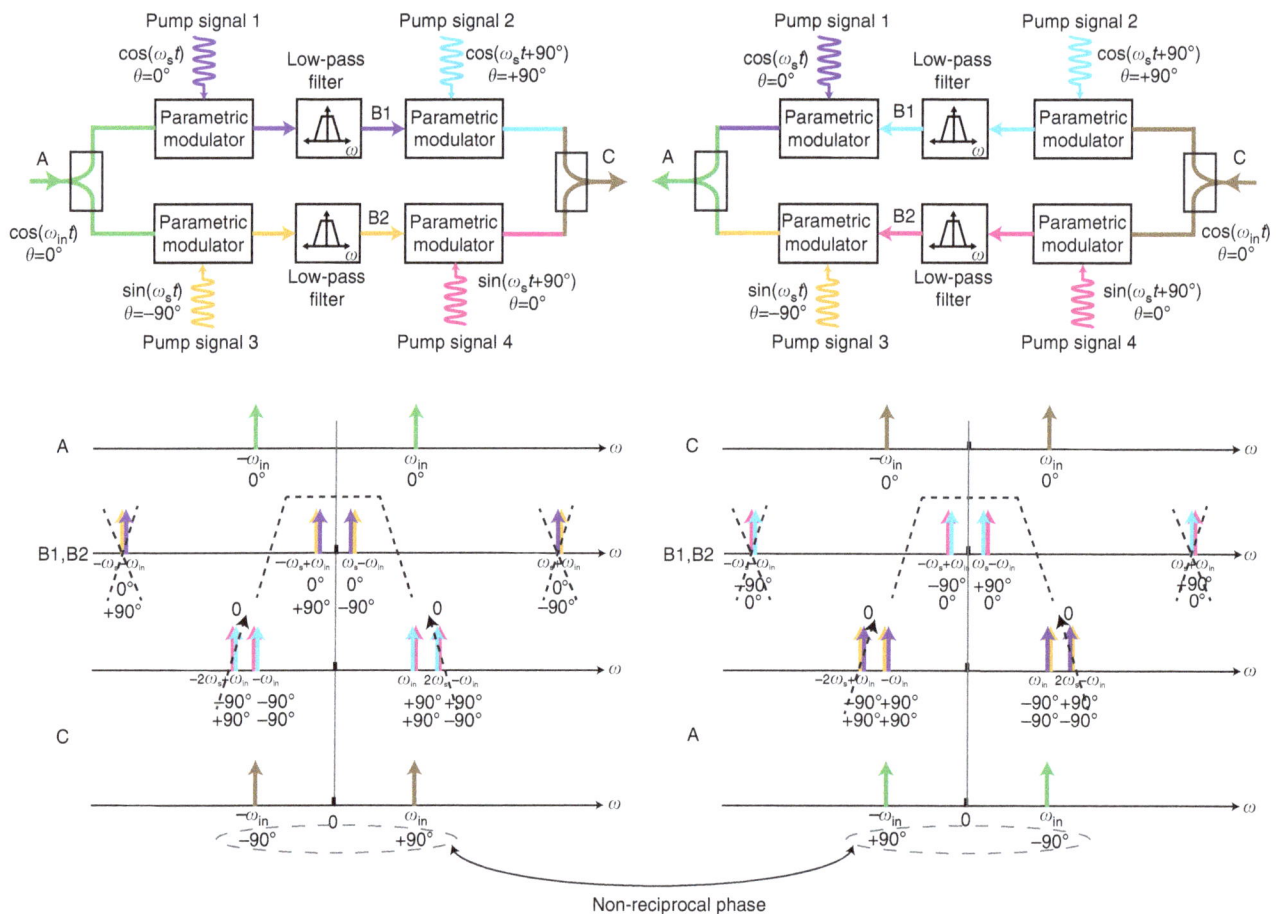

Figure 4 | A frequency up-/down-conversion and filtering-based explanation of phase non-reciprocity in staggered commutated networks. A commutator with at least four paths can be viewed as a reciprocal in-phase and quadrature frequency up-downconverter. The capacitance of the commutated media acts as a low-pass filter that attenuates the up-converted components after the first commutation. As a result, phase non-reciprocity is seen for signals travelling in the forward and reverse directions through a staggered commutated network. It may be verified that in the absence of low-pass filtering, all frequency components at the outputs cancel, leading to zero transmission in either direction, which agrees with the theory and simulations in Fig. 3 that show very weak transmission for high Z_{medium}, low l and high v (that is, low capacitance in the commutated media).

delays in between results in non-reciprocity. In our approach, commutation is essentially amplitude modulation with very high modulation ratio. Staggered commutation results in phase non-reciprocity and the filtering effect that is unique to commutation across media with appreciable capacitance eliminates the need for quarter-wavelength delays, allowing the phase non-reciprocity to be achieved within dimensions as small as $\lambda/1{,}250 \times \lambda/1{,}250$ in our device.

A highly-miniaturized RF circulator with enhanced linearity. A three-port circulator matched at all ports to a reference impedance of Z_0 may be realized by contacting the $3\lambda/4$ transmission line (also of characteristic impedance Z_0) at three points as long as the three ports are spaced $\lambda/4$ apart along the ring circumference, as shown in Fig. 5a. To analyse this structure, we consider the case of a voltage excitation V_{in} at port 1. Using conventional microwave circuit analysis techniques along with the S-parameters of the staggered commutated network in equation (1) for the case of $N \rightarrow \infty$, the various node voltages can be determined to be

$$V_1 = \frac{1}{2}V_{in}, V_2 = \frac{-j}{2}V_{in}, V_3 = 0, V_x = \frac{\sin(\beta l)}{2}V_{in}, V_y = \frac{j\sin(\beta l)}{2}V_{in}, \quad (2)$$

where V_1, V_2 and V_3 are the three port voltages, V_x and V_y are the voltages on the left and right sides of the staggered commutated network, β is the propagation constant of the $3\lambda/4$ ring and l is the circumferential distance between port 3 and the commutated network. Based on these node voltages and by repeating the calculations with excitations at the other two ports, the S-parameter matrix of the structure can be derived to be:

$$S_{circ}(f_s) = \begin{bmatrix} 0 & 0 & -1 \\ -j & 0 & 0 \\ 0 & -j & 0 \end{bmatrix}. \quad (3)$$

This matches the S-parameters of an ideal 3-port circulator. It is interesting to note that the S-parameter matrix does not depend on l, meaning that the position of the staggered commutated network relative to ports 1 and 3 does not have an impact on S_{circ}. However, the voltages seen at the two ends of the staggered commutated network (V_x and V_y) when port 1 is excited are functions of l in equation (2). Interestingly, these voltages become 0 when l is 0. This is a direct consequence of the fact that when l is 0, V_y coincides with port 3 and naturally remains quiet for excitations at port 1 due to the isolation of the circulator (Fig. 5b). The S-parameters of the staggered commutated network force V_x and V_y to have the same magnitude, making V_x also a quiet node (Fig. 5b). As a result, voltage swings across the point parametric modulator are suppressed, yielding very high linearity to excitations at port 1.

A more comprehensive analysis of the circulator may be performed using the S-parameters of the staggered commutated network in equation (1) for the case of finite N. Interestingly, for $l = 0$, we once again arrive at the ideal S-parameter matrix of equation (3) and $V_x = V_y = 0$ for excitations at port 1. This implies that when port 3 coincides with V_y, the presence of a finite number of commutated paths will not limit circulator performance. In the presence of finite switch ON resistance, however, there will be finite voltage swing across the commutating switches for excitations at port 1, leading to finite linearity. However, this linearity to port 1 excitations will be vastly superior to the linearity to excitations at the other ports that do not enjoy this suppression mechanism.

We constructed a prototype at RF using conventional 65 nm CMOS IC technology to implement the electronic commutation across a bank of $N = 8$ capacitors, and three lumped C-L-C networks to miniaturize the $3\lambda/4$ transmission line ring (Fig. 6). Miniaturization is eased by the fact that the $3\lambda/4$ ring is unmodulated, resulting in an overall structure that has a maximum dimension of $\approx \lambda/80$ at the operating frequency of 750 MHz. The capacitors of the C-L-C networks are also incorporated in the IC, leaving the three inductors as the only off-chip components. The commutated capacitors are chosen to be 26 pF each, roughly six times $\frac{1}{2\pi f_s Z_0}$ ($Z_0 = 50\,\Omega$), and are each realized on chip as a pair of $80\,\mu m \times 80\,\mu m$ metal–insulator–metal capacitors. The total size of the staggered commutated network is $\sim 320\,\mu m \times 320\,\mu m$ ($\lambda/1{,}250 \times \lambda/1{,}250$). The control signals for the staggered commutation are generated using on-chip clock generation circuitry that is described in additional detail in the Methods section.

Experimental results. S-parameter measurements of the three-port circulator were performed using a measurement setup described in the Methods section. In the absence of commutation, with a pair of transistor switches on either side of the capacitor bank permanently closed, the circuit is perfectly reciprocal (Fig. 7a,d,g). In this configuration the high reciprocal isolations from port 3 to ports 1 and 2 are seen because port 3 is shunted to ground by one of the 26-pF capacitors. The quarter-wave transmission lines from port 1 to the staggered commutated network and from port 2 to port 3 transform this near-short-circuit impedance to open circuits at ports 1 and 2, effectively disconnecting port 3 from the rest of the circuit and resulting in a reciprocal structure that exhibits low-loss transmission between port 1 and port 2. Similarly, another reciprocal configuration without commutation is depicted, where all switches are permanently open. Here, simple circuit analysis reveals that port 1 is effectively disconnected from the rest of the circuit, which now exhibits low-loss reciprocal transmission between ports 2 and 3.

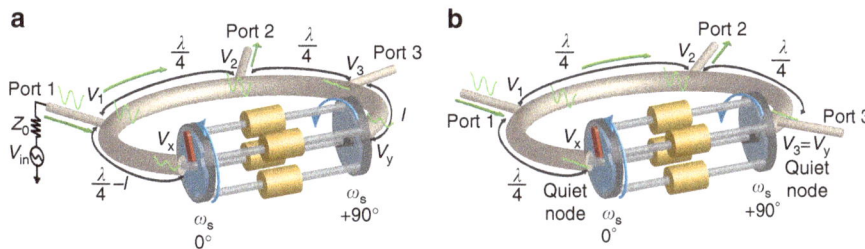

Figure 5 | Circulator architecture. (a) A 3-port circulator may be realized by adding ports at three points along the ring as long as the circumferential spacing between the ports is $\lambda/4$. The S-parameters of such a structure match that of an ideal three-port circulator and are independent of l, the circumferential distance between port 3 and the staggered commutated network. **(b)** However, setting l to 0 suppresses the voltages across the commutated network for excitations at port 1, significantly enhancing the linearity of the structure to port 1 excitations.

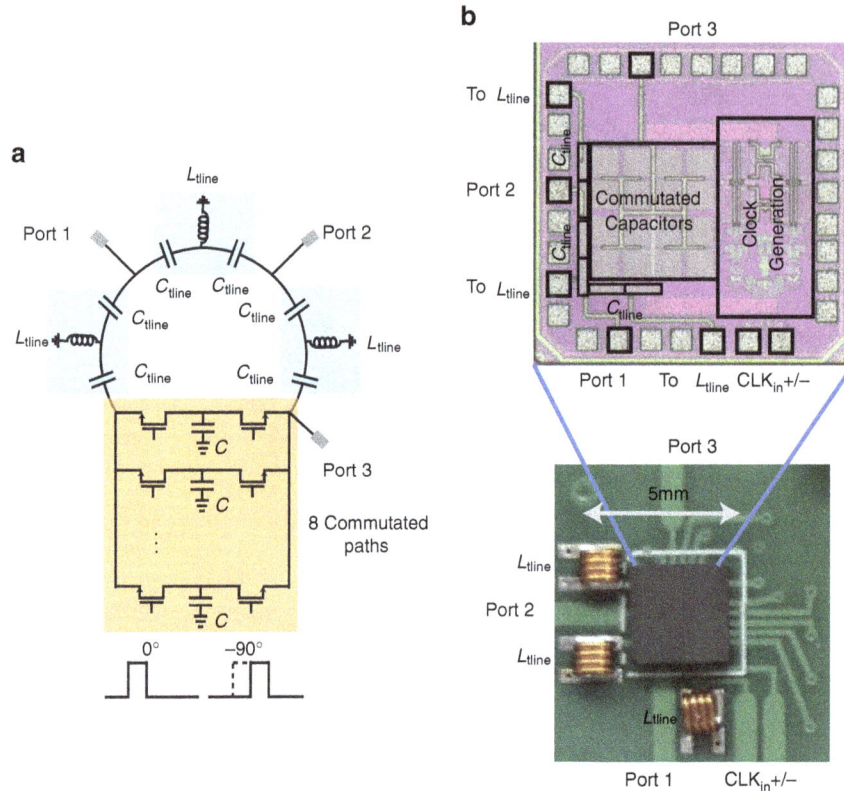

Figure 6 | RF CMOS IC implementation of the circulator. (a) A simplified circuit diagram of the circulator is shown. Electronic commutation across a bank of $N = 8$ capacitors is performed using reciprocal, passive transistor-based switches without direct-current bias. The staggered commutated network enables miniaturization of the unmodulated $3\lambda/4$ ring using three C-L-C sections. **(b)** The microphotograph of the fabricated IC is shown along with a close-up photograph of the fabricated printed circuit board with the IC housed in a quad-flat no-leads (QFN) package and interfaced with the off-chip inductors. The largest dimension of the prototype is 5 mm or $\lambda/80$ at the operating frequency of 750 MHz.

Under commutation at a frequency of 750 MHz with staggering for clockwise circulation, strong non-reciprocity is measured (Fig. 7b,e,h) with low-loss transmission in the direction of circulation (S_{21}, S_{32} and S_{13} are -1.7, -1.7 and -3.3 dB, respectively, at 750 MHz) and strong isolation in the reverse direction (S_{12}, S_{23} and S_{31} are -9.6, -10.4 and -17.4 dB, respectively, at 750 MHz). This represents, on an average, an order of magnitude of non-reciprocity between any two ports and is limited by the imperfect impedance matching at the third port (as is the case with all circulators) caused due to parasitics at the chip-package-board interfaces. If the impedance at the third port is slightly tuned, non-reciprocity of 40–50 dB or four to five orders of magnitude is seen in S_{12}, S_{23} and S_{31} at 750 MHz with negligible impact on transmission in the circulation direction. It should be noted that independent tuning is required at the three ports to achieve very high non-reciprocity in all paths, given the inherent asymmetry in the circulator structure. Figure 7c,f,i depict the simulated S-parameters under commutation, demonstrating an excellent match to the measurements.

Although the analysis in this study has restricted itself to response at the commutation frequency, it is noteworthy that a filtering profile is observed in S_{32} and S_{13} across frequency due to the comb-filter functionality inherent to commutation across a capacitor bank. The LPTV analysis in Supplementary Note 1 may be extended to determine the response of the staggered commutated network and the circulator across frequency, and confirms this observed (and simulated) filtering profile.

Another unique feature of the fabricated prototype is its real-time reconfigurability. By changing the staggering between $+90°$ and $-90°$, the direction of circulation can be altered, as is seen in Fig. 8a. The frequency of operation of the circulator can

also be tuned by changing the frequency of commutation within the limits dictated by the bandwidth of the $3\lambda/4$ transmission line ring. In Fig. 8b, four to five orders of magnitude (40–50 dB) of non-reciprocity in S_{31} is maintained across 700–800 MHz.

In Fig. 8c, we present experimental evidence of the enhanced linearity to excitations at port 1. Two-tone intermodulation distortion tests were performed on the prototype for transmission from port 1 to port 2, and from port 2 to port 3. The input-referred third-order intercept point (IIP3) for transmission from port 1 to port 2 is $+27.5$ dBm (≈ 560 mW), a remarkable number for a 65 nm CMOS IC implementation and nearly two orders of magnitude higher than that from port 2 to port 3 ($+8.7$ dBm or 7.4 mW) due to the suppression of the signal across the point parametric modulator. Aside from intermodulation distortion caused due to third-order non-linearity, other spurious effects to consider include image signals due to quadrature mismatch and modulation feedthrough. The level of the measured spurious image signals produced due to quadrature mismatch for port 1 to port 2 transmission is 51 dB below the main signal. This level of image rejection is high enough so that it is not a serious issue when compared with the third-order intermodulation distortion produced for reasonable port 1 power levels. The measured modulation feedthrough at port 2 is at -57 dBm. Although these levels are already quite low, they can be cancelled further using integrated calibration circuits that are commonly implemented in CMOS RF ICs.

Passive LTI systems have a noise figure (NF) that is equal to their loss. Passive LPTV systems can exhibit noise folding. The measured NF for port 2 to port 3 transmission is 4 dB, higher than the 1.7 dB loss. Simulations indicate that 2 dB degradation arises from modulation path phase noise due to a poor

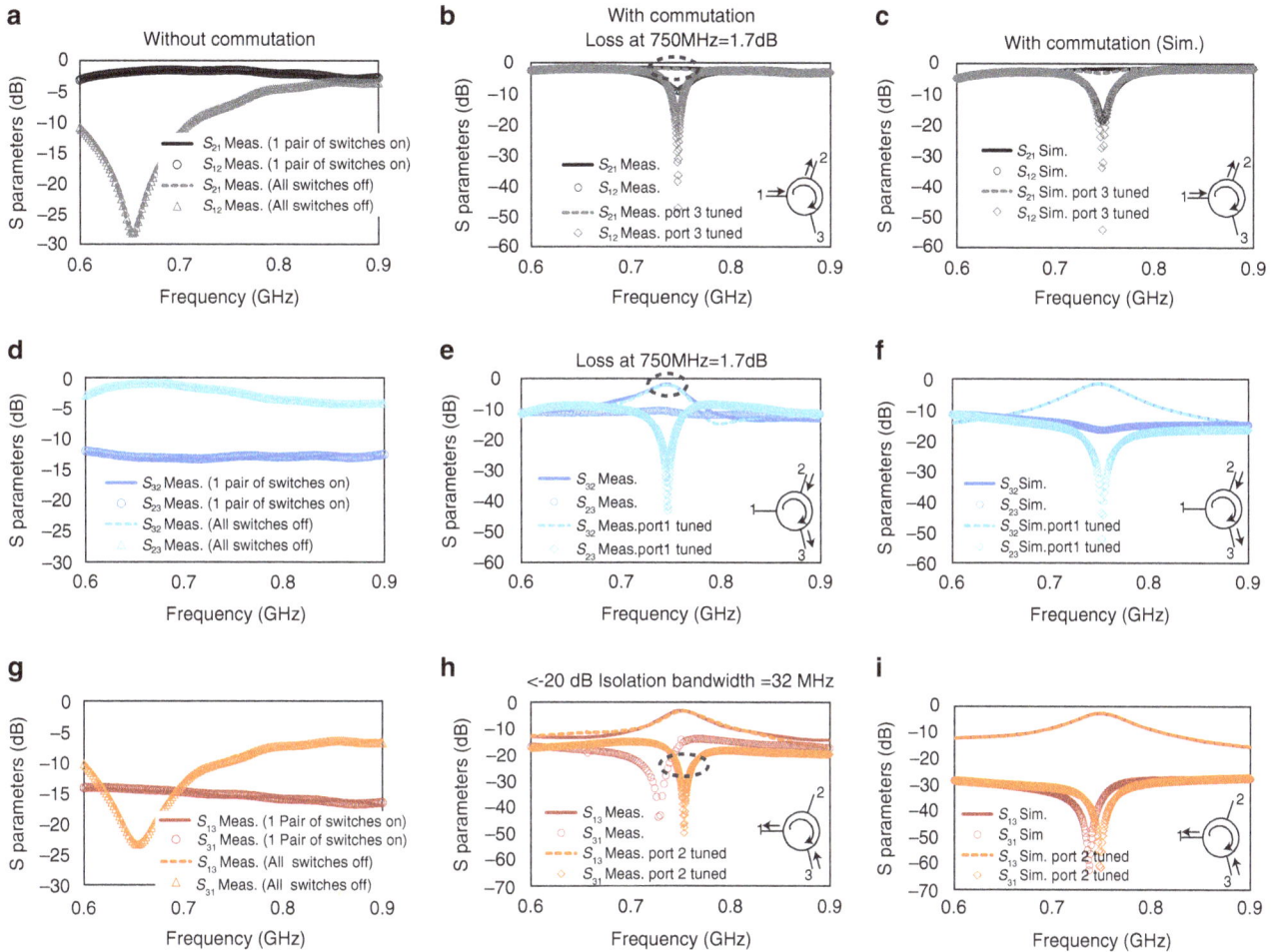

Figure 7 | Circulator S-parameter measurements. Measured circulator S-parameters (**a**), (**d**) and (**g**) without commutation and (**b**), (**e**) and (**h**) with commutation are shown, as well as (**c**), (**f**) and (**i**) simulated S-parameters that show an excellent match to the measurements. Under staggered commutation for clockwise circulation, low loss transmission in the circulation direction (S_{21}, S_{32} and S_{13} are -1.7, -1.7 and $-3.3\,dB$, respectively) and an order-of-magnitude isolation in the reverse direction (S_{12}, S_{23} and S_{31} are -9.6, -10.4 and $-17.4\,dB$, respectively) are measured at the commutation frequency of 750 MHz. When the third port is slightly tuned, non-reciprocity of 40–50 dB is measured in S_{12}, S_{23} and S_{31} at 750 MHz with negligible impact on transmission in the circulation direction. The $-20\,dB$ isolation bandwidth in S_{31} after tuning is 32 MHz or 4.3%.

implementation of the modulation path phase shifter. A better implementation of the modulation path phase shifter restores the NF to 2 dB in simulation, close to the 1.7 dB loss level and indicative of only minor NF degradation due to harmonic noise folding.

Discussion

The ability to integrate magnetic-free passive linear non-reciprocal components in CMOS has the potential to revolutionize RF communications. The CMOS circulator described in this study exhibits extremely low insertion loss ($<2\,dB$) and strong non-reciprocity, is compact and features reconfiguration capabilities. Furthermore, the enhanced linearity to port 1 is particularly useful for full-duplex or simultaneous transmit-and-receive communication and radar applications[3,4], where a high-power transmitter (port 1) and a highly sensitive receiver (port 3) operate simultaneously at the same frequency, and must be interfaced with a shared antenna (port 2). In such a configuration, S_{21} (transmitter to antenna loss), S_{32} (antenna to receiver loss) and S_{31} (transmitter to receiver isolation) are the most important performance metrics, where our device performs particularly well. Full-duplex communication is drawing significant interest for emerging 5G communication networks[43]

due to its potential to double network capacity compared with half-duplex communication at the physical layer while offering numerous other benefits at the network layer. There has been active research on fully integrated CMOS transceiver ICs supporting full duplex[44–48]. The ability to include the circulator with the transceiver on the same CMOS IC would significantly reduce the cost and form factor, and enhance the performance of full-duplex systems. The filtering profile in S_{32} is also very useful in protecting the highly sensitive receiver from interference signals outside the frequency band of operation.

For such applications, the ability to reconfigure the device between reciprocal and non-reciprocal operation is also very useful. We demonstrated reconfigurable modes of operation in the absence of commutation where the structure becomes reciprocal with low-loss transmission between the antenna (port 2) and the transmitter (port 1), and the antenna and the receiver (port 3). In other words, the prototype can be reconfigured between operation as a non-reciprocal circulator for full-duplex communication and a reciprocal low-loss transmit/receive switch for half-duplex communication.

Although the design of this prototype is deliberately asymmetric with respect to the three ports to prioritize linearity to port 1 excitations for communication applications, a symmetric design may also be envisioned where a non-reciprocal

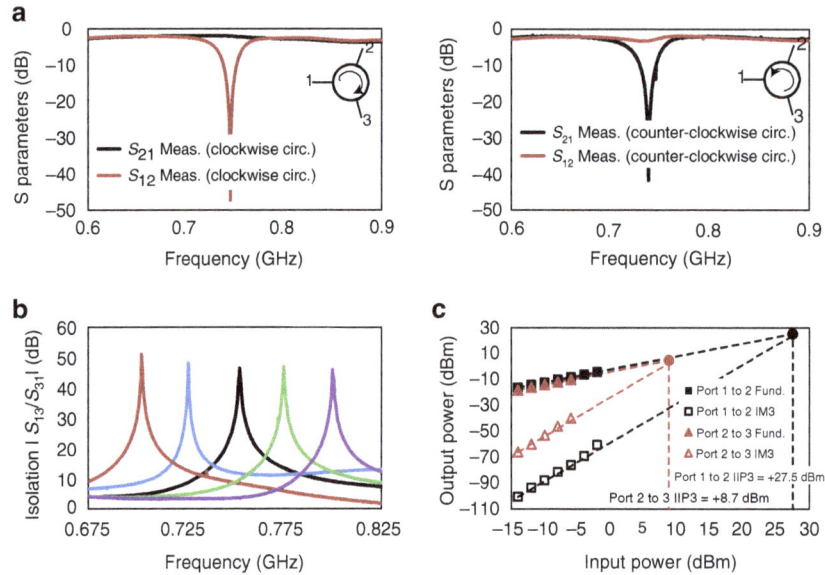

Figure 8 | Experimental evidence of reconfigurability and enhanced linearity to port 1 excitations. (**a**) The direction of circulation can be altered by changing the staggering between $+90°$ and $-90°$. (**b**) The frequency of operation of the circulator can be tuned by changing the commutation frequency within the limits dictated by the bandwidth of the $3\lambda/4$ transmission-line ring. Here we present the non-reciprocity in S_{31} across different commutation frequencies ranging from 700 to 800 MHz. In each case, tuning of the port 2 impedance is exploited to achieve 40–50 dB isolation at the commutation frequency. (**c**) Measured two-tone linearity for transmission from port 1 to port 2 and from port 2 to port 3 are shown when configured for clockwise circulation. Nonlinear systems exhibit intermodulation distortion products when excited with two sinusoidal signals. The input-referred third-order intercept point (IIP3) represents the (extrapolated) input power of each of the two tones at which the third-order intermodulation products (IM3) at the output are as powerful as the fundamental signals. The IIP3 for transmission from port 1 to port 2 is $+27.5$ dBm (≈ 560 mW), nearly two orders of magnitude higher than that from port 2 to port 3 ($+8.7$ dBm or 7.4 mW), owing to the suppression of the signal across the point parametric modulator for port 1 excitations.

phase component is incorporated in all three arms, that is, between all three ports. Such a symmetric circulator structure would enable extension of the concept to non-reciprocal metamaterials that support topologically protected wave propagation modes[49].

The bandwidth of the circulator's isolation is limited by the matching between the frequency responses of the $3\lambda/4$ ring and the staggered commutated network. Currently, the -20 dB isolation bandwidth in S_{31} after tuning is 32 MHz or 4.3% of the operating frequency, useful for commercial cellular LTE and WiFi applications. Dispersion engineering techniques to match their responses over a wider frequency range, thus enhancing the isolation amount versus bandwidth trade-off, represent an interesting topic for future research.

Commutated networks are known to have response at harmonics of the commutation frequency. Response at harmonics can lead to susceptibility to interference near harmonic frequencies and also leads to noise folding. Although we have seen that the effect of noise folding on the NF is very small in this circulator structure, a formal analysis of the harmonic response and noise of the overall circulator structure, as well as techniques to suppress the same, also represent an interesting topic for future investigation.

It is also interesting to consider the application of these concepts to other domains where a high-quality switch is available, such as optical waves. Compact optical switches with one to two orders of magnitude ON/OFF transmission ratio[50,51] open the door to optical non-reciprocity and isolation through commutation-based parametric modulation. The nanosecond-scale switching speed implies GHz-range commutation frequencies, much smaller than the optical carrier frequency, which can be accomodated by commutating across high-Q optical filters that eliminate one of the modulation sidebands, similar to the low-pass filter effect used in our prototype.

Methods

IC implementation details. A block diagram of the IC and component values are provided in Supplementary Fig. 3 and Supplementary Table 1, respectively. The circulator consists of three lumped C-L-C sections that miniaturize the $3\lambda/4$ transmission line and a staggered electronically commutated network of eight capacitors. All components, except for the three inductors, are integrated on the CMOS IC. The switches of the commutated network are implemented using 65 nm CMOS transistors and are driven with two sets of eight non-overlapping clock signals with 12.5% duty cycle. These clock signals are generated from two differential (0/180°) input clocks that run at four times the desired commutation frequency. A divide-by-2 frequency-divider circuit generates four quadrature clocks with 0°/90°/180°/270° phase relationship. These four clock signals drive two parallel paths for the two sets of switches. One of the paths features a programmable phase shifter that allows for arbitrary staggering between the two commutating switch sets. This enables switching between $+90°$ and $-90°$ staggering, which allows dynamic reconfiguration of the circulation direction. The phase shifter also allows for fine tuning of the staggered phase shift, to optimize the transmission loss in the circulation direction and isolation in the reverse direction. This feature has been exploited in the measurements shown in the main text. After phase shifting, another divide-by-2 circuit and a non-overlapping 12.5% duty-cycle clock generation circuit create the clock signals that control the commutating transistor switches.

Experimental setups. Diagrams of the experimental setups are provided in Supplementary Figs 4 and 5, respectively. A list of the equipment used in the two setups is provided in Supplementary Table 2. A 180° hybrid is used to generate two differential (0°/180°) signals from a signal generator to drive the clock inputs of the implemented circulator. A voltage bias signal is added to the clock signals through bias tees to drive the clock inputs. A two-port vector network analyser is used to measure the S-parameters of the circulator two ports at a time, whereas the third port is terminated with a variable impedance tuner to tune the port impedance. For the input-referred third-order intercept point linearity test, a two-way power combiner is used to combine two sinusoidal signals generated from two additional signal generators and feed the input port under consideration. A spectrum analyser is used to monitor the fundamental signals and the third-order intermodulation distortion products of the circulator at the output port under consideration, whereas the third port is terminated with a 50-Ω termination.

References

1. Jalas, D. *et al.* What is—and what is not—an optical isolator. *Nat. Photonics* **7**, 579–582 (2013).
2. Goldsmith, A. *Wireless Communications* (Cambridge Univ. Press, 2005).

3. Bharadia, D., McMilin, E. & Katti, S. Full duplex radios. *ACM SIGCOMM Computer Communication Review* vol. 43, 375–386 (ACM, 2013).

4. Sabharwal, A. *et al.* In-band full-duplex wireless: Challenges and opportunities. *IEEE J. Select. Areas Commun.* **32**, 1637–1652 (2014).

5. Pozar, D. M. *Microwave Engineering* (Addison-Wesley, 1990).

6. Bi, L. *et al.* On-chip optical isolation in monolithically integrated non-reciprocal optical resonators. *Nat. Photonics* **5**, 758–762 (2011).

7. Shoji, Y., Mizumoto, T., Yokoi, H., Hsieh, I. -W. & Osgood, R. M. Magneto-optical isolator with silicon waveguides fabricated by direct bonding. *Appl. Phys. Lett.* **92**, 071117 (2008).

8. Adam, J., Davis, L., Dionne, G. F., Schloemann, E. & Stitzer, S. Ferrite devices and materials. *IEEE Trans. Microw. Theory Techn.* **50**, 721–737 (2002).

9. Adam, J. *et al.* Monolithic integration of an X-band circulator with GaAs MMICs. *IEEE MTT-S International Microwave Symposium Digest* **1**, 97–98 (1995).

10. Oliver, S., Zavracky, P., McGruer, N. & Schmidt, R. A monolithic single-crystal yttrium iron garnet/silicon X-band circulator. *IEEE Microw. Guided Wave Lett.* **7**, 239–241 (1997).

11. Tanaka, S., Shimomura, N. & Ohtake, K. Active circulators—the realization of circulators using transistors. *Proc. IEEE* **53**, 260–267 (1965).

12. Wang, Z. *et al.* Gyrotropic response in the absence of a bias field. *Proc. Natl Acad. Sci. USA* **109**, 13194–13197 (2012).

13. Kodera, T., Sounas, D. L. & Caloz, C. Artificial Faraday rotation using a ring metamaterial structure without static magnetic field. *Appl. Phys. Lett.* **99**, 031114 (2011).

14. Kodera, T., Sounas, D. & Caloz, C. Magnetless nonreciprocal metamaterial (MNM) technology: application to microwave components. *IEEE Trans. Microw. Theory Techn.* **61**, 1030–1042 (2013).

15. Mirzaei, A. & Darabi, H. Reconfigurable RF front-ends for cellular receivers. *2010 IEEE Comp. Semiconduct. Integr. Circuit Symp. (CSICS)*, 1–4 (IEEE, 2010).

16. Carchon, G. & Nanwelaers, B. Power and noise limitations of active circulators. *IEEE Trans. Microw. Theory Techn.* **48**, 316–319 (2000).

17. Peng, B. *et al.* Parity-time-symmetric whispering-gallery microcavities. *Nat. Phys.* **10**, 394–398 (2014).

18. Fan, L. *et al.* An all-silicon passive optical diode. *Science* **335**, 447–450 (2012).

19. Soljačić, M., Luo, C., Joannopoulos, J. D. & Fan, S. Nonlinear photonic crystal microdevices for optical integration. *Opt. Lett.* **28**, 637–639 (2003).

20. Gallo, K., Assanto, G., Parameswaran, K. R. & Fejer, M. M. All-optical diode in a periodically poled lithium niobate waveguide. *Appl. Phys. Lett.* **79**, 314–316 (2001).

21. Mahmoud, A. M., Davoyan, A. R. & Engheta, N. All-passive nonreciprocal metastructure. *Nat. Commun.* **6**, 8359 (2015).

22. Doerr, C. R., Chen, L. & Vermeulen, D. Silicon photonics broadband modulation-based isolator. *Opt. Express* **22**, 4493–4498 (2014).

23. Galland, C., Ding, R., Harris, N. C., Baehr-Jones, T. & Hochberg, M. Broadband on-chip optical non-reciprocity using phase modulators. *Opt. Express* **21**, 14500–14511 (2013).

24. Lira, H., Yu, Z., Fan, S. & Lipson, M. Electrically driven nonreciprocity induced by interband photonic transition on a silicon chip. *Phys. Rev. Lett.* **109**, 033901 (2012).

25. Kang, M., Butsch, A. & Russell, P. S. J. Reconfigurable light-driven opto-acoustic isolators in photonic crystal fibre. *Nat. Photonics* **5**, 549–553 (2011).

26. Yu, Z. & Fan, S. Complete optical isolation created by indirect interband photonic transitions. *Nat. Photonics* **3**, 91–94 (2008).

27. Qin, S., Xu, Q. & Wang, Y. Nonreciprocal components with distributedly modulated capacitors. *IEEE Trans. Microw. Theory Techn.* **62**, 2260–2272 (2014).

28. Zanjani, M. B., Davoyan, A. R., Mahmoud, A. M., Engheta, N. & Lukes, J. R. One-way phonon isolation in acoustic waveguides. *Appl. Phys. Lett.* **104**, 081905 (2014).

29. Sounas, D. L., Caloz, C. & Alù, A. Giant non-reciprocity at the subwavelength scale using angular momentum-biased metamaterials. *Nat. Commun.* **4**, 2407 (2013).

30. Fleury, R., Sounas, D. L., Sieck, C. F., Haberman, M. R. & Alu, A. Sound isolation and giant linear nonreciprocity in a compact acoustic circulator. *Science* **343**, 516–519 (2014).

31. Estep, N. A., Sounas, D. L., Soric, J. & Alu, A. Magnetic-free non-reciprocity and isolation based on parametrically modulated coupled-resonator loops. *Nat. Phys.* **10**, 923–927 (2014).

32. Estep, N., Sounas, D. & Alu, A. On-chip non-reciprocal components based on angular momentum biasing. *2015 IEEE MTT-S Int. Microw. Symp. (IMS)*, 1–4 (IEEE, 2015).

33. Kamal, A., Clarke, J. & Devoret, M. H. Noiseless non-reciprocity in a parametric active device. *Nat. Phys.* **7**, 311–315 (2011).

34. Busignies, H. & Dishal, M. Some relations between speed of indication, bandwidth, and signal-to-random-noise ratio in radio navigation and direction finding. *Proc. IRE* **37**, 478–488 (1949).

35. LePage, W. R., Cahn, C. R. & Brown, J. S. Analysis of a comb filter using synchronously commutated capacitors. *Trans. Am. Inst. Electric. Eng. I Commun. Electron.* **72**, 63–68 (1953).

36. Ghaffari, A., Klumperink, E., Soer, M. & Nauta, B. Tunable high-Q N-path band-pass filters: modeling and verification. *IEEE J. Solid State Circuits* **46**, 998–1010 (2011).

37. Andrews, C. & Molnar, A. Implications of passive mixer transparency for impedance matching and noise figure in passive mixer-first receivers. *IEEE Trans. Circuits Syst. I Regul. Papers* **57**, 3092–3103 (2010).

38. Reiskarimian, N. & Krishnaswamy, H. Design of All-Passive Higher-Order CMOS N-Path Filters. *2015 IEEE RFIC Symp.*, 83–86 (IEEE, 2015).

39. Thomas, C. & Larson, L. Broadband synthetic transmission-line N-path filter design. *IEEE Trans. Microw. Theory Techn.* **63**, 3525–3536 (2015).

40. Gharpurey, R. Linearity enhancement techniques in radio receiver front-ends. *IEEE Trans. Circuits Syst. I Regul. Papers* **59**, 1667–1679 (2012).

41. Tyagi, S. *et al.* An advanced low power, high performance, strained channel 65nm technology. *2005 IEEE Int. Electron Devices Meet. Tech. Digest* 245–247 (2005).

42. Forbes, T., Ho, W.-G. & Gharpurey, R. Design and analysis of harmonic rejection mixers with programmable LO frequency. *IEEE J. Solid State Circuits* **48**, 2363–2374 (2013).

43. Hong, S. *et al.* Applications of self-interference cancellation in 5G and beyond. *IEEE Commun. Mag.* **52**, 114–121 (2014).

44. Zhou, J., Chuang, T.-H., Dinc, T. & Krishnaswamy, H. Receiver with >20MHz bandwidth self-interference cancellation suitable for FDD, co-existence and full-duplex applications. *2015 IEEE Int. Solid State Circuits Conf. (ISSCC)* 342–343 (IEEE, 2015).

45. van den Broek, D. -J., Klumperink, E. & Nauta, B. A self-interference-cancelling receiver for in-band full-duplex wireless with low distortion under cancellation of strong TX leakage. *2015 IEEE Int. Solid State Circuits Conf. (ISSCC)* 344–345 (IEEE, 2015).

46. Zhou, J., Chuang, T.-H., Dinc, T. & Krishnaswamy, H. Integrated wideband self-interference cancellation in the RF domain for FDD and full-duplex wireless. *IEEE J. Solid State Circuits* **50**, 3015–3031 (2015).

47. Yang, D., Yuksel, H. & Molnar, A. A wideband highly integrated and widely tunable transceiver for in-band full-duplex communication. *IEEE J. Solid State Circuits* **50**, 1189–1202 (2015).

48. Dinc, T., Chakrabarti, A. & Krishnaswamy, H. A 60 GHz same-channel full-duplex CMOS transceiver and link based on reconfigurable polarization-based antenna cancellation. *2015 IEEE RFIC Symp.* 31–34 (2015).

49. Khanikaev, A. B., Fleury, R., Mousavi, S. H. & Alù, A. Topologically robust sound propagation in an angular-momentum-biased graphene-like resonator lattice. *Nat. Commun.* **6**, 8260 (2015).

50. Lira, H. L. R., Manipatruni, S. & Lipson, M. Broadband hitless silicon electro-optic switch for on-chip optical networks. *Opt. Express* **17**, 22271–22280 (2009).

51. Vlasov, Y., Green, W. M. J. & Xia, F. High-throughput silicon nanophotonic wavelength-insensitive switch for on-chip optical networks. *Nat. Photonics* **2**, 242–246 (2008).

Acknowledgements

This work was supported by the DARPA ACT programme under Grant FA8650-14-1-7414. We acknowledge useful discussions with Dr Troy Olsson, Dr Ben Epstein.

Author contributions

N.R. and H.K. developed the concept. N.R. conducted the theoretical analysis and numerical simulations, designed the device and performed the experiments. H.K. directed and supervised the project. N.R. and H.K. wrote the paper.

Additional information

26

Bioabsorbable polymer optical waveguides for deep-tissue photomedicine

Sedat Nizamoglu[1,2], Malte C. Gather[1,3], Matjaž Humar[1,4], Myunghwan Choi[1,5], Seonghoon Kim[6], Ki Su Kim[1], Sei Kwang Hahn[7], Giuliano Scarcelli[1,8], Mark Randolph[1,9], Robert W. Redmond[1] & Seok Hyun Yun[1,10]

Advances in photonics have stimulated significant progress in medicine, with many techniques now in routine clinical use. However, the finite depth of light penetration in tissue is a serious constraint to clinical utility. Here we show implantable light-delivery devices made of bio-derived or biocompatible, and biodegradable polymers. In contrast to conventional optical fibres, which must be removed from the body soon after use, the biodegradable and biocompatible waveguides may be used for long-term light delivery and need not be removed as they are gradually resorbed by the tissue. As proof of concept, we demonstrate this paradigm-shifting approach for photochemical tissue bonding (PTB). Using comb-shaped planar waveguides, we achieve a full thickness ($>10\,$mm) wound closure of porcine skin, which represents $\sim$10-fold extension of the tissue area achieved with conventional PTB. The results point to a new direction in photomedicine for using light in deep tissues.

[1] Wellman Center for Photomedicine, Harvard Medical School, Massachusetts General Hospital, 65 Landsdowne Street UP-5, Cambridge, Massachusetts 02139, USA. [2] Department of Electrical and Electronics Engineering, Koc University, Istanbul 34450, Turkey. [3] SUPA, School of Physics and Astronomy, University of St Andrews, St Andrews KY16 9SS, UK. [4] Condensed Matter Department, J. Stefan Institute, Jamova 39, SI-1000 Ljubljana, Slovenia. [5] Global Biomedical Engineering, Sungkyunkwan University, Center for Neuroscience and Imaging Research, Institute for Basic Science, 2066, Seoburo, Jangan, Suwon, Gyeonggi 440-746, Korea. [6] Graduate School of Nanoscience and Technology, Korea Advanced Institute of Science and Technology, 291 Daehakro, Yuseong, Daejeon 305-701, Korea. [7] Department of Materials Science and Engineering, Pohang University of Science and Technology, San 31, Hyoja-dong, Nam, Pohang, Kyungbuk 790-784, Korea. [8] Department of Bioengineering, University of Maryland College Park, College Park, Maryland 20742, USA. [9] Division of Plastic Surgery and Harvard Medical School, Massachusetts General Hospital, 40 Blossom Street, Boston, Massachusetts 02114, USA. [10] Harvard–MIT Health Sciences and Technology, 77 Massachusetts Avenue, Cambridge, Massachusetts 02139, USA. Correspondence and requests for materials should be addressed to S.H.Y. (email: syun@mgh.harvard.edu).

Absorption and scattering of light in tissue results from fundamental light–matter interactions[1] and have enabled a variety of powerful optical techniques for therapy and imaging[1-9]. However, these interactions are also problematic as they limit the penetration of light in tissues. The 1/e penetration depth is only 1–2 mm at most for visible light[10], and even by using near-infrared light or upconversion nanoparticles[11,12] one can only increase the 1/e depth to ~3 mm. Overcoming the limited penetration has been a challenge in almost all applications of light in photomedicine. Traditional optical fibres and endoscopes can bring light close to the surface of the target organ[13-16], but delivering the light further to the target region within the tissue remains a problem.

Several biomaterials have previously been considered to form implantable photonic devices. Silk fibres[17,18] have been used to fabricate patterned films[19], diffraction gratings[20], printed waveguides[21], laser substrates[22] and implantable reflectors[23]. Light-guiding hydrogel waveguides have been developed for sensing and therapies[24-28]. Synthetic polymers, such as poly (D,L-lactide-co-glycolide) (PLGA) and poly(L-lactic acid) (PLA), are widely used for conventional implantable devices and injectable products, for example, for resorbable sutures, but have not been explored for optical applications. These synthetic polymers exhibit advantageous properties for implantable optics, although so far they have mostly been used in an opaque form. However, these polymers can be made transparent and then have a typical refractive index of ~1.47, adequate for guiding and delivering light in tissue. (The refractive index of the human skin is 1.38–1.44 (ref. 29).) The mechanical rigidity and flexibility of polymers can be tuned by varying their molecular weight and the degree of crosslinking. Moreover, polymers offer a wide range of biodegradation half-times, from 1 min to over a year[30]. Lactide and glycolide-based copolymers are degraded by ester hydrolysis and break into monomers, which are further degraded and resorbed in the body with minimal systemic toxicity. Therefore, we reason that these polymers are attractive candidates for making resorbable optical devices.

Here we propose a method of delivering light into living systems using bioabsorbable photonic waveguides for photo-medicine. We describe the design, fabrication and characterization of waveguides made from several different transparent polymers with different mechanical and degradation properties. Further, we demonstrate an application to photochemical tissue bonding, for which a thin flexible waveguide comb is inserted into a skin incision model to deliver light uniformly along the full thickness (>10 mm) of the skin tissue. This allows watertight crosslinking between the tissues to be formed *in situ*; the waveguide need not be removed from the site after the procedure, as it is eventually biodegraded and absorbed by the tissue. This paradigm-shifting approach is expected to impact a variety of applications in the field of photomedicine, such as health monitoring, controlled drug release and chronic photodynamic therapy.

Results

Biopolymer waveguides. To make slab waveguides, we first fabricated transparent polymer films (Fig. 1a) using melt pressing, solvent casting and ultraviolet-induced crosslinking techniques (see Methods). The thickness of the films was typically in a range from 200 to 800 μm to achieve an optimal combination of mechanical rigidity/flexibility and appropriate optical extraction efficiency depending on specific applications (Fig. 1b). The fabricated films were then cut into specific shapes by high-precision laser cutting (Fig. 1c–f). A conventional silica-based jacketed multimode fibre was pigtailed to the polymer device by

using epoxy glue to couple light into the device (Fig. 1d–f). In applications, the pigtail fibre remains outside the tissue and is removed after use, and the polymer device stays in the tissue during the operation and is resorbed thereafter.

To investigate the waveguide-assisted light delivery in tissues, we prepared full-thickness sections of bovine skin and imaged lateral light scattering when laser light at 532 nm was delivered to just the tissue surface (epi-illumination) or launched into a waveguide embedded between two tissue sections (Fig. 2a). Without a waveguide, laser light penetrates less than 5 mm; however, with the waveguide ($1 \times 1 \times 23$ mm), bright emission was observed along the entire region of the tissue containing the waveguide (Fig. 2a). The intensity distribution along the depth indicated that the 10-dB penetration length, where the optical intensity is reduced to 10%, was extended from 7 to >23 mm (Fig. 2b). Two-dimensional ray optics simulations of our structures yielded results that matched the experimental data (Fig. 2c–d).

Light extraction from a planar waveguide. The amount of light scattered from the waveguide as a function of depth can be easily approximated. Let $P_{out}(z)$ be the amount of light extracted from the waveguide per unit length along the depth, z. This output power distribution function is equal to the amount of optical loss in the waveguide (Fig. 2e), which can be expressed as:

$$p_{out}(z)dz = -dP_{in}(z) = \sigma(z)P_{in}(z)dz, \qquad (1)$$

where dz denotes the thickness of a waveguide section, $P_{in}(z)$ is the optical power varying in z in the waveguide, and $\sigma(z)$ is the loss coefficient of the waveguide. Note that the depth profile $P_{out}(z)$ has dimension of $W\,m^{-1}$. The general solution of equation (1) is given by:

$$P_{in}(z) = P_{in}(0)\exp\left[-\int_0^z \sigma(z)dz\right]; \qquad (2)$$

$$P_{out}(z) = P_{in}(0)\sigma(z)\exp\left[-\int_0^z \sigma(z)dz\right]. \qquad (3)$$

Here $P_{in}(0)$ is the initial power at $z=0$. The above equations describe how the intensity profiles inside and outside the waveguide are determined from the loss function $\sigma(z)$. As an inverse problem, an arbitrary axial intensity distribution, $P_{out}(z)$, can be obtained in principle if the waveguide provides the (non-negative) z-dependent loss coefficient

$$\sigma(z) = P_{out}(z)/\left(P_{in}(0) - \int_0^z P_{out}(z)dz\right). \qquad (4)$$

A physical interpretation of this equation is that the loss coefficient at z is equivalent to a ratio of the extracted power to the total power in the waveguide. As an example, for uniform power distribution in tissue over a depth of L, one may wish to have $P_{out}(z) = a$, where a is a constant and $z = [0, L]$. From equation (3), this requires $\sigma(z) = a/(P_{in}(0) - az)$ and $P_{in}(0) \geq aL$.

In practice, the specific loss function of a waveguide can be controlled by design, such as the shape and structure, choice of material and surface quality. One of the most convenient tuning parameters is the cross-section of the waveguide. For smaller cross-sections, optical rays propagating in the waveguide bounce off the polymer/tissue interface more frequently, thus experiencing higher optical loss and, therefore, more rapid extraction into the tissue. To test this, we fabricated PLA slab waveguides from a 240-μm-thick substrate but with different waveguide widths, and measured the propagation loss by a cutback technique. To investigate the effect of different refractive index (n) of the surrounding, we tested waveguides in three different

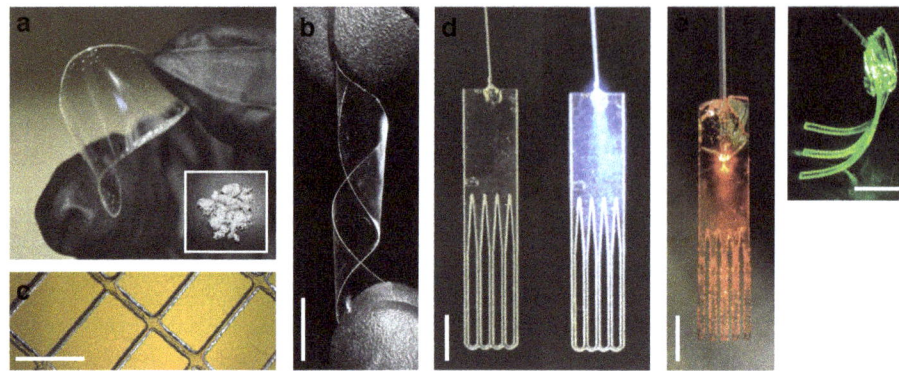

Figure 1 | Various biopolymer films and planar waveguide demonstrations. (a) Transparent and flexible poly(L-lactic acid) biopolymer film obtained by melt-pressing technique. Inset: biopolymers in their initial powder form before film formation. **(b)** A twisted bio-film before laser cutting. **(c)** Laser cutting allows for a simple fabrication of films with well-controlled meshes. **(d)** A fibre-coupled waveguide array with an unstructured top region uniformly coupling the light to the array of thin waveguides. Left, off state; right, light on. Only the lower part of the devices is to be embedded into tissue, the top structure including the pigtail fibre is cut off and removed after use. **(e)** Red light coupled to a waveguide array. **(f)** A polyethylene glycol hydrogel waveguide array carrying green laser light. Scale bars, 10 mm.

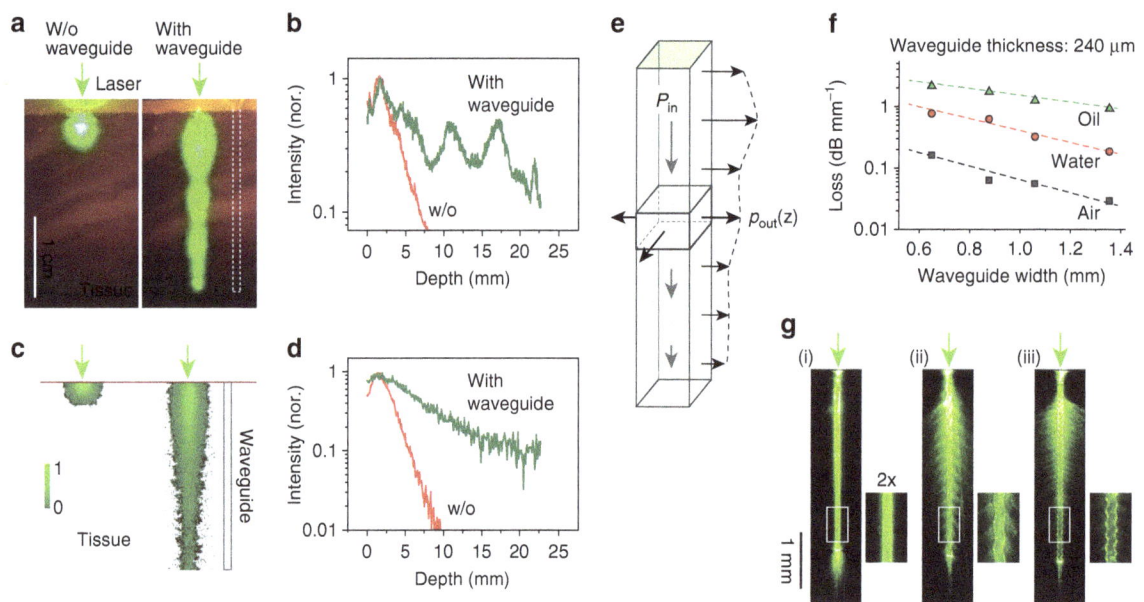

Figure 2 | Propagation of light in biopolymer-based planar waveguides. (a) Microscopic image of the light in bovine tissue (thickness ∼2 mm) taken before (left panel) and after (right panel) insertion of the waveguide. Left panel: an external light source of a green laser is directly applied to the surface of the tissue. Right panel: a green laser is coupled to a waveguide implanted into the tissue. Scale bar, 10 mm. **(b)** Profile of the light decay in the tissue with (green line) and without (w/o; red line) the waveguide. **(c)** Two-dimensional ray simulations of light propagation using the same geometry as for experimental measurement in **a**. **(d)** Profile of the simulated light decay. **(e)** Schematic representation of the light extraction from a planar waveguide. $P_{out}(z)$ corresponds to the light extraction profile from the waveguide, and $P_{in}(z)$ corresponds to input per unit length. **(f)** Measured propagation loss of the waveguides in air, water and oil for increasing waveguide widths and at a constant height of 240 µm. ($N=1$ for each medium; R^2 is 0.979, 0.938 and 0.906 in oil, water and air, respectively. **(g)** Waveguides with different surface patterns enable tuning the light extraction profile; here, in air. Light is coupled to the waveguide by focusing a laser to the top end of the waveguide. Scale bar, 1mm. nor., normalized.

environments: in air ($n=1$), water ($n=1.33$) and oil ($n=1.5$), respectively. Our measurements showed that in all cases, the optical intensity in the waveguide decreased exponentially with the propagation distance, as expected from equation (2), and the loss coefficients decreased linearly with the waveguide width (Fig. 2f). For a width of 650 µm, the loss coefficient was 0.16 dB mm^{-1} or $\sigma=0.37$ mm^{-1} in air, but it increases to 0.76 dB mm^{-1} in water and 2 dB mm^{-1} in oil (Fig. 2f). Most soft tissues[31] have a refractive index of 1.33–1.51. Therefore, the optical loss *in vivo* would fall somewhere between that of water and oil. A loss of 1 dB mm^{-1} may be adequate for an application where the majority (90%) of input optical energy is to be delivered through a tissue section over a depth of 10 mm.

To illuminate over thicker tissues, waveguides with lower loss would be necessary. One way to decrease the propagation loss is to increase the width; for 1.35 mm the loss coefficient in oil is 1.3 dB mm^{-1}, rather than 2 dB mm^{-1} for the 650 µm discussed above (Fig. 2f). Increasing the thickness of the waveguide also reduces the loss. We fabricated PLA slab waveguides with a thickness of 440 µm and obtained consistently lower loss than for the 240-µm-thick waveguides. For example, for a width of 580 µm, the loss coefficients were 0.15 dB mm^{-1} in air, 0.62 dB mm^{-1} in water and 1.5 dB mm^{-1} in oil.

Another practical method to control the loss function is modulation of the sidewall patterns by introducing bending losses on straight waveguides. We fabricated three PLA waveguides with flat and sine-wave sidewall patterns with different periods ($\Lambda = 0.23$ mm and 0.11 mm, respectively). Images of the light scattering patterns for the surface structures show drastically modified attenuation profile of the guided light and the axial profile of the extracted light intensity owing to the enhanced field deformation and curvature losses (Fig. 2g).

Biodegradation of polymer optical waveguides. We investigated the optical and mechanical properties of various polymers in an H-shaped waveguide over 24 h when immersed into phosphate-buffered solution (PBS) (Fig. 3a). Poly(vinylpyrrolidinone) (PVP), a water-soluble polymer, dissolves in PBS within minutes. Silk deforms after several minutes and swells within hours but does not fully dissolve. PLGA and PLA do not change shape in 24 h although transparency degrades modestly over the course of several hours.

To assess biodegradation of bulk polymer waveguides *in vivo*, we embedded transparent pieces of PLGA 50:50 pieces (1×5 mm $\times$ 500 μm) subcutaneously in live mice. The gross appearance of the implants and the surrounding tissue were examined at different time points after implantation. On day 6, we found the shape and transparency of the implant largely intact, but on Day 17, significant erosion of shape and transparency was apparent with significant loss in material volume (Fig. 3b). Visual inspection and histology indicated no sign of inflammation at the implantation site (Fig. 3c). On day 35, the implant was invisible to the naked eye. Molecular weight measurements by gel permeation chromatography showed that the molecular weight of the material decreased exponentially over time, which indicates the cleavage of polymer bonds by hydrolysis (Fig. 3d). The relative amount of GA in PLGA increases the biodegradation rate. We also tested the biodegradation of PVP and found that subcutaneously implanted PVP dissolved within 1 h *in vivo* (Fig. 3e). These results show that the biodegradability of bulk polymer waveguides varies in a large range and may be optimized for specific application by choosing appropriate polymers.

Light delivery to deep tissue for photochemical tissue bonding. To demonstrate the effectiveness of light-delivering waveguides in medicine, we investigated PTB, which is a dye-assisted photochemical technique that induces crosslinking between wound surfaces. Among several applications of PTB, we tested skin wound enclosure. In this technique, Rose Bengal dye is applied to the wound and excited by green light to generate reactive species that induce covalent crosslinking between the collagen molecules on both sides of the tissues in the wound[32].

To increase the depth to which PTB is effective in the tissue and achieve full-thickness skin bonding, we fabricated a PLA device with three waveguides. Light in the conventional multi-mode fibre is directly coupled to the slab waveguide with an efficiency of $\sim 50\%$, measured by a cutback technique. Each waveguide was tapered and the uniform region was 1 mm in width, 440 μm in thickness and 10 mm in length and has corrugated edges for optimal extraction of optical energy over the entire length of 10 mm (Fig. 4a). The number of waveguides can be adjusted to the size of the wound. The interspacing between the waveguides was 2.2 mm, slightly larger than the intrinsic optical penetration depth of the skin, which was optimized for efficient PTB in the inter-waveguide tissue space. We note that the thickness is comparable to the diameters (200–300 μm) of most typical absorbable sutures. For PTB, a full-thickness incision was made in the middle of the porcine skin, and the comb waveguides were inserted into the incision (Fig. 4b).

Before the tissue-bonding experiment, we cut a thick piece of porcine skin tissue into two halves. After applying Rose Bengal dye onto the exposed surface of one half, a waveguide was placed and 532-nm laser light at 1 W was launched to the pigtail fibre. The optical image shows a uniform distribution of optical energy in the tissue around the waveguides (Fig. 4c). The optical intensity at the tissue interface over an area of $\sim 1 \times 1$ cm^2 is therefore 1 W cm^{-2}, which complies with the maximum permissible exposure (1–4 W cm^{-2}) of skin *in vivo* to prolonged laser irradiation at 400–700 nm (ref. 33). To measure the time-lapse photoactivation of the dye, we imaged the cross-section at various time points after laser irradiation. The two pieces of tissue

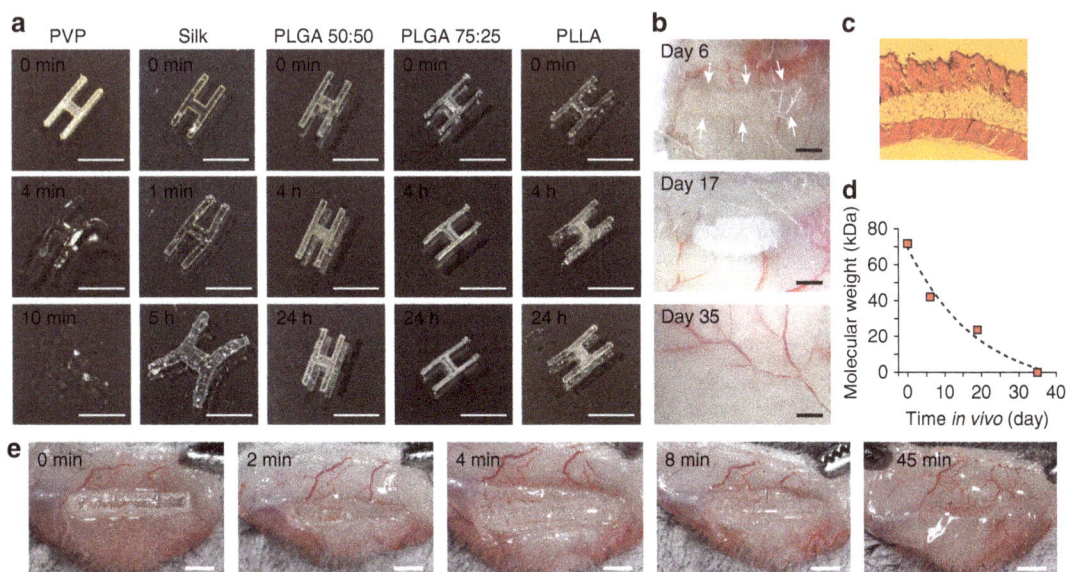

Figure 3 | Biodegradation of biopolymer-based waveguides. (**a**) Short-term behaviour of bio-polymer films are shown by immersing H-shaped waveguide samples into PBS. PVP dissolves within minutes. Silk deforms within minutes and shows considerable swelling after a few hours. PLGA and PLA largely retain their physical structure for more than 24 h. Scale bars, 5 mm. (**b**) Time-dependent *in vivo* degradation of 1×5 mm PLGA 50:50 polymer subcutaneously inserted into a mouse after 6, 17 and 35 days. No inflammation around the waveguide is observed. Scale bar, 1 mm. (**c**) Histology image of skin tissue in the region where the polymer was implanted. (**d**) Average molecular weight of polymer chains in implanted bio-film as a function of degradation time ($N = 1$). Line, fit to exponential decay ($R^2 = 0.93$). (**e**) Dissolution of PVP waveguide *in vivo*.

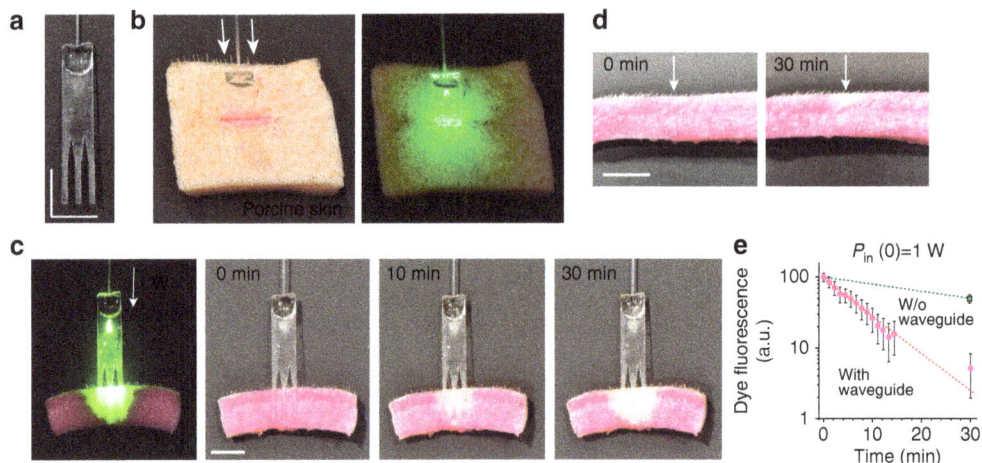

Figure 4 | Light delivery to deep tissue. (**a**) Hybrid biopolymer-based waveguide bundle engineered for optimal light delivery to the tissue. (**b**) Waveguide inserted in porcine skin incision previously stained with Rose Bengal dye (left) and illuminated through a fibre by a green laser (right). (**c**) Cross-section of the photoactivated area of dyed porcine skin after different times of illumination using a biopolymer-based waveguide bundle. The presence of photobleaching down the entire depth of the tissue interface after 30 min of illumination indicates efficient light delivery throughout the entire depth of the sample. (**d**) Identical porcine skin sample treated with surface illumination (that is, without (w/o) using a waveguide). Note that even after 30 min of continuous illumination, only the dye close to the top surface has bleached. (**e**) The amount of remaining intact dye measured by red fluorescence (averaged over whole area of the waveguide, $N = 1$) decreases exponentially with time. The error bars represent s.d. across the area. Scale bars, 10 mm in **a,c** and **d**.

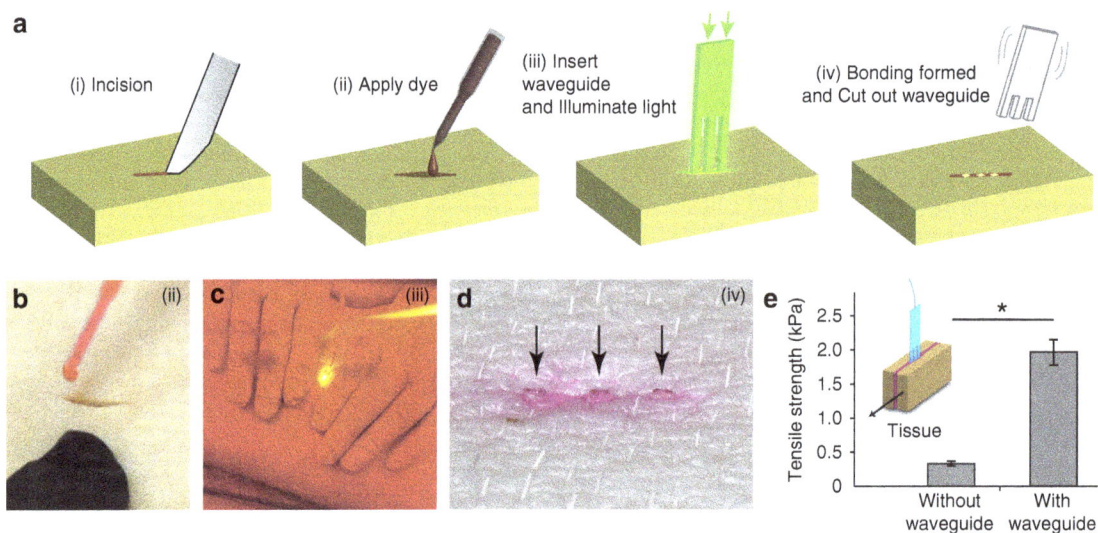

Figure 5 | Waveguide-assisted photochemical tissue bonding. (**a**) Schematic of the experimental procedure. (i) A full-thickness skin incision is performed. (ii) Rose Bengal dye is applied to the surfaces of a wound in porcine skin, and the excess solution is removed. (iii) A waveguide is inserted into the wound, and light is delivered to the waveguide through an optical fibre. (iv) After the procedure, the waveguide is trimmed away close to the skin surface. (**b-d**) Photographs showing steps (ii), (iii) and (iv). The arrows in **d** indicate the position of the waveguide inside the tissue. (**e**) Shear tensile strength of PTB bonds formed *ex vivo* in porcine skin. PTB bonds formed using a waveguide show more than fivefold increased bonding strength compared with conventional superficial illumination of the skin. *P value < 0.01 in t-test ($N = 3$). The errors bars represent the s.d.

were brought into contact during the light delivery to simulate the actual situation of PTB. The time-lapse images of the tissue cross-section showed photobleaching of the dye in the superficial area of the skin after 10 min, but almost complete bleaching occurred over a large area around the entire implant after 30 min (Fig. 4c). Photobleaching results from degradation of the dye over multiple excitations. We note that complete photobleaching is not always required for effective PTB.

For comparison, we illuminated the surface of a piece of porcine skin tissue directly with an unfocused 532-nm laser beam with a diameter of 3 mm at various power levels. We found that photobleaching was limited to a superficial layer at depths of less than 1.5 mm even at a high power of 1 W ($14\,\mathrm{W\,cm^{-2}}$; Fig. 4d).

The total amount of fluorescence from unbleached dye was measured from the region of interest in direct contact with the waveguide. The results indicated an exponential increase of photobleaching over time (Fig. 4e).

Next, we tested waveguide-assisted PTB *in situ* in animals (Fig. 5a). A full-thickness incision was made on the dorsal skin of a pig immediately after being killed (Fig. 5b). After applying Rose Bengal dye to the wound, a polymer waveguide was inserted into the wound. The tissue was approximated to ensure physical contact between tissues and waveguide surface while 532-nm laser light at 1 W was delivered over a duration of 15 min (Fig. 5c). After the irradiation was completed, the exposed part of the waveguide was cut away, leaving the remaining implant within the treated tissue (Fig. 5d). If conventional non-biodegradable fibres had been used, they would have to be removed from the tissue, and this would damage the treated tissue and disrupt the wound closure. Biodegradable polymers, however, will be gradually resorbed and replaced by native tissues if used in an *in vivo* setting. To measure the bond strength between the pieces of tissue, we performed tensometer-based shear tensile measurements (Fig. 5e). Conventional PTB without a waveguide yielded a tensile strength of 0.33 kPa. By contrast, waveguide-assisted PTB showed a much higher strength of 1.94 kPa due to full-thickness photoactivation of porcine skin.

Discussion

To date, biocompatible and biodegradable polymers have been extensively studied and used in many fields of medicine such as cardiovascular medicine[34], orthopedics[35] and neurology[36]. They show minimal systemic toxicity and the rate of resorption of polymeric implants can be tailored to minimize the probability of tissue irritation during degradation. Therefore, they provide a safe material platform to be used for new device applications.

PTB offers a number of advantageous features such as the formation of a water-tight closure across the entire wound interface and minimal scar formation[32,37], compared with wound closure by standard sutures and staples or by cyanoacrylate and fibrin glues. However, until today[32,38], the application of PTB has been limited to superficial wounds with a maximum depth of 1–2 mm. By using biodegradable waveguides, the remit of PTB is substantially expanded and no longer limited in terms of the size of the wound or the type and transparency of the tissue. Moreover, it is known that the efficiency of photochemical PTB is sensitive to oxygen concentration and greatly enhanced in live animals compared with the *ex vivo* tissues used here[39]. We therefore expect a considerably higher tensile strength for *in vivo* full-thickness PTB.

The potential applications of biodegradable polymer waveguides are expected to go far beyond PTB. Implantable waveguides can extend the therapeutic depths in photodynamic and photothermal therapies, enable deep tissue stimulation by low-level light or optogenetic techniques, and offer new strategies for continuous monitoring of medical conditions and sensing diagnostic information. Furthermore, biodegradable graded index lenses or fibres may permit longitudinal imaging of the viability of transplanted organs and monitoring of the healing process after surgery.

In conclusion, we have developed a new class of biocompatible and biodegradable polymer waveguides and validated their effectiveness for inducing photochemical processes in deep tissue. A comb-shaped slab waveguide that was designed and fabricated for optimal light delivery enabled us to demonstrate successful PTB treatment of a full-thickness skin incision (>1 cm deep), which has not been possible by conventional surface illumination and would be impractical with non-biodegradable optical materials. Although the initial focus of this study was on light-delivering devices for PTB, biocompatible and bio-degradable waveguides can be directly applied to other light-based diagnostics, surgery and therapeutics. Therefore, we expect that bioabsorbable waveguides may initiate a new paradigm and find widespread use in medicine.

Methods

Polymer waveguide fabrication. We purchased poly(D,L-lactide-co-glycolide) (Sigma, acid-terminated lactide:glycolide 50:50 with a MW of 38,000–54,000 Da or ester-terminated lactide:glycolide 75:25 with a MW of 76,000–115,000 Da) and poly(lactic acid) (Hycail CML-PLA, MW of 63,000 ± 12,000 Da) as dry powders. The powders were placed onto a glass slide preheated to 230 °C on a hot plate. On melting, another glass slide was used to press the polymer to a thin film. The glass slides were treated beforehand with Sigmacote (Sigma) to prevent sticking of the

polymer to the glass. Glass spacers were used to precisely control the thickness of the films. To cool and separate the film from the glass slides it was immediately submerged into icy water. The poly(vinylpyrrolidinone) (PVP) (Sigma, MW of 360,000 Da) films were prepared by solvent evaporation. PVP was dissolved in water at a concentration of 20 wt%, was poured into a petri dish and dried in air to make the PVP films. Polyethylene glycol hydrogel films were prepared by photopolymerization of the precursor solution containing 80 wt% polyethylene glycol diacrylate (Sigma, MW of 700 Da), 5 wt% 2-hydroxy-2-methylpropiophenone (Sigma) and 15 wt% water using an ultraviolet lamp. Thin films were cut to the desired shape using VersaLaser VLS3.50 cutting/engraving system at 5 mm s^{-1} with a power of 200 mW. To make silk waveguides, the extraction of silk fibroin was performed as described elsewhere[40]. The extracted silk fibroin was concentrated to 20% by dialysis. This solution was dried in waveguide-shaped mould overnight. The waveguides from all the above materials were bonded to a standard multimode optical fibre (200-μm core, 0.48 NA) using an optical adhesive (Norland, NOA 81).

Measurement of waveguide loss. We used the cutback technique to measure the propagation losses of waveguides in air, water and oil. Light from a 635-nm diode laser (Coherent) or a 532-nm miniature diode-pumped Nd:YAG laser (Laser Quantum) is coupled via an objective lens to the waveguide and the light power at the output of the waveguide is measured with a power meter at various lengths. Finally, the propagation loss is calculated according to the measured power levels at different lengths.

Light propagation simulation. The light propagation in waveguides embedded in skin tissue was simulated by using the TracePro ray-tracing software. A light source in the waveguide (that is, influx surface) has a boundary shape of annular and a grid pattern of circular that provide $\sim 10^5$ rays for tracing. The waveguide has a refractive index of 1.476 and the side surface of the waveguides is modelled with a diffuse reflection of 0.92. The tissue has an anisotropy of 0.81, absorption coefficient of 0.145 mm^{-1} and scattering coefficient of 1.5 mm^{-1}. The waveguide in the tissue has dimensions of 1, 0.44 mm and 2 cm. The tissue has a total thickness of 2 mm. The light on the tissue surface is observed 3 mm above it.

Implantation. We used immunocompetent male C57BL6 mice, ages 8–16 weeks, for the implantation study. After anesthesia by intraperitoneal injection of ketamine (100 mg kg^{-1}) and xylazine (10 mg kg^{-1}), the dorsal skin was incised over ~ 1 cm longitudinally and the implant was inserted subcutaneously. The incision was then closed using a 6–0 nylon suture. All animal experiments were performed in compliance with institutional guidelines and approved by the subcommittee on research animal care at the Harvard Medical School.

Histology. A sample of full-thickness skin around the implant was excised and fixed in 4% formalin for over 48 h. The skin sample was frozen-sectioned with a microtome at a thickness of 5 μm and stained with haematoxylin and eosin. The slide was imaged with a bright-field microscope (Olympus).

Photobleaching and PTB experiments. Photobleaching and PTB experiments are done on porcine tissue. The tissues are harvested immediately after the animals were killed and stored at − 20 °C. Immediately before use, the tissues were defrosted in a water bath for 20 min. The waveguide-assisted PTB procedure (in Fig. 5a–c) was demonstrated on a pig immediately after the animal was killed. An incision was made in the tissue and a solution of 0.1% (w/v) Rose Bengal (Aldrich) in PBS was applied to each exposed surface of the opened wound. After 1 min, the excess dye solution was wiped off. The tissue surfaces covered by Rose Bengal were illuminated by a 532-nm continuous-wave KTP-frequency-doubled solid-state laser (LRS-0532-PFH-000500-05, Laserglow Technologies, Canada, 800 mW, 532 nm), at which the Rose Bengal has an extinction coefficient of 25,000 M^{-1} cm^{-1} in PBS. The bonding strength of attached tissues was measured by shear tensile technique using the tensiometer (MTESTQuattro, Admet). The P value of the bonding strengths with and without waveguides is 0.0017 in t-test.

References

1. Richardson, D. S. & Lichtman, J. W. Clarifying tissue clearing. *Cell* **162**, 246–257 (2015).
2. Benedict, E. B. Endoscopy. *N. Engl. J. Med.* **228**, 253–258 (1943).
3. Gora, M. J. *et al.* Tethered capsule endomicroscopy enables less invasive imaging of gastrointestinal tract microstructure. *Nat. Med.* **19**, 238–240 (2013).
4. Dierickx, C. C., Grossman, M. C., Farinelli, W. A. & Anderson, R. R. Permanent hair removal by normal-mode ruby laser. *Arch. Dermatol.* **134**, 837–842 (1998).
5. Haedersdal, M. & Wulf, H. Evidence-based review of hair removal using lasers and light sources. *J. Eur. Acad. Dermatol. Venereol.* **20**, 9–20 (2006).
6. Dougherty, T. J. *et al.* Photodynamic therapy. *J. Natl Cancer Inst.* **90**, 889–905 (1998).

7. Idris, N. M. *et al. In vivo* photodynamic therapy using upconversion nanoparticles as remote-controlled nanotransducers. *Nat. Med.* **18**, 1580–1585 (2012).

8. Huang, X., Jain, P. K., El-Sayed, I. H. & El-Sayed, M. A. Plasmonic photothermal therapy (PPTT) using gold nanoparticles. *Lasers Med. Sci.* **23**, 217–228 (2008).

9. Alvarez-Lorenzo, C., Bromberg, L. & Concheiro, A. Light-sensitive Intelligent Drug Delivery Systems. *Photochem. Photobiol.* **85**, 848–860 (2009).

10. Jacques, S. L. Optical properties of biological tissues: a review. *Phys. Med. Biol.* **58**, R37–R61 (2013).

11. Haase, M. & Schäfer, H. Upconverting nanoparticles. *Angew. Chem. Int. Ed.* **50**, 5808–5829 (2011).

12. Wang, F. *et al.* Simultaneous phase and size control of upconversion nanocrystals through lanthanide doping. *Nature* **463**, 1061–1065 (2010).

13. Sheldon, E. E. *Fiber Endoscope Provided with Focusing Means and Electroluminescent Means* (Google Patents, 1968).

14. Lee, C. M., Engelbrecht, C. J., Soper, T. D., Helmchen, F. & Seibel, E. J. Scanning fiber endoscopy with highly flexible, 1 mm catheterscopes for wide-field, full-color imaging. *J. Biophotonics* **3**, 385–407 (2010).

15. Kim, P. *et al. In vivo* wide-area cellular imaging by side-view endomicroscopy. *Nat. Methods* **7**, 303–305 (2010).

16. Kiesslich, R. *et al.* Identification of epithelial gaps in human small and large intestine by confocal endomicroscopy. *Gastroenterology* **133**, 1769–1778 (2007).

17. Omenetto, F. G. & Kaplan, D. L. A new route for silk. *Nat. Photonics* **2**, 641–643 (2008).

18. Tao, H., Kaplan, D. L. & Omenetto, F. G. Silk materials–a road to sustainable high technology. *Adv. Mater.* **24**, 2824–2837 (2012).

19. Perry, H., Gopinath, A., Kaplan, D. L., Dal Negro, L. & Omenetto, F. G. Nano-and micropatterning of optically transparent, mechanically robust, biocompatible silk fibroin films. *Adv. Mater.* **20**, 3070–3072 (2008).

20. Lawrence, B. D., Cronin-Golomb, M., Georgakoudi, I., Kaplan, D. L. & Omenetto, F. G. Bioactive silk protein biomaterial systems for optical devices. *Biomacromolecules* **9**, 1214–1220 (2008).

21. Parker, S. T. *et al.* Biocompatible silk printed optical waveguides. *Adv. Mater.* **21**, 2411–2415 (2009).

22. Toffanin, S. *et al.* Low-threshold blue lasing from silk fibroin thin films. *Appl. Phys. Lett.* **101**, 091110 (2012).

23. Tao, H. *et al.* Implantable, multifunctional, bioresorbable optics. *Proc. Natl Acad. Sci. USA* **109**, 19584–19589 (2012).

24. Manocchi, A. K., Domachuk, P., Omenetto, F. G. & Yi, H. Facile fabrication of gelatin-based biopolymeric optical waveguides. *Biotechnol. Bioeng.* **103**, 725–732 (2009).

25. Jain, A., Yang, A. H. & Erickson, D. Gel-based optical waveguides with live cell encapsulation and integrated microfluidics. *Opt. Lett.* **37**, 1472–1474 (2012).

26. Choi, M. *et al.* Light-guiding hydrogels for cell-based sensing and optogenetic synthesis *in vivo. Nat. Photonics* **7**, 987–994 (2013).

27. Sykes, E. A., Albanese, A. & Chan, W. C. Biophotonics: implantable waveguides. *Nat. Photonics* **7**, 940–941 (2013).

28. Choi, M., Humar, M., Kim, S. & Yun, S.-H. Step-Index optical fiber made of biocompatible hydrogels. *Adv. Mater.* **27**, 4081–4086 (2015).

29. Ding, H., Lu, J. Q., Wooden, W. A., Kragel, P. J. & Hu, X.-H. Refractive indices of human skin tissues at eight wavelengths and estimated dispersion relations between 300 and 1600 nm. *Phys. Med. Biol.* **51**, 1479–1489 (2006).

30. Ratner, B. D., Hoffman, A. S., Schoen, F. & Lemons, J. E. *Biomaterials Science: an Introduction to Materials in Medicine* 162–164, 2004).

31. Jacques, S. L. Optical properties of biological tissues: a review. *Phys. Med. Biol.* **58**, R37–R61 (2013).

32. Tsao, S. *et al.* Light-activated tissue bonding for excisional wound closure: a split-lesion clinical trial. *Br. J. Dermatol.* **166**, 555–563 (2012).

33. American National Standard. in *American National Standard for Safe Use of Lasers.* (Laser Institute of America, Orlando, FL, USA, 2007).

34. Pilgrim, T. *et al.* Ultrathin strut biodegradable polymer sirolimus-eluting stent versus durable polymer everolimus-eluting stent for percutaneous coronary revascularisation (BIOSCIENCE): a randomised, single-blind, non-inferiority trial. *Lancet* **384**, 2111–2122 (2014).

35. An, Y. H., Woolf, S. K. & R. J., Friedman Pre-clinical *in vivo* evaluation of orthopaedic bioabsorbable devices. *Biomaterials* **21**, 2635–2652 (2000).

36. Bini, T. *et al.* Peripheral nerve regeneration by microbraided poly (L-lactide-co-glycolide) biodegradable polymer fibers. *J. Biomed. Mater. Res. A* **68**, 286–295 (2004).

37. Yao, M., Yaroslavsky, A., Henry, F. P., Redmond, R. W. & Kochevar, I. E. Phototoxicity is not associated with photochemical tissue bonding of skin. *Lasers Surg Med.* **42**, 123–131 (2010).

38. Yang, P., Yao, M., DeMartelaere, S. L., Redmond, R. W. & Kochevar, I. E. Light-activated sutureless closure of wounds in thin skin. *Lasers Surg. Med.* **44**, 163–167 (2012).

39. Verter, E. E. *et al.* Light-initiated bonding of amniotic membrane to cornea. *Invest. Ophthalmol. Vis. Sci.* **52**, 9470–9477 (2011).

40. Rockwood, D. N. *et al.* Materials fabrication from Bombyx mori silk fibroin. *Nat. Protoc.* **6**, 1612–1631 (2011).

Acknowledgements

We thank Orhan Celiker for help in numerical simulation. This work was funded by the U.S. National Institutes of Health (R21EB013761, P41EB015903, R01CA192878), Department of Defense (FA9550-13-1-0068), Bullock-Wellman Fellowships, Marie Curie Career Integration Grant (631679), Marie Curie International Outgoing Fellowship N°627274 within the 7th European Community Framework Programme, Human Frontier Science Program (Young Investigator Grant RGY0074/2013), the Bio & Medical Technology Development Program of the National Research Foundation of Korea (2012M3A9C6049791) and the IT Consilience Creative Program of MKE and NIPA (IITP-2015-R0346-15-1007).

Author contributions

S.N., M.C.G., R.W.R. and S.H.Y. designed the experiments. S.N., M.C.G. and M.H. performed the experiments. M.C., S.K., K.S.K., S.K.H., G.S. and M.R. provided the materials. S.N., M.C.G., M.H. and S.H.Y. wrote the manuscript with input from all the authors.

Additional information

27

All-optical design for inherently energy-conserving reversible gates and circuits

Eyal Cohen[1], Shlomi Dolev[1] & Michael Rosenblit[2]

As energy efficiency becomes a paramount issue in this day and age, reversible computing may serve as a critical step towards energy conservation in information technology. The inputs of reversible computing elements define the outputs and vice versa. Some reversible gates such as the Fredkin gate are also universal; that is, they may be used to produce any logic operation. It is possible to find physical representations for the information, so that when processed with reversible logic, the energy of the output is equal to the energy of the input. It is suggested that there may be devices that will do that without applying any additional power. Here, we present a formalism that may be used to produce any reversible logic gate. We implement this method over an optical design of the Fredkin gate, which utilizes only optical elements that inherently conserve energy.

[1]Department of Computer Science, Ben Gurion University of the Negev, 653 Be'er, Sheva 84105, Israel. [2]Ilze Katz Institute for Nanoscale Science and Technology, Ben Gurion University of the Negev, 653 Be'er, Sheva 84105, Israel. Correspondence and requests for materials should be addressed to E.C. (email: koderex@gmail.com).

One of the greatest challenges in computing is to reduce the energy consumption, not only for the sake of reducing the tremendous amount of electric power used by the computer-based industry (for example, Google, Amazon or Facebook) but also to enable computing devices to operate at a higher frequency without melting as a result of the extra heat. A radical change in computer design may possibly lead to the crucial breakthrough needed for achieving more energy-efficient information processing. One promising direction for a breakthrough is to implement an all-optical[1] universal reversible gate, such as the Fredkin gate[2], and then use the Fredkin gate implementation as a building block in reversible circuits, thus overcoming this significant challenge.

The Fredkin gate is a universal reversible logic gate. Its universality means that any logic operation can be produced using only Fredkin gates. It has been suggested that reversible gates may preserve energy together with the data and reversible devices may be built entirely with energy-conserving elements[3,4]. These elements include a network of directional couplers and controlled nonlinear phased modulators. While the ability to build energy-conserving reversible gates has been discussed, no proof was given for the feasibility of these devices. In addition, no mathematical model has been presented which describes how to design such devices. Therefore, the search for a universal, energy-conserving, reversible logic device is still crucial for reversible computing and its beneficial prospects.

Research in the field of reversible computing is focused on two major aspects which include the implementation of reversible gates on one hand, and their usage for large-scale circuits on the other.

Extensive work has been done trying to utilize the Fredkin gate to perform simple or complex computing primitives. The assumption is that an ideal Fredkin gate is available and used to build efficient circuits. For example, a paradigm named 'Directed Logic'[5] is presented as an energy-efficient computation model used for Boolean gates. Later on, directed logic was used to build automata and circuits[6].

The other major aspect in the research of reversible computing suggests implementations for the Fredkin gate. Publications differ in their technological paradigms. Therefore, we will refer to them as the 'electronic approach', the 'all-optical approach' and the 'hybrid electro-optic approach'. Since electronic and hybrid devices rely on transistors and similar semiconductors paradigms they suffer from inherent energy loss.

Previous publications in the scope of the electronic approach suggested complementary metal-oxide semiconductor (CMOS) implementations[7]. The designs focus on an effort to minimize the energy loss and measure it on every logical state of the gate.

The hybrid electro-optic approach suggests that data and control signals may be partly optical and partly electronic. For example, implementations for 'directed logic' are suggested[8–10], where an electric control may be used for switching the two inputs to the appropriate two outputs. The control is based on manipulation of the resonance of a ring resonator by a silicon p-i-n junction built over the cross-section of the ring[11,12]. This approach requires the conversion of optical data to electronic controls, or electronic signals to optical ones.

The last approach suggests an all-optical approach, where all inputs and controls are optical. For example, it is suggested that light projection may be used to mechanically manipulate the positioning of molecules[13]. This minute manipulation may be very significant if it is done over a ring or a disc resonator. The movement may tune or detune the resonator and, in essence, implement a switch. This design is still not energetically efficient, since the optical energy is transformed into mechanical energy.

Other examples involve utilizing an effect called four-wave mixing (FWM). Under certain conditions in a nonlinear medium, a strong continuous wave signal on one wavelength may switch signals between two other wavelengths. When the continuous wave is absent no manipulation is done, and the signals continue to propagate with negligible modification. This technique is used to implement a Fredkin gate over the wavelength space[14]. A very high intensity is needed for the continuous wave, in order to exhibit the expected wavelength conversion, which also suffers from losses such as harmonics generation. However, it is possible to lower these losses[15] by utilizing coupled-resonator optical waveguides in order to enhance the FWM interaction. The main issue with FWM implementations is that all signals propagate in the same waveguide which requires additional separation, amplification or wavelength conversion in order to cascade the gates.

A different approach is to attempt to utilize quantum optics in order to produce quantum optical gates[16,17]. Although these devices are relatively simple in design, they require single-photon sources. While single-photon sources are becoming available[18], these devices are limited to work only with single-photon sources, making them difficult to integrate with non-quantum optical elements.

It should be noted that from a logic point of view, reversible gates are quantum gates and quantum gates are reversible gates. The two fields share similar concepts and mathematical approaches. However, while quantum gates require quantum implementations, reversible gates are not limited by that requirement, and our formalism and implementation do not refer necessarily to quantum mechanics or quantum implementations.

Here, we present a design for reversible gates built entirely from energy-conserving elements. First, we present our formalism, that utilizes linear and nonlinear unitary transformations in relation to conserving optical elements such as directional couplers. This formalism may be used by designers to design any reversible logic. Next, we use this formalism in order to design a Fredkin gate. We also simulate the different optical elements used such as directional couplers and graphene-induced silicon waveguide showing negligible energy loss. These results are later used to simulate our design of the Fredkin gate. All this effort was made while taking into account further circuit design. For example, making sure that all inputs and outputs are interchangeable without power or wavelength modification. Also, it will permit any cascading design of the circuits.

Results

Optical elements. When considering a computational machine, theoretical values need to be associated with a physical value. In an optical implementation for example, values used will refer to an amplitude of an electric field. The energy hiding in this field is proportional to the square of the magnitude of the electric field. To preserve the overall energy, all values may interact using unitary transformations which preserve the sum over the squares of the magnitudes.

Supplementary Notes 1 and 2 present a few basic available building blocks that manipulate the optical signals in a unitary fashion while Supplementary Fig. 1 demonstrates a representation of the relevant parameters of a general directional coupler. We are going to use these building blocks in order to design our devices. These elements include a linear waveguide which is used to introduce a phase to a signal, and a nonlinear Kerr-induced waveguide, which is used to introduce a nonlinear phase to a signal. Also, we will explain how a directional coupler or a Mach–Zehnder coupler performs a linear unitary transformation on two signals, while an array of such couplers may be used to perform a linear unitary transformation over a set of signals.

Unitary transform formalism. We would like to propose a general theory that allows the design and building of devices that solve a specific reversible logic gate. For example, we will explain the steps needed to design the reversible Fredkin gate. A truth table of the Fredkin gate is given in Table 1. The inputs of the device are considered in a straightforward manner, where each input channel represents a logical input value of the gate. In binary gates, each of these inputs may hold two complex values E_0 and E_1, which are amplitudes that would represent the states 0 and 1, respectively. Similarly, the outputs of the device may also hold the same values as those determined by the gate logic.

Given a state of the gate, we can arrange the inputs for each input channel in a column vector. The different vectors may be arranged side-by-side to build a matrix. This will be the input matrix, which we will denote with A. Similarly, we can arrange the outputs for each output channel in a column vector and arrange these vectors side-by-side to build an output matrix, which is denoted by B.

Our goal is that when the device receives an input that is a column of A, it will produce an output corresponding to an appropriate column of B. The number of columns on both matrices is the number of states the gate supports. The number of rows in A is the number of input channels and the number of rows in B is the number of output channels. Notice that if the gate is reversible, the number of input and output channels is the same, and so are the dimensions of A and B.

If we can find a unitary matrix M that satisfies $B = MA$, we can stop here and build the device using a coupling array that decomposes the matrix. Details on unitary matrix decomposition is given in Supplementary Notes 3 and 5, while Supplementary Fig. 2 demonstrates a coupling array. However, generally, such unitary matrices may not be found, particularly in the case of Fredkin gate, merely because of the fact that the number of columns is greater than the number of rows of A and B. That would mean that the dimensions of M are too small and the degrees of freedom are not enough to solve the equation.

To overcome this problem, we may add rows to both A and B. We denote the matrix added to A with F' and the matrix added to B with F. F is part of the product of the multiplication, while after a small modification it will turn into F', which is then used as part of the multiplier. The transformation from F to F' is done with no loss. Hence, the prime represents a little modification to F.

In other words, we are looking for a unitary matrix M that solves:

$$\begin{pmatrix} B \\ F \end{pmatrix} = M \begin{pmatrix} A \\ F' \end{pmatrix} \qquad (1)$$

A schematic design of the presented setup is given in Fig. 1. The design illustrated presents three inputs, three outputs and four feedback loops with a nonlinear element. Note that all presented elements relate to the matrices A, B, F and F', which hold the electric field over each channel at different states. The row index of the matrices defines the channel and the column

index represents the state. The gate illustrated in Fig. 1a exhibits coupling between seven channels denoted by M. This seven-channel coupling is designed with 21 adjacent couplings illustrated in Fig. 1b. Also, each F channel is manipulated through a nonlinear effect denoted by γ to become F'.

Note that in Fig. 1b, there is no notation for the different channels, it is not described which inputs hold the A channels or the F' channels or which outputs hold the B channels or the F channels. We are free to choose any permutation of these channels as this permutation can be described by a unitary matrix that may be integrated as part of M. A permutation should be chosen such that the decomposition of M will be the easiest to manufacture, taking into account manufacturing properties and limitations.

Two questions are raised by adding the rows of F' and F to A and B. The first regards how many rows need to be added to A or B in the form of F' and F respectively. The answer is that we cannot predict the dimensions of F, and each case should be investigated for its properties. The second regards whether it is possible to have feedback with no nonlinear elements, or is the nonlinearity obligatory.

We can try to use the relation between A, B, F and F' through the unitary matrix M and get:

$$B^\dagger B - A^\dagger A = F'^\dagger F' - F^\dagger F \qquad (2)$$

Note that in this expression M is eliminated. The left-hand side of the equation is determined only by the state representation of an examined gate. Also, note that if the gate is conserving energy for every state, the diagonal of the left-hand side is filled with zeros. Moreover, if some elements of the left-hand side are not zero, this means that a nonlinear interaction must come into play in order to manipulate F to be F', such that this condition is satisfied.

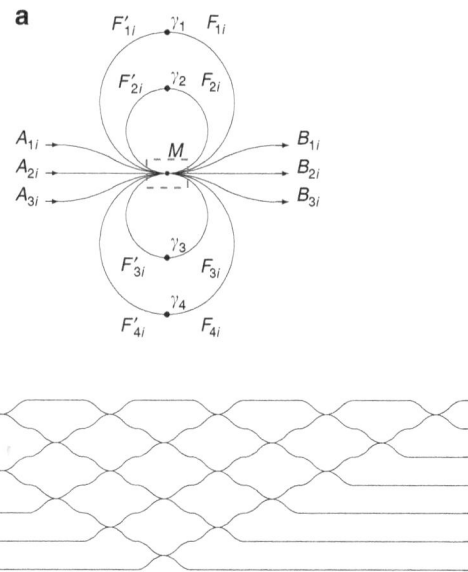

a

Figure 1 | A representation of a three channel gate. (a) This example shows three inputs denoted by A, and three outputs denoted by B. Four more channels are used for feedback, and denoted by F and F'. The coupling is between seven channels and denoted by M and a red broken box. (b) A zoom on the coupling M. This coupling array couples seven channels: the inputs are three A channels and four F' channels. Similarly, the outputs are three B channels and four F channels. This coupling is decomposed to 21 couplings between adjacent channels. When F' is added as an input for the coupling, F is added to the output of the coupling. Each F is then manipulated with an appropriate nonlinear element denoted by γ, which transforms it to F'. Then, F' is fed back to the system. The column index i represents different states.

Table 1 | A truth table of the reversible Fredkin gate.					
C	X₁	X₂	C′	Y₁	Y₂
0	0	0	0	0	0
0	0	1	0	0	1
0	1	0	0	1	0
0	1	1	0	1	1
1	0	0	1	0	0
1	0	1	1	1	0
1	1	0	1	0	1
1	1	1	1	1	1

X_1, X_2 and C represent three input bits, while Y_1, Y_2 and C' represent three output bits.

Note, however, that this equation was derived from equation (1), but it holds less information. A solution for equation (1) is therefore more inclusive and we are going to focus on it.

While different solutions for equation (1) may be found, it does not mean that all these solutions may converge to a desired output given an input. In fact, it only means that the desired states are steady states. These states may also be weak steady states, that might diverge after a short time period. Supplementary Note 4 discusses the issue of convergence and its time period. Supplementary Notes 6 and 7 also discusses different representations for logical data using optics, giving examples of the XOR and Fredkin gates.

Numerical solution for Fredkin gate. The terms and conditions formulated were implemented in Matlab, where numerical solutions were found for the different elements of the Fredkin gate. A solution for equation (1) was found, where M is later decomposed to an array of four 3×3 matrices, while the design utilizes three nonlinear elements. This design is given in Fig. 2. Note that feedback was eliminated with this design, which allows the interaction of pulsed light and not necessarily based on continuous wave. The derivation of this design is further discussed in the methods section. The minimal propagation time through this element is proportional to the sum of the gamma values which was minimized to $S_\gamma = 12.5$. Additional information about the derivation of this solution is given in Supplementary Note 8, while Supplementary Fig. 3 shows an intermediate stage in the decomposition of the design.

Optical elements. Finite-difference-time-domain (FDTD) simulations are done over simple elements that include a symmetric waveguide coupler and a waveguide infused with a Kerr nonlinear element. Subsequently, these elements are used as building blocks for photonic integrated circuits.

All the simulations assumed rectangular waveguides of silicon (Si) layered over glass substrate. The waveguide had a width of

400 nm and height of 220 nm. The simulations were done with a wavelength of $\lambda = 1.5 \, \mu m$, since it exhibits negligible absorption in Si and low bend loss in single-mode waveguides when utilizing the recent advances in CMOS production[19,20]. Note, however, that Si single-mode waveguides produced utilizing these methods still exhibit a relatively high-propagation loss of $0.3 db \, cm^{-1}$. In our discussion, we will remain with typical lengths that are overall smaller than 1 cm. Also, we may consider using different materials, resulting in a lower loss. All simulations refer to the fundamental propagating mode that this waveguide can support. An illustration of the cross-section is given in Fig. 3.

Note that this waveguide profile supports only single-mode propagation. The sources used in the simulation stimulate only this mode, while the monitors that read the corresponding outputs compare their results to this mode. It is assumed that only this mode may propagate and all discussions refer to this mode.

While many simulations, fabrications and tests on directional couplers and Mach–Zehnder interferometers were done and produced significant results, we are still required to produce our own design and results. First, we intend to investigate the losses presented by these elements. Moreover, we choose a specific wavelength and a specific cross-section throughout the design for all waveguides. These waveguides should match the simulations done for the Kerr-induced waveguide. The desired data and parameters may be problematic to obtain based solely on past literature; hence, these elements must be simulated again. Details of these simulations are given in the methods section and Supplementary Note 9. In addition Supplementary Fig. 4 shows the schematic design of the coupler used in our simulations.

A set of directional couplers were simulated to characterize their properties. An example of a propagation process in our design of the directional coupler is given on Supplementary Fig. 5. The results show a loss lower than 0.3% in the energy, or $-0.013 db$ for all the range of elements simulated. Supplementary Fig. 6 illustrates the derived coupling values as a function of the bend radius.

This element is later used to build a Mach–Zehnder interferometer by integrating two 50% directional couplers. The results for the Mach–Zehnder interferometer show a loss lower than 0.3% in the energy. An example of a propagation process in our design of the Mach–Zehnder coupler is given on Supplementary Fig. 7.

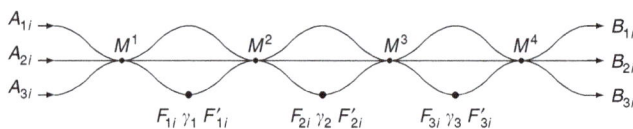

Figure 2 | An implementation of the Fredkin gate where feedbacks are eliminated. This design uses four coupling nodes. All nodes couple three channels. Three channels are used for nonlinear manipulation and denoted by F and F'. Each F is then manipulated with an appropriate nonlinear element denoted by γ, which transforms it to F'. The column index i represents different states.

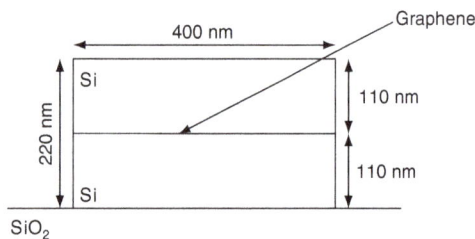

Figure 3 | An illustration of a waveguide cross-section. The waveguide is rectangular silicon over a glass substrate. The width of the waveguide is 400 nm and its height is 220 nm. When needed, a graphene sheet is included into the waveguide centred at the middle with a height (or thickness) of 1 nm. The graphene is included only when the Kerr effect is needed, while it is not included otherwise.

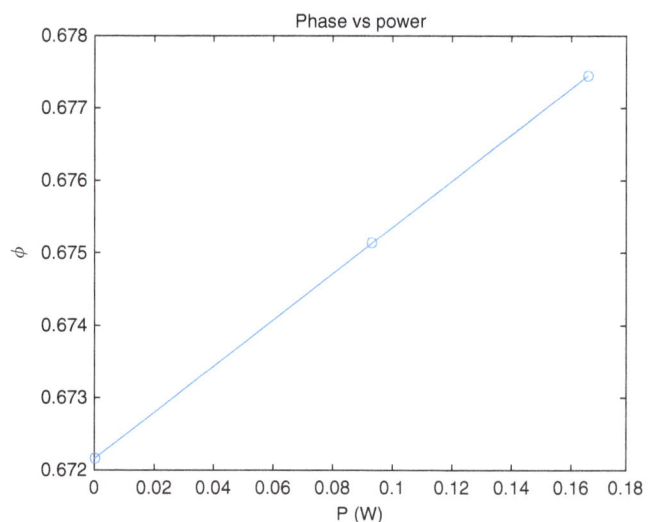

Figure 4 | Phase as a function of the power. A minor phase response is observed when the power propagating through the waveguide is increased. The waveguide includes graphene with a thickness of 1 nm. The length of the Kerr element is 5 μm.

A graphene infused waveguide was also simulated. The length chosen for the Kerr element simulation was $L = 5\,\mu m$. The results show a negligible loss, lower than 0.07% in the energy, or $-0.003db$. The phase response is given in Fig. 4. It shows a relatively high response giving:

$$\frac{\gamma}{L} = 6.31 \times 10^3 \left[W^{-1}m^{-1}\right] \qquad (3)$$

As explained earlier, a sum of the nonlinear values was calculated as $S_\gamma = 12.5$. This result corresponds to a length of $L = 2\,mm$ and a converging time of $T = 2 \times 10^{-11}$ s when using 1 W sources. A convergence energy $E_c = 2 \times 10^{-11}$ J. Note of course that a faster system convergence may be achieved with

higher power and designing shorter elements. Also, low-power devices may be produced with longer elements at the expense of the system convergence time.

Integrated circuit. The elements investigated in the FDTD simulation were now used for an integrated circuit simulation. The allows incorporating results acquired with 'Lumerical FDTD Solutions' for a specific basic optical element. These elements may be inter-connected into a simulation of one integrated device. It also provides many built in optical elements such as sources, detectors, Mach–Zehnder coupler, phase and delay lines which may be used together with customized scripted elements, and other integrated compiled elements.

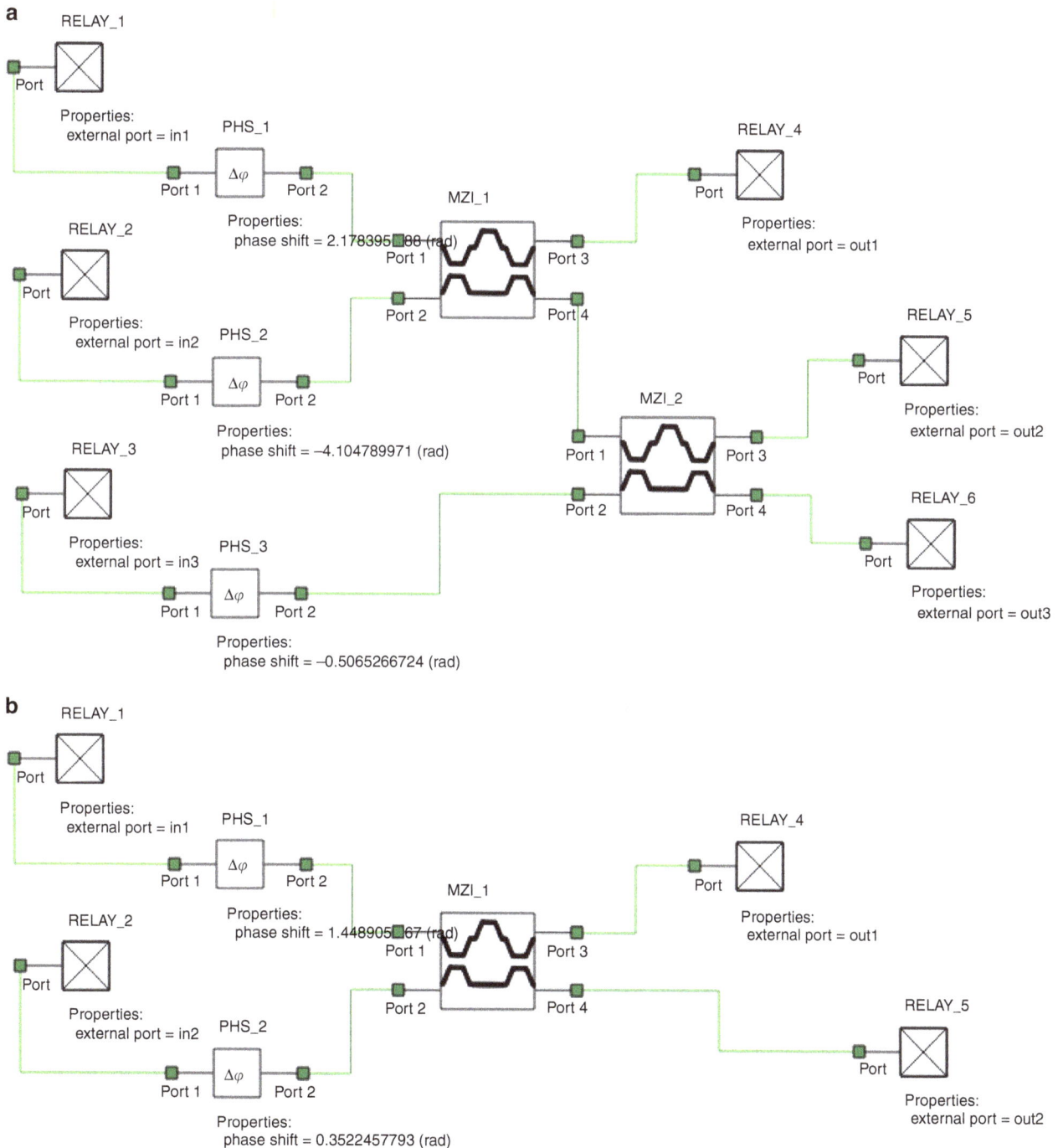

Figure 5 | Compiled elements that perform a unitary transformation. (a) An expansion of the compiled element 'mzi-comp' that includes two MZI couplers and three phase elements. (b) An expansion of the compiled element 'mzi-ex' that includes one MZI coupler and two phase elements.

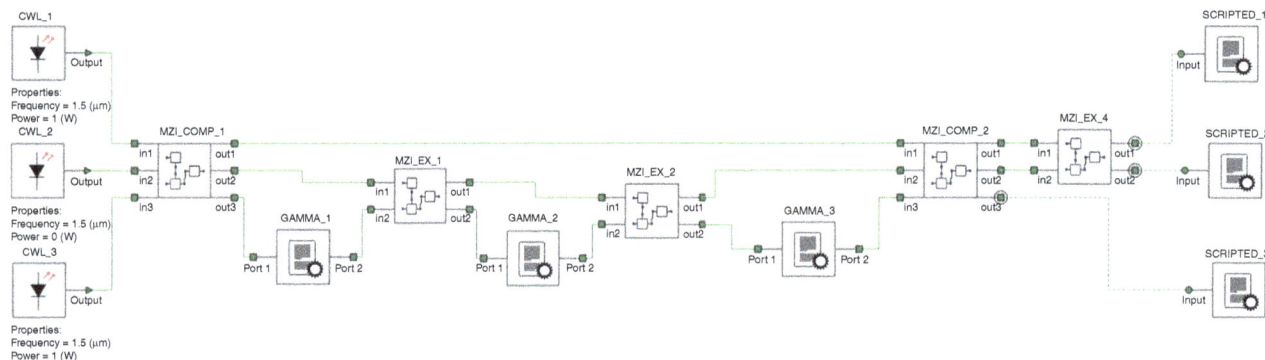

Figure 6 | Schematics of the device suggested. This scheme uses seven Mach–Zehnder couplers, and phase shift elements. This scheme uses compiled elements 'mzi-comp' and 'mzi-ex' in order to simplify the display of the design. Also, the nonlinear Kerr elements were created and are labelled 'gamma'.

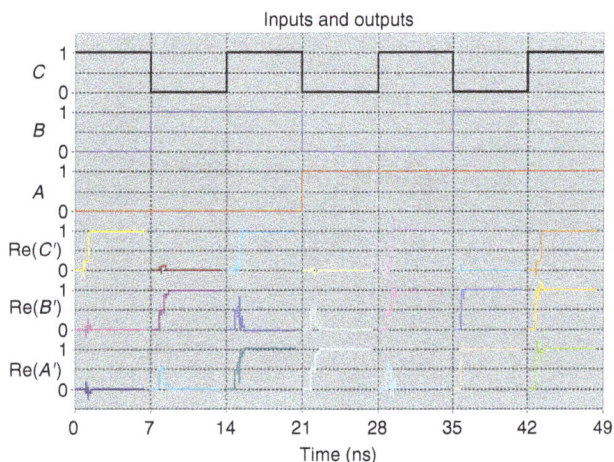

Figure 7 | Timing sequences of the Fredkin gate. Input values and output results are presented across time. The three input values A, B and C are shown together with the real values of the outputs A', B' and C'. As expected from a Fredkin gate, the value of the output C' is very similar to the value of the input C. When C = 0, the values of A' and B' are similar to the values of A and B respectively. When C = 1, the values of A' and B' are similar to the values of B and A, respectively. Note the time delay of the outputs as a result of the propagation delay over the optical elements.

First, in order to simplify the scheme of the design we compile and script a 2×2 unitary transformation by integrating a Mach–Zehnder coupler with phase elements into one compiled element titled 'mzi-ex'. The script selects the proper values for the coupler and phase elements such that the compiled element performs a user defined unitary transformation. Similarly, we compile and script a 3×3 unitary transformation by integrating two Mach–Zehnder couplers with phase elements into one compiled element titled 'mzi-comp'. These elements are displayed in Fig. 5. All couplings were achieved with Mach–Zehnder coupler elements with two 50% couplers and typical phase lengths of 15 μm.

The schematics of the Fredkin gate device are given in Fig. 6. These schematics include the compiled elements from Fig. 5. Three sources were used as input with the power values of zero or 1 W representing the logical 0 and 1. The nonlinear Kerr elements were scripted to produce a phase shift proportional to the power. Output values were collected with scripted elements (titled 'scripted') allowing us to collect phase information together with the intensity.

The integrated circuit was tested with all possible inputs of the Fredkin gate. First, a simulation was done using ideal elements with no loss. The results show almost no deviation from the desired output.

Next, losses were introduced, according to the results of the FDTD simulations, where 0.3% of energy loss was applied over all Mach–Zehnder interferometers. The result of the integrated circuit simulation are displayed on Fig. 7 which displays the timing sequence for the different possible signals of the Fredkin gate where A, B and C are three input signals and A', B' and C' are the three output signals. These results exhibit a small deviation from the desired values. The highest error introduced to the results occurred on the $ABC = 101$ state, where the normalized result of the B' channel was $E_{B'} = 0.961 - 0.044i$ instead of the value 1, which is a 6% deviation over the complex plane.

Discussion

A formalism for the design of reversible circuits was presented. This formalism utilizes only unitary transformations. Linear unitary transformations were represented by a unitary matrix. The definition of unitary transformations was extended to nonlinear transformations by maintaining that the sum over all the squares of the magnitudes is preserved.

The formalism suggested provides a theoretical condition over the reversible gate truth table, to determine if the specific logic may be implemented only by linear elements, or whether it is required to use nonlinear elements.

Optical elements that maintain these unitary properties were presented and tested. A directional coupler or a Mach–Zehnder coupler are used to represent a 2×2 unitary matrix transformation. An array of these couplers may be used for a general unitary matrix transformation.

Also, a Kerr-induced waveguide was simulated. By using graphene as a Kerr material, the effect becomes very strong and shorter elements may be used. This element provides the nonlinear unitary transformation by inducing self phase modulation. Still, since Kerr interactions are usually weak the length needed for these elements is relatively high compared with other elements in the system. This length will determine the propagation time needed in order to achieve the required result.

Simulation over all optical elements show a very low loss, which is almost negligible. This demonstrates the superiority of optical elements in energy efficiency.

We used the formalism presented in order to determine the needed elements for a Fredkin gate. We tried to minimize the length of the nonlinear elements by minimizing the sum over the nonlinear values S_γ. We also managed to eliminate feedback loops in the design, allowing the usage of pulsed light and not only continuous waves.

The integrated circuit simulation provided a promising result. The 6% deviation will still allow a cascading usage of the device in larger scale logic devices. Assuming the worst-case scenario where a 6% deviation is accumulated in a geometric progression, it is still possible to cascade the device five times before corrections with thresholds are needed.

This 6% deviation in the amplitude translates to a 7.5% energy loss. If we modify the device to support short pulses, a 1 W source with a pulse time of 50 fs is 10 wavelengths long, and it holds an energy of 50 fJ. The loss in that case is lower than 4 fJ, where we may assume that most of it probably propagates in the wrong channel rather than heating the system. A lower power source may be used to lower the loss even further; however, this will demand a longer propagation in the nonlinear element. Hence a lower power and lower loss may be achieved at the expense of a longer delay.

Also, this deviation may be lowered by modifying the numerical solution. This simulation was intended to assess the deviation of the output from an ideal device given losses. The results showed quasi-random deviation from the desired output. However, the numerical solution considered only ideal elements, and it may be modified to consider losses, such that the outputs will exhibit the same, yet minimal, power loss. These in turn may be used for other Fredkin gates that consider a lower input power as the normalized logical values.

We presented an optical architecture for reversible logic that does not dissipate heat and, while converged to a state, does not waste energy. The new mathematical framework provided here can be implemented for any reversible logic gate. The design was based on three basic energy-conserving elements: linear phase; coupling; and nonlinear manipulation.

The design is not limited to conventional optics, as different physical phenomena may be used. For example, electron optics may be considered. Another physical approach may suggest that the different inputs may be held on the same channel but on different wavelengths. Coupling between different wavelengths is possible through FWM, which was demonstrated efficiently over graphene[21].

The device only utilizes elements that conserve energy, namely waveguides and conserving couplers as linear elements and Kerr effect materials for nonlinear manipulations. The linear coupling may be achieved with different technologies such as a mode-waveguide coupler, or Mach–Zehnder coupling. The Kerr effect was chosen since it is energy-conserving, is widely available with different materials, and is easily formulated. However, the Kerr effect may be replaced by a different nonlinear, energy-conserving effect. Different effects and materials may exhibit significantly stronger interactions, which would allow lower energy loss and higher working frequency.

While converging, the wasted energy is not transformed to heat but rather is removed from the system through the output ports. This energy may be reused. For example, it may be redirected back to induce population inversion which may be used at a laser source.

For ease of manufacturing, a small number of couplings should be used. In our case, the proposed Fredkin gate utilized only seven couplings. A relatively long time is needed to allow propagation through the nonlinear element. This is a result of the extremely weak interaction of the Kerr effect. To achieve better performance, a shorter wavelength should be used. Better performance may be achieved by the discovery and usage of materials with stronger Kerr interactions. Stronger Kerr interactions may be available with future investigation of graphene. It may even be possible to use a totally different effect for the nonlinear interaction.

The example given for the Fredkin gate used a binary representation for each input or output channel. It may be possible to design gates where each channel may hold more than two values. These values may also be complex.

Note that states of the device were defined by a complex value for the electric field and the Kerr effect was dependent on a continuous wave. Future research may try to redesign the device such that it may support short pulses of light. This may be done by carefully choosing the lengths of the elements, or the optical lengths, and allowing signals that interact with each other to arrive and interact with each other at the appropriate point in time and space. By using pulses, however, special attention should be given to the shape of the pulse when dispersion and the Kerr effect may destroy the shape and introduce undesired noise. It may be possible that a soliton solution may be found while considering the presence of these effects.

Although our device introduces long time delay, a device that supports pulses may not necessarily suffer from bandwidth loss. This is because the device may be used asynchronously, while interaction with the pulses does not modify the shape of the pulse.

This device may serve as an intermediate step towards reversible optical quantum computing. The next step may be to design the devices that incorporate energy-efficient elements together with elements that are governed by quantum effects, while supporting one-photon interaction together with two-photon, three-photon or four-photon interactions.

Another approach may use the formalism and design described and integrate it with a semiconductor optical amplifier only to compensate for the minor optical losses. This may lead to a simpler design while maintaining a very low-power consumption.

Methods

Fabrication feasibility. When we deal with the available building blocks, we have to remember that we want to build devices with negligible loss. In other words, the sum of the intensity over all inputs should be nearly the same as the sum of the intensity over all outputs.

Note that negligible loss is possible to maintain providing the advances in fabrication. An 'ultra-high-Q' optical resonator has been presented[22], featuring a design and fabrication process that managed to produce a toroid microcavity resonator on a chip with a Q factor in excess of 100 million. This suggests that loss is negligible and was eliminated by the advanced fabrication methods described.

The formalism presented deals with an ideal theory behind the device. Several parameters and assumptions were chosen and are given in Supplementary Note 10. In the results section, we tested different optical elements, to challenge these assumptions. We tested the feasibility of physically implementing this theory and ultimately demonstrated that these assumptions are almost entirely correct.

Numerical solution. Matlab was used to solve equation 1. A solver was used to find the parameters of the matrix M, under the conditions that it is a unitary matrix. The matrix M was also confined to the convergence conditions given in Supplementary Note 4. The values of F are dependent on the values of M. The values of F' were chosen according to the relation $F'_{jk} = F_{jk} \exp[i\gamma_j |F_{jk}|^2]$, where γ_j is the scaled Kerr coefficient. This solver also provided the values of γ_j while optimizing the solutions such that the sum $\sum_j \gamma_j$ will be minimized.

Another Matlab script was used to decompose the matrix M into arrays of 2×2 unitary transformations. These values were later used in the simulation of the integrated device.

Optical simulations. Simulations were done in two stages. First, the different optical elements were tested using 'Lumerical FDTD solutions'. The simulations included directional couplers, Mach–Zehnder couplers, and Kerr elements. Details about the FDTD simulations of the different optical elements is given in Supplementary Note 9.

Later, a simulation of the integrated circuit using 'Lumerical Interconnect'. 'Lumerical Interconnect' allows incorporating results acquired with 'Lumerical FDTD Solutions' for a specific basic optical element. 'Lumerical Interconnect' also allows the scripting of its elements which was necessary for elements such as general unitary couplers, Kerr elements and the different detectors.

References

1. Caulfield, H. J. & Dolev, S. Why future supercomputing requires optics. *Nat. Photon.* **4,** 261–263 (2010).
2. Fredkin, E. & Toffoli, T. Conservative logic. *Int. J. Theor. Phys.* **21,** 219–253 (1982).
3. Shamir, J., Caulfield, H. J., Micelli, W. & Seymour, R. J. Optical computing and the fredkin gates. *Appl. Opt.* **25,** 1604–1606 (1986).
4. Zavalin, A. I., Shamir, J., Vikram, C. S. & Caulfield, H. J. Achieving stabilization in interferometric logic operations. *Appl. Opt.* **45,** 360–365 (2006).

5. Hardy, J. & Shamir, J. Optics inspired logic architecture. *Opt. Express.* **15**, 150–165 (2007).

6. Anter, A., Dolev, S. & Shamir, J. *Optical Supercomputing.* pp 92–104 (Springer, 2013).

7. Vasudevan, D. P., Lala, P. K., Di, J. & Parkerson, J. P. Reversible-logic design with online testability. *IEEE Trans. Instrum. Meas.* **55**, 406–414 (2006).

8. Tian, Y. *et al.* Proof of concept of directed or/nor and and/nand logic circuit consisting of two parallel microring resonators. *Opt. Lett.* **36**, 1650–1652 (2011).

9. Xu, Q. & Soref, R Reconfigurable optical directed-logic circuits using microresonator-based optical switches. *Opt. Express* **19**, 5244–5259 (2011).

10. Zhang, L. *et al.* Simultaneous implementation of xor and xnor operations using a directed logic circuit based on two microring resonators. *Opt. Express* **19**, 6524–6540 (2011).

11. Xu, Q., Manipatruni, S., Schmidt, B., Shakya, J. & Lipson, M. 12.5 Gbit/s carrier-injection-based silicon micro-ring silicon modulators. *Opt. Express* **15**, 430–436 (2007).

12. Xu, Q., Schmidt, B., Pradhan, S. & Lipson, M. Micrometre-scale silicon electro-optic modulator. *Nature* **435**, 325–327 (2005).

13. Roy, S., Sethi, P., Topolancik, J. & Vollmer, F. All-optical reversible logic gates with optically controlled bacteriorhodopsin protein-coated microresonators. *Adv. Opt. Technol.* **2012**, 727206 (2012).

14. Gui, C., Wang, J., Xiang, C., Liu, Y. & Li, S. *Asia Communications and Photonics Conference* AS3G–AS33 (Optical Society of America, 2012).

15. Morichetti, F. *et al.* Travelling-wave resonant four-wave mixing breaks the limits of cavity-enhanced all-optical wavelength conversion. *Nat. Commun.* **2**, 296 (2011).

16. Tasca, D. S., Gomes, R. M., Toscano, F., Ribeiro, P. H. S. & Walborn, S. P. Continuous-variable quantum computation with spatial degrees of freedom of photons. *Phys. Rev. A.* **83**, 052325 (2011).

17. Milburn, G. J. Quantum optical fredkin gate. *Phys. Rev. Lett.* **62**, 2124 (1989).

18. Lukishova, S. G. *et al.* Organic photonic bandgap microcavities doped with semiconductor nanocrystals for room-temperature on-demand single-photon sources. *J. Mod. Opt.* **56**, 167–174 (2009).

19. Cardenas, J. *et al.* Low loss etchless silicon photonic waveguides. *Opt. Express* **17**, 4752–4757 (2009).

20. Vlasov, Y. & McNab., S. Losses in single-mode silicon-on-insulator strip waveguides and bends. *Opt. Express* **12**, 1622–1631 (2004).

21. Gu, T. *et al.* Regenerative oscillation and four-wave mixing in graphene optoelectronics. *Nat. Photon.* **6**, 554–559 (2012).

22. Armani, D. K., Kippenberg, T. J., Spillane, S. M. & Vahala, K. J. Ultra-high-q toroid microcavity on a chip. *Nature* **421**, 925–928 (2003).

Acknowledgements

We thank Joseph Shamir from the Technion—Israel Institute of Technology for discussions; S.D. is partially supported by the Rita Altura Trust Chair in Computer Sciences, and the Israel Science Foundation (grant 428/11).

Author contributions

E.C. led the design of the formalism, the design of the simulations and the writing of the manuscript, in all of these processes, the other two authors were also consistently contributing ideas, guidance and supervision.

Additional information

28

Metal-induced rapid transformation of diamond into single and multilayer graphene on wafer scale

Diana Berman[1], Sanket A. Deshmukh[1], Badri Narayanan[1], Subramanian K.R.S. Sankaranarayanan[1], Zhong Yan[2], Alexander A. Balandin[2], Alexander Zinovev[3], Daniel Rosenmann[1] & Anirudha V. Sumant[1]

The degradation of intrinsic properties of graphene during the transfer process constitutes a major challenge in graphene device fabrication, stimulating the need for direct growth of graphene on dielectric substrates. Previous attempts of metal-induced transformation of diamond and silicon carbide into graphene suffers from metal contamination and inability to scale graphene growth over large area. Here, we introduce a direct approach to transform polycrystalline diamond into high-quality graphene layers on wafer scale (4 inch in diameter) using a rapid thermal annealing process facilitated by a nickel, Ni thin film catalyst on top. We show that the process can be tuned to grow single or multilayer graphene with good electronic properties. Molecular dynamics simulations elucidate the mechanism of graphene growth on polycrystalline diamond. In addition, we demonstrate the lateral growth of free-standing graphene over micron-sized pre-fabricated holes, opening exciting opportunities for future graphene/diamond-based electronics.

[1]Center for Nanoscale Materials, Argonne National Laboratory, 9700 South Cass Avenue, Argonne, Illinois 60439, USA. [2]Department of Electrical and Computer Engineering, Materials Science and Engineering Program, Bourns College of Engineering, University of California—Riverside, Riverside, California 92521, USA. [3]Materials Science Division, Argonne National Laboratory, 9700 South Cass Avenue, Argonne, Illinois 60439, USA. Correspondence and requests for materials should be addressed to A.V.S. (email: sumant@anl.gov).

The exceptional electrical, thermal chemical, mechanical and optical properties of graphene are continuing to make strides in many fields[1-4]. The fabrication of devices[5] at the wafer scale for applications particularly related to the electrical and optical properties of graphene relies on a method that can produce large-area single-crystal graphene directly on dielectric substrates. The existing methods based on silicon carbide (SiC)[6-9] and hexagonal boron nitride (h-BN)[10,11] provide direct means of growing high-quality single-crystal graphene on these substrates. Another recently demonstrated method of wafer-scale growth of single-crystal graphene on germanium Ge (111) (refs 12,13) is also promising.

The use of diamond as a substrate for supporting graphene is very appealing, as it offers several advantages and unique properties such as low trap density for charges, high energy for optical phonons, chemical inertness and high thermal conductivity[14,15]. In our previous studies on exfoliated graphene transferred onto ultrananocrystalline diamond (UNCD) thin films and single-crystal diamond[16], we demonstrated that diamond offers an excellent platform for graphene by displaying a breakdown current density reaching up to $10^9 \, \mathrm{A \, cm^{-2}}$ due to the excellent heat dissipation provided by the diamond under layer. It was demonstrated that replacing silicon dioxide (SiO_2) with diamond-like carbon allows one to improve radio-frequency characteristics on graphene transistors[17]. However, diamond-like carbon is an amorphous material with thermal conductivity $K = 0.2 - 3.5 \, \mathrm{W \, m^{-1} \, K^{-1}}$ at room temperature[4], which is a very low value even when compared with SiO_2. Synthetic diamond would be an ideal substrate for graphene devices in terms of its thermal properties, radio-frequency characteristics and compatibility with silicon complementary metal-oxide semiconductor technology. Although the technological benefits of having active material—graphene and substrate—diamond channels consisting of the same element (carbon) are tremendous, achieving this in practice remains a major synthesis and fabrication challenge.

Transition metals, such as nickel[18] and iron[19], have been used to demonstrate growth of multilayer and single-layer graphene on diamond (001) and (111) surfaces, respectively, by annealing them at high temperatures. However, in both cases, the resulting film still contains metals at the graphene–diamond interface, therefore not suitable for direct device fabrication. In addition, the size of graphene is limited by the size of the single-crystal diamond.

In this letter, we report on a process for graphene synthesis, based on rapid thermal annealing of a UNCD film in the presence of a metal catalyst, such as Ni. A distinguishing feature of our process is the elimination of the need for any transfer process. More importantly, we also demonstrate interesting features of lateral growth of suspended graphene over micron-sized holes where graphene appears to be single domain as demonstrated later with selected area electron diffraction technique performed on the graphene transferred from various places onto transmission electron microscope (TEM) grid. This is very important for practical applications, where devices could be fabricated over suspended graphene. In addition, the graphene can be grown selectively just by patterning the deposition of metal at micron scale.

Results

Graphene on UNCD films and their properties. The rapid thermal annealing process (Fig. 1a) presented here is capable of producing graphene on polycrystalline diamond sample and Raman spectroscopy analysis carried out at different locations clearly indicates graphene coverage all over the 4-inch diameter wafer (Fig. 1b–g). We choose UNCD as a substrate as it offers

unique properties that no other diamond film provides due to its small grain size (2–5 nm) and high-volume fraction of grain boundaries. A 50-nm Nickel film was deposited on the UNCD surface by e-beam evaporation. As a result, during thermal annealing process (Fig. 1a), nickel diffuses rapidly through the UNCD grain boundaries as compared with through UNCD grains and sits at the bottom of the UNCD/Si interface. Due to the rapid diffusion of Ni during the temperature ramp, only a small amount of Ni is left over on the top in the form of islands (which also eventually diffuses in diamond), which act as nucleation centres for the graphene growth. The continuous supply of carbon from the UNCD underneath enables the rapid lateral growth of graphene at the wafer scale. The graphene films produced by this method were characterized using X-ray photoemission spectroscopy (XPS) demonstrating its high-quality nature through the clear carbon signature without any indication of the presence of nickel on the surface after the annealing process (Fig. 2). Cross-section scanning electron microscopy (SEM) corroborates the complete nickel segregation to the bottom of the UNCD layer (Supplementary Fig. 1). Also, detailed analysis of the carbon peak in the XPS high-resolution scan (Fig. 2b) demonstrates the presence of only sp^2 bonded carbon on the surface, hence indicating the high-purity nature of the obtained graphene, without any modification (such as oxidation).

Surface analysis of the produced wafer indicates growth of a continuous graphene film (Supplementary Figs 2 and 3). The SEM (Fig. 3a) and atomic force microscopy (AFM) (Fig. 3b) images demonstrate the surface morphology of the as-grown film, displaying a smooth surface (root mean square roughness $R_q = 0.8 \, \mathrm{nm}$) with only occasional folds of the graphene layer (areal density, defined as the area under the wrinkles divided by the total scanning area, is $< 0.5\%$). We believe that these folds as seen on the AFM image (Fig. 3b) may also contribute to the defect peak D (at $\sim 1,335 \, \mathrm{cm^{-1}}$) in the Raman signature[20] (Fig. 3c) along with underlying UNCD. The presence of the folds could be attributed to uneven distribution of stress as a consequence of the fast annealing process.

To produce graphene films with different thicknesses, we varied the temperature of the process from 800 °C to up to 1,000 °C (Fig. 3c). It is important to note subtle differences in the Raman spectra of the single-layer graphene directly grown on UNCD versus single-layer graphene, grown initially on copper by chemical vapour deposition and subsequently transferred onto the UNCD surface. Due to substrate effects caused by the underlying UNCD film[15], the Raman signature shows the presence of a defect peak D (at $\sim 1,335 \, \mathrm{cm^{-1}}$) and low-intensity ratio second order defect peak of 2D (at $\sim 2,650 \, \mathrm{cm^{-1}}$) to G (at $\sim 1,585 \, \mathrm{cm^{-1}}$) bands $I_{2D}/I_G = 0.6$ in comparison with that of a single-layer graphene ($I_{2D}/I_G = 2$) (refs 21,22), grown on copper foil (Fig. 3c). Moreover, significant blue shifts on both, G and 2D bands, as well as an increased 2D-band full-width half maximum are observed for these graphene films on UNCD. This effect may occur due to strong interfacial bonding between graphene and UNCD, which produces strain that manifests itself as the shift observed in the 2D bands. A similar effect has been observed for graphene grown on SiC substrates[23]. For annealing temperatures as low as 800 °C, the Raman signature indicates significantly reduced number of graphene layers[24,25], whereas for annealing at higher temperatures (1,000 °C), the resulting graphene films are multilayered and display a smoother morphology (root mean square roughness decreases from 1.43 nm for single-layer graphene at 800 °C down to 0.8 nm for multilayer graphene at 1,000 °C). Therefore, the thickness of the graphene film can be tuned by varying the annealing temperature. However, it is important to mention that we managed to achieve either single or multilayer graphene only, due to constraints on our experimental setup,

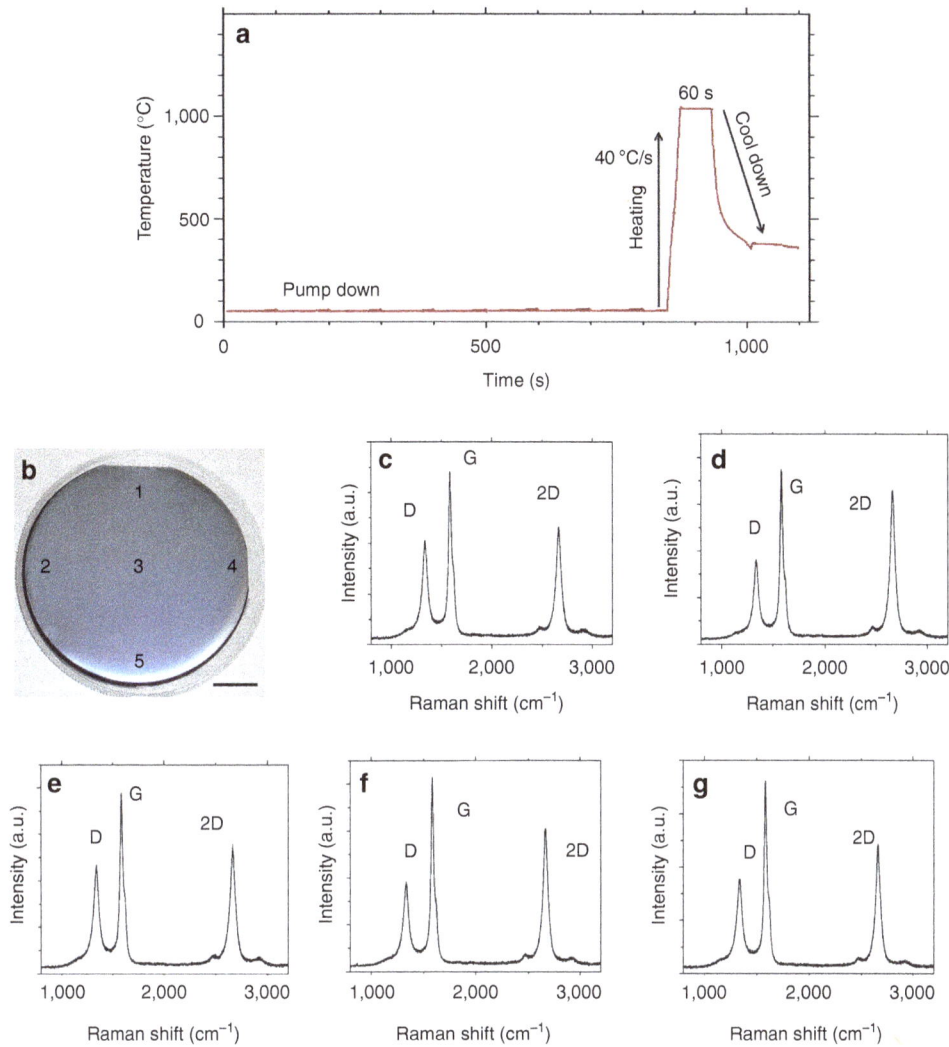

Figure 1 | Wafer-scale growth of graphene on diamond. (**a**) Recipe for the thermal annealing process shows rapid heating of the sample's temperature up to 1,000 °C, keeping the system at high temperature for 60 s followed by rapid cooling down to room temperature. (**b**) A photograph of the 4-inch diameter wafer covered with graphene after the annealing process is shown and the Raman analysis of the sample at five random spots is presented. (**c**) Corresponds to spot 1, (**d**) to spot 2, (**e**) to spot 3, (**f**) to spot 4 and (**g**) to spot 5. Scale bar, 2 cm (**b**).

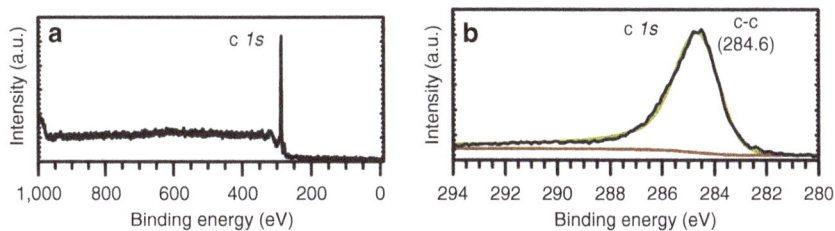

Figure 2 | XPS analysis of the graphene on diamond sample. (**a**) The full survey XPS scan of the sample demonstrates the presence of prominent carbon C $1s$ peak on the surface, indicating absence of residual nickel on the surface as well as no oxygen presence. (**b**) detailed carbon C $1s$ peak confirms sp^2 hybridization of graphene through presence of C–C bond peak highlighted in green. Black colour indicates the collected XPS data and brown highlights the background spectra.

mainly involving the fast, precise control of temperature ramping. The quality of the graphene films grown on UNCD was also evaluated by measuring the sheet resistance using the four-point probe method. The sheet resistance of the obtained graphene/UNCD films varied from 0.09 Ω per square for thick (20–30 nm) graphene samples grown at 1,000 °C to 3.1 Ω per square for single-layer graphene grown at 800 °C. This is the lowest value of sheet resistance reported so far for multilayer, bilayer and

doped free-standing single-layer graphene[26–28], which we attribute to the high quality of the produced graphene.

We have also fabricated graphene top-gate field-effect transistors. The current–voltage (I–V) characteristics of the top-gated devices were measured using a semiconductor parameter analyzer (Fig. 4). The linearity of the source-drain current versus source-drain voltage (I_{ds} versus V_{ds}) curve (Fig. 4a) confirms the high-quality Ohmic contact. The V-shape of source-drain current as a function of the gate bias (I_{ds}–V_g) curve (Fig. 4b) is characteristic for graphene and indicates that the charge carrier type can be switched from electrons to holes by tuning the gate bias. The calculated electron mobility at room temperature $\sim 2{,}000\,cm^2\,V^{-1}\,s^{-1}$ with carrier density of $3.5 \times 10^{12}\,cm^{-2}$ indicates suitability for radio-frequency applications.

In addition to producing high-quality planar sheets of graphene directly onto UNCD surfaces, free-standing graphene films were successfully grown over holes made in diamond (Fig. 5 and Supplementary Fig. 4). Figure 5a demonstrates that

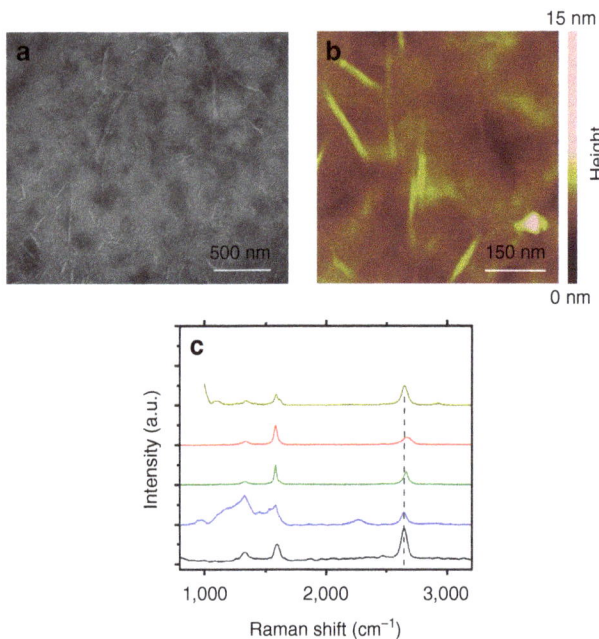

Figure 3 | Characterization of graphene on UNCD sample. (a) The SEM and **(b)** AFM images of grown graphene indicate the formation of the continuous uniform film with occasional folds occurring; **(c)** Raman signature of graphene films produced at different temperatures indicates the variation in the number of grown layers (green line highlighting single layer grown at 800 °C and red line highlighting multilayer grown at 1,000 °C). The Raman signature of graphene grown at 800 °C and then thermal-release tape transferred onto silicon dioxide substrate (dark yellow) confirms the single-layer thickness of graphene. The Raman spectra of single-layer graphene grown on copper (black line) and then transferred on the UNCD film (blue line) are provided for reference. Dashed line highlights the position of 2D peak.

annealing of the Ni/UNCD sample resulted in three holes being completely covered with graphene while one remaining hole was only partially covered. Lateral growth of graphite layers under a thermal gradient was theoretically predicted by Ye et al.[29]. Here, however, we present an experimental evidence of such successful lateral graphene growth. We should emphasize, that the successful growth of free-standing graphene is limited by the size of the hole and the presence of nickel in close proximity. Also, we demonstrate that just by patterning the nickel film, a selective growth of graphene can be obtained at predefined locations on diamond (Supplementary Fig. 5). X-ray diffraction analysis (Supplementary Fig. 6) showed predominantly Ni (111) crystal plane orientation of the evaporated nickel films on UNCD.

To determine the quality of the free-standing graphene films, samples were transferred to TEM grid. The selected area electron diffraction pattern demonstrates the single-domain graphene layer (Fig. 5d,e). Detailed Raman study of the graphene on the hole indicates lowering the intensity of the defect peak D (at $\sim 1{,}335\,cm^{-1}$) and increasing the intensity of 2D (at $\sim 2{,}650\,cm^{-1}$) peak due to the reduced coupling with the UNCD underneath, though the signature does not correspond yet to the single-layer graphene due to shallow shape and small size (500 nm) of the hole (Fig. 6) as compared with the Raman spot size (1.2 μm). In addition, to confirm the single-layer nature of the graphene, we have been able to transfer some part of the graphene on a SiO_2-coated Si substrate (although with some difficulty) using a thermal release tape and the corresponding Raman signature confirms single-layer signature of graphene (Supplementary Fig. 2).

Atomic-scale mechanism of the graphene formation on diamond. To gain insights into the atomic-scale mechanisms underlying the formation of graphene on diamond on rapid annealing Ni/polycrystalline UNCD films, we turn to reactive molecular dynamic (MD) simulations (Supplementary Note 1). Previous studies have successfully employed MD simulations to investigate the atomic-scale processes governing the nucleation and subsequent growth of graphene and carbon nanotubes on transition metal surfaces;[30–32] a comprehensive review on this topic has been recently published by Elliot and co-workers[33]. In those studies, the carbon atoms were deposited on transition metal surfaces (usually Ni) at various rates and were assumed to arise by decomposition of hydrocarbons for example, in chemical vapour deposition (CVD). The Ni/UNCD films used in the present study, however, pose hitherto unanswered questions related to interaction between Ni films and UNCD grain boundaries, as well as associated atomic transport phenomena;

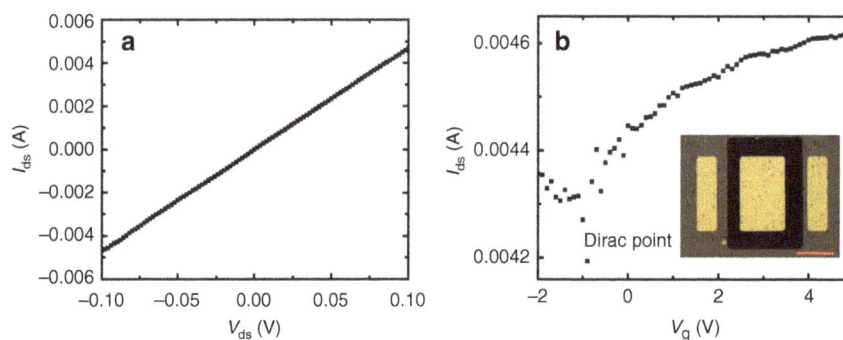

Figure 4 | Top gate field-effect transistor fabricated on graphene on diamond. (a) Two terminal current–voltage characteristics I_{ds} versus V_{ds} at zero gate bias. **(b)** Three-terminal measurement source-drain current versus top-gate bias. Drain voltage is fixed at 0.1 V. Scale bar, 100 μm. The dark part is a dielectric layer of HfO_2 with the size of 200 μm by 300 μm. The gate electrode is 200 μm in width and 120 μm in length. The source and drain electrodes are 200 μm in width and 50 μm in length.

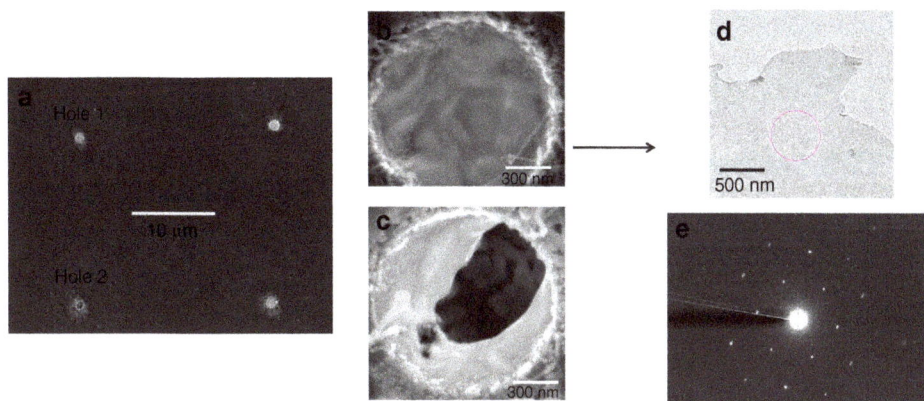

Figure 5 | Lateral growth of single domain free-standing graphene on diamond. (**a**) Successful growth of graphene on four holes is presented. (**b**) Full coverage of hole 1 is demonstrated. (**c**) Graphene partially grown on the hole 2 confirms free-standing nature. (**d**) TEM image confirms a single-domain graphene growth with (**e**) selected area electron diffraction (SAED) pattern indicating diffraction of single-domain graphene film. The circle in **d** highlights the area where SAED pattern was recorded.

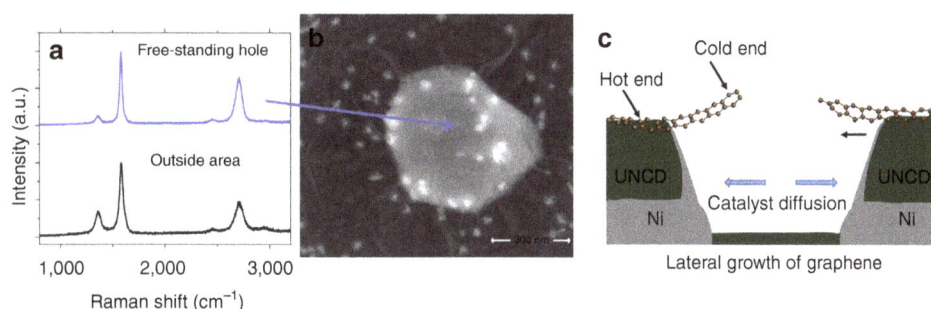

Figure 6 | Growth mechanism of free-standing graphene. (**a**) Changes in the intensities of D and 2D peaks for Raman signature of the graphene on the hole (from $I_D/I_G = 0.3$ and $I_{2D}/I_G = 0.4$ outside the hole to $I_D/I_G = 0.1$ and $I_{2D}/I_G = 0.6$) indicate growth of the free-standing graphene. (**b**) The Raman signature of graphene grown directly on UNCD (outside area) is provided for the reference. (**c**) The schematic of mechanism of free-standing graphene growth is presented.

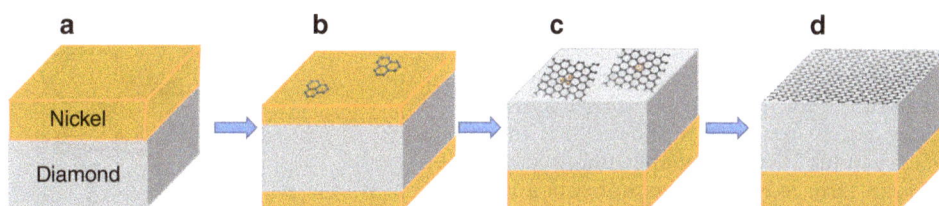

Figure 7 | Schematic of the Ni (111)-facilitated graphene growth. Subsequent growth procedure of graphene starting from initial Ni/UNCD (**a**) configuration with following nucleation (**b**) and increase in domain size (**c**) of graphene up to full coverage (**d**).

such an understanding is crucial to gain precise control over synthesis of monolayer/few-layer graphene from Ni/UNCD films. To investigate these phenomena, we performed two sets of simulations. First, we anneal Ni (111) slab placed on top of a typical grain boundary (GB), namely, Σ13 twist (100) in UNCD at various temperatures 1,200–1,600 K (Supplementary Fig. 7). Here we focus on the counter-diffusion of carbon (C) and Ni at the Ni/UNCD interface, and the evolution of the structure at the interface. The concurrence of these events eventually leads to the nucleation of graphene islands on the free surface. In the second set, we investigate the diffusion of C atoms on free Ni surface, which assists rapid lateral growth of the formed graphene nuclei. For this set, we deposited several C atoms at random locations on Ni (111) and Ni (001) surfaces ensuring identical C coverage. These two sets of simulations provide a holistic picture of the

atomic-scale events leading to the nucleation and growth of graphene on annealing of Ni/UNCD films (Fig. 7).

Discussion

Our first set of simulations on Ni (111)/UNCD interface capture the sequence of atomic scale events leading to graphene nucleation on Ni (111). Initially, Ni atoms penetrate into the UNCD GB, which, in turn, introduces structural disorder near the GB, and eventually leads to rapid amorphization of UNCD (Supplementary Movie 1). Concurrently, the C atoms from the amorphous region dissolve into the Ni thin film; this dissolution continues until the Ni film gets saturated with C atoms. At the annealing temperatures employed in this study (1,200–1,800 K), the solubility of C in Ni is known to be ~1–2% (ref. 34), which is

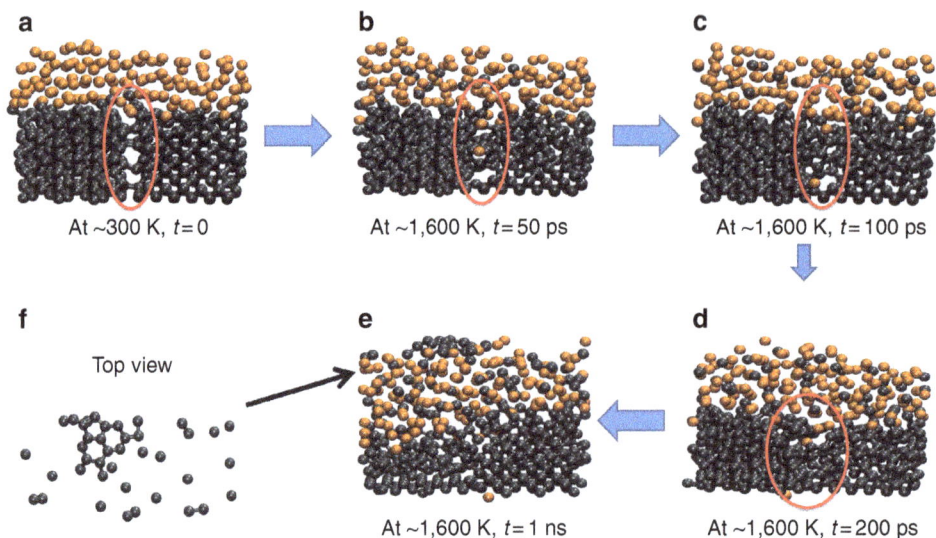

Figure 8 | Carbon diffusion through the diamond grain boundaries mechanism. Carbon diffusion through the nickel film and eventual segregation of nickel through diamond grain boundaries is highlighted through MD simulations: (**a**) initial configuration shows nickel film on two diamond grains, (**b**) heating up to 1,600 K for 50 ps indicates segregation of nickel atoms through the grain boundary, (**c**) at 100 ps one nickel atom diffuses all the way down to the bottom of diamond, (**d**) more nickel atoms segregate at 200 ps and top layer of diamond starts to graphitize, (**e**) indicates the formation of graphitic layer on top of nickel, and (**f**) top view depicts the formation of graphene rings. Carbon atoms are represented by dark grey colour, while nickel atoms by dark orange colour. Red oval highlights the grain boundary.

well described by the reactive force field (ReaxFF) used in this study (Supplementary Note 2)[35]. Beyond the solubility limit, the C atoms proximal to the free Ni surface precipitate out, and eventually result in graphene-like C-rings (Fig. 8). Once these rings (nuclei) form, they grow rapidly in the lateral dimensions through diffusion of C atoms over the free Ni surface (Fig. 9). These events involving C super-saturation of Ni films, and precipitation of C on Ni surfaces are analogous to earlier MD reports on formation of graphene on transition metal surfaces during CVD synthesis[16,36–41]. We note that the rate of nucleation of these graphene-like C rings is governed mainly by the concurrent counter-diffusion of Ni and C, and the consequent amorphization of the Ni/UNCD interface.

We ascertain that the presence of GBs in UNCD assists the nucleation of graphene on free Ni surfaces (Supplementary Note 3). MD simulations using Ni(111) on single-crystal diamond (Supplementary Fig. 8) conclusively showed that the kinetics of Ni dissolution into the UNCD is severely impaired in the absence of GB. This, in turn, restricts the amorphization of the UNCD within 1–2 layers of the Ni/UNCD interface within MD timescales (~ 2 ns), and limits the C flux into Ni slab to a few atoms. Thus, in the absence of UNCD GB, the transport of C to the free Ni surface is negligible.

Our MD simulations (Supplementary Note 4) of surface diffusion of C atoms from random locations on Ni surfaces (Fig. 9) indicate that the diffusion of C over a Ni surface occurs over timescales (in the order of tens of ps) that are much faster than those for nucleating graphene islands (in the order of ns). This result is valid irrespective of the crystallographic orientation of Ni surface. Such a distinct difference between the kinetics of nucleation and growth phases is ideal for forming single-domain graphene sheets. The growth kinetics of graphene is particularly favourable on the Ni (111) surface (Supplementary Fig. 9) owing to low diffusion barrier for C on Ni (111) of 0.5 eV (ref. 42). In addition, the Ni atoms on (111) plane are arranged on a close-packed triangular lattice (Fig. 9c), which closely imitates the honeycomb structure of graphene. This commensurability between Ni (111) substrate and graphene enables the growth of graphene nuclei into large domains, which are mostly uniform

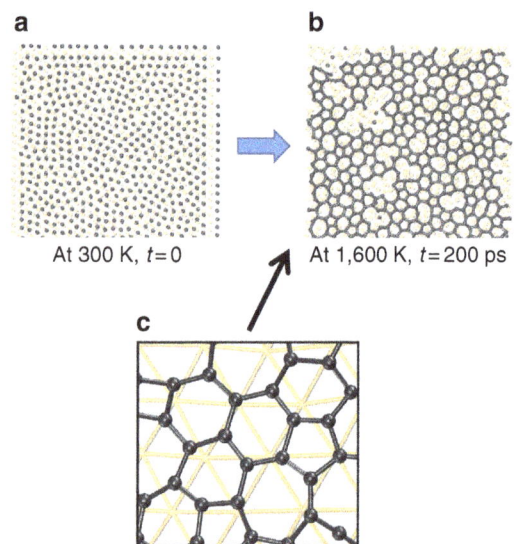

Figure 9 | Growth of graphene film on Ni (111) surface. Growth of uniform graphene film on Ni (111) surface is highlighted. (**a**) Initially, amorphous carbon is placed on Ni (111) film. (**b**) After heating up to 1,600 K for 200 ps formation of ordered graphene rings on the nickel surface is observed with (**c**) high-magnification image indicating lattice configuration of graphene on nickel. Carbon atoms are represented by dark grey colour, while nickel atoms by pale orange colour.

and defect-free. Our simulations suggest that the diffusivity of C through the Ni film is much faster than the rate of depletion of Ni at the Ni–GB interface, thereby ensuring that the (111) preferred orientation of the free surface is preserved.

On the Ni (111) surface, we observe another interesting behaviour exhibited by the C atoms. Initially, the C atoms diffuse rapidly on the Ni (111) surface and bind to neighbouring C atoms in the form of rings, which may either be graphitic (six-membered) or defective (non six-membered, for example, in

Stone–Waals defects). Once a continuous largely planar network of the C atoms is formed, the re-arrangement of C atoms continues, which results in healing of defects (that is, conversion of defective rings into graphitic ones). This re-arrangement can be attributed to the template provided by the underlying Ni (111), and does not occur over Ni (001) (Supplementary Fig. 9). Consequently, the monolayer graphene that forms on Ni (111) surface features the least number of defects, and covers nearly the entire substrate (Fig. 9). We note that these results are robust; the effect of the Ni surface orientation on graphene growth is unaffected by the temperature schedule and the highest temperature achieved during the annealing. The kinetics of re-organization of C-atoms to reduce the defects, as expected, is slow at lower temperatures. These simulation results are in good qualitative agreement with experimental observations. Overall, the general picture that emerges from our experimental results and well-supported by simulations is that graphene nucleation and growth occurs on Ni (111) surface and is much faster than Ni diffusion into UNCD GBs, therefore graphene growth occurs first followed by complete diffusion of Ni into the UNCD bottom leaving intact a uniform graphene layer on top of the UNCD.

In summary, we have demonstrated a process of growing high-quality wafer-scale graphene film directly onto an insulating substrate by rapid thermal annealing of diamond film in the presence of a metal catalyst. This is very promising as our process does not involve transferring of graphene over other substrate. The tuning of the graphene thickness is achieved through the variation of the annealing temperature such that, at lower temperatures (800 °C) the obtained graphene film consists of a single layer, while increasing the annealing temperature to up to 1,000 °C results in multilayer graphene. Reactive MD simulations elucidate an intriguing kinetically controlled mechanism of graphene nucleation and growth facilitated by Ni-induced amorphization of diamond. This fabrication scheme can be easily extended to form suspended graphene membranes directly on diamond. This opens up more avenues to exploit full advantages of intrinsic properties of suspended graphene. Thus, the graphene-on-diamond platform is promising for the development of high performance, energy-efficient nanoelectronic devices.

Methods

Fabrication procedure. UNCD film is grown by a microwave plasma-enhanced CVD (MPCVD) process using an Ar-rich/CH_4 chemistry (Ar (99%)/CH_4 (1%)) on silicon substrate[43]. The thickness of UNCD films varied from 200 nm up to 300 nm, measured with Spectroscopic Reflectometer SR 300.

A 50-nm nickel film was deposited on the UNCD surface by e-beam evaporation using a Kurt J Lesker PVD250 e-beam evaporator with a base pressure of $\sim 10^{-8}$ torr and a deposition rate of 5 Å s^{-1}, which, along with the low base pressure, provides high-purity nickel film (source: pellets 99.995% pure). The thickness of the nickel film was controlled by quartz crystal monitor pre-calibrated before the deposition process using a lithographic patterning, thus guaranteeing <1% thickness variation during the deposition. After this, the samples were processed in a rapid thermal annealing system (ramping speed $40° \text{s}^{-1}$) in a temperature range between 800 and 1,000 °C for 60 s while flowing 500 s.c.c.m. of the forming gas mixture (5% of H_2 and 95% of N_2). This procedure proved to be effective for growing uniform graphene layers on UNCD surfaces.

Free-standing graphene. The free-standing graphene samples were obtained using the specific procedure. Initially, SEM ion milling was used to make holes, 500 nm deep, on UNCD films and then 50 nm Ni film was evaporated. The holes had diameters varying from 300 nm up to 1 µm. Afterwards, to produce lateral growth of graphene over the holes, RTA annealing steps were performed following the same procedure as described for the plain UNCD sample.

Characterization. XPS analysis was performed by a home-built X-ray photo-electron spectrometer, which includes a hemispherical electron energy analyzer of 0.9 eV energy resolution and a non-monochromated Mg K-alpha soft X-ray source at 1,253 eV.

Raman spectroscopy analysis has been performed by an Invia Confocal Raman Microscope using the red laser light (wavelength $\lambda = 633$ nm) to confirm the

formation of a graphene layer on the UNCD surface. The intensity and position of the characteristic G and 2D graphene peaks, as well as the full width at half maximum of the 2D peak in the Raman spectra show the variation in the number of graphene layers grown at different temperatures. The presence of defect peak D can be explained by graphene folds clearly seen in SEM and AFM images (Fig. 3).

SEM images were obtained using a FEI Nova 600 Nanolab dual-beam microscope with focused ion beam used for making the cross-section of the grown layers. To protect the sample, two platinum films are deposited on the surface before making the cross-section. Energy-dispersive X-ray spectroscopy analysis of the layers confirmed the position of nickel layer (as shown in Supplementary Fig. 1) before and after annealing, and carbon layer, confirming nickel segregation inside the UNCD and thus producing direct graphene deposition on the diamond surface.

AFM measurements were performed to demonstrate the three-dimensional structure of the grown layers. For this purpose, the images were acquired by an AFM Veeco Microscope at ambient air conditions (relative humidity $\sim 30\%$) using a n-doped silicon tip in tapping mode.

TEM studies were performed on two different types of samples: grown graphene layer and the cross-section of graphene grown on UNCD. In the first case, the sample was prepared by using thermal release tape to transfer the top layer on the TEM 300 mesh copper grid. In the second case, the sample was prepared using focused ion beam at the SEM instrument, when the sample was cut at the cross-section and then attached to the TEM lift-out grid using focused ion-beam milling. Observation of the single-domain graphene is performed using TEM JEOL JEM-2100F.

X-ray diffraction analysis was performed with Bruker D2 Phaser Diffractometer to demonstrate the crystal orientation of the grown nickel films on UNCD wafer.

Graphene on diamond-based transistors. To test the electron mobility of the CVD-grown graphene on UNCD substrates, we fabricated a number of graphene top-gate field-effect transistors. The top-gate dielectric layer that consists of 20-nm-thick HfO_2 was grown by atomic layer deposition. The electron beam lithography technique was used to define the source, drain and gate electrodes. Ti/Au films with thickness of 10 nm/100 nm were deposited by electron beam evaporation to form the metal contact to the devices.

The current–voltage (I–V) characteristics of the top-gated devices were measured using a semiconductor parameter analyzer. First, the source-drain current versus source-drain voltage was measured at zero gate bias with two terminal measurements. Second, for the source-drain current as a function of the gate bias curve, the source-drain voltage was kept at 0.1 V while the gate bias swept from −2 to 5 V. The Dirac point of the tested device is around −1 V. The carrier mobility of electrons and holes can be extracted using the Drude formula (equation (1)):

$$\mu_{FE} = \frac{g_m}{C_g V_{ds}} \frac{L_g}{W} \qquad (1)$$

where $g_m = \frac{dI_{ds}}{dV_g}$ is the transconductance, $C_g = \frac{\varepsilon_r \varepsilon_0}{d}$ is the gate capacitance per area, where the relative permittivity of HfO_2 is 25. L_g and W are the length and width of the gate, respectively. The electron mobility at $\sim 2,000 \text{ cm}^2 \text{V}^{-1}\text{s}^{-1}$ was obtained at −0.5 V gate bias. The carrier density was $3.5 \times 10^{12} \text{ cm}^{-2}$ at this point. The maximum gate bias applied was +5 V and carrier density was $\sim 4.2 \times 10^{13} \text{ cm}^{-2}$.

Molecular dynamic simulations. We employed reactive MD simulations to investigate the atomic-scale events leading to the formation of graphene domains on the surface of diamond during thermal annealing of thin films comprising of Ni and polycrystalline UNCD. To explore the nucleation and growth of graphene in this complex system within timescales afforded by MD simulations, we performed two sets of simulations.

First, we probe the atomic-scale events leading to the nucleation of graphene (few six-membered C rings) on the free surface on annealing a system consisting of Ni placed onto a UNCD bicrystal containing a typical twist GB. The methods used to generate the computational supercell are detailed in Supplementary Note 1. The thermal annealing was performed at various temperatures in the range 1,200–1,600 K. The conclusions from all these simulations are identical, except as expected, the frequency of occurrence of these events decreases at lower temperatures. The events leading up to the nucleation, as detailed in the main manuscript, consist of penetration of Ni into UNCD GB causing disorder near the GB, propagation of this disorder leading to amorphization of UNCD, transport of C from amorphous region to the Ni film, supersaturation of Ni with carbon, and precipitation of C on the free surface, which forms carbon rings (assisted by commensurability of Ni (111) and graphene).

In the second set, we study the fast kinetics of growth of graphene nucleus (that occurs over tens of ps). Here we deposit carbon atoms at random locations on the free surface of a Ni slab and monitor the formation of C rings on annealing at different temperatures. We have also studied the effect of crystallographic orientation of the Ni slab.

Additional details on the simulations can be found in Supplementary Note 4.

References

1. Bonaccorso, F., Sun, Z., Hasan, T. & Ferrari, A. C. Graphene photonics and optoelectronics. *Nat. Photon.* **4**, 611–622 (2010).
2. Geim, A. K. Graphene status and prospects. *Science* **324**, 1530–1534 (2009).
3. Novoselov, K. S. *et al.* A roadmap for graphene. *Nature* **490**, 192–200 (2012).
4. Balandin, A. A. Thermal properties of graphene and nanostructured carbon materials. *Nat. Mater.* **10**, 569–581 (2011).
5. Liao, L. *et al.* High-speed graphene transistors with a self-aligned nanowire gate. *Nature* **467**, 305–308 (2010).
6. Berger, C. *et al.* Ultrathin epitaxial graphite: 2D electron gas properties and a route toward graphene-based nanoelectronics. *J. Phys. Chem. B* **108**, 19912–19916 (2004).
7. de Heer, W. A. *et al.* Epitaxial graphene. *Solid State Commun.* **143**, 92–100 (2007).
8. Emtsev, K. V. *et al.* Towards wafer-size graphene layers by atmospheric pressure graphitization of silicon carbide. *Nat. Mater.* **8**, 203–207 (2009).
9. Kim, J. *et al.* Layer-resolved graphene transfer via engineered strain layers. *Science* **342**, 833–836 (2013).
10. Son, M., Lim, H., Hong, M. & Choi, H. C. Direct growth of graphene pad on exfoliated hexagonal boron nitride surface. *Nanoscale* **3**, 3089–3093 (2011).
11. Tang, S. *et al.* Nucleation and growth of single crystal graphene on hexagonal boron nitride. *Carbon* **50**, 329–331 (2012).
12. Lee, J.-H. *et al.* Wafer-scale growth of single-crystal monolayer graphene on reusable hydrogen-terminated germanium. *Science* **344**, 286–289 (2014).
13. Wang, G. *et al.* Direct growth of graphene film on germanium substrate. *Sci. Rep.* **3**, 2465 (2013).
14. Auciello, O. *et al.* Are diamonds a MEMS' best friend? *Microw. Mag IEEE* **8**, 61–75 (2007).
15. Sumant, A. V., Auciello, O., Carpick, R. W., Srinivasan, S. & Butler, J. E. Ultrananocrystalline and nanocrystalline diamond thin films for MEMS/NEMS applications. *MRS Bull.* **35**, 281–288 (2010).
16. Yu, J., Liu, G., Sumant, A. V., Goyal, V. & Balandin, A. A. Graphene-on-diamond devices with increased current-carrying capacity: carbon sp2-on-sp3 technology. *Nano Lett.* **12**, 1603–1608 (2012).
17. Wu, Y. *et al.* High-frequency, scaled graphene transistors on diamond-like carbon. *Nature* **472**, 74–78 (2011).
18. García, J. M. *et al.* Multilayer graphene grown by precipitation upon cooling of nickel on diamond. *Carbon NY* **49**, 1006–1012 (2011).
19. Cooil, S. P. *et al.* Iron-mediated growth of epitaxial graphene on SiC and diamond. *Carbon NY* **50**, 5099–5105 (2012).
20. Gupta, A. K., Nisoli, C., Lammert, P. E., Crespi, V. H. & Eklund, P. C. Curvature-induced D-band Raman scattering in folded graphene. *J. Phys. Condens. Matter.* **22**, 334205 (2010).
21. Lenski, D. R. & Fuhrer, M. S. Raman and optical characterization of multilayer turbostratic graphene grown via chemical vapor deposition. *J. Appl. Phys.* **110**, 013720 (2011).
22. Malard, L. M., Pimenta, M. A., Dresselhaus, G. & Dresselhaus, M. S. Raman spectroscopy in graphene. *Phys. Rep.* **473**, 51–87 (2009).
23. Wang, Y. y. *et al.* Raman studies of monolayer graphene: the substrate effect. *J. Phys. Chem. C* **112**, 10637–10640 (2008).
24. Ferrari, A. C. Raman spectroscopy of graphene and graphite: disorder, electron–phonon coupling, doping and nonadiabatic effects. *Solid State Commun.* **143**, 47–57 (2007).
25. Ferrari, A. C. & Basko, D. M. Raman spectroscopy as a versatile tool for studying the properties of graphene. *Nat. Nanotechnol.* **8**, 235–246 (2013).
26. Bae, S. *et al.* Roll-to-roll production of 30-inch graphene films for transparent electrodes. *Nat Nanotechnol.* **5**, 574–578 (2010).
27. Iacopi, F. *et al.* A catalytic alloy approach for graphene on epitaxial SiC on silicon wafers. *J. Mater. Res.* **30**, 609–616 (2015).
28. Liu, Y. P. *et al.* 'Quasi- freestanding' graphene- on- single walled carbon nanotube electrode for applications in organic light- emitting diode. *Small* **10**, 944–949 (2014).
29. Ye, J. & Ruoff, R. S. Graphite fountain: modeling of growth on transition metals under a thermal gradient. *J. Appl. Phys.* **114**, 023516 (2013).
30. Shibuta, Y. & Maruyama, S. Molecular dynamics simulation of formation process of single-walled carbon nanotubes by CCVD method. *Chem. Phys. Lett.* **382**, 381–386 (2003).
31. Neyts, E. C., Shibuta, Y., van Duin, A. C. T. & Bogaerts, A. Catalyzed growth of carbon nanotube with definable chirality by hybrid molecular dynamics-force biased Monte Carlo simulations. *ACS Nano* **4**, 6665–6672 (2010).
32. Neyts, E. C., van Duin, A. C. T. & Bogaerts, A. Changing chirality during single-walled carbon nanotube growth: a reactive molecular dynamics/Monte Carlo study. *J. Am. Chem. Soc.* **133**, 17225–17231 (2011).
33. Elliott, J. A., Shibuta, Y., Amara, H., Bichara, C. & Neyts, E. C. Atomistic modelling of CVD synthesis of carbon nanotubes and graphene. *Nanoscale* **5**, 6662–6676 (2013).
34. Singleton, M. & Nash, P. The C-Ni (carbon-nickel) system. *Bull. Alloy Phase Diagr.* **10**, 121–126 (1989).
35. Mueller, J. E., van Duin, A. C. T. & Goddard, W. A. Development and validation of ReaxFF reactive force field for hydrocarbon chemistry catalyzed by nickel. *J. Phys. Chem. C* **114**, 4939–4949 (2010).
36. Herrera, J. E., Balzano, L., Borgna, A., Alvarez, W. E. & Resasco, D. E. Relationship between the structure/composition of Co-Mo catalysts and their ability to produce single-walled carbon nanotubes by CO disproportionation. *J. Catal.* **204**, 129–145 (2001).
37. Suiter, P. W., Flege, J. I. & Er, E. A. S. Epitaxial graphene on ruthenium. *Nat. Mater.* **7**, 406–411 (2008).
38. Nikolaev, P. *et al.* Gas-phase catalytic growth of single-walled carbon nanotubes from carbon monoxide. *Chem. Phys. Lett.* **313**, 91–97 (1999).
39. Reina, A. *et al.* Large area, few-layer graphene films on arbitrary substrates by chemical vapor deposition. *Nano Lett.* **9**, 30–35 (2009).
40. Keun Soo, K. *et al.* Large-scale pattern growth of graphene films for stretchable transparent electrodes. *Nature* **457**, 706–710 (2009).
41. Amara, H., Bichara, C. & Ducastelle, F. Understanding the nucleation mechanisms of carbon nanotubes in catalytic chemical vapor deposition. *Phys. Rev. Lett.* **100**, 056105 (2008).
42. Hofmann, S., Csanyi, G., Ferrari, A. C., Payne, M. C. & Robertson, J. Surface diffusion: the low activation energy path for nanotube growth. *Phys. Rev. Lett.* **95**, 036101 (2005).
43. Sumant, A. V. *et al.* Toward the ultimate tribological interface: surface chemistry and nanotribology of ultrananocrystalline diamond. *Adv. Mater.* **17**, 1039–1045 (2005).

Acknowledgements

The help in the TEM data collection by Yuzi Liu is greatly appreciated. Use of the Center for Nanoscale Materials was supported by the US Department of Energy, Office of Science, Office of Basic Energy Sciences, under contract no. DE-AC02-06CH11357. This research used resources of the National Energy Research Scientific Computing Center, which is supported by the Office of Science of the US Department of Energy under contract no. DE-AC02-05CH11231.This research used resources of the Argonne Leadership Computing Facility at Argonne National Laboratory, which is supported by the Office of Science of the US Department of Energy under contract no. DE-AC02-06CH11357. XPS study was supported by the U.S. Department of Energy, Office of Science, Basic Energy Sciences, Materials Sciences and Engineering Division.

Author contributions

A.V.S. conceived the idea, helped in the data analysis of experimental results and directed the project. D.B. performed the experiments and analysed the data. S.A.D., B.N. and S.K.R.S.S. devised and performed the molecular dynamics simulations and performed all the related data analysis. S.K.R.S.S. guided the simulation effort. Z.Y. and A.A.B. performed device fabrication and characterization. A.Z. guided XPS analysis. D.R. performed the evaporation of Ni films. All the authors, including A.V.S., D.B., S.A.D., B.N., S.K.R.S.S. and A.A.B., contributed in discussing the results and composing the manuscript.

Additional information

29

Enhanced electronic properties in mesoporous TiO$_2$ via lithium doping for high-efficiency perovskite solar cells

Fabrizio Giordano[1], Antonio Abate[1], Juan Pablo Correa Baena[2], Michael Saliba[3], Taisuke Matsui[4], Sang Hyuk Im[5], Shaik M. Zakeeruddin[1], Mohammad Khaja Nazeeruddin[3], Anders Hagfeldt[2] & Michael Graetzel[1]

Perovskite solar cells are one of the most promising photovoltaic technologies with their extraordinary progress in efficiency and the simple processes required to produce them. However, the frequent presence of a pronounced hysteresis in the current voltage characteristic of these devices arises concerns on the intrinsic stability of organo-metal halides, challenging the reliability of technology itself. Here, we show that n-doping of mesoporous TiO$_2$ is accomplished by facile post treatment of the films with lithium salts. We demonstrate that the Li-doped TiO$_2$ electrodes exhibit superior electronic properties, by reducing electronic trap states enabling faster electron transport. Perovskite solar cells prepared using the Li-doped films as scaffold to host the CH$_3$NH$_3$PbI$_3$ light harvester produce substantially higher performances compared with undoped electrodes, improving the power conversion efficiency from 17 to over 19% with negligible hysteretic behaviour (lower than 0.3%).

[1] Laboratory of Photonics and Interfaces, Institute of Chemical Sciences and Engineering, School of Basic Sciences, Ecole Polytechnique Fédérale de Lausanne (EPFL), Lausanne CH-1015, Switzerland. [2] Laboratoire des sciences photomoléculaires, Institute of Chemical Sciences and Engineering, School of Basic Sciences, Ecole Polytechnique Fédérale de Lausanne (EPFL), Lausanne CH-1015, Switzerland. [3] Group for Molecular Engineering of Functional Materials, Institute of Chemical Sciences and Engineering, School of Basic Sciences, Ecole Polytechnique Fédérale de Lausanne (EPFL), Lausanne CH-1015, Switzerland. [4] Advanced Research Division, Materials Research Laboratory, Panasonic Corporation, 1006, Kadoma, Kadoma City, Osaka 571-8501, Japan. [5] Department of Chemical Engineering, Kyung Hee University, 1732 Deogyeong-daero, Giheung-gu, Yongin-si, Gyeonggi-do 446-701, Republic of Korea. Correspondence and requests for materials should be addressed to F.G. (email: fabrizio.giordano@epfl.ch) or to M.G. (email: michael.graetzel@epfl.ch).

Perovskite-based solar cells (PSCs) have made impressive strides in just a few years with maximum power conversion efficiencies (PCEs) jumping from 3.8% (ref. 1) in 2009 to 20.1% (ref. 2) in 2015. Even though further improvements are still expected[3], such rapid progress is unprecedented for any photovoltaic (PV) material. For instance, silicon, GaAs, CIGS and CdTe required decades to fully realize their potential as solar cells[4].

Perovskites comprise a large family of crystalline materials, where the most commonly used for solar cells have an ABX_3 chemical composition containing an organic cation A, such as methylammonium (MA) or formamidinium (FA)[5,6], a divalent metal B, such as Pb or Sn[7,8], and a halide X, such as Br or I. These organic–inorganic perovskites can be processed by a large number of techniques ranging from spin coating[5], dip coating[9], two-step interdiffusion[10], chemical vapour deposition[11], spray pyrolysis[12], atomic layer deposition[13], ink-jet printing[14], to thermal evaporation[15,16]. The PV performances have been attributed to their outstanding optoelectronic properties such as remarkably high absorption over the visible spectrum[7], charge carrier diffusion lengths in the micrometre-range[17–19] implying a sharp optical band edge, and a tuneable band gap from 1.1 to 2.3 eV by interchanging the above cations[2,20], metals[21,22] and/or halides[23].

Recently, Jeon et al. achieved one of the highest certified PCEs of 17.9% by using the mixed halide and cation formulation, $(FAPbI_3)_{0.85}(MAPbBr_3)_{0.15}$ (ref. 24). Their record solar cell architecture contains a mesoporous TiO_2 layer, which is infiltrated by a liquid perovskite precursor solution forming the solid perovskite film after subsequent annealing. Mesoporous TiO_2 has been widely used for high-surface-area electrodes, in optoelectronic applications[25] and, in particular, in dye-sensitized solar cells (DSSCs)[26], where they have been demonstrated to collect and transport electrons photoinjected from a surface-adsorbed sensitizer. One strategy to improve DSSCs was to enhance the electron transport of the mesoporous TiO_2 making use of substiutional dopants[27–37]. Also, lithium intercalation has been employed to lower the conduction band edge of TiO_2 facilitating electron injection and transport in the mesoporous TiO_2 (refs 38–41). By analogy, it may be expected that the introduction of n-dopants would also enhance the performance of PSCs. However, so far very few studies have examined the doping effect on the electron transport within the mesoporous TiO_2 scaffold employed as PCSs[42]. This may be attributed to the perovskites already being excellent charge transporting materials inherently suited for electron conduction in high-efficiency solar cells[17]. This makes it challenging to investigate the effects of enhanced charge transport in mesoporous TiO_2 as this improvement would only have a discernible impact for embodiments when the perovskite is operating at its limit.

Therefore, at first glance, it may be assumed that highly efficient PSCs could be achieved by depositing a perovskite film directly on a thin TiO_2 compact layer, which would effectively work as an electron-selective contact[15]. However, flat junction PSCs prepared on a compact TiO_2 suffer from relatively poor charge collection efficiency under steady-state forward voltage bias[43]. Although Wojciechowski et al. have shown significant improvement of this architecture by modifying the TiO_2-perovskite flat heterojunction with fullerene derivatives[44], the steady-state PCE of TiO_2-based flat PSCs is still substantially lower than those based on mesoporous TiO_2. The planar PSC architecutres are also plagued by severe hysteresis in the J–V curve rendering it very difficult to determine their solar to electric PCEs. For this reason, none of the TiO_2-based flat PSC architectures have been certified so far. Recently, Guillén et al. demonstrated that the charge collection in a PSC with mesoporous TiO_2 involves two separate electron transport pathways: one through the perovskite, and one through the mesoporous TiO_2 (refs 45,46). Following this, Heo et al. reported that surface deposition by sintering of bis(trifluoromethane)sulfonimide lithium salt (Li-TFSI) onto mesoporous TiO_2 results in substantial improvements in the PCE and suppression of the hysteresis.

In this work, we show that mesoporous TiO_2 can be n-doped in a facile and effective way by a similar lithium salt surface treatment. We show that the Li-doping enables faster electron transport within the mesoporous TiO_2 electrodes and demonstrate that PSCs prepared on such electrodes achieve substantially higher performances compared with undoped electrodes, improving PCEs from 17 to 19.3% with negligible hysteresis behaviour (lower than 0.3%).

Results

X-ray photoelectron spectroscopy analysis. We applied the lithium ion surface treatment of the meso-porous TiO_2 layer via spin coating of a LiTFSI solution (see the Methods section). After the deposition and solvent evaporation, the substrates were sintered at 450 °C for 30 min. The introduction of Li^+ ions by the thermal diffusion modifies the surface of the particles. Together with the doping effect of the TiO_2 an overlayer of LiO_2 or LiOH could be formed[47]. Interestingly, also the formation of spinel structures like $Li_4Ti_5O_{12}$ was observed with a synthetic procedure similar to the one employed here[48].

We used X-ray photoelectron spectroscopy (XPS) to study the elemental composition of the Li-treated and untreated TiO_2 after sintering. No traces of sulphur or fluorine from the LiTFSI precursor for the treated sample were detected and the high-resolution spectra for C 1s showed no difference between the Li-treated and untreated samples as seen in Supplementary Fig. 1. The O 1s spectra in Fig. 1a,c show that Li-treated TiO_2 (Fig. 1a) has a more pronounced shoulder at the higher energy of the main peak compared with the untreated TiO_2 (Fig. 1c). The deconvolution of this signal reveals a second small peak at 531.2 eV for the untreated sample and a much more pronounced one for the Li-treated one that has been previously assigned to the oxygen interaction with the lithium[47]. The Ti 2p spectra can be found in Supplementary Fig. 2, and no difference was detected for the treated and untreated samples. On the right hand side of Fig. 1, the spectrum from 50 to 65 eV for the Li-treated TiO_2 (Fig. 1b) electrodes shows a peak corresponding to the Ti 3s and a weak, but distinct, signal for Li 1s. In the untreated TiO_2 (Fig. 1d),

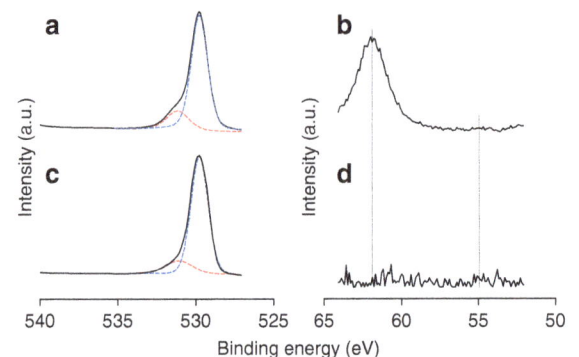

Figure 1 | X-ray photoelectron spectroscopy. Doped and undoped mesoporous TiO_2 layers for the O 1s peaks Li doped (**a**) and the undoped control (**c**), Ti 3s and Li 1s peaks slightly visible at 55 eV for Li-doped (**b**) and the signal for the undoped TiO_2 here shown as reference, which reveals the absence of the peaks related to Ti 3s (dashed line at 61.5 eV) and Li 1s (**d**), dashed line at 54.9 eV.

no such signals, measured over a series of samples, could be detected in this energy region and it is here shown as noise. The Ti $3s$ peak at 61.5 eV reveals the presence of Ti^{3+} for Li-treated electrodes and thus indicates that the Li^+ treatment induces a partial reduction of Ti^{4+} to Ti^{3+} within the TiO_2 lattice[39,49], which is not seen for the untreated electrode. Pathak et al. demonstrated that a small amount of species with valency $+3$ can passivate the electronic defects or trap states that originate from oxygen vacancies within the TiO_2 lattice[37]. Accordingly, the Li^+ doping mechanism is consistent with a passivation of electronic trap states resulting in improved charge transport properties and thus in better performing mesoporous TiO_2 electrodes[50].

Charge extraction and electron transport analysis. To study the impact of the Li^+ doping on the electronic states and the charge transport within the TiO_2, we prepared solid-state DSSCs using Li^+-doped mesoporous TiO_2 as electron transporting layer. DSSCs were prepared according to the previously reported procedures[51].

In the DSSC field, charge extraction is a well-established light-assisted technique, which qualitatively draws the density of state distribution below the TiO_2 conduction band[52]. In Fig. 2a, we report the charge extracted from the DSSCs at open circuit condition as a function of the open circuit voltage. At the same open circuit voltage, the devices employing $Li:TiO_2$ hold significantly less charges than the undoped TiO_2 (for example at 0.46 V, $Li:TiO_2$ and undoped TiO_2 holds 44.3 and 154.3 nC, respectively). This suggests that the Li^+ doping reduces the concentration of sub-bandgap states in the TiO_2 supporting our previous conjecture that a partial reduction of Ti^{4+} to Ti^{3+} passivates the trapping states associated with oxygen vacancies within the TiO_2 lattice.

From the same DSSCs, we extracted the charge transport time constant by intensity-modulated photocurrent spectroscopy[45,53,54]. The cell was biased at short circuit under light and the time constants were measured for different light intensities. In Fig. 2b, we compare the transport time constants for DSSCs prepared with Li^+-treated and untreated TiO_2 electrodes. Li^+-treated devices display up to over one order of magnitude faster charge transport than the untreated devices over the whole range of current densities. To explain this trend, we note that charge transport in DSSCs is controlled by the electron transport in the TiO_2 (ref. 41). In particular, it has been shown

that the electronic transport in the TiO_2 is limited by the temporary localization of electrons within sub-bandgap states, which can be passivated with different doping mechanisms[53,55]. Our results indicate that Li^+ doping also reduces the concentration of sub-bandgap states (Fig. 2a) and improves the electronic transport within mesoporous TiO_2 electrodes (Fig. 2b). As we observe the presence of Li^+ ions within the TiO_2 lattice (Fig. 1), we can regard this method as an effective way of n-doping via a facile post treatment of the mesoporous TiO_2 films. Even though such doping strategies may be effective for improving DSSC performance, they may not necessarily benefit PSCs, as the latter already accomplish fast charge transport within the perovskite layer[17]. To further elucidate the effect of Li doping on the electron transport in mesoscopic perovskite solar cells, we performed light intensity-modulated photocurrent spectroscopy also on PSC[56-58].

In Fig. 3a,b, we report the intensity modulated photocurrent spectroscopy (IMPS) spectra at different light intensities for PSCs prepared with mesoporous TiO_2 films with and without the Li^+-doping. The Nyquist plot of the untreated sample shows two distinct semicircles related to two different transport processes. In the treated sample, on the contrary, the slow time component (semicircle on the right) merges with the faster component (semicircle on the left) confirming that the TiO_2 transport time is faster than that one observed in the control. These processes are resolved in frequency in Fig. 3b, where the imaginary part of IMPS is plotted against the frequency. The electron transport time for a given pathway equals the inverse of the frequency of the corresponding peak[53]. In the case of the untreated TiO_2 films, we observe two distinct peaks at 10^3 and 10^4–10^5 Hz. Guillén et al. correlated these peaks with two separate electron transport pathways, running in parallel, involving the mesoporous TiO_2 and the perovskite, respectively[45]. The peak at low frequency (10^3 Hz) is directly linked to the electron transport in TiO_2 and it shifts to higher frequency (faster transport) as the photocurrent density increases, as revealed also by the trends in Fig. 2b. By contrast, the high-frequency peak (10^4–10^5 Hz) is related to the charge transport through the perovskite and does not shift significantly with increasing the photocurrent density[45]. Therefore, at higher current density, the TiO_2 and the perovskite peaks overlap at around 10^4–10^5 Hz. Interestingly, for the Li^+-doped PSCs, the TiO_2 peak at low frequency does not show up even for extremely low current density, such as $0.07\,mA\,cm^{-2}$. As Li^+-doping of TiO_2 enables faster electron injection and transport in DSSCs, we infer from the lack of a low-

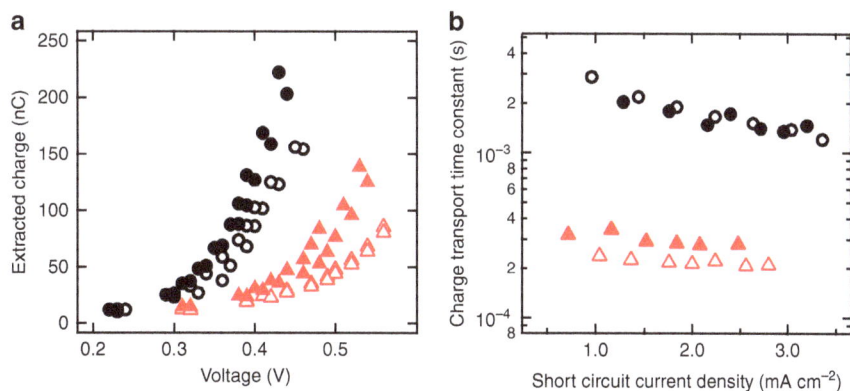

Figure 2 | Charge extraction and electron transport time constant. (a) Charge extracted at open circuit as function of the voltage for no treated samples (black) and Li-doped samples (red) and **(b)** charge transport lifetime as function of the short circuit current density for solid-state dye-sensitized solar cells prepared without and with Li-doped mesoporous TiO_2 (black and red markers, respectively). Two samples per condition are shown: open triangle, sample 1 Li-doped; closed triangle, sample 2 Li-doped; open circle, sample 1 no treated; closed circle, sample 2 no treated. For the charge extraction measurement **(a)**, each sample was measured twice.

a

b

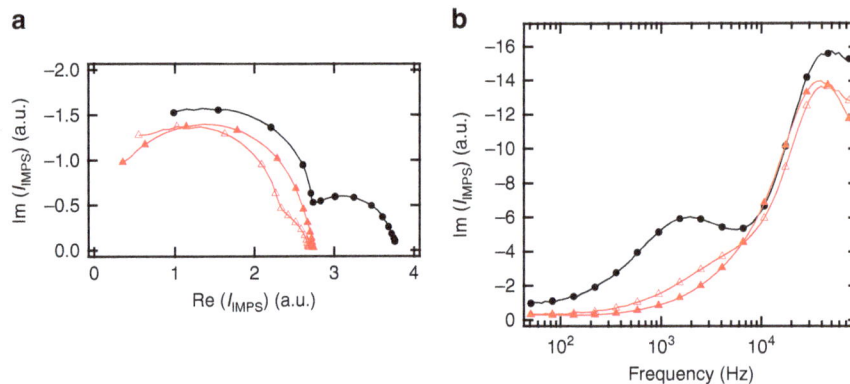

Figure 3 | IMPS spectra. Nyquist plot (**a**) for PSCs prepared with no treated TiO_2 electrode (in black at 0.15 mA cm^{-2}) and with Li-doped TiO_2 electrode at 0.07 mA cm^{-2} (open triangle) and 0.26 mA cm^{-2} (closed triangle). (**b**) Imaginary component of the same IMPS shown in **a** versus the frequency.

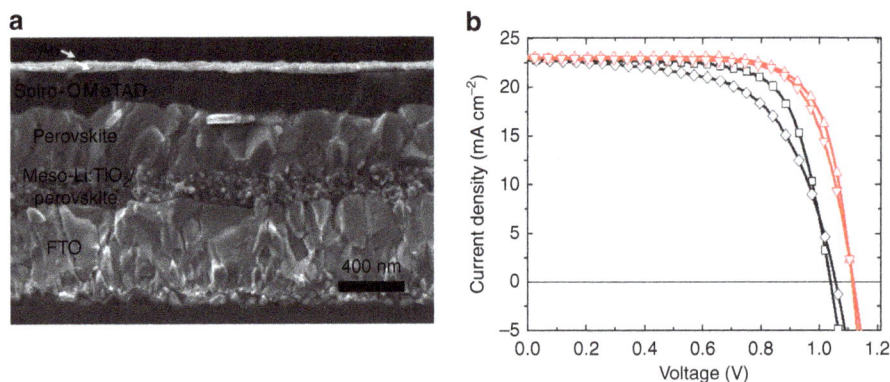

a

b

Figure 4 | Morphology and J–V curves of the solar cells. (**a**) SEM cross-sectional image of the device. (**b**) Current density–voltage curves of the solar cells with and without the Li$^+$ doping (red and black curve, respectively) collected under AM1.5 simulated sun light. Devices were masked with a black metal aperture of 0.16 cm^2 to define the active area. The curves were recorded scanning at 0.01 V s^{-1} from forward bias to short circuit condition and *vice versa* with no device preconditioning such as light soaking or holding at forward voltage bias. Legend: black diamond, control forward scan; black square, control reverse scan; red down-pointing triangle, Li$^+$-doped forward scan; red up-pointing triangle, Li$^+$-doped reverse scan.

frequency TiO_2 peak for the Li$^+$-treated PSCs (Fig. 3) that the electron transport through the TiO_2 occurs at a similar rate as through the perovskite. This result also suggests that in PSCs employing doped TiO_2 further improvements may originate from applying hole transporting materials with a higher hole mobility than the commonly used spiro-MeOTAD.

PV characterization. In Fig. 4a, we show the stack architecture used in this study composed of fluorinated tin oxide (FTO)/compact-TiO_2/mesoporous-TiO_2/perovskite/Spiro MeOTAD/gold. An X-ray diffraction pattern of the perovskite employed in solar cell fabrication is shown in Supplementary Fig. 3, where the peroskvite structure is detected. Figure 4b shows a typical current density–voltage (J–V) characteristic that we measured for PSCs with and without Li$^+$-doped mesoporous TiO_2. In Table 1, we summarize the device performance parameters, as extracted from the J–V curves in Fig. 4b, and the light intensity measured during each J–V scan. We note that the difference between the backward and the forward scans is significantly larger in the control than in the Li$^+$-doped device. The short circuit currents are quite similar, whereas the Li$^+$-doped device showed significantly larger open circuit voltage (about 60 mV) and 0.11 units higher fill factor. The current mismatch between the measured current density and the current density calculated by the incident photon-to-electron conversion efficiency (IPCE) over the Solar AM1.5 G spectrum was less than 2% (Supplementary Figs 4 and 5). The

Table 1 | Solar cell performance parameters.

	Scan direction	J_{sc} (mA cm^{-2})	V_{oc} (V)	FF	PCE (%)	Light intensity (mW cm^{-2})
Li$^+$	Backward	23.0	1.114	0.74	19.3	98.1
doped	Forward	23.1	1.118	0.72	19.0	
Control	Backward	22.7	1.038	0.72	17.1	99.4
	Forward	22.7	1.056	0.61	14.7	

FF, fill factor; PCE, power conversion efficiency.
Short circuit photocurrent (J_{sc}), open circuit voltage (V_{oc}), FF extracted from the data in Fig. 4.

overall power conversion efficiency of devices employing the Li$^+$-doped TiO_2 electrodes was systematically higher than the devices employing the undoped scaffold (Supplementary Fig. 6). This result is consistent with the fact that the Li$^+$-doping decreases the number of deep traps, which act as recombination centres and induces faster charge transport within the TiO_2, improving the open circuit voltage and fill factor, respectively.

Discussion

In summary, we demonstrated a doping mechanism that allowed preparing mesoporous TiO_2 electrodes with superior electron properties. The doping can be accomplished with a facile post treatment of the mesoporous TiO_2 making use of lithium salts to induce a partial reduction of Ti^{4+} to Ti^{3+} within the TiO_2 lattice and passivating electronic defect states acting as nonradiative recombination centres. We exploit the Li^{+}-doped mesoporous TiO_2 electrodes to improve the maximum power conversion efficiency of perovskite solar cells from 17% to over 19%, which is comparable to the highest values reported in the literature.

Methods

Substrate preparation and Li-doping TiO2.
Nippon Sheet Glass of $10\,\Omega\,sq^{-1}$ was cleaned by sonication in 2% Hellmanex water solution for 30 min. After rinsing with deionized water and ethanol, the substrates were further cleaned with ultraviolet ozone treatment for 15 min. Then, 30-nm TiO_2 compact layer was deposited on FTO via spray pyrolysis at 450 °C from a precursor solution of titanium diisopropoxide bis(acetylacetonate) in anhydrous ethanol. After the spraying, the substrates were left at 450 °C for 45 min and left to cool down to room temperature. Then, mesoporous TiO_2 layer was deposited by spin coating for 20 s at 4,000 r.p.m. with a ramp of $2,000\,r.p.m.\,s^{-1}$, using 30 nm particle paste (Dyesol 30 NR-D) diluted in ethanol to achieve 150- to 200-nm-thick layer. After the spin coating, the substrates were immediately dried at 100 °C for 10 min and then sintered again at 450 °C for 30 min under dry air flow.

Li-doping of mesoporous TiO_2 was accomplished by spin coating a 0.1 M solution of Li-TFSI in acetonitrile. The solution was prepared freshly before the application in nitrogen atmosphere. 150 µl were poured on $1.4 \times 2.4\,cm^2$ substrate. After 5 s of loading time, the spinning programme started with an acceleration of $1,000\,r.p.m.\,s^{-1}$ to a final speed of 3,000 r.p.m., the substrate was left spinning for 30 s. Both Li^{+}-doped and undoped electrodes were completed with a second calcination step at 450 °C for 30 min. After cooling down to 150 °C, the substrates were immediately transferred in a nitrogen atmosphere glove box for the deposition of the perovskite films.

DSSC preparation procedure.
Glass substrates for solid-state DSSCs were prepared following the same procedure used for PSC. 900 nm of mesoporous TiO_2 (Dyesol 30 nrd ethanol diluited) were deposited by spin coating at 4,000 r.p.m. After sintering, the substrates were cooled down to 70 °C and immersed in 0.1 mM solution of dye (Y123) in 1:1 mixture of acetonitrile and *tert*-butyl alcohol for 30 min. After the dyed films were rinsed in abundant acetonitrile, the hole conductor was applied by spincoating at 2,000 r.p.m. for 20 s. The hole transport composition and the following steps to complete the DSSCs were identical to what used for PSCs.

Perovskite precursor solution and film preparation.
The perovskite films were deposited from a precursor solution containing FAI (1 M), PbI_2 (1.1 M), MABr (0.2 M) and $PbBr_2$ (0.2 M) in anhydrous dimethylformamide/ dimethylsulphoxide (4:1 (v:v)) solution. The perovskite solution was spin coated in a two-step programme at 1,000 and 4,000 r.p.m. for 10 and 30 s, respectively. During the second step, 100 µl of clorobenzene was poured on the spinning substrate 15 s prior the end of the programme. The substrates were then annealed at 100 °C for 1 h in nitrogen-filled glove box.

We note that the perovskite precursor solution for this paper was prepared with a different composition from what reported by Jeon et al., who used an equimolar amount of FAI and PbI_2 to achieve a certified PCE of 17.9% with the mixed halide and cation formulation, $(FAPbI_3)_{0.85}(MAPbBr_3)_{0.15}$ (ref. 24). Interestingly, we observed a systematic improvement in PCE moving away from the equimolar concentration for FAI and PbI_2 towards 10 mol% lower stoichiometric amount of FAI.

Hole transporting layer and top electrode.
After the perovskite annealing, the substrates were cooled down for few minutes and a spirofluorene-linked methoxy triphenylamines (spiro-OMeTAD, from Merck) solution (70 mM in chlorobenzene) was spun at 4,000 r.p.m. for 20 s. The spiro-OMeTAD was doped with bis(trifluoromethylsulfonyl)imide lithium salt (Li-TFSI, from Aldrich), tris(2-(1H-pyrazol-1-yl)-4-*tert*-butylpyridine)- cobalt(III) tris(bis(trifluoromethylsulfonyl)imide) (FK209, from Dyenamo) and 4-*tert*-Butylpyridine (TBP, from Aldrich)[40,41,59]. The molar ratio of additives for spiro-OMeTAD was: 0.5, 0.03 and 3.3 for Li-TFSI, FK209 and TBP, respectively. Finally, 70–80 nm of gold top electrode was thermally evaporated under high vacuum.

PV device testing.
The solar cells were measured using a 450-W xenon light source (Oriel). The spectral mismatch between AM1.5G and the simulated illumination was reduced by the use of a Schott K113 Tempax filter (Präzisions Glas & Optik GmbH). The light intensity was calibrated with a Si photodiode equipped with an IR-cutoff filter (KG3, Schott) and it was recorded during each measurement. Current–voltage characteristics of the cells were obtained by applying an external voltage bias while measuring the current response with a

digital source meter (Keithley 2400). The voltage scan rate was $10\,mV\,s^{-1}$ and no device preconditioning was applied before starting the measurement, such as light soaking or forward voltage bias applied for long time. The starting voltage was determined as the potential at which the cells furnishes 1 mA in forward bias, no equilibration time was used. The cells were masked with a black metal mask (0.16 cm²) to estimate the active area and reduce the influence of the scattered light. The devices were characterized 2 days after their preparation.

Charge extraction technique.
Charge extraction measurement were performed with Autolab potentiostat PGSTAT30 driven by NOVA software. The procedure for the charge extraction comprised four steps. First, the cell was kept for 10 s in dark at short circuit. At this stage, the carriers eventually accumulated in the intrinsic capacitances of the device were discharged. Then, the potential was brought to open circuit and the light was switched on for an equilibration time of 10 s. The light was then switched off and the open circuit voltage decay was monitored for a defined decay time T_d. In the last step, the cell was brought back to short circuit condition from V_{TD} (voltage at time T_d) and the discharge current was measured. The integration over the time (starting from T_d) of this current gave the value of charge stored at the voltage V_{TD}.

References

1. Kojima, A., Teshima, K., Shirai, Y. & Miyasaka, T. Organometal halide perovskites as visible-light sensitizers for photovoltaic cells. *J. Am. Chem. Soc.* **131**, 6050–6051 (2009).
2. Yang, W. S. *et al.* High-performance photovoltaic perovskite layers fabricated through intramolecular exchange. *Science* **348**, 1234–1237 (2015).
3. Park, N. G. Perovskite solar cells:an emerging photovoltaic technology. *Mater. Today* **18**, 65–72 (2015).
4. NREL Efficiency chart http://www.nrel.gov/ncpv/images/efficiency_chart.jpg (2015).
5. Lee, M. M., Teuscher, J., Miyasaka, T., Murakami, T. N. & Snaith, H. J. Efficient hybrid solar cells based on meso-superstructured organometal halide perovskites. *Science* **338**, 643–647 (2012).
6. Eperon, G. E. *et al.* Formamidinium lead trihalide: a broadly tunable perovskite for efficient planar heterojunction solar cells. *Energy Environ. Sci.* **7**, 982–988 (2014).
7. Hao, F., Stoumpos, C. C., Cao, D. H., Chang, R. P. & Kanatzidis, M. G. Lead-free solid-state organic-inorganic halide perovskite solar cells. *Nat. Photonics* **8**, 489–494 (2014).
8. Noel, N. K. *et al.* Lead-free organic–inorganic tin halide perovskites for photovoltaic applications. *Energy Environ. Sci.* **7**, 3061–3068 (2014).
9. Burschka, J. *et al.* Sequential deposition as a route to high-performance perovskite-sensitized solar cells. *Nature* **499**, 316–319 (2013).
10. Xiao, Z. *et al.* Efficient, high yield perovskite photovoltaic devices grown by interdiffusion of solution-processed precursor stacking layers. *Energy Environ. Sci.* **7**, 2619–2623 (2014).
11. Chen, Q. *et al.* Planar heterojunction perovskite solar cells via vapor-assisted solution process. *J. Am. Chem. Soc.* **136**, 622–625 (2013).
12. Barrows, A. T. *et al.* Efficient planar heterojunction mixed-halide perovskite solar cells deposited via spray-deposition. *Energy Environ. Sci.* **7**, 2944–2950 (2014).
13. Sutherland, B. R. *et al.* Perovskite thin films via atomic layer deposition. *Adv. Mater.* **27**, 53–58 (2015).
14. Wei, Z., Chen, H., Yan, K. & Yang, S. Inkjet printing and instant chemical transformation of a CH3NH3PbI3/nanocarbon electrode and interface for planar Perovskite solar cells. *Angew. Chem.* **126**, 13455–13459 (2014).
15. Liu, M., Johnston, M. B. & Snaith, H. J. Efficient planar heterojunction perovskite solar cells by vapour deposition. *Nature* **501**, 395–398 (2013).
16. Malinkiewicz, O. *et al.* Perovskite solar cells employing organic charge-transport layers. *Nat. Photonics* **8**, 128–132 (2014).
17. Stranks, S. D. *et al.* Electron-hole diffusion lengths exceeding 1 micrometer in an organometal trihalide perovskite absorber. *Science* **342**, 341–344 (2013).
18. Xing, G. *et al.* Long-range balanced electron- and hole-transport lengths in organic-inorganic CH3NH3PbI3. *Science* **342**, 344–347 (2013).
19. Dong, Q. *et al.* Electron-hole diffusion lengths > 175 µm in solution-grown CH3NH3PbI3 single crystals. *Science* **347**, 967–970 (2015).
20. Pellet, N. *et al.* Mixed-organic-cation Perovskite photovoltaics for enhanced solar-light harvesting. *Angew. Chem. Int. Ed.* **53**, 3151–3157 (2014).
21. Ogomi, Y. *et al.* CH3NH3Sn x Pb (1−x) I3 Perovskite solar cells covering up to 1060 nm. *J. Phys. Chem. Lett.* **5**, 1004–1011 (2014).
22. Stoumpos, C. C., Malliakas, C. D. & Kanatzidis, M. G. Semiconducting tin and lead iodide perovskites with organic cations: phase transitions, high mobilities, and near-infrared photoluminescent properties. *Inorg. Chem.* **52**, 9019–9038 (2013).
23. Noh, J. H., Im, S. H., Heo, J. H., Mandal, T. N. & Seok, S. I. Chemical management for colorful, efficient, and stable inorganic–organic hybrid nanostructured solar cells. *Nano Lett.* **13**, 1764–1769 (2013).

24. Jeon, N. J. *et al.* Compositional engineering of perovskite materials for high-performance solar cells. *Nature* **517**, 476–480 (2015).

25. Crossland, E. J. W. *et al.* Mesoporous TiO_2 single crystals delivering enhanced mobility and optoelectronic device performance. *Nature* **495**, 215–219 (2013).

26. O'Regan, B. & Gratzel, M. A low-cost, high-efficiency solar cell based on dye-sensitized colloidal TiO2 films. *Nature* **353**, 737–740 (1991).

27. Ko, K. H., Lee, Y. C. & Jung, Y. J. Enhanced efficiency of dye-sensitized TiO_2 solar cells (DSSC) by doping of metal ions. *J. Colloid Interface Sci.* **283**, 482–487 (2005).

28. Fabregat-Santiago, F. *et al.* High carrier density and capacitance in TiO_2 nanotube arrays induced by electrochemical doping. *J. Am. Chem. Soc.* **130**, 11312–11316 (2008).

29. Lee, S. *et al.* Nb-doped TiO_2: a new compact layer material for TiO_2 dye-sensitized solar cells. *J. Phys. Chem. C* **113**, 6878–6882 (2009).

30. Lü, X. *et al.* Improved-performance dye-sensitized solar cells using Nb-doped TiO_2 electrodes: efficient electron injection and transfer. *Adv. Funct. Mater.* **20**, 509–515 (2010).

31. Nah, Y. C., Paramasivam, I. & Schmuki, P. Doped TiO_2 and TiO_2 nanotubes: synthesis and applications. *Chemphyschem.* **11**, 2698–2713 (2010).

32. Zhang, X., Liu, F., Huang, Q.-L., Zhou, G. & Wang, Z.-S. Dye-sensitized W-doped TiO_2 solar cells with a tunable conduction band and suppressed charge recombination. *J. Phys. Chem. C* **115**, 12665–12671 (2011).

33. Zhang, X., Wang, S.-T. & Wang, Z.-S. Effect of metal-doping in TiO_2 on fill factor of dye-sensitized solar cells. *Appl. Phys. Lett.* **99**, 113503 (2011).

34. Duan, Y. *et al.* Sn-doped TiO_2 photoanode for dye-sensitized solar cells. *J. Phys. Chem. C* **116**, 8888–8893 (2012).

35. Cho, I. S. *et al.* Codoping titanium dioxide nanowires with tungsten and carbon for enhanced photoelectrochemical performance. *Nat. Commun.* **4**, 1723 (2013).

36. Leijtens, T. *et al.* Overcoming ultraviolet light instability of sensitized TiO_2 with meso-superstructured organometal tri-halide perovskite solar cells. *Nat. Commun.* **4**, 2885 (2013).

37. Pathak, S. K. *et al.* Performance and stability enhancement of dye-sensitized and Perovskite solar cells by Al doping of TiO_2. *Adv. Funct Mater.* **24**, 6046–6055 (2014).

38. Kopidakis, N., Benkstein, K. D., van de Lagemaat, J. & Frank, A. J. Transport-limited recombination of photocarriers in dye-sensitized nanocrystalline TiO_2 solar cells. *J. Phys. Chem. B* **107**, 11307–11315 (2003).

39. Olson, C. L., Nelson, J. & Islam, M. S. Defect chemistry, surface structures, and lithium insertion in anatase TiO_2. *J. Phys. Chem. B.* **110**, 9995–10001 (2006).

40. Abate, A. *et al.* Lithium salts as 'redox active' p-type dopants for organic semiconductors and their impact in solid-state dye-sensitized solar cells. *Phys. Chem. Chem. Phys.* **15**, 2572–2579 (2013).

41. Abate, A. *et al.* Protic ionic liquids as p-dopant for organic hole transporting materials and their application in high efficiency hybrid solar cells. *J. Am. Chem. Soc.* **135**, 13538–13548 (2013).

42. Kim, D. H. *et al.* Niobium doping effects on TiO_2 mesoscopic electron transport layer-based Perovskite solar cells. *ChemSusChem* **8**, 2392–2398 (2015).

43. Xing, G. *et al.* Interfacial electron transfer barrier at compact TiO_2/ $CH_3NH_3PbI_3$ heterojunction. *Small* **11**, 3606–3613 (2015).

44. Wojciechowski, K. *et al.* Heterojunction modification for highly efficient organic–inorganic Perovskite solar cells. *ACS Nano* **8**, 12701–12709 (2014).

45. Guillén, E., Ramos, F. J., Anta, J. A. & Ahmad, S. Elucidating transport-recombination mechanisms in perovskite solar cells by small-perturbation techniques. *J. Phys. Chem. C* **118**, 22913–22922 (2014).

46. Heo, J. H. *et al.* Hysteresis-less mesoscopic $CH_3NH_3PbI_3$ perovskite hybrid solar cells by introduction of Li-treated TiO_2 electrode. *Nano Energy* **15**, 530–539 (2015).

47. Södergren, S. *et al.* Lithium intercalation in nanoporous anatase TiO_2 studied with XPS. *J. Phys. Chem. B* **101**, 3087–3090 (1997).

48. Bouattour, S. *et al.* Li-doped nanosized TiO_2 powder with enhanced photocatalytic acivity under sunlight irradiation. *Appl. Organomet. Chem.* **24**, 692–699 (2010).

49. Westermark, K. *et al.* Determination of the electronic density of states at a nanostructured TiO_2/Ru-dye/electrolyte interface by means of photoelectron spectroscopy. *Chem. Phys.* **285**, 157–165 (2002).

50. Cappel, U. B. *et al.* Characterization of the interface properties and processes in solid state dye-sensitized solar cells employing a perylene sensitizer. *J. Phys. Chem. C* **115**, 4345–4358 (2011).

51. Abate, A. *et al.* An organic 'Donor-Free' dye with enhanced open-circuit voltage in solid-state sensitized solar cells. *Adv. Energy Mater.* **4**, 1400166 (2014).

52. Barnes, P. R. *et al.* Interpretation of optoelectronic transient and charge extraction measurements in dye-sensitized solar cells. *Adv. Mater.* **25**, 1881–1922 (2013).

53. Dloczik, L. *et al.* Dynamic response of dye-sensitized nanocrystalline solar cells: characterization by intensity-modulated photocurrent spectroscopy. *J. Phys. Chem. B.* **101**, 10281–10289 (1997).

54. Krüger, J., Plass, R., Grätzel, M., Cameron, P. J. & Peter, L. M. Charge transport and back reaction in solid-state dye-sensitized solar cells: a study using intensity-modulated photovoltage and photocurrent spectroscopy. *J. Phys. Chem. B.* **107**, 7536–7539 (2003).

55. Han, Y. S. & Kim, J. T. Enhanced performance of dye-sensitized solar cells with surface-treated titanium dioxides. *Mol. Cryst. Liq. Cryst.* **565**, 138–146 (2012).

56. Marinova, N. *et al.* Light harvesting and charge recombination in CH3NH3PbI3 Perovskite solar cells studied by hole transport layer thickness variation. *ACS Nano* **9**, 4200–4209 (2015).

57. Lee, Y. H. *et al.* Unraveling the reasons for efficiency loss in Perovskite solar cells. *Adv. Funct. Mater.* **25**, 3925–3933 (2015).

58. Chen, Q. *et al.* Controllable self-induced passivation of hybrid lead iodide perovskites toward high performance solar cells. *Nano Lett.* **14**, 4158–4163 (2014).

59. Abate, A., Staff, D. R., Hollman, D. J., Snaith, H. J. & Walker, A. B. Influence of ionizing dopants on charge transport in organic semiconductors. *Phys. Chem. Chem. Phys.* **16**, 1132–1138 (2014).

Acknowledgements

M.G. thanks the financial support from SNSF-NanoTera (SYNERGY), Swiss Federal Office of Energy (SYNERGY), CCEM-CH in the 9th call proposal 906: CONNECT PV and the SNSF NRP70 'Energy Turnaround' and GRAPHENE project supported by the European Commission Seventh Framework Programme under contract 604391 is gratefully acknowledged. A.A. has received funding from the European Union's Seventh Framework Programme for research, technological development and demonstration under grant agreement no 291771. We thank Aisin Cosmos R&D Co., Ltd, Japan for financial support.

Author contributions

F.G. proposed the experiment. F.G., A.A., J.P.C.B., M.S. and T.M. prepared the PSC devices. J.P.C.B. performed the microscopy analysis and analysed the XPS data. F.G., A.A. and J.P.C.B. prepared the SSDSC devices, performed IMPS, charge extraction and analysed the data. All the authors contributed to the analysis of the data and the writing of the paper. M.G. supervised the experiment.

Additional information

Permissions

List of Contributors

Yicheng Zhao, Jing Wei, Heng Li, Yin Yan, Wenke Zhou, Dapeng Yu and Qing Zhao
State Key Laboratory for Mesoscopic Physics, School of Physics, Peking University, Beijing, 100871, China Collaborative Innovation Center of Quantum Matter, Beijing, 100084, China

Maria Ibáñez
Department of Chemistry and Applied Biosciences, Institute of Inorganic Chemistry, ETH Zürich, Vladimir Prelog Weg 1, CH-8093 Zurich, Switzerland
Laboratory for Thin Films and Photovoltaics, Empa-Swiss Federal Laboratories for Materials Science and Technology, Dübendorf, Überlandstrasse 129, CH-8600 Dübendorf, Switzerland
Advanced Materials Department, Catalonia Energy Research Institute - IREC, Sant Adria de Besos, Jardins de les Dones de Negre n.1, Pl. 2, 08930 Barcelona, Spain

Laura Piveteau and Maksym V. Kovalenko
Department of Chemistry and Applied Biosciences, Institute of Inorganic Chemistry, ETH Zürich, Vladimir Prelog Weg 1, CH-8093 Zurich, Switzerland
Laboratory for Thin Films and Photovoltaics, Empa-Swiss Federal Laboratories for Materials Science and Technology, Dübendorf, Überlandstrasse 129, CH-8600 Dübendorf, Switzerland

Zhishan Luo, Silvia Ortega, Doris Cadavid, Oleksandr Dobrozhan and Yu Liu
Advanced Materials Department, Catalonia Energy Research Institute - IREC, Sant Adria de Besos, Jardins de les Dones de Negre n.1, Pl. 2, 08930 Barcelona, Spain

Aziz Genc
Department of Advanced Electron Nanoscopy, Catalan Institute of Nanoscience and Nanotechnology (ICN2), CSIC and The Barcelona Institute of Science and Technology, Campus UAB, Bellaterra, 08193 Barcelona, Spain

Maarten Nachtegaal
Paul Scherrer Institute, 5232 Villigen PSI, Switzerland

Mona Zebarjadi
Department of Mechanical and Aerospace Engineering, Rutgers University, 98 Brett Rd, Piscataway, New Jersey 08854-8058, USA

Jordi Arbiol
Department of Advanced Electron Nanoscopy, Catalan Institute of Nanoscience and Nanotechnology (ICN2), CSIC and The Barcelona Institute of Science and Technology, Campus UAB, Bellaterra, 08193 Barcelona, Spain Institució Catalana de Recerca i Estudis Avanc¸ats, ICREA, Passeig de Lluís Companys, 23 08010 Barcelona, Spain

Andreu Cabot
Advanced Materials Department, Catalonia Energy Research Institute - IREC, Sant Adria de Besos, Jardins de les Dones de Negre n.1, Pl. 2, 08930 Barcelona, Spain Institució Catalana de Recerca i Estudis Avanc¸ats, ICREA, Passeig de Lluís Companys, 23 08010 Barcelona, Spain

Hao Chen, Harley T. Johnson and Kimani C. Toussaint Jr
Department of Mechanical Science and Engineering, University of Illinois Urbana-Champaign, Urbana, Illinois 61801, USA

Abdul M. Bhuiya and Qing Ding
Department of Electrical and Computer Engineering, University of Illinois Urbana-Champaign, Urbana, Illinois 61801, USA

Takahiro Nagata, Seungjun Oh, Yutaka Wakayama and Takashi Sekiguchi
International Center for Materials Nanoarchitectonics, National Institute for Materials Science, 1-1 Namiki, Ibaraki 305-0044, Japan

Kentaro Watanabe
International Center for Materials Nanoarchitectonics, National Institute for Materials Science, 1-1 Namiki, Ibaraki 305-0044, Japan
Graduate School of Engineering Science, Osaka University, 1-3 Machikaneyama-cho, Osaka 560-8531, Japan

Yoshiaki Nakamura
Graduate School of Engineering Science, Osaka University, 1-3 Machikaneyama-cho, Osaka 560-8531, Japan

János Volk
MTA EK Institute of Technical Physics and Materials Science, Konkoly Thege M. ut 29-33, Budapest 1121, Hungary

Vladislav E. Demidov and Boris Divinskiy
Institute for Applied Physics and Center for Nanotechnology, University of Muenster, Correnstrasse 2-4, Muenster 48149, Germany

Sergej O. Demokritov
Institute for Applied Physics and Center for Nanotechnology, University of Muenster, Correnstrasse 2-4, Muenster 48149, Germany
M.N. Miheev Institute of Metal Physics of Ural Branch of Russian Academy of Sciences, Yekaterinburg 620041, Russia

Sergei Urazhdin and Ronghua Liu
Department of Physics, Emory University, Atlanta, Georgia 30322, USA

Andrey Telegin
M.N. Miheev Institute of Metal Physics of Ural Branch of Russian Academy of Sciences, Yekaterinburg 620041, Russia

Sergio Boixo, Alireza Shabani, Sergei V. Isakov, Vasil S. Denchev, Masoud Mohseni and Hartmut Neven
Google, Venice, California 90291, USA

Vadim N. Smelyanskiy
Google, Venice, California 90291, USA. 2 NASA Ames Research Center, Moffett Field, California 94035, USA

Mark Dykman
Department of Physics and Astronomy, Michigan State University, East Lansing, Michigan 48824, USA

Anatoly Yu. Smirnov
D-Wave Systems Inc., Burnaby, British Columbia, Canada V5C 6G9

Mohammad H. Amin
D-Wave Systems Inc., Burnaby, British Columbia, Canada V5C 6G9
Department of Physics, Simon Fraser University, Burnaby, British Columbia, Canada V5A 1S6

Yun-Pil Shim
Laboratory for Physical Sciences, College Park, Maryland 20740, USA. 2 Department of Physics, University of Maryland, College Park, Maryland 20742, USA

Charles Tahan
Laboratory for Physical Sciences, College Park, Maryland 20740, USA

D. Pierangeli, M. Ferraro and E. DelRe
Dipartimento di Fisica, Università di Roma 'La Sapienza', Rome 00185, Italy

F. Di Mei and G. Di Domenico
Center for Life Nano Science@Sapienza, Istituto Italiano di Tecnologia, Rome 00161, Italy

C.E.M. de Oliveira and A.J. Agranat
Department of Applied Physics, Hebrew University of Jerusalem, Jerusalem 91904, Israel

Peng You, Zhike Liu, Helen L.W. Chan and Feng Yan
Department of Applied Physics, The Hong Kong Polytechnic University, Hung Hom, 999077 Kowloon, Hong Kong

Qidong Tai
Department of Applied Physics, The Hong Kong Polytechnic University, Hung Hom, 999077 Kowloon, Hong Kong
Institute for Interdisciplinary Research and Key Laboratory of Optoelectronic Chemical Materials and Devices of Ministry of Education, Jianghan University, 430056Wuhan, China

Hongqian Sang and Chenglong Hu
Institute for Interdisciplinary Research and Key Laboratory of Optoelectronic Chemical Materials and Devices of Ministry of Education, Jianghan University, 430056Wuhan, China

Fabrizio Torricelli
Department of Information Engineering, University of Brescia, via Branze 38, Brescia 25123, Italy
Department of Electrical Engineering, Eindhoven University of Technology, Groene Loper 19, PO Box 513, Eindhoven 5600MB, The Netherlands

Luigi Colalongo and Zsolt Miklós Kovács-Vajna
Department of Information Engineering, University of Brescia, via Branze 38, Brescia 25123, Italy

Daniele Raiteri and Eugenio Cantatore
Department of Electrical Engineering, Eindhoven University of Technology, Groene Loper 19, PO Box 513, Eindhoven 5600MB, The Netherlands

Gary Bulman, Phil Barletta, Jay Lewis and Nicholas Baldasaro
RTI International, Electronics and Applied Physics Division, Research Triangle Park, North Carolina 27709, USA

Michael Manno, Avram Bar-Cohen and Bao Yang
Department of Mechanical Engineering, University of Maryland, College Park, Maryland 20742, USA

Yang Zhou, Lu You, ShiweiWang, Lei Chang, Le Wang, Peng Ren and Junling Wang
School of Materials Science and Engineering, Nanyang Technological University, Block N4.1-02-24, 50 Nanyang Avenue, Singapore 639798, Singapore

Zhiliang Ku and Hongjin Fan
School of Physical and Mathematical Sciences, Nanyang Technological University, Singapore 639798, Singapore

Daniel Schmidt and Andrivo Rusydi
Singapore Synchrotron Light Source, National University of Singapore, 5 Research Link, Singapore 117603, Singapore

Liufang Chen and Guoliang Yuan
Department of Materials Science and Engineering, Nanjing University of Science and Technology, Nanjing 210094, China

Lang Chen
Department of Physics, South University of Science and Technology of China, Shenzhen 518055, China

Antonio Ortiz-Ambriz and Pietro Tierno
Department of Structure and Constituents of Matter, University of Barcelona, Avinguda Diagonal 647, 08028 Barcelona, Spain
Institute of Nanoscience and Nanotechnology, IN2UB, Universitat de Barcelona, 08028 Barcelona, Spain

Ariel Epstein, Joseph P.S. Wong and George V. Eleftheriades
The Edward S. Rogers Department of Electrical and Computer Engineering, University of Toronto, Toronto, Ontario, Canada M5S 2E4

Sho C. Takatori and John F. Brady
Division of Chemistry and Chemical Engineering, California Institute of Technology, Pasadena, California 91125, USA

Jan Vermant
Department of Materials, ETH Zürich, Vladimir-Prelog-Weg 5, Zürich 8093, Switzerland

Raf De Dier
Department of Materials, ETH Zürich, Vladimir-Prelog-Weg 5, Zürich 8093, Switzerland
Department of Chemical Engineering, KU Leuven, Leuven B-3001, Belgium

Fenggong Wang, Fan Zheng, Ilya Grinberg and Andrew M. Rappe
The Makineni Theoretical Laboratories, Department of Chemistry, University of Pennsylvania, Philadelphia, Pennsylvania 19104-6323, USA

Steve M. Young
Center for Computational Materials Science, United States Naval Research Laboratory, Washington, DC 20375, USA

Dehui Li and Chih-Yen Chen
Department of Chemistry and Biochemistry, University of California, 607 Charles E. Young Drive East, Los Angeles, California 90095, USA

Gongming Wang and Xiangfeng Duan
Department of Chemistry and Biochemistry, University of California, 607 Charles E. Young Drive East, Los Angeles, California 90095, USA
California Nanosystems Institute, University of California, Los Angeles, California 90095, USA

Hung-Chieh Cheng, Hao Wu and Yuan Liu
Department of Materials Science and Engineering, University of California, Los Angeles, California 90095, USA

Yu Huang
California Nanosystems Institute, University of California, Los Angeles, California 90095, USA
Department of Materials Science and Engineering, University of California, Los Angeles, California 90095, USA

Colin Ophus, Jim Ciston and Peter Ercius
National Center for Electron Microscopy, Molecular Foundry, Lawrence Berkeley National Laboratory, 1 Cyclotron Road, Berkeley, California 94720, USA

Jordan Pierce, Tyler R. Harvey, Jordan Chess and Benjamin J. McMorran
Department of Physics, University of Oregon, 1585 E 13th Avenue, Eugene, Oregon 97403, USA

Cory Czarnik
Gatan Inc., 5794 W Las Positas Boulevard, Pleasanton, California 94588, USA

Harald H. Rose
Department of Physics, Center for Electron Microscopy, Ulm University, Albert Einstein-Allee 11, 89069 Ulm, Germany

W. He
Nano Science and Technology Program, Hong Kong University of Science and Technology, Clear Water Bay, Kowloon, Hong Kong

H. Song, Y. Su and P. Tong
Department of Physics, Hong Kong University of Science and Technology, Clear Water Bay, Kowloon, Hong Kong

L. Geng and H.B. Peng
Division of Life Science, Hong Kong University of Science and Technology, ClearWater Bay, Kowloon, Hong Kong

B.J. Ackerson
Department of Physics, Oklahoma State University, Stillwater, Oklahoma 74078, USA

Georg Kaltenboeck, Marios D. Demetriou, Scott Roberts and William L. Johnson
Keck Engineering Laboratories, California Institute of Technology, Pasadena, California 91125, USA

C. Carrétéro, V. Garcia, S. Fusil, A. Barthélémy and M. Bibes
Unité Mixte de Physique, CNRS, Thales, Univ. Paris-Sud, Université Paris-Saclay, 91767 Palaiseau, France

D. Sando
Unité Mixte de Physique, CNRS, Thales, Univ. Paris-Sud, Université Paris-Saclay, 91767 Palaiseau, France
School of Materials Science and Engineering, University of New South Wales, Sydney 2052, Australia

Yurong Yang and L. Bellaiche
Department of Physics and Institute for Nanoscience and Engineering, University of Arkansas, Fayetteville, Arkansas 72701, USA

E. Bousquet and Ph. Ghosez
Theoretical Materials Physics, Université de Liége, B-5, B-4000 Sart-Tilman, Belgium

D. Dolfi
Thales Research and Technology France, 1 Avenue Augustin Fresnel, 91767 Palaiseau, France

Kent A.G. Fisher, Jean-Philippe W. MacLean and Kevin J. Resch
Institute for Quantum Computing and Department of Physics and Astronomy, University of Waterloo, 200 University Avenue West, Waterloo, Ontario, Canada N2L 3G1

Duncan G. England and Philip J. Bustard
National Research Council of Canada, 100 Sussex Drive, Ottawa, Ontario, Canada K1A 0R6

Benjamin J. Sussman
National Research Council of Canada, 100 Sussex Drive, Ottawa, Ontario, Canada K1A 0R6
Department of Physics, University of Ottawa, 150 Louis Pasteur, Ottawa, Ontario, Canada K1N 6N5

Juliane Simmchen and Jaideep Katuri
Max-Planck-Institut für Intelligente Systeme, Heisenbergstrasse 3, D-70569 Stuttgart, Germany

William E. Uspal, Mihail N. Popescu and Mykola Tasinkevych
Max-Planck-Institut für Intelligente Systeme, Heisenbergstrasse 3, D-70569 Stuttgart, Germany

IV. Institut für Theoretische Physik, Universität Stuttgart, Pfaffenwaldring 57, D-70569 Stuttgart, Germany

Samuel Sánchez
Max-Planck-Institut für Intelligente Systeme, Heisenbergstrasse 3, D-70569 Stuttgart, Germany.
Institut de Bioenginyeria de Catalunya (IBEC), Baldiri I Reixac 10-12, 08028 Barcelona, Spain
Institució Catalana de Recerca i Estudis Avancats (ICREA), Pg. Lluís Companys 23, 08010 Barcelona, Spain

Negar Reiskarimian and Harish Krishnaswamy
Department of Electrical Engineering, Columbia University, 1300 South West Mudd, 500 West 120th Street, New York, New York 10027, USA

Ki Su Kim and Robert W. Redmond
Wellman Center for Photomedicine, Harvard Medical School, Massachusetts General Hospital, 65 Landsdowne Street UP-5, Cambridge Massachusetts 02139, USA.

Sedat Nizamoglu
Wellman Center for Photomedicine, Harvard Medical School, Massachusetts General Hospital, 65 Landsdowne Street UP-5, Cambridge, Massachusetts 02139, USA
Department of Electrical and Electronics Engineering, Koc University, Istanbul 34450, Turkey

Malte C. Gather
Wellman Center for Photomedicine, Harvard Medical School, Massachusetts General Hospital, 65 Landsdowne Street UP-5, Cambridge, Massachusetts 02139, USA
SUPA, School of Physics and Astronomy, University of St Andrews, St Andrews KY16 9SS, UK

Matjaž Humar
Wellman Center for Photomedicine, Harvard Medical School, Massachusetts General Hospital, 65 Landsdowne Street UP-5, Cambridge, Massachusetts 02139, USA
Condensed Matter Department, J. Stefan Institute, Jamova 39, SI-1000 Ljubljana, Slovenia

Myunghwan Choi
Wellman Center for Photomedicine, Harvard Medical School, Massachusetts General Hospital, 65 Landsdowne Street UP-5, Cambridge, Massachusetts 02139, USA
Global Biomedical Engineering, Sungkyunkwan University, Center for Neuroscience and Imaging Research, Institute for Basic Science, 2066, Seoburo, Jangan, Suwon, Gyeonggi 440-746, Korea

Giuliano Scarcelli
Wellman Center for Photomedicine, Harvard Medical School, Massachusetts General Hospital, 65 Landsdowne Street UP-5, Cambridge, Massachusetts 02139, USA
Department of Bioengineering, University of Maryland College Park, College Park,Maryland 20742, USA

Mark Randolph
Wellman Center for Photomedicine, Harvard Medical School, Massachusetts General Hospital, 65 Landsdowne Street UP-5, Cambridge, Massachusetts 02139, USA
Division of Plastic Surgery and Harvard Medical School, Massachusetts General Hospital, 40 Blossom Street, Boston, Massachusetts 02114, USA

Seonghoon Kim
Graduate School of Nanoscience and Technology, Korea Advanced Institute of Science and Technology, 291 Daehakro, Yuseong, Daejeon 305-701, Korea

Sei Kwang Hahn
Department of Materials Science and Engineering, Pohang University of Science and Technology, San 31, Hyoja-dong, Nam, Pohang, Kyungbuk 790-784, Korea

Seok Hyun Yun
Wellman Center for Photomedicine, Harvard Medical School, Massachusetts General Hospital, 65 Landsdowne Street UP-5, Cambridge, Massachusetts 02139, USA

Harvard–MIT Health Sciences and Technology, 77 Massachusetts Avenue, Cambridge, Massachusetts 02139, USA

Eyal Cohen and Shlomi Dolev
Department of Computer Science, Ben Gurion University of the Negev, 653 Béer, Sheva 84105, Israel

Michael Rosenblit
Ilze Katz Institute for Nanoscale Science and Technology, Ben Gurion University of the Negev, 653 Béer, Sheva 84105, Israel

Diana Berman, Sanket A. Deshmukh, Badri Narayanan, Subramanian K.R.S. Sankaranarayanan, Daniel Rosenmann and Anirudha V. Sumant
Center for Nanoscale Materials, Argonne National Laboratory, 9700 South Cass Avenue, Argonne, Illinois 60439, USA

Zhong Yan, Alexander A. Balandin
Department of Electrical and Computer Engineering, Materials Science and Engineering Program, Bourns College of Engineering, University of California – Riverside, Riverside, California 92521, USA

Alexander Zinovev
Materials Science Division, Argonne National Laboratory, 9700 South Cass Avenue, Argonne, Illinois 60439, USA

Fabrizio Giordano, Antonio Abate, Shaik M. Zakeeruddin and Michael Graetzel
Laboratory of Photonics and Interfaces, Institute of Chemical Sciences and Engineering, School of Basic Sciences, Ecole Polytechnique Fédérale de Lausanne (EPFL), Lausanne CH-1015, Switzerland

Juan Pablo Correa Baena and Anders Hagfeldt
Laboratoire des sciences photomoléculaires, Institute of Chemical Sciences and Engineering, School of Basic Sciences, Ecole Polytechnique Fédérale de Lausanne (EPFL), Lausanne CH-1015, Switzerland

Michael Saliba and Mohammad Khaja Nazeeruddin
Group for Molecular Engineering of Functional Materials, Institute of Chemical Sciences and Engineering, School of Basic Sciences, Ecole Polytechnique Fédérale de Lausanne (EPFL), Lausanne CH-1015, Switzerland

Taisuke Matsui
Advanced Research Division, Materials Research Laboratory, Panasonic Corporation, 1006, Kadoma, Kadoma City, Osaka 571-8501, Japan

Sang Hyuk Im
Department of Chemical Engineering, Kyung Hee University, 1732 Deogyeong-daero, Giheung-gu, Yongin-si, Gyeonggi-do 446-701, Republic of Korea

Index

www.ingramcontent.com/pod-product-compliance
Lightning Source LLC
Chambersburg PA
CBHW081710240326
41458CB00156B/4272